MODERN ASPECTS OF ELECTROCHEMISTRY

No. 38

Modern Aspects of Electrochemistry

Modern Aspects of Electrochemistry, No. 37:

- The kinetics of electrochemical hydrogen entry into metals and alloys.
- The electrochemistry, corrosion, and hydrogen embrittlement of unalloyed titanium. This important chapter discusses pitting and galvanostatic corrosion followed by a review of hydrogen embrittlement emphasizing the formation of hydrides and their effect on titanium's mechanical properties.
- Oxidative electrochemical processes of organics, introduces an impressive model that distinguishes active (strong) and non-active (weak) anodes.
- Comprehensive discussions of fuel cells and Carnot engines; Nernst law; analytical fuel cell modeling; reversible losses and Nernst loss; and irreversible losses, multistage oxidation, and equipartition of driving forces. Includes new developments and applications of fuel cells in trigeneration systems; coal/biomass fuel cell systems; indirect carbon fuel cells; and direct carbon fuel cells.
- Exploration of the catalytic effect of trace anions in outer-sphere heterogeneous charge-transfer reactions.
- Results of the experimental and theoretical investigations on bridging electrolyte-water systems as to thermodynamic and transport properties of aqueous and organic systems. Revised version of chapter four in Number 35.

Modern Aspects of Electrochemistry, No. 36:

- The study of electrochemical nuclear magnetic resonance (EC-NMR); in situ both sides of the electrochemical interface via the simultaneous use of 13C and 193Pt NMR
- Recent impressive advances in the use of rigorous ab initio quantum chemical calculations in electrochemistry.
- Fundamentals of ab initio calculations, including density functional theory (DFT) methods, help to understand several key aspects of fuel cell electrocatalysis at the molecular level.
- The development of the most important macroscopic and statistical thermodynamic models that describe adsorption phenomena on electrodes
- Electrochemical promotion and recent advances of novel monolithic designs for practical utilization.
- New methods for CT analysis and an explanation for the existing discrepancy in Li diffusivity values obtained by the diffusion control CT analysis and other methods.

LIST OF CONTRIBUTORS

A. J. APPLEBY
238 Wisenbaker Engineering
Research Center
Texas A&M University,
College Station, TX 77843-3402

ROLANDO GUIDELLI
Dept. of Chemistry,
Florence University,
Via della Lastruccia 3
50019 Sesto Fiorentino, Firenze,
Italy

ARJAN HOVESTAD
Faculty of Chemical Engineering,
Eindhoven University of
Technology,
Eindhoven, The Netherlands

LEONARD J. J. JANSSEN
Faculty of Chemical Engineering,
Eindhoven University of
Technology,
Eindhoven, The Netherlands

JOACHIM MAIER
Max-Planck-Institut für
Festkörperforschung,
Heisenbergstraße 1, 70569
Stuttgart, Germany

N. D. NIKOLIĆ
ICTM–Institute of
Electrochemistry,
Belgrade,
Serbia and Montenegro

K. I. POPOV
Faculty of Technology and
Metallurgy,
University of Belgrade,
Serbia and Montenegro

Z. RAKOČEVIĆ
Vinča Institute of Nuclear
Sciences,
Belgrade,
Serbia and Montenegro

WOLFGANG SCHMICKLER
Abteilung Elektrochemie,
Ulm University,
89069 Ulm, Germany

PAUL STONEHART
Stonehart Associates Inc.
17 Cottage Read,
Madison, Connecticut
06443-1220 U.S.A.

DOUGLAS WHEELER
UTC (United Technologies
Corporation) Fuel Cells,
195 Governor's Highway, South
Windsor, Connecticut
06074 U.S.A.

MODERN ASPECTS OF ELECTROCHEMISTRY

No. 38

Edited by

B. E. CONWAY

University of Ottawa
Ottawa, Ontario, Canada

C. G. VAYENAS

University of Patras
Patras, Greece

RALPH E. WHITE

University of South Carolina
Columbia, South Carolina

and

MARIA E. GAMBOA-ADELCO

Managing Editor
Superior, Colorado

Kluwer Academic / Plenum Publishers
New York, Boston, Dordrecht, London, Moscow

The Library of Congress cataloged the first volume of this title as follows:

Modern aspects of electrochemistry, no. [1]
  Washington Butterworths, 1954–
  v. illus., 23 cm.
  No. 1–2 issued as Modern aspects series of chemistry.
  Editors: No 1– J.O'M. Bockris (with B. E. Conway, No. 3–)
  Imprint varies: no. 1, New York, Academic Press.—No. 2, London, Butterworths.
    1. Electrochemisty—Collected works. I. Bockris, John O'M. ed. II. Conway, B. E. ed.
(Series: Modern aspects series of chemistry)

QD552.M6   54-12732 re

ISBN: 0-306-48703-9
eISBN: 0-306-48704-7

© 2005 Kluwer Academic/Plenum Publishers, New York
233 Spring Street, New York, New York 10013

http://www.wkap.nl/

10 9 8 7 6 5 4 3 2 1

A C.I.P. record for this book is available from the Library of Congress

Preface

This volume comprises six chapters on aspects of fundamental and applied electrochemical science that will be of interest both to researchers in the basic areas of the subject and to those involved in aspects of electrochemical technologies.

Chapter 1 is the first part of a 2-part, major contribution by Joachim Maier on Solid State Electrochemistry: Thermodynamics and Kinetics of Charge Carriers in Solids. Part 2 will follow in volume 39 to be published in year 2005. This contribution reviews modern concepts of the equilibria involving charge carriers in solids in terms of concentrations of defects in solids and at grain-boundaries, including doping effects. Complementarily, kinetics of charge transfer and ion transfer are treated in some detail in relation to conductance, kinetics of surface processes and electrode-kinetics involving solid-state processes. This chapter will be of major interest to electrochemists and physicists in the semiconductor field and that involving ionic solids.

In the second chapter, Appleby presents a detailed discussion and review in modern terms of a central aspect of electrochemistry: Electron Transfer Reactions With and Without Ion Transfer. Electron transfer is the most fundamental aspect of most processes at electrode interfaces and is also involved intimately with the homogeneous chemistry of redox reactions in solutions. The subject has experienced controversial discussions of the role of solvational interactions in the processes of electron transfer at electrodes and in solution, especially in relation to the role of "Inner-sphere" versus "Outer-sphere" activation effects in the act of electron transfer. The author distils out the essential features of electron transfer processes in a *tour de force* treatment of all aspects of this important field in terms of models of the solvent (continuum and molecular), and of the activation process in the kinetics of electron transfer reactions, especially with respect to the applicability of the Franck-Condon principle to the time-scales of electron transfer and solvational excitation. Sections specially devoted to hydration of the proton and its heterogeneous transfer, coupled with electron transfer in cathodic H deposition and H_2 evolution, are important in this chapter.

Chapter 3, by Rolando Guidelli, deals with another aspect of major fundamental interest, the process of electrosorption at electrodes, a topic central to electrochemical surface science: Electrosorption Valency and Partial Charge Transfer. Thermodynamic examination of electrochemical adsorption of anions and atomic species, e.g. as in underpotential deposition of H and metal adatoms at noble metals, enables details of the state of polarity of electrosorbed species at metal interfaces to be deduced. The bases and results of studies in this field are treated in depth in this chapter and important relations to surface - potential changes at metals, studied in the gas-phase under high-vacuum conditions, will be recognized. Results obtained in this field of research have significant relevance to behavior of species involved in electrocatalysis, e.g. in fuel-cells, as treated in chapter 4, and in electrodeposition of metals.

In chapter 4, Stonehart (a major authority in the field of H_2 fuel-cell technology and its fundamental aspects) writes, with co-author Wheeler, on the topic of : Phosphoric Acid Fuel-Cells (PAFCs) for Utilities: Electrocatalyst Crystallite Design, Carbon Support, and Matrix Materials Challenges. This contribution reviews, in detail, recent information on the behavior of very small Pt and other alloy electrocatalyst crystallites used as the electrode materials for phosphoric acid electrolyte fuel-cells.

A materials - science aspect of metal electrodes is treated in chapter 5, by Nicolić and Radočević: Nanostructural Analysis of Bright Metal Surfaces in Relation to Their Reflectivity. This is an area of practical importance in electrochemistry applied, e.g., to metal finishing. Details of how reflectivity and electron microscopy can be usefully applied in this field are given with reference to many examples.

Finally, in chapter 6, another direction of applied electrochemistry is treated by Hovestad and Janssen; Electroplating of Metal Matrix Composites by Codeposition of Suspended Particles. This is another area of metals materials-science where electroplating of a given metal is conducted in the presence of suspended particles, e.g. of Al_2O_3, BN, WC, SiC or TiC, which become electrodeposited as firmly bound occlusions. Such composite deposits have improved physical and electrochemical properties. Process parameters, and mechanisms and models of the codeposition processes are described in relation to bath compositions, particle compositions, micrographs and dispersion hardening conditions.

B. E. Conway

University of Ottawa
Ottawa, Ontario, Canada

R. E. White

University of South Carolina
Columbia, South Carolina

C. Vayenas,

University of Patras
Patras, Greece

Contents

Chapter 1

SOLID STATE ELECTROCHEMISTRY I: THERMODYNAMICS
AND KINETICS OF CHARGE CARRIERS IN SOLIDS

Joachim Maier

I. Introduction 1
II. Solids vs. Liquids 2
III. Point Defects as Charge Carriers 5
VI. Equilibrium Concentration of Charge Carriers in the Bulk 10
  1. Defect Reactions 10
  2. Equilibrium Defect Concentrations in Pure Compounds 13
  3. Doping Effects 22
  4. Frozen-in Defect Chemistry 32
  5. Defect-Defect Interactions 36
V. Defect Chemistry at Boundaries 48
  1. Space Charge Profiles and Capacitances 48
  2. Space Charge Conductance 54
  3. Interfacial Defect Thermodynamics 70
  4. Non-Trivial and Trivial Size Effects: Nanoionics 75
VI. Kinetics of Charge Transfer and Transport 84
  1. Three Experimental Situations 84
  2. Rates, Fluxes and Driving Forces 87
    (*i*) Transport, Transfer and Reaction 87
    (*ii*) Transport in Terms of Linear Irreversible
        Thermodynamics and Chemical Kinetics 89
  2. Bulk Processes 95
    (*i*) Electrical Conduction: Mobility, Conductivity and
        Random Walk 95
    (*ii*) Mobility 96
    (*iii*) Tracer Diffusion 103

(*iv*) Chemical Diffusion 106
4. Correlations and Complications 114
    (*i*) Conductivity: Static and Dynamic Correlation Factors..114
    (*ii*) Chemical Diffusion and Conservative Ensembles..........117
    (*iii*) Diffusion Involving Internal Boundaries.......................128
5. Surface Kinetics 133
6. Electrode Kinetics 145
7. Generalized Equivalent Circuits...147
8. Solid State Reactions..149
9. Non-linear Processes 152
Acknowledgment 161
Symbols..162
Indices 163
References 164

Chapter 2

ELECTRON TRANSFER REACTIONS WITH AND WITHOUT ION
TRANSFER

A. J. Appleby

  I. Introduction...175
 II. The Franck-Cordon Principle and Electron Transfer 176
    1. Historical Development...176
    2. Inner and Outer Sphere Concepts......................................179
    3. Dielectric Continuum Theory...180
    4. Activation via the Dielectric Continuum.............................185
    5. Inner Sphere Rearrangement With "Flow of Charge"...........186
    6. Inner Sphere Rearrangement and Force Constants.............187
    7. The Franck-Condon Approximation 193
    8. The Tafel Slope...194
III. Interaction of Ions with Polar Media to Dielectric Saturation...196
    1. The "Electrostatic Continuum"...196
    2. The Electrostatic Gibbs Energy in a Continuous Polar
       Medium...198
    3. The Approach To Dielectric Saturation..............................202
    4. Polarization..202
    5. The Static Dielectric Constant of Water 205

6. The Dielectric Constant at High Field Strengths.................207
7. Inadequacies of the Booth Theory.................................212
8. The Dielectric Constant as a Function of Displacement 212
9. The Ion-"Continuum" Interaction in Polar Liquids............219
10. Induction Effects...221
11. Polarization of ut Type...221
12. Quadrupole or Multipole Effects 224
13. At Dielectric Saturation 225
14. Summary of Ion-Solvent Interactions 227
IV. The Inner Sphere(s)...228
1. Multivalent Cations as Trikisoctahedra.....................228
2. Inner and Second Sphere Energies...........................230
3. Dipole Nearest Neighbors..235
4. Initial Simulation Results for First and Second
   Trikisoctahedron Shells...238
5. The Third and Fourth Shells 241
6. "Bottom-Up" Modeling of the Third Shell 244
7. Simplified Modeling 244
8. The "Continuum" Energy.......................................246
9. Solvation Energy Estimates 248
V. The Structure and Energies of Liquid Water and
   Solvated H_3O 251
1. Preliminary Approaches...251
2. H_3O^+ Solvation..259
VI. Solvation and Charge Transfer 262
1. FC Redox Processes...262
2. FC Proton Transfer 265
3. The Inertial Term...267
VII. Non-FC Charge Transfer 269
1. Water Molecule Rearrangement in Solvation Shell
   Assembly..269
2. Non-FC Redox Electron Transfer 270
3. Proton Transfer and Other ECIT Processes 274
4. Tafel Plots for Redox ECET and ECIT Processes 280
5. Linear Tafel and Brønsted Slopes 283
VIII. Conclusions..285
   Appendix...286
1. Applicability of $\varepsilon_A = 1 + 4n_o\alpha_{T,A}$ 286
2. An Appropriate Model for Water Molecule Orientation......288
   References...292

Chapter 3

ELECTROSORPTION VALENCY AND PARTIAL CHARGE
TRANSFER

Rolando Guidelli and Wolfgang Schmickler

I. Introduction 303
II. The Electrosorption Valency 304
   1. Some Thermodynamic Considerations 308
   2. The Extrathermodynamic Contributions to the
      Electrosorption Valency 314
   3. The Partial Charge Transfer Coefficient of Lorenz
      and Salié 316
   4. Hard Sphere Electrolyte Model for Specific Adsorption 322
   5. Experimental Procedures for the Determination of the
      Electrosorption Valency 324
III. The Partial Charge Transfer 333
   1. Extrathermodynamic Estimate of the Partial Charge
      Transfer Coefficient from the Electrosorption Valency........333
   2. Partial Charge Transfer in Terms of the Anderson-Newns
      Model 343
   3. Dipole Moment and Electrosorption Valency 347
      (*i*) Adsorption from the Gas Phase 347
      (*ii*) Adsorption from an Electrolyte Solution 348
      (*iii*) Relation to the Electrosorption Valency 349
      (*iv*) Dipole Moments in the Hard-Sphere Electrolyte
      Model 351
IV. Electrosorption Valency and Partial Charge Transfer Coefficient
   in Self-assembled Thiol Monolayers 352
   1. Self-assembled Alkanethiol Monolayer on Mercury............352
   2. The Integral Electrosorption Valency of Metal-Supported
      Thiol Monolayers 355
   3. Absolute Potential Difference Between Mercury and an
      Aqueous Phase..358
   4. Electrosorption Valency of Three Mercury-Supported Thiol
      Monolayers 361
V. Conclusions 365
   References 366

Chapter 4

PHOSPHORIC ACID FUEL-CELLS (PAFCs) FOR UTILITIES:
ELECTROCATALYST CRYSTALLITE DESIGN, CARBON
SUPPORT, AND MATRIX MATERIALS CHALLENGES

Paul Stonehart and Douglas Wheeler

I. Introduction...373
II. The Role of Electrocatalysis in Phosphoric Acid
Fuel-Cells (PAFCs)...374
III. Specific and Mass Activities for Oxygen Reduction on
Platinum in Phosphoric Acid 375
IV. The Stonehart Theory of Crystallite Separation........382
V. Calculation for Hemispherical Diffusion to a Crystallite..........385
VI. Development of Platinum Alloy Oxygen-Reduction
Electrocatalysts 390
VII. The Matrix for PAFCs 400
VIII. Carbon Electrocatalyst Supports...............................404
IX. Development of Alloy Electrocatalysts for Hydrogen-
Molecule Oxidation 414
X. Conclusions...420
Acknowledgments..421
References...422

Chapter 5

NANOSTRUCTURAL ANALYSIS OF BRIGHT METAL
SURFACES IN RELATION TO THEIR REFLECTIVITY

N. D. Nicolić, Z. Radočević and K. I. Popov

I. Introduction...425
II. Real Systems..426
1. Limiting Cases 427
(*i*) Silver Mirror Surface 427
(*ii*) Metal Surfaces...430

2. Systems in Metal Finishing 432
    (*i*) Polished Copper Surfaces 432
    (*ii*) Copper Coatings 439
    (*iii*) Zinc Coatings..450
    (*iv*) Nickel Coatings 459
    (*v*) Discussion of Presented Results and the Model 463
Acknowledgments..472
References 472

Chapter 6

ELECTROPLATING OF METAL MATRIX COMPOSITES BY
CODEPOSITION OF SUSPENDED PARTICLES

Arjan Hovestad and Leonard J. J. Janssen

I. Introduction...475
II. Properties and Applications 477
  1. Dispersion Hardening..477
  2. Wear Resistance 479
    (*i*) Abrasion Resistance...480
    (*ii*) Lubrication...480
  3. Electrochemical Activity...481
    (*i*) Corrosion Resistance 481
    (*ii*) Electrocatalysis 482
III. Process Parameters...483
  1. Particle Properties 483
    (*i*) Particle Material..484
    (*ii*) Particle Size 488
    (*iii*) Particle Shape 489
  2. Bath Composition..490
    (*i*) Bath Constituents 490
    (*ii*) pH..493
    (*iii*) Additives 494
    (*iv*) Aging...498
  3. Deposition Variables 498
    (*i*) Particle Bath Concentration 499
    (*ii*) Current Density...500

(*iii*) Electrolyte Agitation 504
(*iv*) Temperature 506
IV. Mechanisms and Models.. 507
1. Early Mechanisms... 508
2. Empirical Models.. 508
(*i*) Guglielmi 509
(*ii*) Kariapper and Foster 512
(*iii*) Buelens and Celis *et al.* 513
(*iv*) Hwang and Hwang.. 516
3. Advanced Models 316
(*i*) Valdes 518
(*ii*) Guo *et al* 519
(*iii*) Fransaer *et al* 521
(*iv*) Hovestad *et al* 523
List of Symbols 526
References.. 529

INDEX 533

MODERN ASPECTS OF ELECTROCHEMISTRY

No. 38

1

Solid State Electrochemistry I: Thermodynamics and Kinetics of Charge Carriers in Solids

Joachim Maier

Max-Planck-Institut für Festkörperforschung, Heisenbergstraße 1, 70569 Stuttgart, Germany. E-mail: s.weiglein@fkf.mpg.de

I. INTRODUCTION

This contribution deals with thermodynamics and kinetics of charge carriers in solids in the case of zero or non-zero electrical or chemical driving forces. It does not intend to repeat well-known electrochemical principles, however, it intends to underline the special situation in solids by, on one hand, emphasizing characteristic aspects due to the solid nature, but on the other hand, stressing the common and generalizing aspects of the picture whenever it appears necessary. This also implies that specific solid state aspects (such as structural details, anisotropies or strain effects) are neglected whenever their influence is not indispensable for the understanding.

The present contribution does not include solid state techniques to measure electrochemical parameters, nor does it consider applications of solid state electrochemistry. Such topics will be dealt with in a second part which will appear separately.[1]

Modern Aspects of Electrochemistry, Number 38, edited by B. E. Conway *et al.* Kluwer Academic/Plenum Publishers, New York, 2005.

In the traditional electrochemical community solids have a long tradition, however, almost exclusively from the stand-point of electronic processes. Careful studies have been performed in order to understand the behavior of metallic or semi-conductive electrodes in contact to liquid electrolytes in great detail. Ionic processes are mostly considered to occur "outside" the solid, i.e., on its surfaces or in the fluid phases adjacent. More recently, in particular in conjunction with applications such as lithium batteries or fuel cells the ionic conductivity and the appearance of mixed ionic electronic conductivity within the solid state is more and more appreciated, but even in "modern" electrochemical textbooks one hardly finds any specific mention of the ionic charge carriers in solids, which are the ionic point defects. On the other hand, such considerations are in the focus of entire solid state communities which, however, vastly concentrated on solid electrolytes. Recently and again triggered by electrochemical applications such as batteries, fuel cells, sensors or chemical filters, interfacial processes are more and more getting to the fore in the field of solid state ionics. Hence, treatments that try to combine and generalize relevant aspects and concepts are desired. The contribution does not aim to be exhaustive, it concentrates on issues being particularly relevant in this context. It can partly rely on and refer to a variety of excellent existing reviews[2-22] and in particular on a comprehensive monograph that appeared recently.[23,24]

II. SOLIDS VS. LIQUIDS

If a liquid is reversibly cooled below the melting point, the solid, i.e., a three-dimensional giant molecule ("3D polymer") forms, characterized by a long range order. So-formed crystals exhibit strict periodicity. While, in the case of liquids, structural configurations fluctuate in space and time, glasses as "frozen liquids" or other amorphous solids show also fluctuations but essentially only with respect to the position coordinate.

There is neither a sharp demarcation in the treatment of solids compared to molecular units, nor is there a strict demarcation in the treatment of different bonding types. It is Schrödinger's (or Dirac's) equation that describes the bonding situation in all cases. Nonetheless, it proves meaningful to use "ad hoc" classifications.

In some cases the bond strength between the "monomers" (intermolecular forces) is weak and molecular crystals are formed, in others the intermolecular forces are not at all saturated within the monomer and no real distinction between inter- and intramolecular forces can be made, as is the case in typical ionic crystals, covalent crystals or metal crystals. In the latter two cases the orbitals of the bonding partners severely overlap forming comparatively wide bands in which the electrons are delocalised. In the case of metals (e.g. Na) the topmost non-empty band is only partially filled also in the thermal ground state, and the outer electrons are nearly freely mobile. In the case of typical covalent crystals (such as Si), at $T = 0$ K the topmost non-empty band is completely filled, the electrons therein immobile, and the bands of higher lying unoccupied levels separated by a substantial gap (which corresponds to the energy distance between the bonding and anti-bonding level in the two-atom problem). Those electrons can also be thought to be situated between the bonding partners. In the case of ionic crystals (e.g. NaCl) the bonding electrons are affiliated with orbitals of the electronegative partners only. Hence orbitals of the anionic partner usually form the highest occupied band while orbitals of the cationic partner form the lowest unoccupied band. (This is at least so in the case of main group elements. In a variety of transition metal oxides, e.g., both valence and conduction bands refer to the cationic states.) Owing to poor interaction of the atoms of the same kind the bands are quite narrow. (In this context it is also worthy of mention that — according to Mott — too large a distance between atoms prevents sufficient orbital overlap and delocalization.[25,26])

As well known from semiconductor physics, in non-metals electrons are, at finite temperatures, excited from the highest occupied band to the lowest unoccupied band to form excess electrons in the conduction band and electron holes in the valence band. Owing to long-range order each crystal possesses a certain amount of "free" electronic carriers. The mixed conductor which exhibits both ionic and electronic conductivity, will play an important role in this text, since it represents the general case, and pure ionic and electronic (semiconductors) conductors follow as special cases.

Under the conditions we will refer to here (the yield strength is not reached), a potentially applied stress does not change the form (no plastic deformation) during the electrochemical performance or measurements, rather is elastically supported. (The viscosity is virtually infinite; only at very high temperatures and/or extreme driving forces a creep of the solid becomes possible.) There is no convective flow, and diffusive transport is the decisive mechanism of mass transport. Owing to this, solid-solid contacts frequently exhibit a low degree of perfection.

Under usual conditions at least one sublattice is very rigid and—in the case of interest (in particular when dealing with solid ion conductors)—one sublattice exhibits a significant atomic mobility. The selectivity of the conductivity (cf. also the selective solubility of foreign species) is indeed a characteristic feature of solids.

A paramount role is played by point defects;[1,2] the configurational entropy gained by their presence allows them to be present even in thermodynamical equilibrium; they have their counterpart in liquids and will be considered explicitly in the next chapter.

Further characteristic solid state properties following from the high bond strength compared to kT, are the anisotropies of structure and properties (transport coefficients e.g. are tensors) as well as the occurrence of higher-dimensional defects such as dislocations and internal boundaries.

Grain boundaries are a specific feature of the solid state and are typical non-equilibrium defects. Their metastability, however, is often so high and their influence can be so great that it is very necessary to consider them explicitly. In the following text we will refer to them as frozen-in structure elements with zero mobility. Dislocations are usually more mobile and can often be healed out (if they are not constituting grain boundaries); we will neglect them in the following.

While external boundaries are in principle necessary because of mass conservation, their amount, nature and arrangement, and hence the shape, are usually not in equilibrium. (As long as we can ignore the energetic influence of intersections of surfaces, the Wulff-shape represents the equilibrium shape of crystals.[27,28] The Wulff-shape is characterized by a ratio of surface tension and orthogonal distance from the center that is common to every surface plane; as it is only rarely established, we will also consider the shape as frozen-in.) The occurrence of internal interfaces leads to the fact that upon the (usually strictly periodic) atomic structure we have to superimpose the micro-structure which is of rather fuzzy periodicity – if we may use this term at all. Lastly let us come back to the point that sluggish kinetics prevent many materials from adopting the ordered state: In inorganic glasses or many polymers the atomic structure is amorphous. Then, as already stated, structural fluctuations occur in space but under usual conditions not in time. Let us start with the atomic structure and the ionic charge carriers therein.

III. POINT DEFECTS AS CHARGE CARRIERS

The ionic charge carriers in ionic crystals are the point defects.[1,2,23,24] They represent the ionic excitations in the same way as H_3O^+ and OH^- ions are the ionic excitations in water (see Fig. 1). They represent the chemical excitation upon the perfect crystallographic structure in the same way as conduction electrons and holes represent electronic excitations upon the perfect valence situation. The fact that the perfect structure, i.e., ground structure, of ionic solids is composed of charged ions, does not mean that it is ionically conductive. In AgCl regular silver and chloride ions sit in deep Coulomb wells and are hence immobile. The occurrence of ionic conductivity requires ions in interstitial sites, which are mobile, or vacant sites in which neighbors can hop. Hence a "superionic" dissociation is necessary, as, e.g. established by the Frenkel reaction:

$$Ag_nCl_n + Ag_{n'}Cl_{n'} \underset{\leftarrow}{\overset{\rightarrow}{\rightleftharpoons}} (Ag_{n+1}Cl_n)^+ + (Ag_{n'-1}Cl_{n'})^- \qquad (1)$$

or more concisely in a "molecular notation"

$$2\,AgCl \underset{\leftarrow}{\overset{\rightarrow}{\rightleftharpoons}} Ag_2Cl^+ + \square Cl^- \qquad (2)$$

where it is indicated that a vacancy is formed within the cluster from which Ag^+ is removed.

A major difference of the crystalline state compared with the liquid state, is that the sites are crystallographically defined and need to be conserved in such reactions. In defect nomenclature this crystallographic site is usually given as a lower index, where i denotes the interstitial site which is occupied in $(Ag_{n+1}Cl_n)^+$, or referring to Eq. (2) in the Ag_2Cl^+ unit. In the rock-salt structured AgCl where the Ag^+ ions regularly occupy all octahedral interstices of the close-packed

Joachim Maier

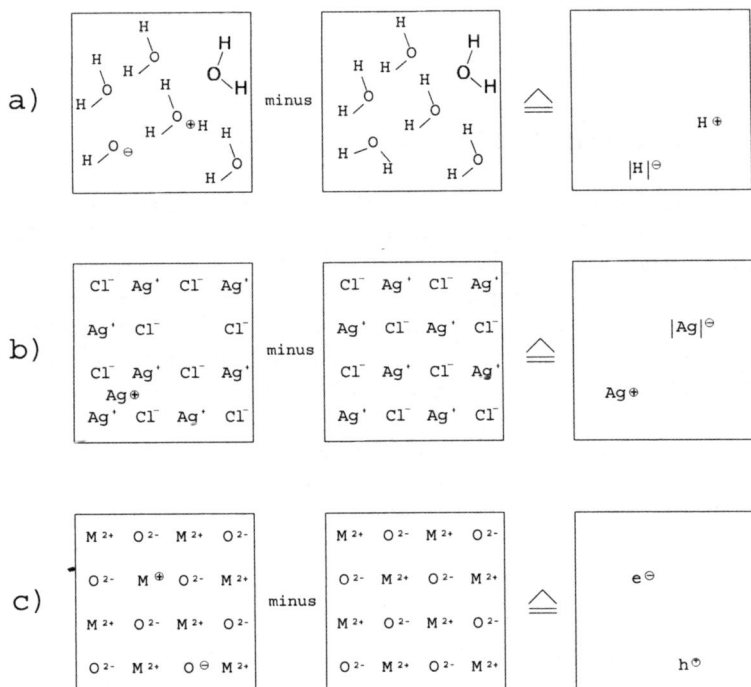

Figure 1. a) If the compositionally unperturbed structure (chemical ground structure) is substracted from the real structure, the point defects shown on the right remain. Naturally each is surrounded by a distorted region (effective radius of the point defect) which affects at least the immediate neighborhood. In the case of fluid phases (see above) this procedure can only be regarded as an instantaneous picture. Owing to the absence of defined sites no distinction is made between various types of defect reactions as is done in the solid state. b) Frenkel disorder is sketched in the second row. c) Third row shows the case of purely electronic disorder. Here the charge carriers are assumed to be localized for the sake of clarity.[20]

chlorine lattice ($n' = 6$), the interstitial site is the tetrahedral interstice ($n = 4$). Instead of writing $\square$ it is customary to use the symbol $\vee$ (for vacancy).[2]

Since the above disorder reaction is restricted to the silver sublattice we can condense Eq. (2) to the even more concise form:

$$Ag^+_{Ag^+} + v^0_i \; \underset{\leftarrow}{\overset{\rightarrow}{}} \; Ag^+_i + v^0_{Ag^+} \tag{3}$$

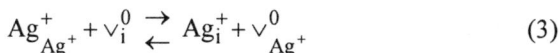

As already considered, the absolute charges do not really matter, the regular Ag-ion, $Ag^+_{Ag^+}$, is not mobile in contrast to the silver vacancy $v^0_{Ag^+}$. Hence it is advantageous to refer to relative charges (the structure element $M^{z^\bullet}_{S_{z^r}}$ has the relative charge $z^\bullet - z^r$). Omitting the absolute charges leads to the Kröger-Vink formulation[29]

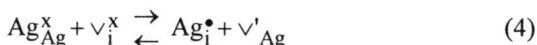

$$Ag^x_{Ag} + v^x_i \; \underset{\leftarrow}{\overset{\rightarrow}{}} \; Ag^\bullet_i + v'_{Ag} \tag{4}$$

Here the cross denotes the relative charge zero, the prime ' the relative charge -1 and the dot · the relative charge +1. Even though it is the special notation that creates quite an activation barrier to deal with point defect chemistry, it proves extremely helpful, once being familiar with it. Equation (4) denotes the so-called Frenkel reaction in structural element formulation, i.e., in terms of particles that actually constitute the real crystal. Unfortunately, thermodynamics demand strictly speaking the treatment of so-called building elements, elements that you can add to the perfect crystal.[30] The chemical potential of a vacant site, e.g. measures the increase in Gibbs energy on adding a vacancy. However, adding the structure element vacancy (v'_{Ag}) requires the removal of a lattice constituent (Ag^x_{Ag}). This leads to a complete relative description in which the Frenkel reaction is described by

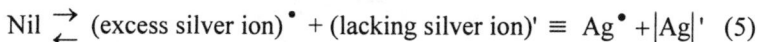

$$\text{Nil} \; \underset{\leftarrow}{\overset{\rightarrow}{}} \; (\text{excess silver ion})^\bullet + (\text{lacking silver ion})' \equiv Ag^\bullet + |Ag|' \tag{5}$$

(On the r.h.s. the defects are written in Schottky's building element notation,[30] also used in Fig. 1.) Equation (5) is obtained by subtracting 2AgCl from Eq. (2) or bringing the regular constituents in Eq. (4) to the right hand side: Obviously (excess silver ion)$^\bullet \equiv (Ag^\bullet_i - Ag^x_{Ag})$ and (lacking silver ion)' $\equiv V'_{Ag} - Ag^x_{Ag}$ are the entities for which chemical potentials can be rigorously defined.

The treatment in terms of excess and lacking particles is the key to defect chemistry. The similarity with "aqueous chemistry" is obvious: We end up with an analogous "disorder" equation (as a snap-shot at a given time) with H_3O^+ and OH^- as "defects" when removing all H_2O from the autoprotolysis reaction:

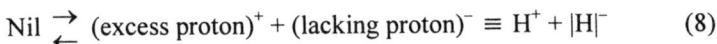

$$2\ H_2O \underset{\leftarrow}{\overset{\rightarrow}{\rightleftharpoons}} H_3O^+ + OH^- \tag{6}$$

$$H_2O \underset{\leftarrow}{\overset{\rightarrow}{\rightleftharpoons}} H^+ + OH^- \tag{7}$$

$$Nil \underset{\leftarrow}{\overset{\rightarrow}{\rightleftharpoons}} (\text{excess proton})^+ + (\text{lacking proton})^- \equiv H^+ + |H|^- \tag{8}$$

(compare Eq. 6 with Eq. 2; Eq. 7 with Eqs. 3 and 4; and Eq. 8 with Eq. 5). Here we still can use absolute charges since the ground state is uncharged.

The analogous situation is met if we excite an electron from the ground state (electron in the valence band, VB) to an energetically higher position (in the conduction band, CB), e.g. in silicon, according to

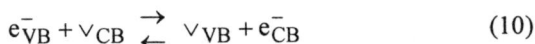

$$Si_{Si} + Si_{Si} \underset{\leftarrow}{\overset{\rightarrow}{\rightleftharpoons}} Si^+_{Si} + Si^-_{Si} \tag{9}$$

$$e^-_{VB} + V_{CB} \underset{\leftarrow}{\overset{\rightarrow}{\rightleftharpoons}} V_{VB} + e^-_{CB} \tag{10}$$

the symbol V denotes here the electronic vacancy (hole). Equation (10) reads in relative charges

$$e^x_{VB} + V^x_{CB} \underset{\leftarrow}{\overset{\rightarrow}{\rightleftharpoons}} e'_{CB} + V^{\bullet}_{VB} \tag{11}$$

or simpler in building element notation ($e' \equiv e'_{CB} - v^x_{CB}$,
$h^{\bullet} \equiv v^{\bullet}_{VB} - e^x_{VB} \equiv |e|^{\bullet}$)

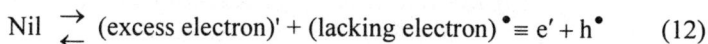

$$Nil \underset{\leftarrow}{\rightarrow} (excess\ electron)' + (lacking\ electron)^{\bullet} \equiv e' + h^{\bullet} \qquad (12)$$

Equation (12) is independent of any detailed knowledge of the band structure. In AgCl where the valence band is formed by the Cl-p orbitals and the conduction band by the Ag-d orbitals, the detailed formulation of Eq. (12) would read

$$Ag^+ + Cl^- \underset{\leftarrow}{\rightarrow} Ag^0 + Cl^0 \qquad (13)$$

while in PbO, to mention a main group metal oxide (see Fig. 1c), we would have to write

$$Pb^{2+} + O^2 \underset{\leftarrow}{\rightarrow} Pb^+ + O^- \qquad (14)$$

In the case of transition metal compounds such as $LaMnO_3$ d-level splitting can be so pronounced that both valence and conduction band are derived from d-orbitals. For $LaMnO_3$, e.g. Eq. (12) has to be interpreted as a redox disproportionation

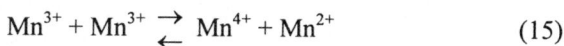

$$Mn^{3+} + Mn^{3+} \underset{\leftarrow}{\rightarrow} Mn^{4+} + Mn^{2+} \qquad (15)$$

Note that generally bands correspond to hybrids, and attributing bands to certain elements is an approximation. Unlike the electronic "energy" level distribution, the distribution of ionic "energy" levels in crystals (the meaning of which we will consider in the next section) is discrete. In the electronic case we face bands comprising a manifold of narrowly neighbored levels, so that we better speak of a continuous density of states.

In a fluid system such as water the energy levels can be conceived as smeared out on spatial and time average, while in amorphous solids the structure fluctuations occur in space but (approximately) not in time (see e.g. Refs.[31-34]).

IV. EQUILIBRIUM CONCENTRATION OF CHARGE CARRIERS IN THE BULK

1. Defect Reactions

A major difference between crystals and fluids refers to the necessity of distinguishing between different sites. So the autoprotolysis in water could, just from a mass balance point of view, also be considered e.g. as a formation of a OH^- vacancy and a H^+ vacancy. In solids such a disorder is called Schottky disorder (S) and has to be well discerned from the Frenkel disorder (F). In the densely packed alkali metal halides in which the cations are not as polarizable as the Ag^+, the formation of interstitial defects requires an unrealistically high energy and the dominating disorder is thus the Schottky reaction

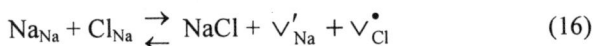

$$Na_{Na} + Cl_{Na} \underset{\leftarrow}{\overset{\rightarrow}{}} NaCl + V'_{Na} + V^{\bullet}_{Cl} \tag{16}$$

which reads in building element notation

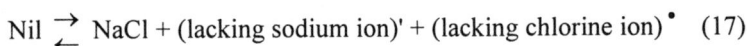

$$\text{Nil} \underset{\leftarrow}{\overset{\rightarrow}{}} NaCl + (\text{lacking sodium ion})' + (\text{lacking chlorine ion})^{\bullet} \tag{17}$$

In particular if the anions are as small as F^-, e.g. in CaF_2, we can have Frenkel disorder in the anion sublattice, which is also referred to as anti-Frenkel-disorder ($\overline{F}$),

$$F_F + V_i \underset{\leftarrow}{\overset{\rightarrow}{}} F'_i + V^{\bullet}_F \tag{18}$$

Equation (18) reads in building element notation

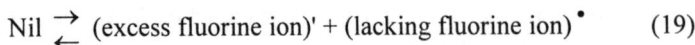

$$\text{Nil} \underset{\leftarrow}{\overset{\rightarrow}{}} (\text{excess fluorine ion})' + (\text{lacking fluorine ion})^{\bullet} \tag{19}$$

In addition, there should also be the possibility of having the so-called anti-Schottky reaction ($\overline{S}$) in materials with comparatively loose structures, such as orthorhombic PbO

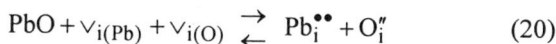

$$PbO + v_{i(Pb)} + v_{i(O)} \underset{\leftarrow}{\overset{\rightarrow}{}} Pb_i^{\bullet\bullet} + O_i'' \tag{20}$$

or in terms of building elements

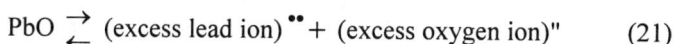

$$PbO \underset{\leftarrow}{\overset{\rightarrow}{}} (excess\ lead\ ion)^{\bullet\bullet} + (excess\ oxygen\ ion)'' \tag{21}$$

More complex internal defect reactions involve association reactions or doping effects; they will be considered later. (For more extensive treatments of defect chemistry see Ref.[2])

All these reactions leave the 1:1 composition, the "Dalton composition"[23], unchanged. However, according to the phase rule, stoichiometric variations are possible by tuning the chemical potential of one component, e.g. the chlorine partial pressure over AgCl or the oxygen partial pressure over PbO, (or generally n-1 component partial pressures in a n-component multinary, e.g. the O_2 and SrO partial pressures over $SrTiO_3$). The interaction with oxygen may be formulated as:

$$\frac{1}{2}O_2 + v_O^{\bullet\bullet} + 2e' \underset{\leftarrow}{\overset{\rightarrow}{}} O_O \tag{22}$$

meaning that the oxygen is incorporated on a vacant site as O^{2-} which requires the annihilation of two conduction electrons. Equally the interaction could have been formulated by assuming that two regular electrons are annihilated (and $2h^{\bullet}$ would be formed), or we could have incorporated oxygen ions into interstitial sites. In equilibrium one formulation suffices, since the others follow by reaction combinations. It is advisable to use a formulation in terms of majority carriers. Evidently, Eq. (22) involves the ionic and electronic defects, changes the oxide $MO_{1+\delta}$ into $MO_{1+\delta+\Delta\delta}$ and implies a variation of the mean oxidation numbers. (In ternary systems such stoichiometric changes can occur without involving the electronic budget, e.g. the incorporation of H_2O in NaOH.)

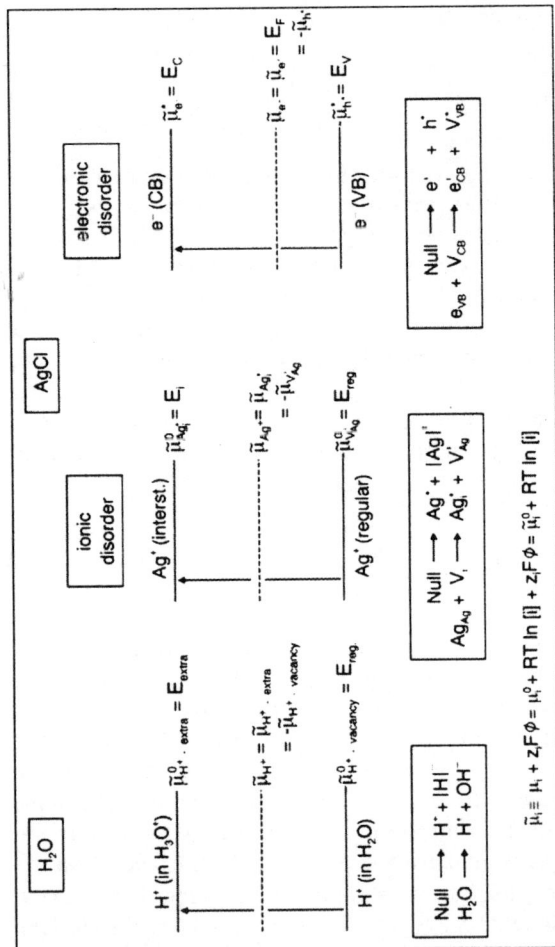

Figure 2. Electronic and ionic disorder in ionic solids and in water in "physical" (top) and "chemical" language (bottom).[20] The coupling of the ionic and electronic Fermi levels takes place via the chemical potential of the neutral components (here: $\tilde{\mu}_{Ag^+} + \tilde{\mu}_{e^-} = \mu_{Ag}$) (see Fig. 3). A further relevant example could be the breaking-up of an ion pair in a polymer.

For a given material all the defect reactions considered so far occur to a greater or lesser extent. It is the problem of the next section to show how the equilibrium concentrations of all defects (including electronic carriers) can be calculated.

2. Equilibrium Defect Concentrations in Pure Compounds

Figure 2 translates the charge carrier formation reaction into an "energy level" diagram for various systems. In fact these levels refer to standard chemical potentials or (in the case of the "Fermi-levels") to full chemical potentials (see e.g. Refs.[3,35]). As long as —in pure materials— the gap remains large compared to RT, the Boltzmann-form of the chemical potential of the respective charge carrier (defect) is valid,

$$\mu_j = \mu_j^\circ + RT \ln (c_j / c_j^\circ) \tag{23}$$

independent of the charge carrier situation and also independent of the form of the energy level distributions. If there is a smearing out of the levels or even a band of levels, μ_j° refers to an effective level, such as the band edges when dealing with the electronic picture. Equation (23) results from a simple combinatorial analysis by assuming a random configurational entropy, and it is the configurational entropy which is the reason why defective states of atomic dimensions (such as vacancies, interstitial sites, excess electrons, holes, H_3O^+, OH^- etc.) are important in thermal equilibrium, i.e., exhibiting non-zero concentrations. Figure 3 shows the coupling of the ionic level picture and the electronic level picture for AgCl via the thermodynamic relation $\mu_{Ag^+} + \mu_{e^-} = \mu_{Ag}$ and hence via the precise position in the phase diagram.

Let us consider an elemental crystal first (with defect d). If N_d identical defects are formed in such a crystal of N identical elements, a local free enthalpy of Δg_d^0 is required to form a single defect, and if interactions can be neglected, the Gibbs energy of the defective crystal (G_p refers to the perfect crystal) is

$$G = G_p + N_d \Delta g_d^\circ - k_B T \ln \left(\frac{N}{N_d} \right) \tag{24}$$

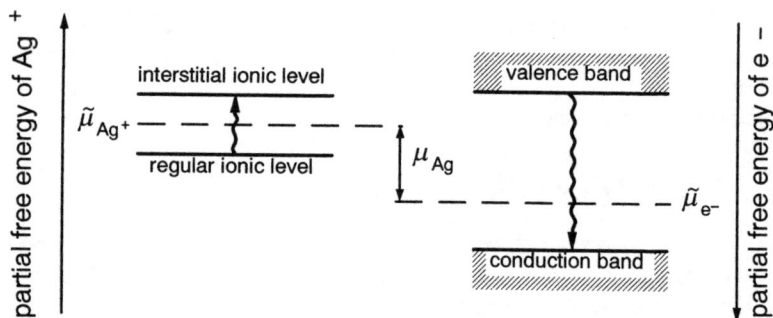

Figure 3. Coupling of ionic and electronic levels in AgCl.[36,37] (Reprinted from J. Maier, "Defect chemistry and ion transport in nanostructured materials. Part II. Aspects of nanoionics." *Solid State Ionics*, **157**, 327-334. Copyright © 2003 with permission from Elsevier.)

(For a description of how Δg^0 can be atomistically computed, the reader is referred to the literature.[18,38-40]) Owing to the infinitely steep decrease of the configurational entropy term (cf. last term in Eq. 24) there is a minimum in G (see Fig. 4) that occurs in our elemental crystal if the

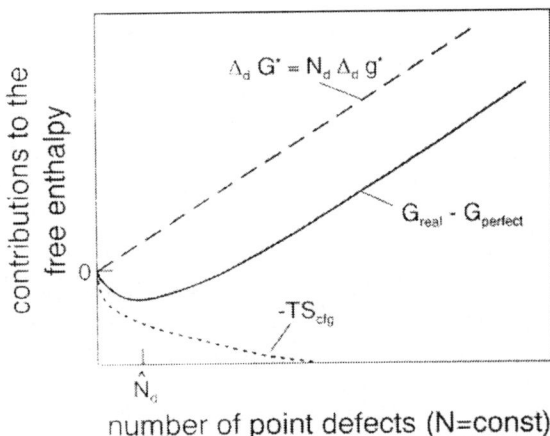

Figure 4. Contributions to the free enthalpy of the solid by defect formation with a constant total number of sites.

chemical potential of d, i.e., $\mu_d \equiv \partial G / \partial(N_d / N_m)$, vanishes, which is given by

$$\mu_d = \mu_d^o + RT \ln \frac{N_d}{N - N_d} \simeq \mu_d^o + RT \ln \frac{N_d}{N} = \mu_d^o + RT \ln \frac{n_d}{n} \qquad (25)$$

Equation (25) follows after differentiation and application of Stirling's formula ($n \equiv N/N_m$, $n_d \equiv N_d/N_m$, N_m = Avogadro's number; $\mu^o = N_m \Delta g^o_d$). It is worthy to note that the strict result (see l.h.s. of Eq. 25) which is formally valid also for higher concentrations, is of the Fermi-Dirac type. This is due to the fact that double occupancy is forbidden and hence the sites are exhaustible similar as it is the case for the quantum states in the electronic problems.

If we consider a situation in which the levels are broadened to a more or less continuous zone the Fermi-Dirac form given by Eq. (25) l.h.s. is only valid, if we attribute N_d and μ_d^o to an infinitely small level interval (ranging from E_d to $E_d + dE_d$); then in order to obtain the total concentration, the molar density of states D (E_d) has to be considered, and the result for n_d follows by integration:

$$n_d = \int_{zone} \frac{D(E_d)\,dE_d}{1 + \exp \dfrac{|E_d - \mu_d|}{RT}} \qquad (26)$$

Note that μ_d (Fermi-level) is independent of E_d in equilibrium. Equation (26) has been discussed for a variety of state densities (delta function for point defects in crystals, Gaussian function for charge carriers in an amorphous matrix, parabolic for electronic carriers in ideal band) in Ref.[31] The above considerations can be obviously only applied to situations in which carriers are statistically distributed, i.e., particles in almost empty zones or vacancies (holes) in an almost full zone. The T-dependence of the occupation refers to the thermal excitation from the top occupied zone to the lowest empty zone. In the systems of interest the Fermi-level (μ_d) lies in between these two zones with the absolute distance between μ_d and the neighbored zone edges being great compared to RT. If, without restriction of generality, the almost empty zone is considered and the lower edge of this zone designated as E_d', for all levels more distant from μ_d than E_d' the unity in the denominator of Eq. (26) can be neglected, leading to

$$n_d \approx \bar{n}_d(T)\exp-\left|E'_d - \mu_d\right|/RT \qquad (27)$$

whereby the effective value is given by

$$\bar{n}_d(T) = \int_{zone} dE_d \left\{ D(E_d)\exp-\left|E_d - E'_d\right|/RT \right\} \qquad (28)$$

and is usually only a weak function of temperature.[32] Equation (28) as Eq. (26) ignores changes of the parameters with occupation (in semiconductors this is called rigid band model).

As anticipated above, the Boltzmann-form of the chemical potential results. If n in Eq. (25) is identified with $\bar{n}_d$, the effective value μ_d° is given by E'_d. As long as we refer to dilute conditions, freedom of normalization is left with respect to the concentration measure (see Eq. 23).

At this point we do not want to treat the case of high concentrations, since then we also have to include interactions. However, we want to draw the reader's attention to a peculiar point: Thermodynamically μ_d refers to the defect as a building element for which an activity term $N_d/(N-N_d)$ results at high concentrations. Frequently, in the literature structure elements (defective and regular ones) are used in the thermodynamic treatment.[29] Even though not correct a priori, it leads to correct results if they are defined properly, since $x_d \equiv N_d/N$ is the concentration of the defective structural unit and $1 - x_d$ is the concentration of the regular constituent (replaced by the defective one) owing to mass conservation. By applying Boltzmann approximations to both we automatically end up with the Fermi-Dirac correction when we later formulate mass action laws. This is also well known when dealing with Langmuir adsorption. The adsorbed species has the activity $\theta/(1-\theta)$ which can also be expressed as the mass action ratio of occupied (θ) and free sites ($1 - \theta$).

Strictly speaking dealing with charged particles requires taking account of the electrochemical potential (already used in Figs. 2 and 3)

$$\tilde{\mu}_d = \mu_d + z_d F\phi \qquad (29)$$

which includes the electrical potential ϕ.[41] However, when we deal with bulk problems where ϕ is constant or generally with reactions that are electroneutral at given position, this term cancels.

If one repeats the above statistical derivation for the formation of a pair of defects or for more complicated reactions of the type

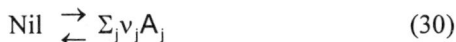

$$\text{Nil} \underset{\leftarrow}{\overset{\rightarrow}{\rightleftharpoons}} \Sigma_j \nu_j A_j \tag{30}$$

Equation (29) is valid for the individual defects as long as interactions and non-random configurational effects can be neglected; hence, mass action laws result

$$\Sigma_j \nu_j \mu_j = 0 = \Sigma_j \nu_j \mu_j^\circ + RT \ln \Pi_j (c_j / c_j^\circ)^{\nu_j} \tag{31}$$

This set of reactions and mass action laws has to be coupled with the electroneutrality reaction

$$\Sigma_j z_j c_j = 0 \tag{32}$$

which is a special case of Poisson's equation for a homogeneous material. Since the mass action laws (dilute) linearly combine the logarithms of the concentrations, we only arrive at simple solutions, if the electroneutrality equation reduces to a proportionality (in the following this case is called Brouwer condition[42]) which is necessarily fulfilled if two (oppositely charged) carriers are in the majority.

Let us explicitly consider an oxide MO and assume that only oxygen defects can be formed to a perceptible extent, then we have to consider the set of internal and external equations given by Table 1. (Note that one disorder reaction is redundant in Table 1.) The free parameters are obviously (i) the temperature which enters the mass action constants (see Table 1) and (ii) the partial pressure P_{X_2} which we also simply term P. (The hydrostatic pressure p is assumed to be constant.) For Brouwer (and Boltzmann) conditions we find results of the type (the index r labels the reaction)

$$c_j(T, P) = \alpha_j^{\beta_j} P^{N_j} \Pi_r K_r(T)^{\gamma_{rj}} \tag{33}$$

Table 1
Simple Defect Chemistry of $M_{1+\delta}X$

Reaction (r)	Mass Action Law
Internal	
Schottky (S)	$[V_X^{\bullet}][V_M'] = K_S$
Frenkel (F)	$[M_i^{\bullet}][V_M'] = K_F$
Anti-Frenkel ($\bar{F}$)	$[V_X^{\bullet}][X_i'] = K_{\bar{F}}$
Anti-Schottky ($\bar{S}$)	$[M_i^{\bullet}][X_i'] = K_{\bar{S}}$
Band-Band (B)	$[h^{\bullet}][e'] = K_B$

$$K_r(T) \propto \exp\left[+\frac{\Delta_r S_m^0}{R}\right]\exp\left[-\frac{\Delta_r H_m^0}{RT}\right]$$

External

Reaction with the gas phase (X) $P_{X_2}^{-1/2}[V_X^{\bullet}]^{-1}[e']^{-1} = K_X$

Electroneutrality Condition

$$[V_X^{\bullet}] + [M_i^{\bullet}] + [h^{\bullet}] = [V_M'] + [X_i'] + [e'](\pm C)$$

(N_j, γ_{rj}, α_j, β_j being simple rational numbers.) As the Brouwer conditions are only valid within certain parameter windows (belonging to the P_{O_2}-T-plane) and change from window to window, Eq. (33) is only a sectional solution and N_j, γ_{rj}, α_j, β_j change from window to window. Equation (33) defines van't Hoff-diagrams characterized by

$$-\frac{\partial \ln c_j}{\partial 1/RT} = \sum_r \gamma_r \Delta_r H^o \qquad (34)$$

For not too extended T-ranges $\Delta_r H^o$ is constant and straight lines are observed in the ln c vs. 1/T representation. Also straight lines are observed if we plot ln c_j vs. ln P defining Kröger-Vink diagrams with the characteristic slopes

$$\frac{\partial \ln c_j}{\partial \ln P} = N_j \tag{35}$$

In most examples we will refer to the conductivity σ_j as a convenient measure of c_j. Since $\sigma_j \propto c_j$ with the proportionality factor containing the mobility $u_j(T)$, the temperature dependence (see Eq. 34) of σ_j will also include the migration enthalpy (cf. Section VI.3.$ii.$), while the P- (and later the C-) dependence (Eqs. 35 and 43) is the same.

Figure 5 displays the internal acid-base and redox chemistry within the phase width of MO. At low P_{O_2} vacant oxygen ions ($v_O^{\bullet\bullet}$) and

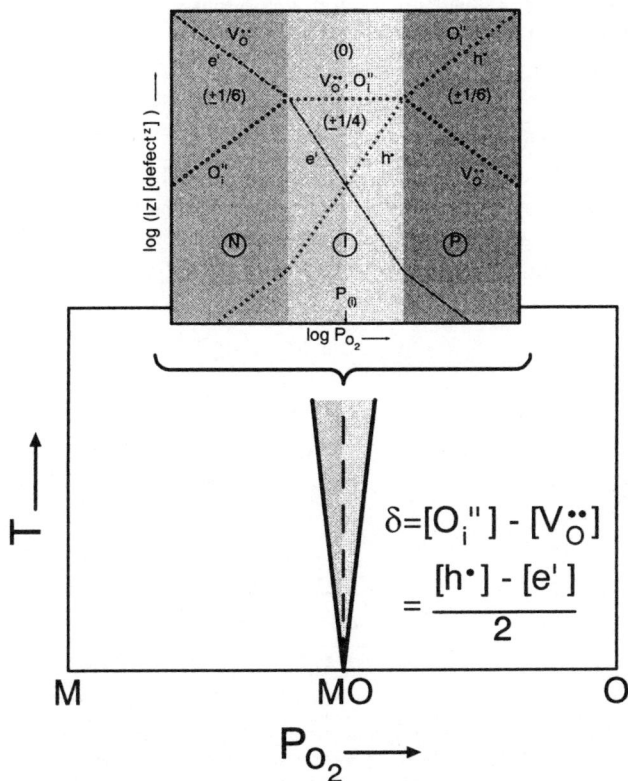

Figure 5. Internal redox and acid-base chemistry in a solid $MO_{1+\delta}$ within the phase width at a given temperature.

reduced   states   (e′)   are   in   the   majority   according   to
$2[v_O^{\bullet\bullet}] \approx [e′] >> [O_i''], [h^\bullet]$ (N-regime); oxygen interstitials (O_i'')
and oxidized states ($h^\bullet$) dominate for very high P_{O_2} : $2[O_i''] \approx [h^\bullet]$
$>> [e′], [v_O^{\bullet\bullet}]$ (P-regime). At intermediate P_{O_2} usually $[v_O^{\bullet\bullet}] \approx [O_i'']$
$>> [e′], [h^\bullet]$ (I-regime; the alternative $[e′] \approx [h^\bullet] >> [v_O^{\bullet\bullet}], [O_i'']$ is less
usual). Exactly at the Dalton-composition, i.e., at $P_{O_2}^{(i)}$ where $2[O_i''] =$
$2[v_O^{\bullet\bullet}] = [e′] = [h^\bullet]$, the "non-stoichiometry" δ in $MO_{1+\delta}$ (i.e., $[O_i'']$ -
$[v_O^{\bullet\bullet}] = (1/2)[h^\bullet] - (1/2)[e′])$ is precisely zero. For $\delta >> 0$ we meet a p-
conductor (because the mobility of $h^\bullet$ is usually much greater than for
O_i''), for $\delta << 0$ we meet (for analogous mobility reasons) an
n-conductor, while at $\delta \approx 0$ mixed conduction is expected (only if the
ionic concentrations are much larger than the electronic ones, pure ion
conduction is expected). Since there are limits of realizing extreme
P_{O_2} –values and also limits with respect to the phase stability
(formation of higher oxides, lower oxides or of the metal), usually not
the entire diagram is observed.

Figure 6 gives three examples. SnO_2 is an oxygen deficient
material[43,44] and the solution of Eq. (22) with $2[v_O^{\bullet\bullet}] = [e′]$ directly
leads to

$$[e′] = 2[v_O^{\bullet\bullet}] \propto P_{O_2}^{-1/6} \qquad (36)$$

(The slopes of -1/4 at lower T stem from doping effects, see below.)
La_2CuO_4 is an example of a p-type conduction for which $[h^\bullet] = 2[O_i''']$;
it follows that

$$[h^\bullet] = 2[O_i''] \propto P_{O_2}^{+1/6} \qquad (37)$$

Figure 6. Three experimental examples of Kröger-Vink diagrams in a pure oxide MO
with ideal defect chemistry: SnO_2 as n-type conductor, PbO as mixed conductor and
La_2CuO_4 as p-type conductor.[45] (Reprinted from J. Maier, "Ionic and Mixed Conductors
for Electrochemical Devices," *Radiat. Eff. Defects Solids*, **158**, 1-10. Copyright ©2003
with permission from Taylor & Francis.)

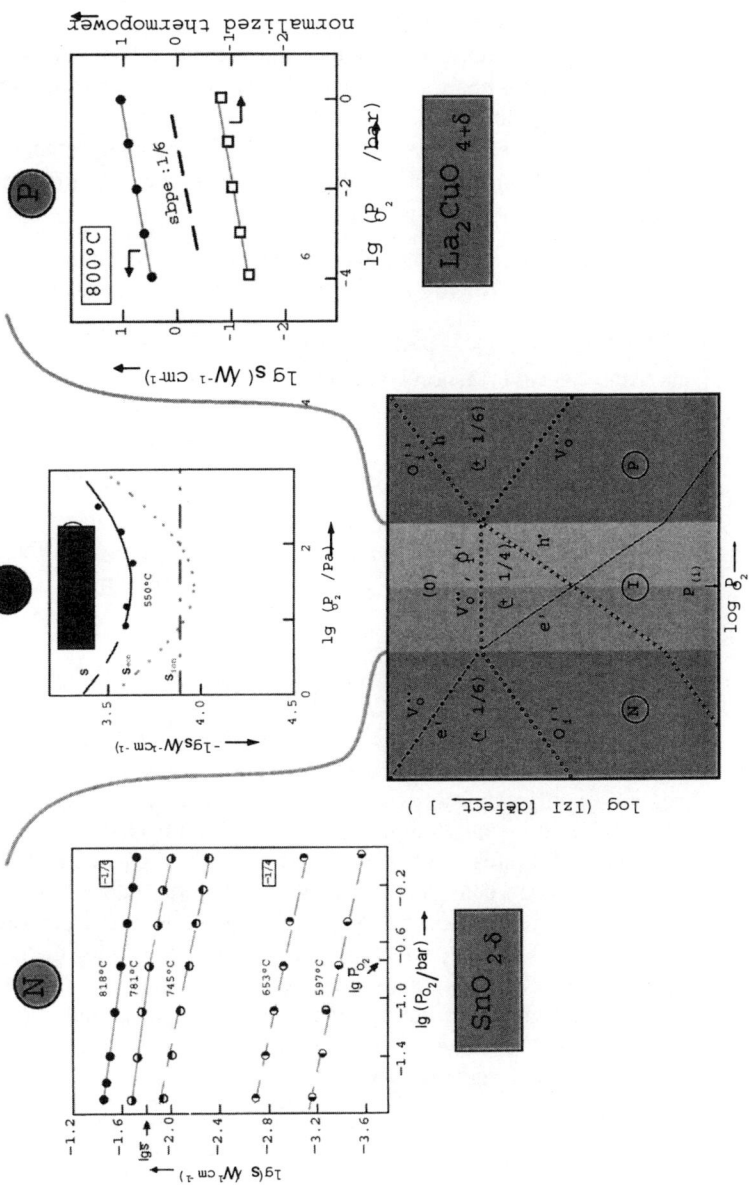

In this case the increased $[h^\bullet]$ at high P_{O_2} leads to superconductivity at very low temperatures.[46] (The fact that we are facing a ternary does not change the picture, as long as metal defects are negligible.)

PbO is an example of I-regime and exhibits mixed conduction,[47,48] σ_{ion} is constant since the decisive ionic defect concentration ($[O_i^{''}]$) is determined by the square of the mass action constant of the ionic disorder reaction. (In PbO very probably the counter-defect to $O_i^{''}$ is $Pb_i^{\bullet\bullet}$ rather than $V_O^{\bullet\bullet}$; the results, however, are not different then). Owing to Eq. (22), $[e'] \propto P_{O_2}^{-1/4}$, $[h^\bullet] \propto P_{O_2}^{+1/4}$.

A related example that also shows predominant ionic disorder, is AgCl.[49] Here the role of P_{O_2} is played by the chlorine partial pressure. Analogously $[V'_{Ag}] = [Ag_i^\bullet] = \sqrt{K_F}$, $[e'] \propto P_{Cl_2}^{-1/2}$, $[h^\bullet] \propto P_{Cl_2}^{+1/2}$ (where now the decisive disorder reaction is the Frenkel reaction of the Ag sublattice; the exponent 1/2 results since the effective charges are ± 1). Please note that in these examples with overwhelming ion disorder, e.g. $[Ag_i^\bullet] \approx [V'_{Ag}] \approx \sqrt{K_F}$, the equality sign refers to a *relative* constancy. Of course any Ag incorporated as Ag^+ involves equal *absolute* changes in the ionic and electronic budget. Thus, the more precise formulation would be $\ln [Ag_i^\bullet] \approx \ln [V'_{Ag}]$, i.e., approximate equality in terms of chemical potential.

3. Doping Effects

At low enough temperatures the defect concentration predicted by the above considerations fall below the impurity limit. In such an extrinsic regime impurity defects — if charged — have to be considered in the electroneutrality equation. In turn, purposeful introduction of impurities is a powerful means of tuning charge carrier chemistry. Since the impurities are considered to be immobile under measurement conditions, their concentration only appears in the electroneutrality equation. Such doping effects are quite characteristic for the solid state in that virtually only the foreign cation or anion is soluble, while usually both ions can be dissolved in a fluid phase and a net charge effect is not achieved. If AgCl contains Cd^{2+} impurities,[49] the electroneutrality equation reads

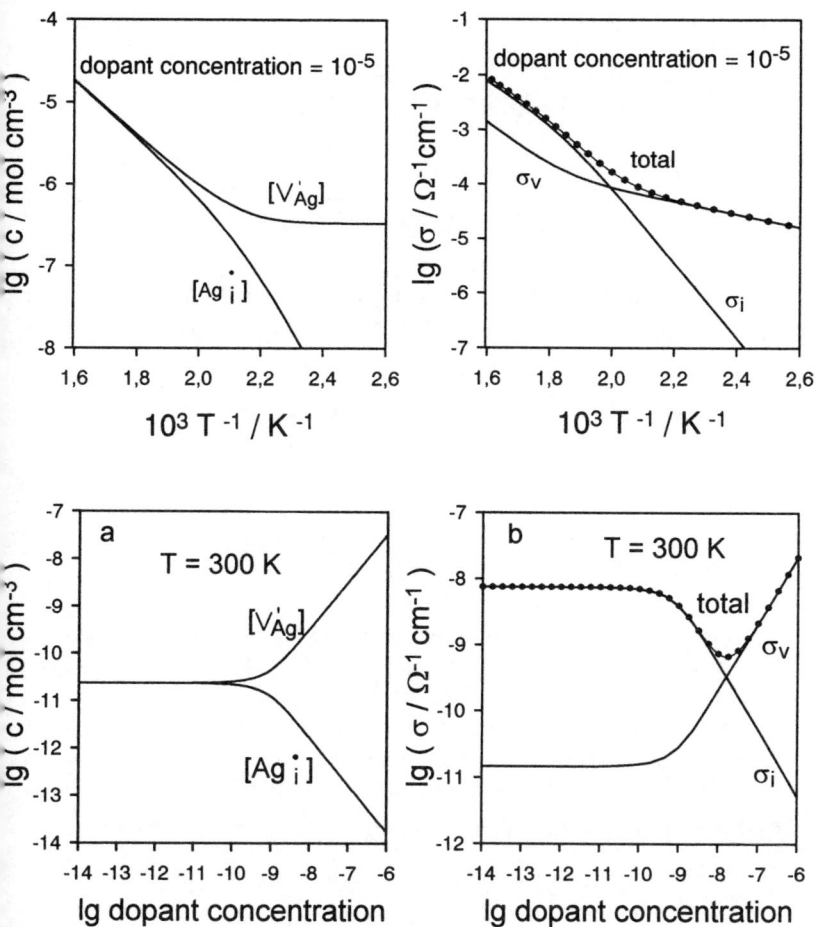

Figure 7. Defect concentrations (l.h.s.) and conductivities (r.h.s.) calculated according to Eq. (39) as a function of temperature for a fixed Cd-content (top row) and as a function of Cd-content for a fixed temperature (bottom row).[24]

$$[Ag_i^\bullet] + [Cd_{Ag}^\bullet] = [\vee_{Ag}']$$ (38)

Coupling with the Frenkel equation, $[Ag_i^\bullet]\,[\vee_{Ag}'] = K_F$ leads to

$$[\vee_{Ag}'] = C/2 + \sqrt{C^2/4 + K_F}$$ (39a)

$$[Ag_i^\bullet] = -C/2 + \sqrt{C^2/4 + K_F}$$ (39b)

where C stands for $[Cd_{Ag}^\bullet]$. Let us again consider Brouwer conditions. For $C \ll \sqrt{K_F}$ the intrinsic result $[\vee_{Ag}^\bullet] = [Ag_i^\bullet] = \sqrt{K_F}$ is obtained. For $C \gg \sqrt{K_F}$ power laws are found that is immediately obvious by neglecting $[Ag_i^\bullet]$ already in the electroneutrality equation

$$[\vee_{Ag}^\bullet] = C$$ (40)

$$[Ag_i^\bullet] = K_F C^{-1}$$ (41)

Figure 7 shows the solutions obtained by the more accurate Eq. (39) for both concentrations and conductivities. As soon as the dopant content becomes appreciable, it correspondingly increases the concentration of the oppositely charged defect which depresses the active counter-defect via mass action. In AgCl the mobility of the silver interstitials exceeds that of the vacancies leading to a minimum in the overall ion conductivity approximately at $\sqrt{K_F u_i / u_v}$ as can be readily verified ($u_{i,v}$: mobility of $Ag_i^\bullet$, $\vee_{Ag}'$).

Figures 8 and 9 show the temperature dependence in pure samples and the doping dependence with regard to positive and negative doping. While the response to Cd^{2+} doping ($Cd_{Ag}^\bullet$) follows exactly the theory (Figs. 8 and 9), S^{2-}-doping (S_{Cl}') suffers from interaction effects (see below), but the absence of a minimum is in qualitative agreement with the fact that an increase of $[Ag_i^\bullet]$ is effected (Fig. 9, l.h.s.). Figures 7 and 8 display the succession of intrinsic ($C \ll \sqrt{K_F}$) and the extrinsic regimes on T decrease. The knee in the $\sigma(T)$ curve corresponds to the

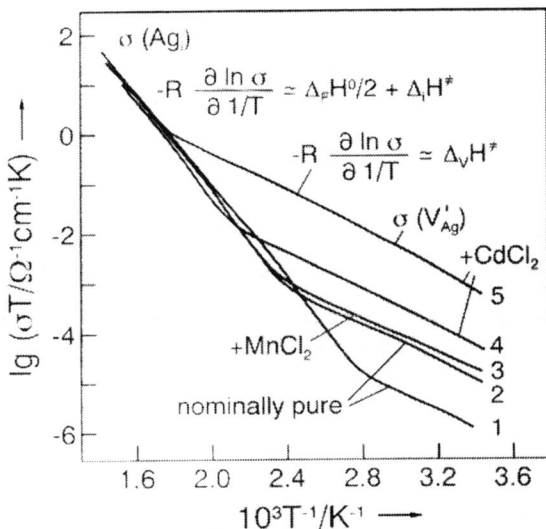

Figure 8. Experimental conductivity data for nominally pure and doped AgCl as a function of 1/T. Here lg (σT) is plotted instead of lg σ to take account of the slight T-dependence of the pre-factor. However, this does not alter the slope noticeably.[50] (Reprinted from J. Corish, P. W. M. Jacobs, "Ionic conductivity of silver chloride single crystals." *J. Phys. Chem. Solids*, **33**, 1799-1818. Copyright © 1972 with permission from Elsevier.)

minimum in the σ(C) curve; the low temperature branch reflects the dominance of the vacancy concentration which is now given by $\sigma \simeq \sigma_v$ = Fu$_v$C. Since the concentration term is constant, the activation energy yields directly the migration energy of the vacancies.

It is clear that, in extension of Eq. (39), under Brouwer and Boltzmann conditions, the result for any carrier concentration must be

(we also replace P^{N_j} by $\Pi_p P_p^{N_{pj}}$ in order to allow for multinary equilibria labeled by p)

$$c_j(T,P,C) = \alpha_j^{\beta_j} \Pi_p P_p^{N_{pj}} C^{M_j} \Pi_r K_r(T)^{\gamma_{rj}} \tag{42}$$

which now also predicts sectionally constant slopes in diagrams of the type log c_j vs. log C according to

$$\frac{\partial \ln c_j}{\partial \ln C} = M_j \tag{43}$$

Let us briefly discuss two further examples. If SnO_2 with the intrinsic electroneutrality equation $2[v_O^{\bullet\bullet}] = [e']$ is doped by Fe or In, the resulting Fe'_{Sn} or In'_{Sn} defects will increase $[v_O^{\bullet\bullet}]$ and depress $[e']$, resulting in a decrease of the electronic conductivity which will now be given by (K_O: mass action constant of reaction Eq. 22)

$$\sigma \approx \sigma_n = u_n F P^{-1/4} C^{-1/2} K_O^{1/2} \tag{44}$$

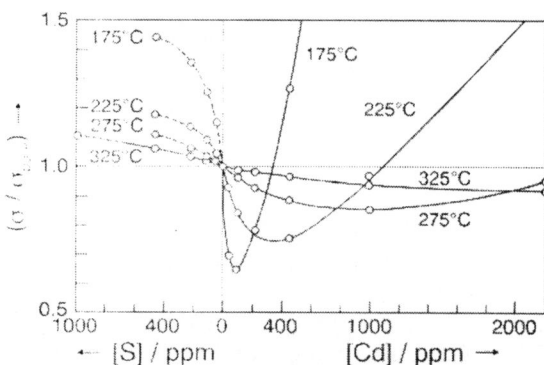

Figure 9. The dependence of the conductivity increase brought about by the impurities, on the S and Cd content of AgCl or AgBr. The solid curves (top) are calculated according to Eq. (39).[49] Reprinted from J. Teltow, Z. physik. Chem. 213-224, 195, Copyright © 1950 with permission from Oldenbourg Wissenschaftsverlag and from J. Teltow, Ann. Physik 63-70, 6, Copyright © 1949 with permission from WILEY-VCH.

This P_{O_2} behavior was seen in Fig. 6 (l.h.s., low T). Higher temperatures and low P_{O_2} favor the intrinsic situation with the slope -1/6. (Similarly also the T-dependence can be exploited.) If we positively dope SnO_2 via higher valent cations or lower valent anions ($Cl_O^{\bullet}$) we increase σ_n and make it P_{O_2} independent ([e$'$] = [$Cl_O^{\bullet}$]) (Fig. 10).

On the contrary, the negative doping of La_2CuO_4 by Sr (Sr'_{La}) increases the hole conductivity (this "LSC" is a superconductor already below 40 K.)[52] According to [Sr'_{La}] = [h$^{\bullet}$] the hole conductivity then becomes independent of P_{O_2}.

Figure 11 shows all relevant defect concentrations at high T as a function of Sr-content. It is visibly that all positively charged defects

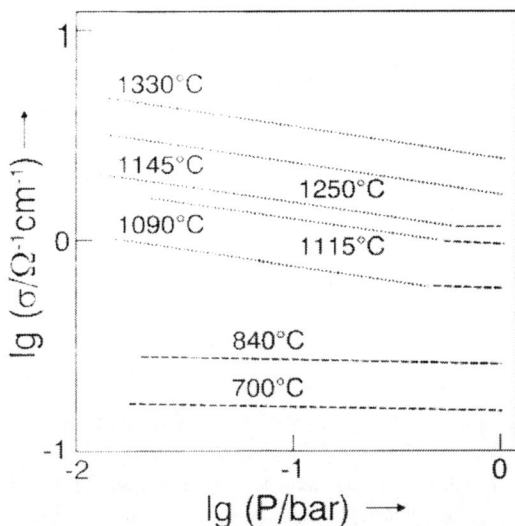

Figure 10. The P_{O_2} dependence of the conduc-tivity[44,45] for positively doped SnO_2 (doped with $SnCl_4$) exhibiting native and impurity-dominated regions.[44] (Reprinted from J. Maier and W. Göpel, *J. Solid St. Chem.* **72**, 293-302. Copy-right © 1988 with permission from Elsevier.)

are increased and all negatively charged defects are depressed in their number. The reason is not electroneutrality alone (this only would demand an overall increase of the positive counter-charge), but also the fact that two oppositely charged carriers appear in the individual disorder reactions.

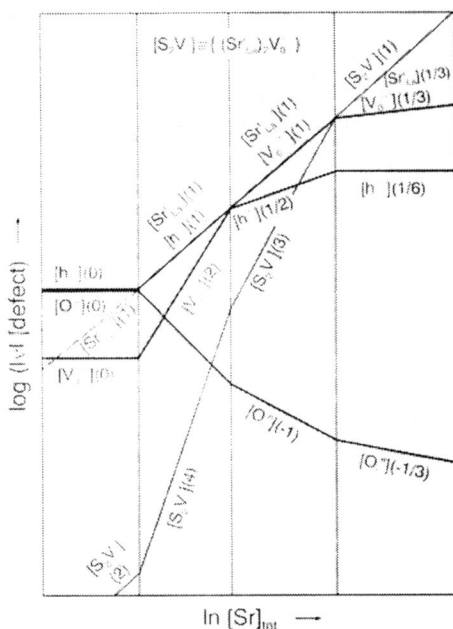

Figure 11. The dependence of defect con-centrations on Sr doping of La_2CuO_4 for a given P_{O_2}. Interactions are not included. Only the formation of one associate $S_2V = (Sr'_{La})_2 V_O^{\bullet\bullet}$ is taken into account (see following section). The figures in brackets give the M values (slopes). The parameter $|v|$ in the ordinate labeling gives the absolute value of the charge in the case of ionized defects; in the case of S_2V, however, the Sr content per formula unit ($|v| = 2$).[53]

Equation (42) highlights the basic parameters via which the defect chemistry can be tuned: these are T, P, C. Owing to their importance let us formulate the following qualitative T-, P-, C-rules which hold for simple defect chemistry:

1. Component activity rule (P-rule). If we increase the partial pressure of the electronegative (-positive) component we increase (decrease) the number of holes and decrease (increase) the number of excess electrons; we increase (decrease) the numbers of all defects which individually increase (decrease) the X to M stoichiometry and decrease (increase) the others. As the other rules, these statements—since compensation effects do not occur—are not trivial in that they would simply follow from conservation laws, but reflect also the individual mass action laws.

2. Temperature rule (T-rule). Temperature increase (decrease) favors endothermic (exothermic) reactions. Since the total T-dependence is determined by a combination of formation energies, the final result is not always obvious. However, usually, the defect concentrations rise with increasing temperature.

3. Rule of (homogeneous) doping (C-rule). If the effective charge of the dopant defect is positive (negative) we increase (decrease) the concentration of all negatively charged defects and decrease (increase) the concentration of all positively charged defects. Again compensation effects (that would be allowed within the electroneutrality equation) do not occur.

Since temperature and partial pressure are often fixed, as far as application is concerned, the most powerful means in modifying a given structure and compound is the homogeneous doping. (In the following we will also see that frozen-in higher dimensional defects as well as frozen-in native defects can act similarly).

Rule 3 can be expressed as

$$\frac{z_j \delta \, c_i}{z \delta \, C} < 0 \tag{45}$$

or more concisely (see Eq. 42)

$$\frac{z_j}{z} M_j < 0 \tag{46}$$

(z: charge number of the dopant.)

Let us briefly consider in this context two important classes of materials. The first are the oxides of fluorite type such as ZrO_2 or CeO_2. They can accommodate high concentrations of lower valent cations. Y_2O_3 doping of ZrO_2[10] leads to the very important YSZ electrolyte. According to

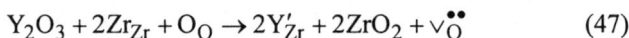

$$Y_2O_3 + 2Zr_{Zr} + O_O \rightarrow 2Y'_{Zr} + 2ZrO_2 + v_O^{\bullet\bullet} \tag{47}$$

oxygen vacancies are introduced enabling the high oxygen ion conductivity at high T. The resulting conductivity is independent of P_{O_2} while the electronic minority species are following $\pm 1/4$ power laws (see Eq. 22). (In spite of the slope being $\pm 1/4$ quite accurately, these materials are far from being ideal, and their discussion belongs into Section IV.5.)

The same is true for negatively doped perovskites. However, here the conductivity is usually predominantly electronic. Iron doping produces primarily Fe'_{Ti} defects enhancing $[v_O^{\bullet\bullet}]$ and $[h^\bullet]$. The transition from n- to p-type is shown in Fig. 12. Further increase of the Fe-content increases $[v_O^{\bullet\bullet}]$ so much that the conductivity behavior around the minimum becomes very flat indicating ionic conductivity. It becomes also flat if we quench the high T situation down to much lower concentrations: the reason for this will be considered in the next section.

It is well known that in such negatively doped oxides H_2O[55-62] can be dissolved via the formation of two hydroxide groups. The "OH"-part occupies an oxygen vacancy while the "H"-part combines with a regular O^{2-} to form an OH^- (i.e., $OH^\bullet_O$):

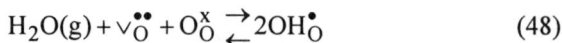

$$H_2O(g) + v_O^{\bullet\bullet} + O_O^x \rightleftarrows 2OH_O^\bullet \tag{48}$$

The $OH^\bullet_O$ defect gives rise to proton conduction according to

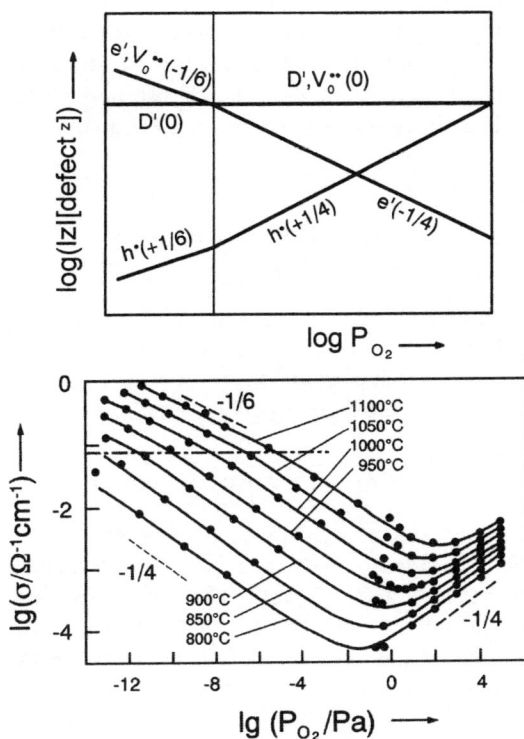

Figure 12. The conductivity of negatively doped SrTiO₃ as a function of oxygen partial pressure. Top: model (D'_: negative doping), Bottom: experimental data from Ref.[54]. (Reprinted from G. M. Choi and H. L. Tuller, "Defect Structure and Electrical Properties of Single-Crystal Ba₀.₀₃Sr₀.₉₇TiO₃." *J. Am. Ceram. Soc.* **71**, 201-205. Copyright © 1988 with permission from the American Ceramic Society.)

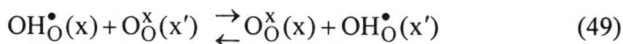

$$OH_O^\bullet(x) + O_O^x(x') \underset{\leftarrow}{\overset{\rightarrow}{\rightleftharpoons}} O_O^x(x) + OH_O^\bullet(x') \tag{49}$$

(x, x' denoting two neighboring sites) as was studied in great detail by experimental and computational methods (see e.g. Refs.[59-61]). Equation

(48) has to be added to the above equations, and P_{H_2O} appears as a parameter in Eq. (42). (The detailed calculation is left to the reader as a useful exercise.) Figures 13a and 13b show the solution in forms of Brouwer diagrams. Hydroxide defects can also be formed by reaction with H_2. If this is the dominant reaction, a great amount of electrons is incorporated

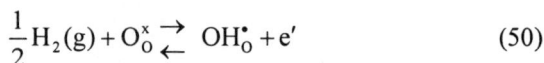

$$\frac{1}{2}H_2(g) + O_O^x \underset{\leftarrow}{\overset{\rightarrow}{\rightleftarrows}} OH_O^{\bullet} + e' \tag{50}$$

and the material will be primarily electronically conducting. (Note that reaction (48) is a pure acid-base reaction.) Thermodynamically speaking one of two incorporation reactions (Eqs. 48 and 50) is redundant in view of the oxygen-hydrogen equilibrium.

4. Frozen-In Defect Chemistry

At very high temperatures dopants such as Fe in $SrTiO_3$ can become mobile and then instead of having a fixed concentration, a reversible segregation reaction enters the game. Different defect kinetic regimes in terms of reversibility of defect concentrations have also to be considered in pure materials: The defect diagrams considered for the ternaries $SrTiO_3$ and La_2CuO_4 already assumed cation defects to be absent or immobile. If they are immobile such as Sr-vacancies in $SrTiO_3$ at temperatures below 1300 K, they enter the electroneutrality equation like an extrinsic dopant and may be called a native dopant.

The same happens with the oxygen sublattice when we typically go below 600 K, then also the oxygen vacancy concentration becomes invariant and only electronic transfer reactions are reversible. The ion defect concentration remains on a high level with unexpectedly high ion conduction at low temperatures (see Fig. 14).[63]

The regime that we just addressed is typically the regime of electronic applications. If one wants to understand how these low temperature concentrations depend on the control parameters that have to be tuned during fabrication or at least how to prepare these reproducibly, one has to look more carefully at the bridge between high and low temperature defect chemistry.[63-67] A very detailed analysis of this important subject is given in Refs.[65,66]. Here only a few points should be highlighted: In order to fix and calculate the low temperature

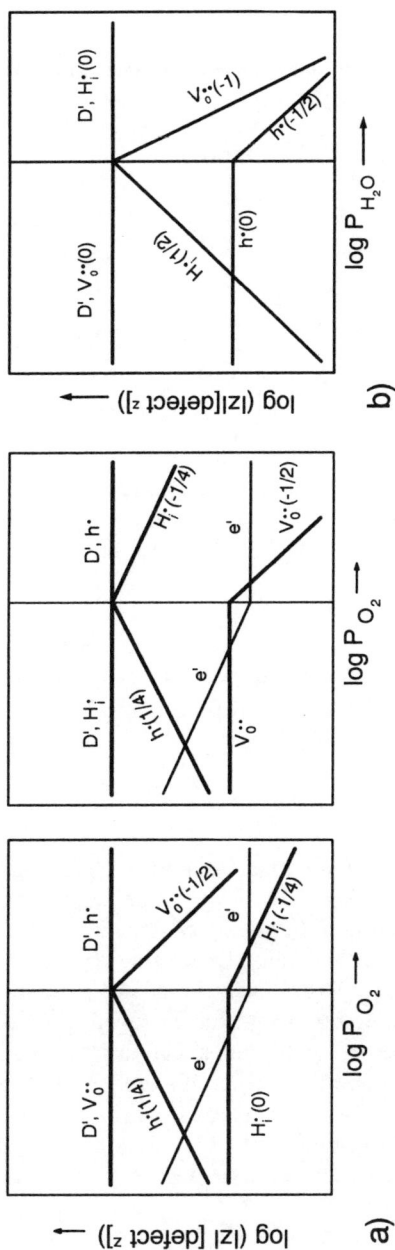

Figure 13: Defect chemistry of water containing perovskites (D': negative dopant) under reversible conditions. Dependence of the defect concentrations on oxygen partial pressure (for different water contents, l.h.s.) and on the partial pressure of water (r.h.s.).[24]

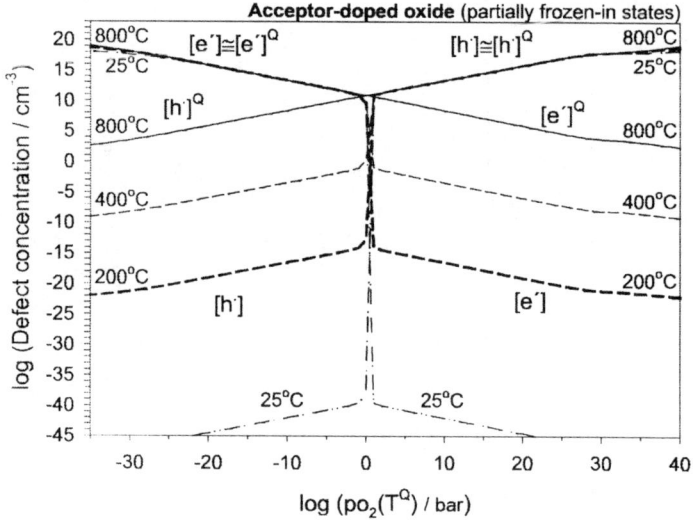

Figure 14. Defect concentrations in an acceptor-doped oxide (shallow acceptor: c_A = 1×10^{18} cm^{-3}), containing minor redox-active acceptor (deep acceptor: c_M = 1×10^{17} cm^{-3}) in partially frozen-in states. The P_{O_2} is the P_{O_2} at which the oxide has been equilibrated at T^Q = 800°C. According to Refs.[65,66]. (Reprinted from K. Sasaki and J. Maier, "Low-temperature defect chemistry of oxides. I. General aspects and numerical calculations." *J. Appl. Phys* **86**, 5422-5433. Copyright © 1999 with permission from the American Institute of Physics.)

defect concentration in a multinary material a multi-step quenching procedure is required.[67] The quenching temperature should be chosen such that the equilibration of the sublattice to be quenched takes as long as one can afford (e.g. a few days). Quenching from a higher T would cause kinetic problems during quenching. At a lower temperature partial equilibrium would not have been achieved. The reason that one cannot simply freeze in an equilibrium situation 1:1 is the occurrence of very different carrier mobilities. Hence if at about 300 K and lower the oxygen vacancy concentration is fixed (sluggish surface reaction), the electrons can still redistribute and react. When freezing one carrier concentration by reducing temperature, one loses a mass action law in

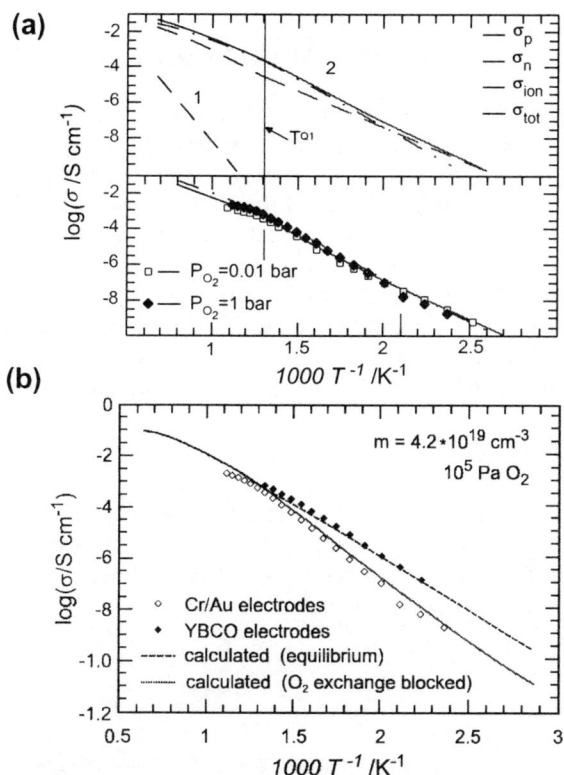

Figure 15. a) Partial (calculated, top) and total (calculated, top and measured, bottom) conductivity of Fe-doped $SrTiO_3$ in the reversible (regime 1) and in the quenched state (regime 2). At $T < T^{Q1}$ the surface reaction is frozen. b) Reversible and quenched path at lower temperatures. The reversible path is taken when the surface is covered by $YBa_2Cu_3O_{6+x}$ which acts as a catalyst of the surface reaction. According to Ref.[68].

the description and gains a conservation law, hence the number of control parameters remains constant. One parameter from the P-term in Eq. (42) (in-situ parameters) is re-shuffled to the C-term (ex-situ parameters). The number of "degrees of freedom" increases owing to the increased deviation from full equilibrium.[67]

Figure 15[68] shows how accurately the defect concentration can also be calculated even under such complicated conditions for the example of Fe-doped $SrTiO_3$, if the quenching procedure is carefully conducted.

5. Defect-Defect Interactions

Defect concentrations can be quite high in solids, and in fact the heavy admixture of a foreign component may lead to significant structural changes and even to new phases. Here we are considering situations in which the dilute limit is no longer fulfilled but severe perturbations are not yet met. Like in liquids we follow two approaches, one being the introduction of associates, which leads to a rescaling of the nature of the defects and allows treatment in terms of the new set in a dilute approximation.[7,69,70] The other is explicitly introducing corrections (activity coefficients) for each species. Of course both approaches can be combined as done in the Bjerrum-Debye-Hückel[71] approach.

(i) Associates

As already seen in Fig. 9, the Ag_2S-doping of AgCl leads to significant deviations from the dilute behavior. Unsurprisingly, the introduced S'_{Cl} and $Ag_i^{\bullet}$ defects, both being very polarizable, exert strong interactions that might be described by:

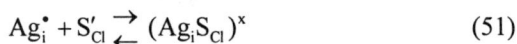

$$Ag_i^{\bullet} + S'_{Cl} \underset{\leftarrow}{\overset{\rightarrow}{}} (Ag_iS_{Cl})^x \tag{51}$$

The associate is formally neutral and drops out of the conduction process. The mass action law

$$K_{ass} = \frac{[(Ag_iS_{Cl})^x]}{[Ag_i^{\bullet}][S'_{Cl}]} \tag{52}$$

introduces a new unknown. The mass conservation requires $C = [S'_{Cl}] + [(AgS_{Cl})^x]$, and combination with the electroneutrality equation and the Frenkel equation allows for the calculation of all the defect concentrations including the free silver defects which are responsible for the conductivity. Again it is helpful to consider the limiting cases (Brouwer approximations). For small K_{ass} the situation is as before, for great K_{ass} almost all sulphur is bound. Since for strong doping silver vacancies can be neglected, the electroneutrality equation reads $[Ag^{\bullet}] = [S'_{Cl}]$ leading to $K_{ass} = C/[Ag_i^{\bullet}]^2$, and hence to $[Ag_i^{\bullet}] = C^{1/2} K_{ass}^{-1/2}$. Unlike the situation in the non-associated cases where $[Ag_i^{\bullet}] = C$, the concentration is now thermally activated again. The situation is inverse for doping with $CdCl_2$ which enhances v'_{Ag} and reduces $Ag_i^{\bullet}$. On strong doping most of the v'_{Ag} defects are bound as $(Cd_{Ag}v_{Ag})^x$ while the free carriers are now thermally activated. Fig. 16 shows this for the case of positive doping ($CdCl_2$).

Other prominent examples are associates such as $(v_O^{\bullet\bullet} Y'_{Zr})^{\bullet}$ or $(Y'_{Zr} v_O^{\bullet\bullet} Y'_{Zr})^x$ in heavily Y_2O_3 doped ZrO_2[10] or analogous centers in heavily Sr-doped La_2CuO_4 (Sr'_{La} instead of Y'_{Zr}) (cf. Fig. 11).[72] (If more and more neighbors have to be considered and/or ordering becomes more and more significant, we leave the comfortable point defect chemical treatment.[6])

Ion pairing is exceedingly the case in ion-conducting polymers. Polar functional groups enable a partial dissociation of the dissolved "salt molecules"[73] according to

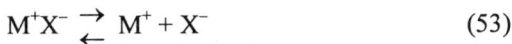

$$M^+X^- \underset{\leftarrow}{\overset{\rightarrow}{}} M^+ + X^- \tag{53}$$

Enhancement of the charge carrier density in such polymers is equivalent to shifting this reaction to the right by varying solvent, solute or adding additional particles (see Section VI.3.$ii.$).

Other variants of ionic defect pairs are vacancy pairs such as $(v_{Na} v_{Cl})^x$ in NaCl or $(Pb_i O_i)^x$ in PbO being precursors of pores or precipitates.

Very important are associations between ionic and electronic carriers. Examples are color centers formed in alkali halides.[18,74] If Na

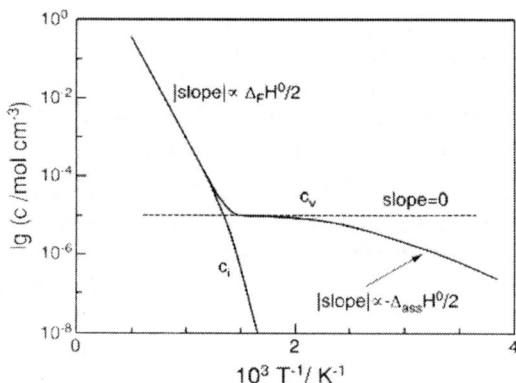

Figure 16. Concentrations of vacancies and interstitial particles in a positively doped Frenkel-disordered material taking account of association between cationic vacancies and dopant ions. The parameters used were $\Delta_{ass}S_m^o = 0, \Delta_F S_m^o = 10R, \Delta_{ass}H_m^o = -40kJ$ / mol and $\Delta_F H_m^o = 200\,kJ\,/\,mol$.[24]

is dissolved in NaCl, a Cl-deficiency is created and e′ are introduced, which tend to strongly associate with the chlorine vacancy according to

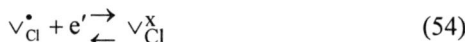

$$V_{Cl}^\bullet + e' \underset{\leftarrow}{\overset{\rightarrow}{}} V_{Cl}^x \tag{54}$$

It is interesting that the electron cloud confined in the octahedron formed by the Na^+ neighbors can be approximately treated as a free electron in a box of such dimensions. The straightforward calculation directly explains the color "ab initio".[18] Trapping of electrons by $V_O^{\bullet\bullet}$ occurs in SnO_2 at lower temperatures and leads to $V_O^\bullet$ and V_O^x. Similarly a trapping of $h^\bullet$ and O_i'' is expected in the high temperature superconductors.[53] In fact trapping according to

$$O_i'' + 2h^\bullet \underset{\leftarrow}{\overset{\rightarrow}{}} O_i' + h^\bullet \underset{\leftarrow}{\overset{\rightarrow}{}} O_i^x \tag{55}$$

occurs in $YBa_2Cu_3O_{6+x}$ already at quite high temperatures owing to the high concentrations. Valence changes of impurity defects such as

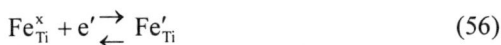

$$Fe_{Ti}^x + e' \underset{\leftarrow}{\overset{\rightarrow}{\rightleftharpoons}} Fe_{Ti}' \tag{56}$$

in $SrTiO_3$ or

$$P_{Si}^\bullet + e' \underset{\leftarrow}{\overset{\rightarrow}{\rightleftharpoons}} P_{Si}^x \tag{57}$$

in silicon also belong into this section.[75] While the equilibrium constant

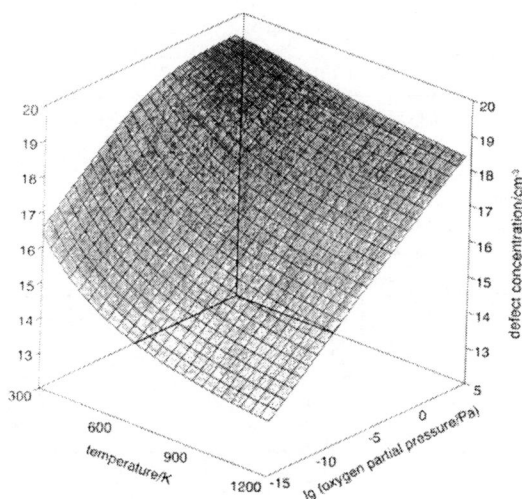

Figure 17. Fe^{4+} ("the color") concentration in Fe-doped $SrTiO_3$ as a function of T and P_{O_2}. The total iron concentration ($\sim 10^{19}/$ cm^3) corresponds to the limiting case for (T, P_{O_2}) $\rightarrow$ (0, ∞).[76] (Reprinted from T. Bieger, J. Maier and R. Waser, "An Optical In-Situ Method to Study Redox-Kinetics in SrTiO$_3$ ", *Ber. Bunsenges. Phys. Chem.*, **97** 1098-1104. Copyright © 1993 with permission from WILEY-VCH Verlag GmbH.)

of the first reaction is reasonably high, the one for the second equation is very small. In a band diagram we speak of a deep donor in the first case and a flat donor in the latter. Hence Fe^{4+} and Fe^{3+} both exist in $SrTiO_3$ (in Fig. 17) whilst phosphorus in silicium is almost exclusively tetravalent with the formal charge +1. This is visualized by defect levels in the band gap of the band diagram (see Fig. 18 r.h.s. bottom)

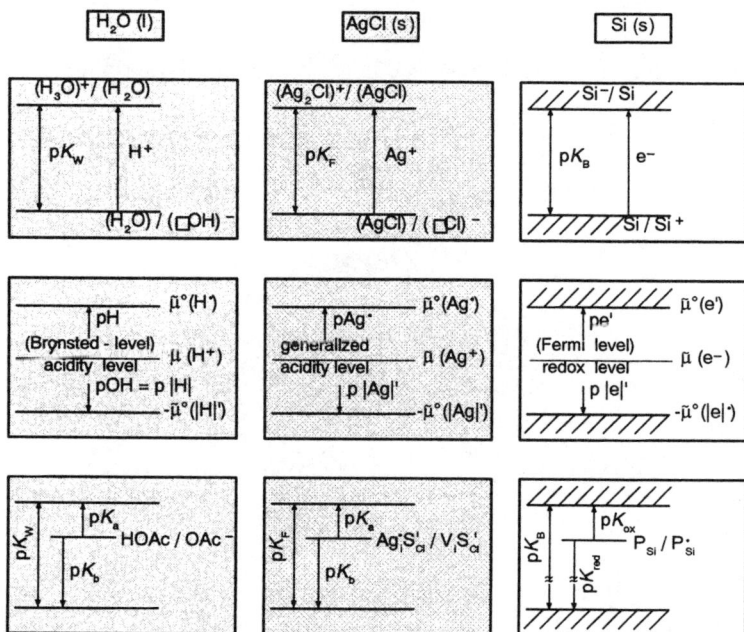

Figure 18. In the same way as the concentration of protonic charge carriers characterizes the acidity (basicity) of water and in the same way as the electronic charge carriers characterize the redox activity, the concentration of elementary ionic charge carriers, that is of point defects, measure the acidity (basicity) of ionic solids, while associates constitute internal acids and bases. The definition of acidity/basicity from the (electro-)chemical potential of the exchangeable ion, and, hence, of the defects leads to a generalized and thermodynamically firm acid-base concept that also allows to link acid-base scales of different solids.[77] (In order to match the decadic scale the levels are normalized by ln 10.) (Reprinted from J. Maier, "Acid-Base Centers and Acid-Base Scales in Ionic Solids." *Chem. Eur. J.* **7**, 4762-4770. Copyright © 2001 with permission from WILEY-VCH Verlag GmbH.)

whose position is characterized as follows: The difference of the iron level to the conduction band is the reaction energy of Eq. (56) hence the difference of $\tilde{\mu}_{e'}^{\circ}$ and $\tilde{\mu}_{e^-(Fe)}^{\circ}$, the latter parameter being identical with $\tilde{\mu}_{Fe'_{Ti}}^{\circ} - \tilde{\mu}_{Fe_{Ti}^x}^{\circ}$, while the distance to the valence band is given by $\tilde{\mu}_{Fe_{Ti}^x}^{\circ} - \tilde{\mu}_{Fe'_{Ti}}^{\circ} - \tilde{\mu}_{h^\bullet}^{\circ}$ and is interrelated with the other value by the band gap $E_g = \tilde{\mu}_{e'}^{\circ} + \tilde{\mu}_{h^\bullet}^{\circ}$. Note that $\tilde{\mu}_{e'}^{\circ}$ and $-\tilde{\mu}_{h^\bullet}^{\circ}$ refer to the edges of conduction and valence band. (Strictly speaking the identification of the energy level with $\tilde{\mu}^{\circ}$ requires appropriate normalization, cf. Section IV.2.)

In the same way as the electron transfer is mapped by the band diagram, the ion transfer of the ion-ion associates can be represented by an "ionic" level diagram (see Fig. 18 center, bottom).[75,77] Figure 18 indicates that, in particular by comparison with the situation of water, the ionic associates play the role of internal acids and bases. It is even possible to transform the Brønsted-concept in its ionotropic generalization to solids by using the point defect concept. In the same way as the numbers of H^+ and OH^- reflect the acidity and basicity in water, the numbers of v'_{Ag} and $Ag_i^\bullet$ reflect the (ionotropic) acidity and basicity in AgCl. An acidity function based on $\tilde{\mu}_{Ag^+}$ which is identical with $\tilde{\mu}(Ag^\bullet) = -\tilde{\mu}(|Ag|')$ in equilibrium also includes the definition of surface acidity. For more details the reader is referred to Ref.[77].

Finally, there are also associates between electronic carriers such as

$$e' + h^\bullet \underset{\leftarrow}{\overset{\rightarrow}{\rightleftharpoons}} (e'h^\bullet) \qquad (58)$$

or

$$2e' \underset{\leftarrow}{\overset{\rightarrow}{\rightleftharpoons}} (e')_2 \qquad\qquad 2h^\bullet \underset{\leftarrow}{\overset{\rightarrow}{\rightleftharpoons}} (h^\bullet)_2 \qquad (59)$$

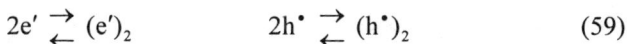

that are worthy of mention. The first are excitons, the second Cooper pairs. While the first are separated by an activation threshold from the ground state and play an important role in non-equilibrium situations, the second associates can be held together e.g. by phonon interactions and only exist at very low temperatures.[32,78]

(ii) Activity Coefficients

According to our splitting the chemical potential in a non-configurational and a configurational term $\mu^{ex} = \mu^{ex(c)} + \mu^{ex(nc)}$, we can split the activity corrections in two parts. The activity coefficients defined by $\mu^{ex} \equiv RT \ln f$ then multiply ($f = f^c \, f^{nc}$). To a certain degree, the introduction of activity coefficients can circumvent the introduction of associates: if we do not distinguish between free and trapped vacancies in AgCl formed upon $CdCl_2$ doping, then the total cation content C would be identical with the (overall) vacancy concentration and we would have to introduce an activity coefficient given by $(K_{ass}C)^{1/2} < 1$ to satisfy the decreased activity. Of course if there are good reasons to introduce associates, it is much better to use these and to try to find activity coefficients for the detailed species. In this way both concepts can be combined (cf. the Bjerrum concept in liquids.[69])

The classical approach to correct charge carrier interactions in liquid systems is the Debye-Hückel theory which is extensively discussed in textbooks.[79] The decisive parameter is the screening length

$$\lambda = \sqrt{\frac{\varepsilon \, RT}{2z^2 F^2 c_\infty}} \tag{60}$$

c_∞ being the defect concentration of the majority carriers for which we assumed the same charge number z.

The activity coefficient is approximately

$$RT \ln f_j^{nc} = -\frac{z_j^2 F^2}{8\pi\varepsilon N_m \lambda} \propto c_j^{1/2} \tag{61}$$

The Debye-Hückel concept fails very soon (see Fig. 19) and probably earlier than in liquid electrolytes (at latest for ~1). Also straightforward higher order corrections do not lead to reliable extensions into the defect range of practical interest where the coarse grained nature cannot be neglected. Accurate approaches that take into

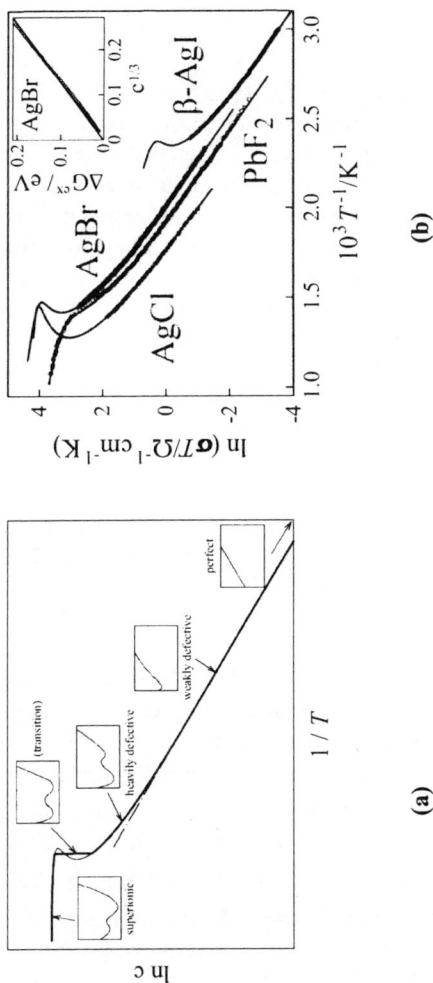

(b)

(a)

Figure 19. High temperature anomalies in the conductivity can be explained with the cube root law (see text). In the insert in (b) literature data[80] for the excess free enthalpy for AgBr are plotted against the cube root of the defect concentration. Even the phase transition temperatures themselves are well predicted.[81,82] (a): Behavior in terms of the energetic situation. (b): Experimental examples and curves fitted according to Eq. (62). (Reprinted from J. Maier and W. Münch, "Thermal Destiny of an Ionic Crystal" *Z. Anorg. Allg. Chem.* **626**, 264-269. Copyright © 2000 with permission from WILEY-VCH Verlag GmbH.)

account the structural feature become soon individual and non-manageable.[83]

In literature[81] an ad-hoc model has been used for the description of the conductivity properties of various binary halides that corrects the chemical potential according to

$$\mu^{\text{ex}} = -J_{\pm}x_{\pm}^{1/3} \qquad (62)$$

($x_{\pm}$: defect mole fraction of the majority carriers).

It is worth noting that cube root corrections have also been used for liquid electrolytes,[84] however, there without real structural connection and without the possibility of testing that over a wide T-range. Supported by Monte Carlo calculations of molten salts[85] which show that their Coulomb energy can be approximated by a Madelung concept (with a somewhat smaller Madelung factor) in spite of the fluctuating distribution, it proposes the interaction energy to be similar to the Madelung energy of a defect superlattice superimposed on the perfect lattice. The calculation then readily gives Eq. (62) with $J_{\pm}$ being

$$J_{\pm} \simeq \frac{2}{3}\frac{\varphi_d}{\varphi}\frac{U}{\varepsilon_r} \qquad (63)$$

(φ, φ_d: Madelung constants of perfect lattice and defect lattice, U, ε_r: lattice energy and dielectric number of the perfect lattice.) The concept was tested by conductivity (see Fig. 19) and specific heat experiments as well as by Monte Carlo and MD simulations and found to work surprisingly well.[86,87]

In contrast to the silver halides for which the use of the static dielectric constant in Eq. (57) yields good agreement, for PbF_2 with its high static dielectric constant a value more typical for this structure ($\varepsilon \approx \varepsilon_{CaF_2}$) had to be used. Since at the small distances involved the full static permittivity is not operative, the neglect of the high static polarizability of the lead ion makes sense. (Similarly the interaction between $V_O^{\bullet\bullet}$ and acceptor dopants in $SrTiO_3$ could be well described by a cube root law with an effective dielectric constant of ~10 instead of the static value which is one order of magnitude greater.[88])

There are a variety of important consequences of Eq. (62). First, the attractive defect interactions lower the effective formation energy according to the implicit function

$$x_{\pm} \equiv [Ag_i^{\bullet}] = [V_{Ag}'] = \exp-\frac{\Delta_F G_m^{\circ} - 2J_{\pm}x_{\pm}^{1/3}}{2RT} \qquad (64)$$

so that the conductivity rises over its values computed "according to Boltzmann". This is indeed found and called the pre-melting regime in ion conductors,[81] and is well described by Eq. (64) for AgCl, AgI and PbF$_2$. Moreover, since a higher defect concentration leads to ever higher values, this avalanche causes a phase transformation.[81,89] In contrast to Fig. 4, now the free enthalpy curve shows two minima (see inserts in Fig. 19, l.h.s.), the second of which becomes lower above the transition temperature (T$_c$) (see Fig. 19). Surprisingly the J-values given above suffice to even predict the phase transformations which are a first order transition from the low defective β-AgI to the completely disordered α-AgI (molten Ag$^+$ sublattice), a first order transition from the low defective AgCl to the totally molten state and a second order phase transition in the case of PbF$_2$ (Fig. 19). In the second case even the high T conductivity was described reasonably well. The precise criterion for the order of the phase transition is given in Ref.[81]. The fact that the T$_c$ corresponds to reality even in the first order cases implies that the transition from the virtual sublattice molten crystal with the low temperature structure to the real structure of the partially or totally molten state does not exhibit great free energy changes. In principle only the upper limit of such a transition can be predicted in this way.

Interestingly also the electron-hole interaction can scale with such a distance law. The pre-melting corresponds then to what is called level narrowing,[90] and the transition to a superionic state (degenerate situation) corresponds to the insulator-metal transition expected in such a case.

These considerations should also be relevant for boundary effects and boundary phase transitions.

Figure 20 shows the "thermal destiny" of an ionic crystal which (given structure, bond strength and permeabilities) from the perfect state at T = 0 up to the superionic transitions.[81,82] Owing to the finite disorder energy (which depends on bond strength and permeabilities), there is a formation of point defects, the concentration of which first increases according to Boltzmann; with increasing concentrations the formation energy is successively lowered leading to an anomalous concentration increase (also determined by bond strength and permeabilities) and in a self-amplified way to the transition into a

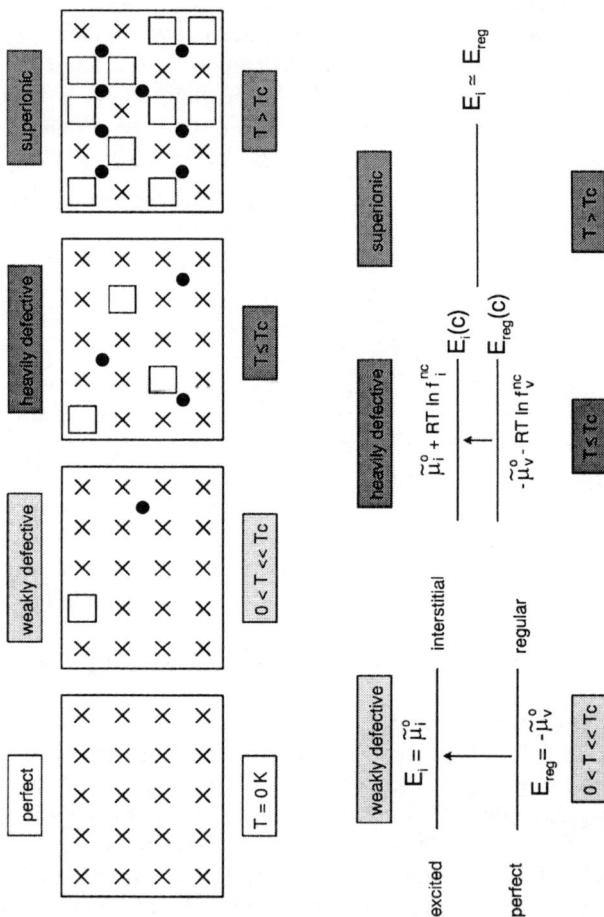

Figure 20. Sketch of the real structure of a Frenkel disordered crystal (only the affected sublattice is shown). The increase in temperature corresponds to an increase in defect concentration. Interaction leads to a narrowing of the spacing of the "energy levels" (electrochemical potentials minus configurational term) and eventually to a transition into the superionic state.[82] (Reprinted from J. Maier and W. Münch, *Z. Anorg. Allg. Chem.*, **626**, 264-269, Copyright © 2000 with permission from WILEY-VCH Verlag GmbH.)

totally disordered state (unless the system evades this by a structural transformation into another state). In master example AgCl the whole T-range also including the above discussed impurity effects can be described by the simple defect chemistry.

For high concentrations $x_\pm$ has to be replaced by $x_\pm/(1 - x_\pm)$ in order to account for the finite number of states. It was already mentioned that this stems from configurational corrections, more specifically from the Fermi-Dirac correction. The results are corrections of the type

$$f_\pm^c(x_\pm) = (1 - x_\pm)^{-1} \qquad (65)$$

for ionic point defects or, in the case of electronic carriers, e.g. for the electrons in the conduction band[3,5] ($\mathfrak{I}_{1/2}$: Fermi-Dirac integral for parabolic state densities, Δ: distance from Fermi-level to band edge normalized with respect to kT)

$$f_{e'}^c = \frac{\pi^{1/2} \exp \Delta}{2 \mathfrak{I}_{1/2}(\Delta)} \quad \text{with} \quad \mathfrak{I}_{1/2}(y) \equiv \int_0^\infty \frac{\tau^{1/2} d\tau}{\exp(\tau + y) + 1} \qquad (66)$$

which is obtained from Eq. (26).

It is noteworthy that these activity coefficients are larger than 1 and lead to an increased activity (cf. Fermi pressure) (see Fig. 21). The opposite is true for a Bose-Einstein statistics ($f_\pm^c = (1 + x_\pm)^{-1}$ instead of Eq. (65), cf. Bose condensation).

In realistic situations the interactions "act back" on the configuration, and the assumption of a random configuration for a situation with energetic interaction is not free from inconsistency. Nevertheless, experience shows that for weak interaction this may be a well-working assumption. Corrections in the configurational entropy are extremely complicated[83] and simple corrections are sought, even if very crude. Site exclusion is such a simple correction: Here it is, e.g. assumed that two positions may both be available with equal probability for the very first defect, but cannot be occupied both; then the binominal expression in Eq. (24) has to be simply multiplied by 2^{N_d}.[91] (In realistic situations the coordination number has to be considered rather than a factor of 2.)

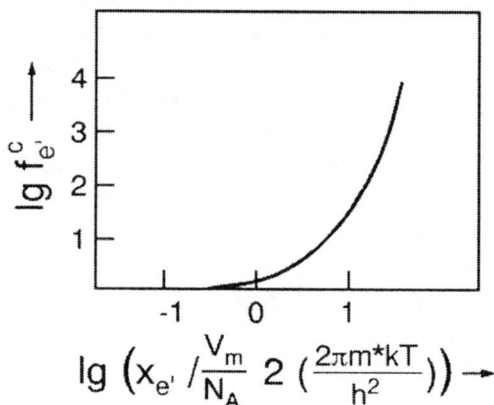

Figure 21. Fermi-Dirac activity coefficient of electronic charge carriers as a function of concentration.[4] (Reprinted from H. Schmalzried, *Solid State Reactions,* Verlag Chemie, Weinheim. Copyright © 1981 with permission from WILEY-VCH Verlag GmbH.)

V. DEFECT CHEMISTRY AT BOUNDARIES

1.   Space Charge Profiles and Capacitances

The treatment has to be changed if we are concerned with interfaces. The electroneutrality equation is only a limiting case valid for the homogeneous bulk (and valid there only on a coarse grained scale). When heterogeneities are introduced, the symmetry break results in space charges. Put differently, owing to the structural change at interfaces, the (free) formation energy and hence μ_d^0 is different there; this not only leads to a changed point defect chemistry in this interfacial core but also to space charge zones in the region adjacent. In particular in materials of interest characterized by a low carrier density but significant mobility in the bulk, the latter phenomenon is of prime importance. The situation is characterized by non-zero electrical potential gradients and electrochemical potentials. In order to describe the electrostatic information, the combination of the two relevant Maxwell-equations (magnetic fields will be neglected), namely $\nabla \cdot (\varepsilon \mathbf{E})$ $= \rho$ and $\nabla \times \mathbf{E} = 0$, i.e., $\mathbf{E} = -\nabla \phi$ ($\mathbf{E}$ being the electric field and $\rho =$ $\Sigma_j z_j F c_j$ the charge density) leads to Poisson's equation (which for the bulk reduces to $\rho = 0$). The equilibrium condition (for random distribution)

$$\nabla \tilde{\mu}_j = \nabla \mu_j + z_j F \nabla \phi = RT \nabla c_j / c_j + \nabla \mu_j^{ex} + z_j F \nabla \phi = 0 \qquad (67)$$

offers the second information.[92] If we consider interaction-free solutions ($\mu^{ex} = 0$) and one-dimensional situations, the concentration profile in the boundary regions is determined by the electrical potential profile (space charge zone) according to

$$\left(\frac{c_j(x)}{c_{j\infty}} \right)^{1/z_j} \equiv \zeta_j^{1/z_j} = \exp - \frac{\phi(x) - \phi_\infty}{RT} F \qquad (68)$$

being independent of j. Combination with Poisson's equation leads to the well-known Poisson-Boltzmann equation[93]

$$\frac{d^2(\phi - \phi_\infty)}{dx^2} = -\frac{F}{\varepsilon} \Sigma_k c_{k\infty} z_k \exp - \left(z_k F \frac{\phi - \phi_\infty}{RT} \right) \qquad (69)$$

For two oppositely charged majority carriers with $z_+ = z_- \equiv z$ the Gouy-Chapman-profile (see textbooks on electrochemistry) results which we write ($\xi \equiv x/\lambda$) here as

$$\zeta_\pm = \left(\frac{1 + \vartheta_\pm \exp - \xi}{1 - \vartheta_\pm \exp - \xi} \right)^2 = \zeta_+^{-1} \qquad (70)$$

where

$$\vartheta_\pm = \frac{\zeta_{\pm0}^{1/2} - 1}{\zeta_{\pm0}^{1/2} + 1} = -\vartheta_\mp \qquad (71)$$

The parameter ϑ varies between -1 and 0 for depletion and between 0 and +1 for accumulation, and plays the role of a degree of influence of the interface.[94]

Figure 22 (r.h.s.) illustrates the contact thermodynamics and its influence on ionic and electronic carrier concentrations. The level bending expresses the variation in the electrical potential, and the constancy of $\tilde{\mu}_{ion}$ and $\tilde{\mu}_{eon}$ the electronic and ionic contact equilibrium.[35] (Note that the electric potential term—as a non-configurational term—is to be included into the "energy levels".) The constancy of the chemical potential of the neutral component is automatically fulfilled (see Fig. 22 r.h.s.).

Two extreme cases are worth mentioning: the first if the concentration variation is small ($\vartheta \to 0$), then we end up with the exponential solution [$\zeta_{j0} \equiv \zeta_j(x = 0)$]

$$\zeta_j = 1 + (\zeta_{j0} - 1) \exp - \xi \qquad (72)$$

and the other for large effects ($\vartheta_1 \to 1$)

$$\zeta_1 = \frac{\zeta_{10}}{(1 + \sqrt{\zeta_{10}} \xi / 2)^2} \qquad (73)$$

where 1 denotes the accumulated carrier. The Gouy-Chapman solutions have been used in liquid electrochemistry for a long time when dealing with the metal/electrolyte contact.[95,96] There space charge zones occur, the charge of which is compensated by adsorbed carriers at the very interface and also by an excess electronic charge at the metal side of the contact.[96] In materials with low carrier concentrations, such as dilute electrolytes (ions) or weakly doped semi-conductors (electrons) this

Figure 22. L.h.s.: Four basic space charge situations involving ionic conductors (here silver ion conductor): a) contact with an isolator, b) contact with a second ion conductor, c) grain boundary, d) contact with a fluid phase. R.h.s.: Bending of "energy levels" and concentration profiles in space charge zones ($\xi = 0$ refers to the interfacial edge).

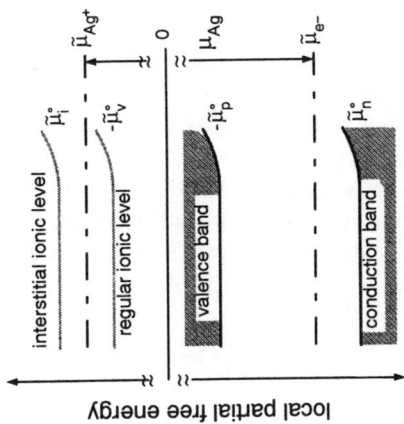

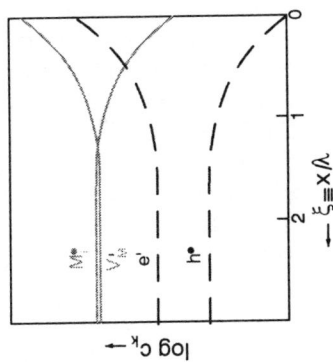

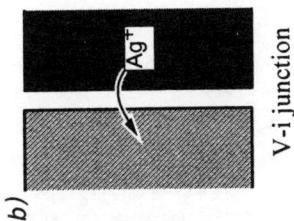

heterogeneous doping

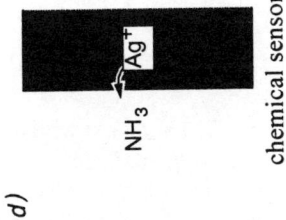

V-i junction

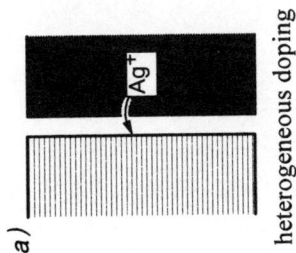

grain boundary engineering

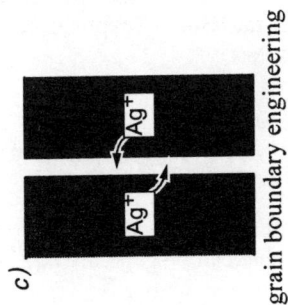

chemical sensors

charge distribution is diffuse.[33] In our more general case we have to consider electronic and ionic profiles in both phases as well as in the core of the boundary. Later we will mainly work[24] with Eq. (73), which in essence ignores the counter defect in the charge density. For the same reason this approximation is—given that large effects are indeed met—quite general in terms of defect chemistry and independent of whether or not the counter-defect is immobile or whether or not its charge coincides with the majority defect. A completely different situation is only met if an immobile dopant forms the majority carrier and the counter defect (2) is depleted. Then a constant dopant profile means a constant charge density, and a Mott-Schottky-profile is obtained[97,98] for the majority counter carrier:

$$\zeta_2 = \exp{-\left|\frac{z_2}{z_1}\right|\left(\frac{x - \lambda^*}{2\lambda}\right)^2} = \exp{-\left|\frac{z_2}{z_1}\right|\left(\frac{\xi - \xi^*}{2}\right)^2} \tag{74}$$

Unlike the Debye length the effective width λ^* depends on the space charge potential:

$$\lambda^* \equiv \sqrt{\frac{2\varepsilon}{z_1 F c_{1\infty}}(\phi_\infty - \phi_0)} \tag{75}$$

From Eqs. (73) and (74) which describe accumulation and depletion for strong effects, the concentrations of the other mobile species are simply accessible through Eq. (68). In liquid electrochemistry these profiles have been used to evaluate electrode/liquid electrolyte capacitances.[79]

In the Gouy-Chapman case the differential capacitance (a: area) follows as

$$C_{sc}/a = \frac{\varepsilon}{\lambda}\cosh\frac{|z|F(\phi_0 - \phi_\infty)}{2RT} \tag{76}$$

which (for small space charge potentials) reduces to $(C_{sc}/a) = \varepsilon / \lambda$, a result that can be generalized[101] to

$$C_{sc}/a = \varepsilon / l_0 \tag{77}$$

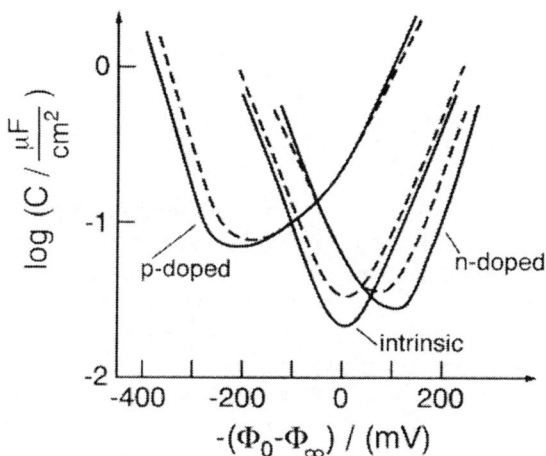

Figure 23. Theoretical space charge capacitance according to Bohnenkamp and Engell as a function of the voltage drop over the boundary $- (\phi_0 - \phi_\infty)$ for intrinsic and doped semiconductors (donor concentration N_D/V $3.5 \times 10^{14} cm^{-3}$, acceptor concentration $N_A/V = 7.5 \cdot 10^{15} cm^{-3}$) at room temperature (dashed line: 45°C).[99]

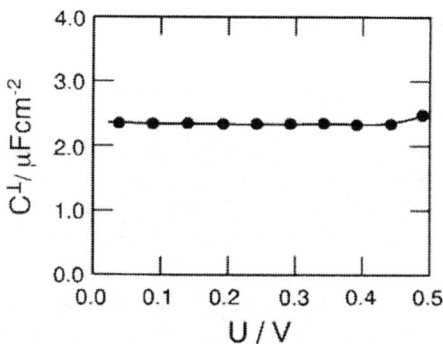

Figure 24. Voltage dependence of the boundary layer capacitance for $C|\alpha$-AgI at 175 °C. From Ref.[102]. (Reprinted from R. D. Armstrong and R. Mason, "Double Layer Capacity Measurements involving Solid Electrolytes." *J. Electroanal. Chem.* **41**, 231-241. Copyright © 1973 with permission from Elsevier.)

where l_0 is the centroid of the charge perturbation. The space charge capacitance in the Mott-Schottky case is given by:

$$C_{sc}/a = \sqrt{\frac{|z|Fm\varepsilon}{2|\phi_0 - \phi_\infty|}} = \frac{\varepsilon}{\lambda^*} \tag{78}$$

Equation (78) has been widely used in semiconductor physics and semiconductor electrochemistry, in particular with respect to the evaluation of important parameters such as the potential of zero-charge (flat-band potential).[71] Equation (76) predicts a minimum in the capacitance vs. potential curve. (The generalization to a doped situation[99] (see Fig. 23) shows that this minimum is sensitive to the dopant concentration.) A minimum has never clearly been observed for solid ionic conductors, what was attributed to interfacial states, saturation and finite size effects.[103,104] The total capacitance also includes contributions from the core layer itself.[71] Figure 24 shows the double layer capacitance of α-AgI for which the Debye-length is negligible. Figure 25 refers to a depletion layer in the mixed conductor SrTiO$_3$. According to Eq. (78) the boundary capacitance changes only weakly with P_{O_2}.

In mixed conductors the "chemical capacitance" plays an important role. This will be extensively considered in Part II[1] (see also Section VI, in particular VI.7.).

2. Space Charge Conductance

The space charge regions naturally lead to severe conductance effects. In solid state electrochemistry such effects can be well measured by using thin films, dispersions, polycrystalline materials and/or by taking advantage of special configurations.[1,23,94] Table 2 compiles conductance effects, parallel and perpendicular to the interface, calculated on the basis of Eqs. (68), (73) and (74) for a variety of situations. Let us first refer to accumulation effects and study the conduction effects along the interfaces. Four basic space charge situations in ionically conducting systems displayed in Fig. 22 (l.h.s.) will be discussed in more detail in the following text.

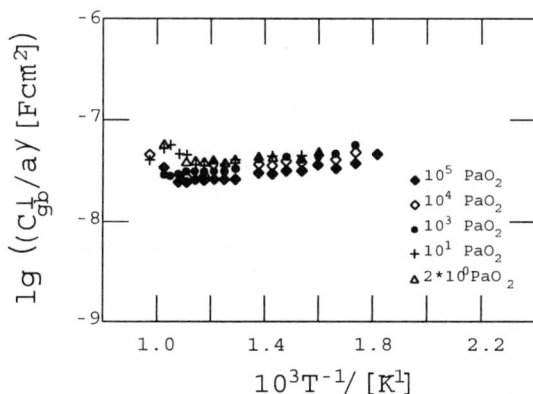

Figure 25. Grain boundary capacitance of a Fe-doped $SrTiO_3$ polycrystal (m_{Fe} = 6.5 x $10^{19} cm^{-3}$), normalized to the electrode surface and measured at various oxygen partial pressures as a function of reciprocal temperature.[100] Typical space charge potentials vary between 300 and 800 mV. (Reprinted from I. Denk, J. Claus and J. Maier, "Electrochemical Investigations of $SrTiO_3$ Boundaries." *J. Electrochem. Soc.* **144**, 3526-3536. (Copyright © 1997 with permission from The Electrochemical Society, Inc.)

Before we do so, let us mention a particular difference between solid and liquid systems which is very important in this context, namely the occurrence of a network of internal boundaries in multiphase or polycrystalline materials. The proportion of boundary regions can be so high that the overall conductivity can be interfacially dominated.

As far as solid state ion conductors are concerned, the interest in boundary effects started with the surprising finding that the overall conductivity in two-phase systems can exceed the conductivities of the bulk phases by several orders of magnitude.[106-109] If e.g. fine Al_2O_3 particles are dispersed in matrices of moderate cation or anion conductors such as LiI, AgCl, AgBr, β-AgI, CuCl, $TlCl$[94] or CaF_2, PbF_2,[110] the ion conductivity—at not too high temperatures—drastically increases with an activation energy close to the migration energy of the mobile ion (see Fig. 26). The more acidic oxides (e.g. SiO_2) are less active in the case of the cation conductors but more active in the case of

Table 2

Possible space charge situations and the respective conductance effects parallel ($\parallel$) or perpendicular ($\perp$) to the interface.[105] (Reprinted from S. Kim, J. Fleig and J. Maier, "Space charge conduction: Simple analytical solutions for ionic and mixed conductors and application to nanocrystalline ceria.", Phys. Chem. Chem. Phys. 2268-2273, 5, Copyright © 2003 with permission from the PCCP Owner Societies.)

Model	$\ln c_j$	For $z_1 = -z_2$	$\ln c_j$	For $2z_1 = -z_2$	$\ln c_j$	For $z_1 = -2z_2$
Gouy-Chapman	a	$(\sigma_m^{\parallel})_1 \propto c_{10}^{1/2}$ $(\rho_m^{\perp})_2 \propto \dfrac{1}{c_{20}^{1/2} c_{2\infty}}$	b	$(\sigma_m^{\parallel})_1 \propto c_{10}^{1/2}$ $(\rho_m^{\perp})_2 \propto \dfrac{1}{(c_{20} c_{2\infty})^{3/4}}$	c	$(\sigma_m^{\parallel})_1 \propto c_{10}^{1/2}$ $(\rho_m^{\perp})_2 \propto \dfrac{1}{c_{2\infty}^{3/2}} \ln\left(\dfrac{c_{2\infty}}{c_{20}}\right)$
	d	$(\sigma_m^{\parallel})_1 \propto \dfrac{1}{c_{2\infty}^{1/2}}(c_{10} c_{1\infty})^{1/2}$ $(\rho_m^{\perp})_2 \propto \dfrac{1}{c_{20}^{1/2} c_{2\infty}}$	e	$(\sigma_m^{\parallel})_1 \propto \dfrac{1}{c_{2\infty}^{1/2}}(c_{10} c_{1\infty})^{1/2}$ $(\rho_m^{\perp})_2 \propto \dfrac{1}{(c_{20} c_{2\infty})^{3/4}}$	f	$(\sigma_m^{\parallel})_1 \propto \dfrac{1}{c_{2\infty}^{1/2}}(c_{10} c_{1\infty})^{1/2}$ $(\rho_m^{\perp})_2 \propto \dfrac{1}{c_{2\infty}^{3/2}} \ln\left(\dfrac{c_{2\infty}}{c_{20}}\right)$
	g	$(\sigma_m^{\parallel})_1 \propto c_{10}^{1/2}$ $(\rho_m^{\perp})_2 \propto \dfrac{1}{(c_{1\infty} c_{20} c_{2\infty})^{1/2}}$	h	$(\sigma_m^{\parallel})_1 \propto c_{10}^{1/2}$ $(\rho_m^{\perp})_2 \propto \dfrac{1}{c_{1\infty}^{1/2} (c_{20}^{3/4} c_{2\infty}^{1/4})}$	i	$(\sigma_m^{\parallel})_1 \propto c_{10}^{1/2}$ $(\rho_m^{\perp})_2 \propto \dfrac{1}{c_{1\infty}^{1/2} c_{2\infty}} \ln\left(\dfrac{c_{2\infty}}{c_{20}}\right)$
	j		k		l	

Table 2
Continuation

Mott-Schottky (graphs m, n)

$$(\sigma_m^\parallel)_1 \propto c_{10}^{1/2}$$

$$(\rho_m^\perp)_2 \propto \frac{1}{(c_{1\infty} c_{20} c_{2\infty})^{1/2}}$$

Combined (graphs o, p)

Column o:

$$(\sigma_m^\parallel)_1 \propto \frac{1}{c_{2\infty}^{1/2}} \frac{c_{10}}{[\ln(c_{10}/c_{1\infty})]^{1/2}}$$

$$(\rho_m^\perp)_2 \propto \frac{1}{c_{2\infty}^{1/2}} \frac{1}{c_{20}[\ln(c_{2\infty}/c_{20})]^{1/2}}$$

$$(\sigma_m^\parallel)_1 \propto c_{10}^{1/2}$$

$$(\rho_m^\perp)_2 \propto \frac{1}{c_{1\infty}^{1/2}} \frac{1}{c_{20}^{3/4} c_{2\infty}^{1/4}}$$

Column p:

$$(\sigma_m^\parallel)_1 \propto c_{10}^{1/2}$$

$$(\rho_m^\perp)_2 \propto \frac{1}{c_{1\infty}^{1/2} c_{2\infty}} \ln\left(\frac{c_{2\infty}}{c_{20}}\right)$$

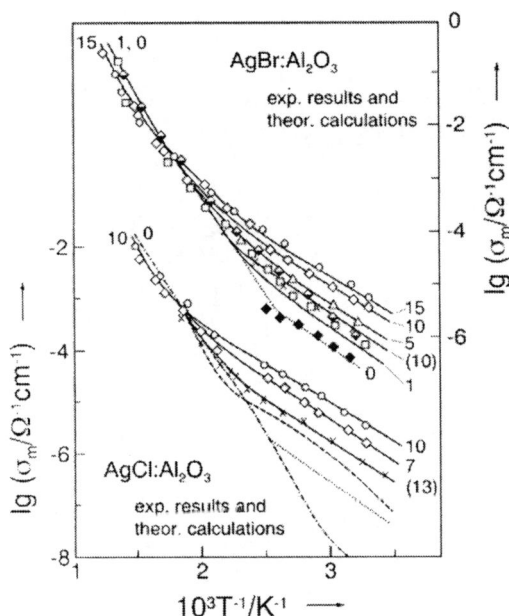

Figure 26. Experimental results (symbols) and theoretical calculations (solid lines) for AgBr:Al$_2$O$_3$ and AgCl:Al$_2$O$_3$ two-phase mixtures. The labels give the volume fraction of Al$_2$O$_3$ as a percentage and refer to an Al$_2$O$_3$ grain size of 0.06μm (if not in brackets) or 0.15μm (if in brackets). The line marked with dashes for AgCl refers to the nominally pure single crystal, the dotted line to the polycrystal and the broken line to a positively doped single crystal (with respect to the knee, cf. Fig 27).[94,109] (Reprinted from J. Maier, "Ionic Conduction in Space Charge Regions." *Prog. Solid St. Chem.* **23**, 171-263. Copyright © 1995 with permission from Elsevier.)

the fluoride ion conductors. γ-Al$_2$O$_3$ proved to be more active than α-Al$_2$O$_3$ which was attributed to the OH-groups on the surface. Making the surface non-polar (using (CH$_3$)$_3$SiCl) nullifies the effect. All these phenomena can be explained by a strong adsorption of the mobile ion towards the surface leaving behind vacant sites and hence refer to case a in Fig. 22a, which in the case of a cation conductor, is described by[108,109]

$$\vee_A + M_M \underset{\leftarrow}{\overset{\rightarrow}{\rightleftharpoons}} M_A^{\bullet} + \vee_M'. \tag{79}$$

A denotes a site at the oxide/halide interface. In principle the superposition of the individual pathways to the overall response can be very complicated. Standard percolation theory starts from two phases of extremely different conductivities, assumes a random distribution and equal particle sizes. In this case the simplest situation is to assume a highly conducting interface between two insulating phases. Then two percolation thresholds have to be considered: the first on the Al_2O_3 poor side where the first interface pathway is formed, and the second on the Al_2O_3 rich side where the last interface pathway is blocked.[111] Even though this is intelligible, the agreement is only qualitative. The assumption of a homogeneous interface conductivity and the neglect of the conductivity of the base material is, however, not the most serious shortcoming: (i) The very small size of the Al_2O_3 particles depresses the first percolation threshold to very small values. (ii) The assumption of random distribution is inconsistent owing to the necessary interfacial interaction and not independent of this. (iii) The grain size of the matrix changes upon increasing the volume fraction of Al_2O_3. As expected from the local energetics, the Al_2O_3 particles populate the grain boundaries and very soon form percolating pathways. On increasing the volume fraction of alumina the grain size of the AgCl particles is reduced and the monolayer situation maintained. Hence, the excess conductivity is approximately proportional to both the specific area (Ω_A) and to the volume fraction of the oxide particles. The measured conductivity (see Table 2, case a) is then given by[94,108,109]

$$\sigma_m = (1 - \varphi_A)\sigma_\infty + \beta_L \Omega_A \varphi_A (2\varepsilon RT)^{1/2} u_1 \sqrt{c_{10}} \tag{80}$$

(For the mechanism proposed by Eq. (79), carrier 1 is the metal vacancy. The parameter β_L is explained in Section V.4., Eq. 90.) Figure 26 shows that Eq. (80) describes the conductivity effects excellently. Since $\sqrt{c_{10}}$ is only weakly activated, the T-dependence of the heterogeneously doped samples is similar to the positively (homogeneously) doped samples (e.g. $CdCl_2$).

Like the homogeneous $CdCl_2$ doping, the heterogeneous Al_2O_3 doping leads to a change in the conduction mechanism in pure silver halides in which the interstitial defects dominate the conductivity. Unlike there, however, we do not only meet a transition in the overall conductivity from v- to i-type, the type of the conductivity changes also locally from the bulk towards the boundary. Similarly TlCl which is an anion conductor intrinsically and exhibits Schottky disorder, can be made cation-conducting if heterogeneously doped by Al_2O_3.[112]

Figure 27 compares the heterogeneous and homogeneous doping situations. In both cases the excess charge introduced generates an enhancement (depression) of the concentration of the oppositely (equally) charged intrinsic defects; the distribution of charge and counter charge, however, is different in both situations. The similarity is also expressed by the rule of heterogeneous doping (Σ-rule)[20] (adding to the three defect chemical rules given in Section IV.3., c.f. Eq. 45.)

$$\frac{z_j \delta c_j}{\delta \Sigma} < 0 \qquad (81)$$

or in words:

(4) If one introduces higher dimensional defects into a system, which attract a positive (negative) charge, the concentrations of all mobile negatively (positively) charged defects will be increased (decreased) in the space charge zone and vice versa. Compensation effects do not occur (Σ: charge density at the interfacial core).

One consequence of the modified defect chemistry being restricted to the boundary is readily seen in Fig. 27. Unlike in the case of homogeneous doping, in heterogeneous doping the transition from interstitial to vacancy type does not show up as a knee in the conductivity curves, since, as soon as the boundary zone becomes less conductive, the bulk which is in parallel, dominates.[113]

The depletion in the silver interstitial concentration has been investigated at the contact $RuO_2/AgCl$, in which the role of RuO_2 corresponds to that of Al_2O_3; however, due to its high electronic conductivity, it can also serve as an electrode and the effect across the space charge zone including the space charge capacitance can be measured.[114]

I:

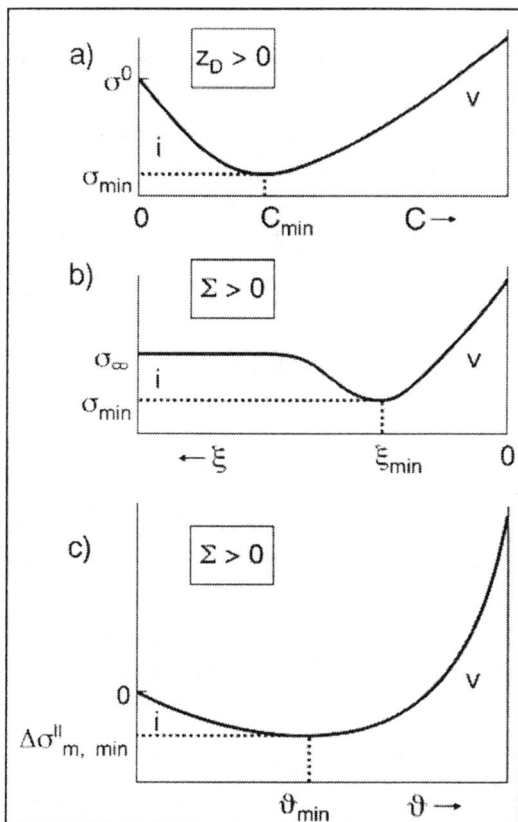

Figure 27. I: a) Comparison of homogeneously (a) and heterogeneously (b,c) doped material (example: AgCl): (a) conductivity as a function of the concentration of a homogeneous positive dopant ($z_D > 0$) of concentration C, (b) local conductivity as a function of the spatial coordinate at positive charge density of the interface core ($\Sigma > 0$) in the case of heterogeneous doping ($\xi \equiv x/\lambda$); $\xi = 0$ denotes the layer adjacent to the interfacial core, (c) integral conductivity increase ($\Delta\sigma_m^I$) as a function of the strength of the interfacial interaction in heterogeneous doping. II.: The dependence of the conductivity on temperature for positively homogeneously and heterogeneously doped material (example: AgCl[94]). The "knee" which is characteristic for the homogeneous case, is by-passed in the heterogeneous case.[24] Reprinted from J. Maier, "Ionic Conduction in Space Charge Regions." *Prog. Solid St. Chem.* **23**, 171-263. Copyright © 1995 with permission from Elsevier.)

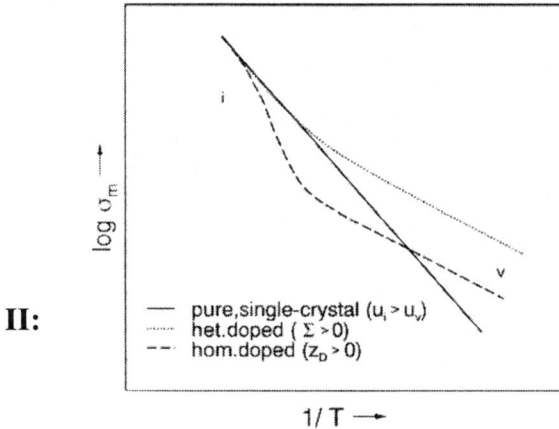

II:

Figure 27. Continuation

A revealing experiment consists in blocking the ionic conduction in AgCl/Al$_2$O$_3$ composites by pure electron conductors. This addresses the extension of the Wagner-Hebb method to inhomogeneous systems[115] (see Part II[1]) and leads to the detection of minority species therein. The results are completely in agreement with the space charge picture (cf. Eq. 68) which predicts by an increased [e'] and a decreased [h•] in the space charge zones. The total information can be comprised by Kröger-Vink diagrams of boundary regions which exhibit the distance from the boundary as a different parameter (see Section V.3., Fig. 35). One point is worthy of note at this stage: If these carriers are also the minority carriers in the core of the boundary, they are—unlike the majority carriers—not important for the establishment of the space charge field, do, however, react on it equally. In this respect it is pertinent to speak of a "fellow traveler effect".[94] As far as electron conductors are concerned, much more attention should be paid to such a foreign control, since the space charge effects of the electrons often control the conductivity, whereas, however, the origin of the space charge potential may be solely due to ionic carrier effects (see e.g. the SrTiO$_3$ below).

Oxide admixtures also proved beneficial (see, e.g. reference[116]) in the case of ion conducting polymers. Even though mobility effects occur, too, the separation of an ion pair in the rather covalent matrix by adsorbing one partner is possible, an effect that can be subsumed under the phenomenon of heterogeneous doping.[37]

The second contact problem of significance is the contact of two ionic (or more generally two mixed) conductors (see Fig. 22, case b).[94,107,117,118] Here a redistribution affects two boundary layers (charge storage in the proper interface is now neglected). Let us consider the contact MX/MX' of two M$^+$ conducting materials in which also electronic carriers (but not X$^-$) may be mobile. The full thermodynamic equilibrium demands the invariance of $\tilde{\mu}_{M^+}(=\tilde{\mu}_{M_i^{\bullet}}=-\tilde{\mu}_{V_{M'}})$, $\tilde{\mu}_{X^-}$ and $\tilde{\mu}_{e^-}(=\tilde{\mu}_{e'}=-\tilde{\mu}_{h^{\bullet}})$. The global bulk thermodynamics characterized by $\nabla\mu_M = 0$, $\nabla\mu_X = 0$, is then automatically fulfilled ($\mu_M = \tilde{\mu}_{M^+}+\tilde{\mu}_{e^-}$; $\mu_X = \tilde{\mu}_{X^-}-\tilde{\mu}_{e^-}$; $\mu_{MX} = \mu_M + \mu_X$) and hence also $\nabla\mu_{MX} = 0 = \nabla\mu_{MX'}$. At the temperatures of interest only the sufficiently mobile ions can redistribute $\nabla\mu_M = 0 \neq \nabla\mu_X$, $\nabla\mu_{M^+} = 0 \neq \nabla\mu_{X^-}$ (i.e., $\nabla\mu_{MX} \neq 0$). Figure. 28 plots the decisive potentials and functions for the contact together with the profiles for the ionic carriers. The behavior of the electronic carriers is not shown. Here we refer to Fig. 22 (r.h.s.) which applies for one side of the contact. The construction of the total energy level diagram for ions and electrons (analogously to Fig. 22 (r.h.s., top)) is left to the reader as useful exercise.[35] Relevant cases in which we assume or observe ion redistribution are the contacts AgI/AgCl and BaF$_2$/CaF$_2$[118,119]

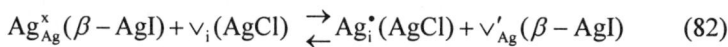

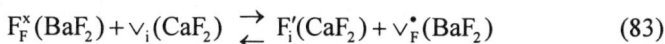

$$Ag_{Ag}^x(\beta - AgI) + v_i(AgCl) \underset{\leftarrow}{\overset{\rightarrow}{\rightleftharpoons}} Ag_i^{\bullet}(AgCl) + v_{Ag}'(\beta - AgI) \qquad (82)$$

$$F_F^x(BaF_2) + v_i(CaF_2) \underset{\leftarrow}{\overset{\rightarrow}{\rightleftharpoons}} F_i'(CaF_2) + v_F^{\bullet}(BaF_2) \qquad (83)$$

(Of course the thermodynamics can also be described in terms of transfer reactions of the individual defects given the coupling of the

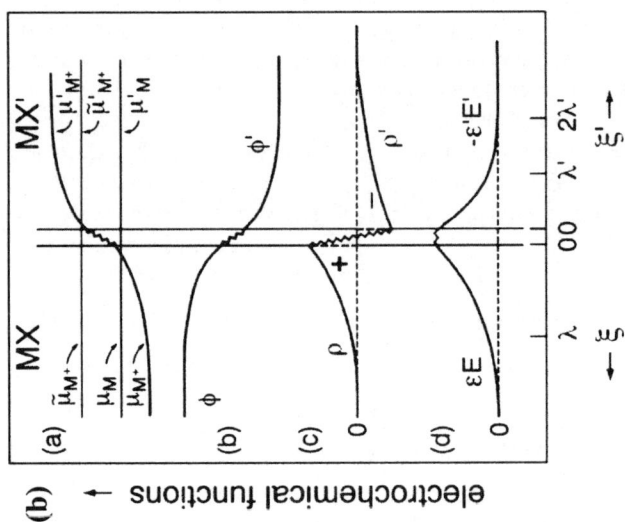

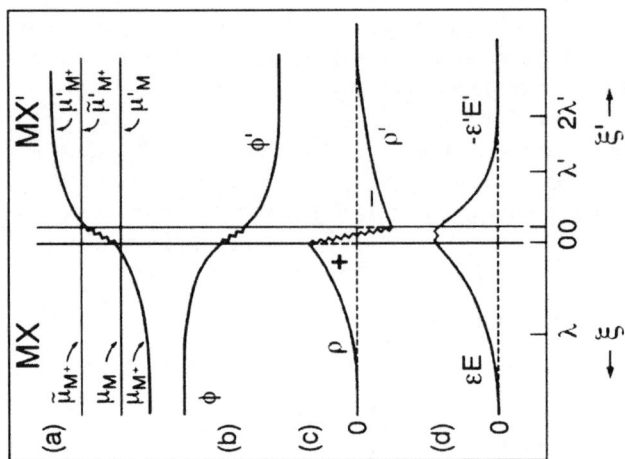

Figure 28. a) Ion redistribution process at the contact MX/MX': concentration effects.[117] b) Variation of the potentials, charge densities, and dielectric displacements at the contact of two Frenkel defect ionic conductors.[94] (Reprinted from J. Maier, "Ionic Conduction in Space Charge Regions." Prog. Solid St. Chem. 171-263, 23, Copyright © 1995 with permission from Elsevier.)

bulk disorder equations.[117]) In this way we can produce v-i transitions which correspond to "ionic p-n transitions". The two boundary concentrations that appear as parameters in the conductance in MX and MX' are simply linked via the ratio of the dielectric constants if we assume large space charge potentials and ignore charge storage between $x = 0$ and $x' = 0$. Equation (82) explains the large conductivity anomalies in the two-phase system AgI/AgCl which is shown in Fig. 29 (l.h.s.). (Here the preparation is made from quenching a high temperature homogeneous mixture into the solubility gap such that we can assume a coexistence of AgI(AgCl) with AgCl(AgI).)

A more elegant example which does not involve the complicated percolation effects[117] of these two phase mixtures, refers to MBE grown heterolayers of BaF_2 and CaF_2.[119] Here owing to the comparatively low preparation temperature almost no cation mixing occurs. The anion redistribution gives rise to a strong excess conductivity due to interfaces, which increases progressively if the interfacial density is increased. In both examples the space charge calculations are in very good agreement with the results.

The third contact is the contact of two chemically identical but differently oriented grains, i.e., a grain boundary (cf. Fig. 30).[94,120] A grain boundary can also separate two similarly oriented but "badly contacted" grains which may imply insufficient sintering (e.g. inclusion of pores) or inclusion of impurities (e.g. OH-groups or glassy phases). In all cases, however, the symmetry break introduced by the grain boundary will change the ion distribution at the expense of an excess charge (see Fig. 30a). Examples are highly conducting grain boundaries in AgCl that can be even locally investigated in Fig. 30b.[121,122] The effect is very related to the above heterogeneous doping in the case of which small second phase particles within the grain boundaries caused similar but enhanced boundary potentials. The conductivity can also be tuned by treatment with Lewis acids and bases. One example is the effect of contamination with NH_3 on the conductivity of AgCl.[94] While the cation adsorption is enhanced by bases, the inverse effect occurs in the case of the anti-Frenkel disordered earth alkaline fluorides. An increased internal adsorption of F^- is achieved here by treatment with Lewis-acids. (Analogously, SiO_2 particles in the grain boundary are more active than the more basic Al_2O_3.) In the case of SbF_5 contaminated CaF_2 grain boundaries[123] the F^- segregation which already occurs in the clean boundaries according to

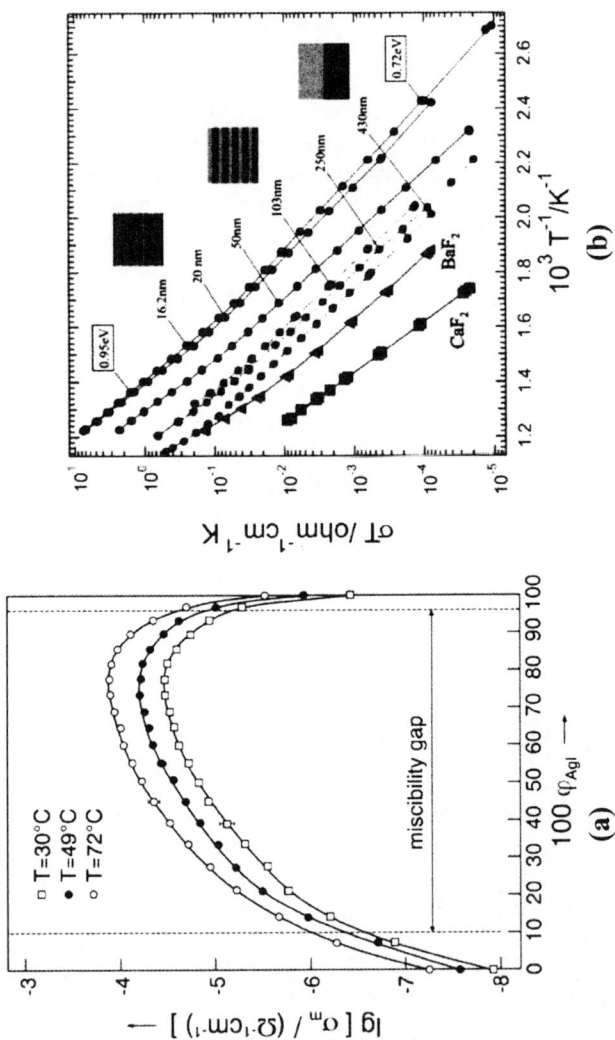

Figure 29. a) Conductivity variation in the system β-AgI-AgCl (φ: volume fraction) for different temperatures.[118] b) Conductivity variation in CaF$_2$-BaF$_2$ heterolayers as a function of temperature for different periods (spacings).[119] (Reprinted from N. Sata, K. Eberman, K. Eberl and J. Maier, "Mesoscopic fast ion conduction in nanometre-scale planar heterostructures." *Nature*, 408, 946-949. Copyright © 2000 with permission from Macmillan Magazines Ltd.).

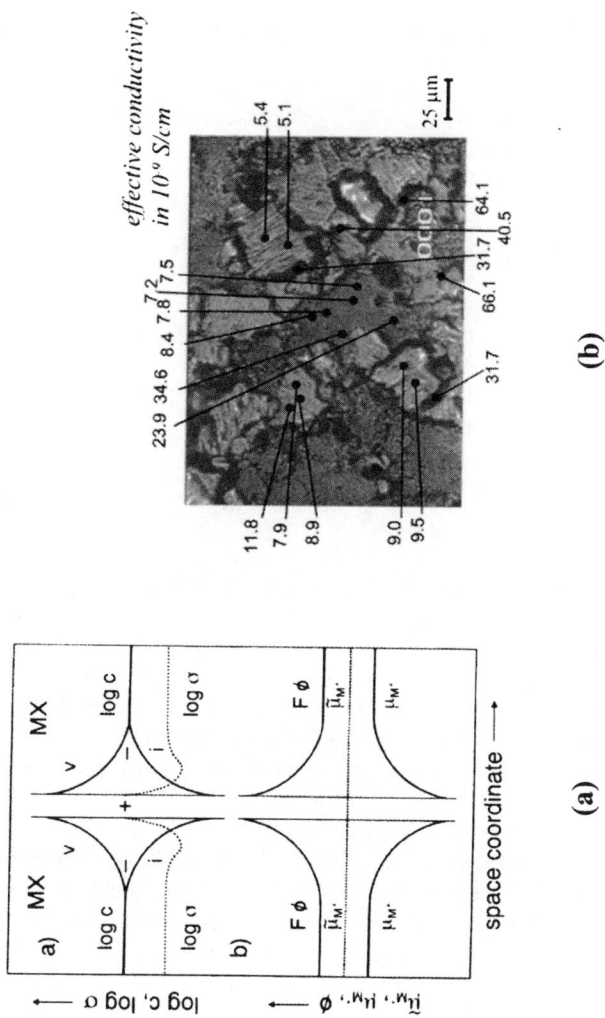

Figure 30. Ionic space charge effects at grain boundaries in Frenkel disordered materials. (a) Theoretical profiles if $u_i >$ u_v.[120] (b) The enhanced grain boundary conductivity can be verified by point electrode impedance spectroscopy.[121] The number given are in units of nS / cm and refer to room temperature.

$$F_F + V_{gb} \overset{\rightarrow}{\underset{\leftarrow}{}} F'_{gb} + V_F^{\bullet} \qquad (84)$$

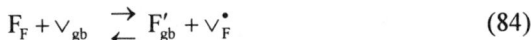

is additionally increased most probably according to

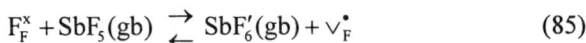

$$F_F^x + SbF_5(gb) \overset{\rightarrow}{\underset{\leftarrow}{}} SbF_6'(gb) + V_F^{\bullet} \qquad (85)$$

These acid-base effects can be directly studied by exposing surfaces of the ion conductors to acid-base active gases, such as CaF_2 to BF_3, or AgCl to NH_3 according to

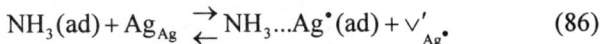

$$NH_3(ad) + Ag_{Ag} \overset{\rightarrow}{\underset{\leftarrow}{}} NH_3...Ag^{\bullet}(ad) + V'_{Ag} \qquad (86)$$

In the latter case, the increased vacancy conductivity can be used to sense ammonia quite selectively.[124,125] This is obviously the analogue to the Taguchi sensor[126] where an electron conductor senses redox active gases, e.g. O_2 by trapping electrons according to

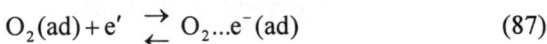

$$O_2(ad) + e' \overset{\rightarrow}{\underset{\leftarrow}{}} O_2...e^-(ad) \qquad (87)$$

The correspondence between ions and electrons is displayed by Fig. 31. Besides its importance for conductivity sensors, surface defect chemistry should be paid much more attention to in terms of heterogeneous catalysis. While in homogeneous catalysis acid-base effects are overwhelmingly studied, the role of the point defects as elementary acid-base centers are not really acknowledged. Examples are the dehydrohalogenation of t-BuCl by homogeneously or heterogeneously doped AgCl[127,128] (see also Refs.[129,130])

Depletion layers with respect to ionic conductors are no less interesting. In acceptor doped $SrTiO_3$,[131-134] CeO_2[135,136] and presumably also in ZrO_2,[137-141] space charge layers have been found to exhibit a positive space charge potential. As a consequence the oxygen vacancies, the relevant ionic carriers, are depleted as well as holes, while e' should be enriched.

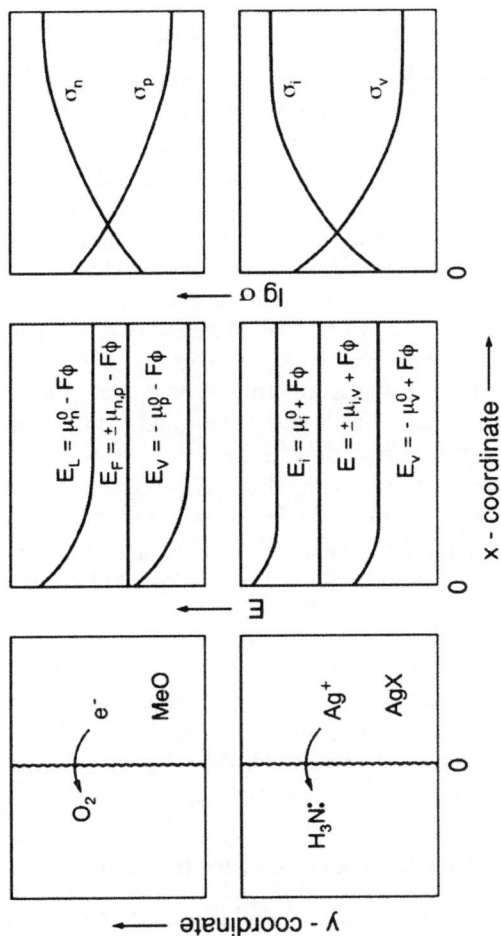

Figure 31. The change in the ionic surface conductivity upon a variation in the partial pressure of the acidbase active gas (bottom) is analogous to the Taguchi sensor which works for redox-active gases.[20]

In the case of the Fe-doped $SrTiO_3$ bicrystal referred to in Fig. 32a (grain boundary parallel to the electrodes), the impedance response shows a clear interfacial response whose capacitance (cf.Mott-Schottky length λ^* (Eq. 75) as well as resistance parameters provide evidence of a hole-depletion zone. Evaluating these parameters as a function of partial pressure, doping content, temperature and bias (see, e.g. Fig. 32a) gives a clear picture of the defect chemistry. The space charge potentials for the boundary under investigation (here Σ5-tilt) obtained are typically of the order of several 100 mV (here 500 mV) almost independent of P_{O_2} and weakly dependent on temperature in agreement with the defect chemical modeling (see Fig. 32b).[100,131]

According to Eq. (68) we now also expect an oxygen vacancy depletion, and in fact a much more severe one according to the double positive charge. The separation of electronic and ionic conductivity by a blocking technique confirmed the predictions in detail (Fig. 33[134]).

The simultaneous depletion of both the relevant ionic and the relevant electronic carriers leads to two further consequences. One is a severe chemical resistance with respect to the kinetics of stoichiometry[142] changes which we will consider later in more detail in Section VI.4.*iii.* Figure 32b shows the good agreement between electrical and chemical experiments in terms of the space charge potential. The second is the finding that a changed $\sigma_{ion}/\sigma_{eon}$ ratio at the boundaries compared to the bulk leads to a bulk polarization if an electrical field is applied,[143] similar to the electrochemical polarization induced by selectively blocking electrodes which will be considered in Part II.[1]

The last consequence to be discussed is the fact that a positive space charge potential does not only lead to a hole depletion but also to an accumulation of excess electrons. Their influence may be observed in $SrTiO_3$ only at very high space charge potentials,[144] but has been clearly seen in nanocrystalline CeO_2[136,159] (space charge potential $\simeq 300$ mV) and will be considered again in Section V.4.

3. Interfacial Defect Thermodynamics

The parameter c_0 measuring the concentration in the first layer adjacent to the core, i.e., in the first layer of the assumed bulk structure, is a convenient parameter for the mathematical description but not a

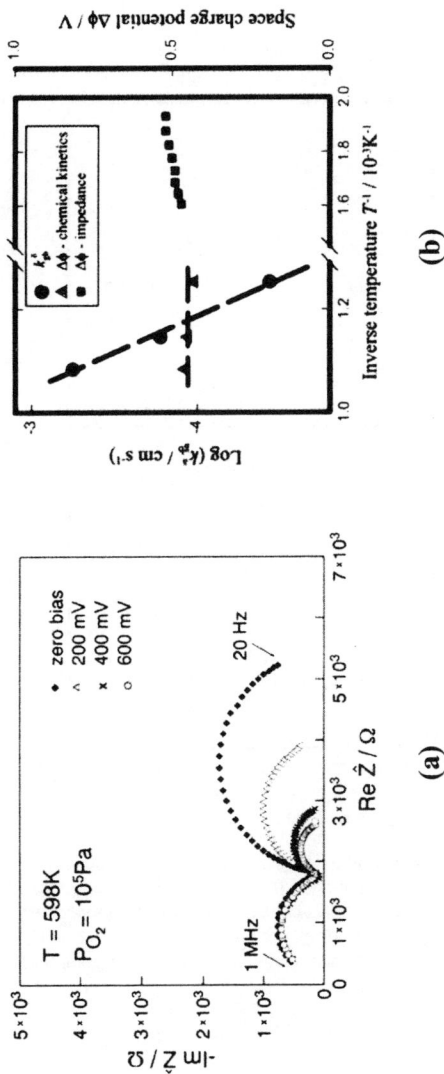

Figure 32. a) Impedance measurements on the O_2, Pt|SrTiO$_3$|SrTiO$_3$|Pt, O_2 as function of the superimposed d. c. bias. Electrodes are parallel to the bicrystal boundary (Σ5 tilt grain boundary, iron content: 2 x 10^{18}cm^{-3}). Both bulk and boundary resistances are predominately electronic resistances.[100] b) Measurements of oxygen diffusion into a SrTiO$_3$ bicrystal (interface perpendicular to transport), effective rate constants, k$_{gb}^\delta$, and space charge potentials, $\Delta\phi$, determined by impedance spectroscopy at low temperatures and by the diffusion experiment at high temperatures.[142] (Reprinted from M. Leonhardt, J. Jamnik and J. Maier, "In situ Monitoring and Quantitative Analysis of Oxygen Diffusion through Schottky-Barriers in SrTiO$_3$ Bicrystal." *Electrochem. Solid-St. Lett.* **2**, 333-335. Copyright © 1999 with permission from The Electrochemical Society, Inc.)

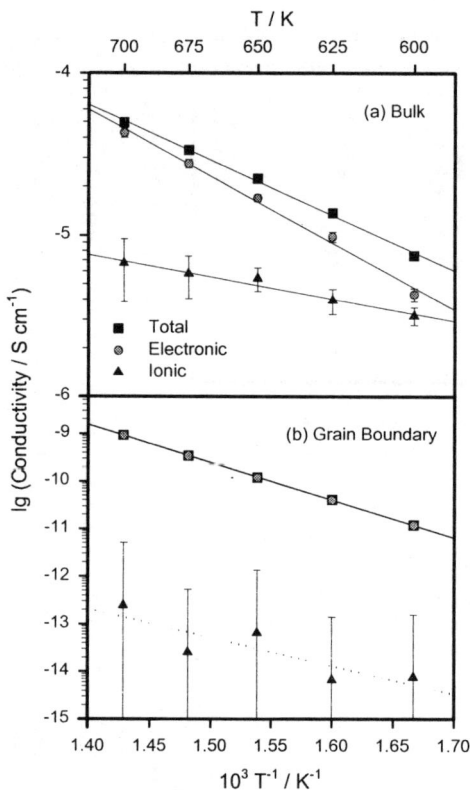

Figure 33. Temperature dependence of total and partial conductivities (a) in bulk and (b) at the grain boundary. Note that the values of the ionic conductivity at the grain boundary are approximate.[134] (Reprinted from X. Guo, J. Fleig and J. Maier, "Sepa-ration of Electronic and Ionic Contributions to the Grain Boundary Conductivity in Acceptor-Doped SrTiO₃." *J. Electrochem. Soc.*, **148**, J50-J53. Copyright © 2001 with permission from The Electrochemical Society, Inc.)

fundamental interfacial materials parameter; the same is true for the surface charge or the space charge potential. A pertinent parameter is the local standard chemical potential of a particle or defect (i.e., its "energy level") in the boundary zone. This leads to the (rigid) core-space charge picture described in Refs.[23,24,108,145] and is illustrated by Fig. 34 for the charging of the boundary core (e.g. a grain boundary).

In the simplest case the perfect structure changes at the boundary in an abrupt way, and the standard chemical potential of the carriers can

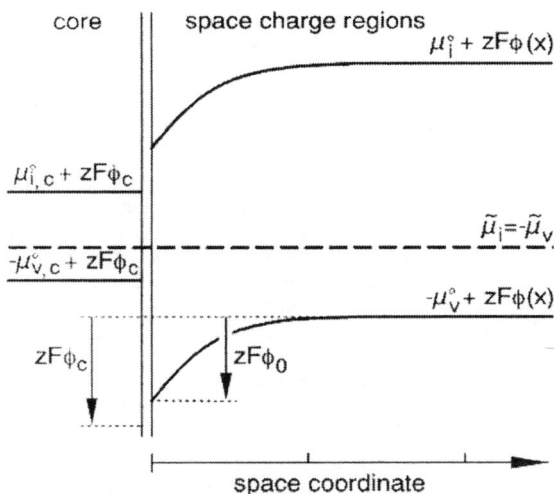

Figure 34. Boundary equilibrium in the level diagram for Frenkel defects (i, $\vee$: e.g. $Ag_i^\bullet$, V'_{Ag}, $z = 1$) or anti-Frenkel defects (i,$\vee$: e.g. O_i'', $V_O^{\bullet\bullet}$, $z = -2$). The electrical potentials cause the levels ($\tilde{\mu}^\circ$) to bend in the space charge zone and to shift in the core in order to satisfy that $\tilde{\mu} = \text{const}$. For the sake of simplicity the electrical bulk potential (ϕ_∞) is set to zero. The index c designates the core region.[145] (Reprinted from J. Jamnik, J. Maier, S. Pejovnik, "Interfaces in solid ionic conductors: Equilibrium and small signal picture." *Solid State Ionics*, **75**, 51-58. Copyright © 1995 with permission from Elsevier.)

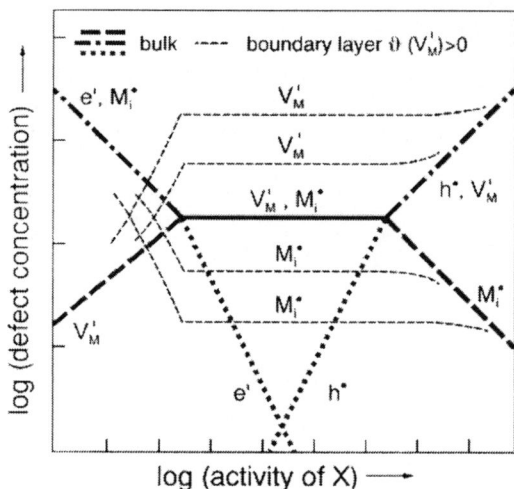

Figure 35. Kröger-Vink diagram of boundary layers for our model substance MX when $\mathcal{G}(v'_M) > 0$. The broken lines refer to the ionic defect concentrations at (two) different distances from the interface. The behavior of the electrons in boundary regions is not shown.The approach to the bulk values (printed boldface) for extreme abscissa values is attributable to the disappearing Debye length. The mirror symmetry on compa-rison of v'_M with $M_i^\bullet$ follows from Eq. (68).[94] (Reprinted from J. Maier, "Ionic Conduction in Space Charge Regions." *Prog. Solid St. Chem.* **23**, 171-263. Copyright © 1995 with permission from Elsevier.)

be described by a step function. Without great loss of generality, let us consider the case of a grain boundary (Fig. 34). The boundary core may be approximately conceived as a region of special structure of the width of one or a few atomic layers, which again for simplicity's sake is considered to be homogeneous. As experiments and capacitances show, this picture is not only simple but in many cases also quite realistic.[146]

The different structure is energetically reflected by a non-zero interfacial tension γ. The terms γa (a: interfacial area) and $\Sigma \mu_k n_k^\Sigma$ add to the bulk value of the free energy (μ_k describes the chemical potential of the neutral component, n_k^Σ the interfacial excess mole number). This knowledge is not sufficient for the defect chemical treatment. For this purpose we need to know the standard chemical potentials (or at least the relevant differences) of the defects also in the core regions. The defect concentrations then follow by a lattice-gas statistical model as indicated in Fig. 34. In more detail this is described for an oxide surface in Refs.[23,24]. There also the P_{O_2} and T dependencies are discussed, which are particularly simple if a power law form (analogous to Eq. 42) is valid for c_0. This then leads to the Kröger-Vink diagram of boundary regions (see Fig. 35). Calculation of the partial free energies ("energy levels") of the point defects for the interfacial layers as well as the elucidation of more advanced core-space charge models is of pronounced significance in this context. The latter point includes questions of structural flexibility of the interfacial core (i.e., variance of $\tilde{\mu}_j^\circ$ in the core layer) as well as more realistic profiles of $\tilde{\mu}_j^\circ$ within the core layer. The situation is more complicated if the interface is curved. For more details see Refs.[23,24,28,147].

4. Non-Trivial and Trivial Size Effects: Nanoionics

So far we essentially discussed the behavior under semi-infinite boundary conditions, i.e., situations that are locally not affected by the presence of neighboring interfaces. This is different if the spacing is very narrow and of the order of the characteristic decay length at semi-infinite interfaces (i.e., the above discussed Debye length).[94,148,149] If the latter is fulfilled or the interface structure itself is changed by the presence of a neighboring interface, the properties change locally and are different from the properties at isolated interfaces. In this case we speak of true size effects.[36,37,150,151] Also in the other cases the overall response may perceptibly depend on the spacing of interfaces simply due to the increased interface-to-volume proportion. This we name trivial size effects. The distinction is precise as long as the interfaces are planar and edges and corners are neglected, as it is the case for thin films, or for polycrystalline materials that can be described by the bricklayer model (cf. reference[152]) which will be explained in more detail below.

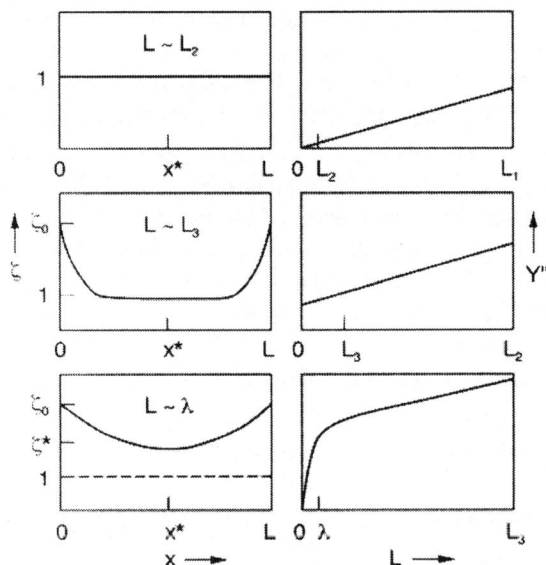

Figure 36. Defect concentration and conductance effects for three different thicknesses $L_1 \gg L_2 \gg L_3$. The mesoscale effect on defect concentration (l.h.s.) discussed in the text, when $L < 4\lambda$, is also mirrored in the dependence of the conductance on thickness (r.h.s.). If the boundary layers "overlap", the interfacial effect previously hidden in the intercept is now resolved. It is presupposed that surface concentration and Debye length do not depend on L. (Both can be violated, c_∞, at sufficiently small L because of interaction effects and exhaustibility of bulk concentrations.)[36,94] (Reprinted from J. Maier, "Defect chemistry and ion transport in nanostructured materials. Part II. Aspects of nanoionics." *Solid State Ionics*, **157**, 327-334. Copyright © 2003 with permission from Elsevier.)

Before dealing with the superposition in polycrystalline materials, let us first consider the conductance of a thin film with an accumulation[148] effect at the interface to the (identical) neighboring phases for different thicknesses and consider Fig. 36. In the top figure the film is so thick that the interfacial effect is not seen; the carrier profile is horizontal and the (parallel) conductance increases linearly with increasing thickness with a zero intercept. At distinctly smaller thicknesses the boundary effects appear and manifest themselves as a non-zero intercept in the conductance behavior; however, the interfaces are still separated. Only if the thickness is of the order of the space charge width or below, the conductance varies in a more sophisticated way according to

$$\Delta\sigma_m^{\parallel} \cdot L \equiv \Delta Y^{\parallel} \simeq 2u_1[2RT\varepsilon(c_{10} - c_1^*)]^{1/2} \qquad (88)$$

(1 denotes the accumulated majority carrier, c^* is the concentration value in the center and can be calculated from L and c_{10} via elliptical integrals of the first kind.) The detailed calculation is given in Ref.[148]. References[94,148,149] give experimental examples for the above situations. Figure 37 shows the same phenomenon in alternative presentations. The l.h.s. figure plots the mean excess film conductivity (i.e., the excess conductance normalized to the film thickness) as a function of the inverse thickness. Now the film thickness decreases from the left to the right. As long as $L > 4\lambda$, the boundary parts approach but are still separated leading to a linear increase of the mean conductivity as non-contributing bulk parts drop out. When L becomes smaller than 4λ, the mean specific value does not remain constant, rather the defect concentration is now locally enhanced (see Fig. 38, l.h.s. bottom). This enhancement can be described by the nano-size factor

$$g(L)- \approx \frac{4\lambda}{L}\left[\frac{c_{10} - c_1^*}{c_{10}}\right]^{1/2} \qquad (89)$$

In the r.h.s. part of the figure the excess conductance is normalized with respect to the space charge width (2λ if $L > 4\lambda$ and L if $L < 4\lambda$) being a measure of the mean conductivity of the space charge zone. This value remains constant in the regime of trivial size effects but increases according to g(L) in the regime of true size effects.

Analogously we have to proceed if we are measuring in the orthogonal direction; the results will be completely different (space charge profile, core effects). Even though the overall effect can be obtained more precisely by a superposition of bulk values and interfacial excess contributions, it is more convenient for large space

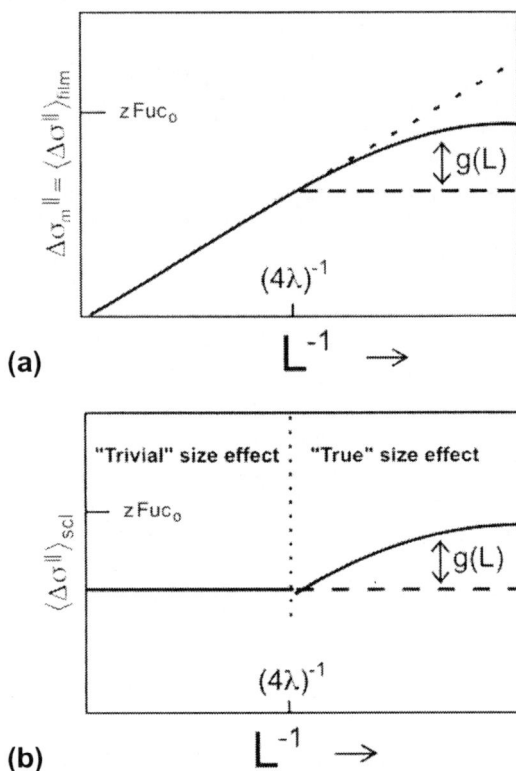

Figure 37. (a) Area related space charge conductance per film thickness representing the measured excess space charge conductivity, as a function of inverse spacing. (b) Area related space charge conductance per boundary thickness representing the mean specific excess space charge conductivity, as a function of inverse spacing.[37] (Space charge width is 4λ for $L \geq 4\lambda$, and L otherwise.) (Reprinted from J. Maier, "Nano-Ionics: Trivial and Non-Trivial Size Effects on Ion Conduction in Solids." *Z. Physik. Chem.* **217**, 415-436. Copyright © 2003 with permission from Oldenbourg Wissenschaftsverlag.)

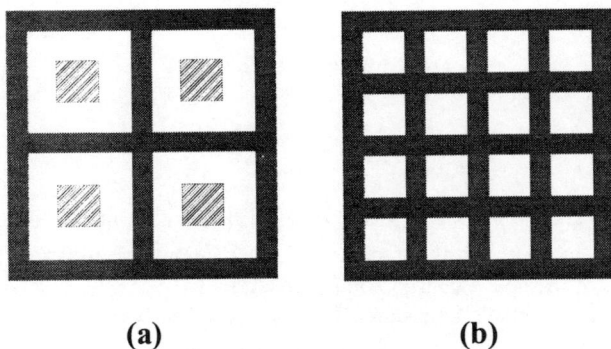

(a)   **(b)**

Figure 38. Section of a polycrystal (bricklayer-model). (a) Grain
size large compared to 4λ. (b) Grain size smaller than 4λ. The
white area refers to the space charge zones, the hatched to the
electroneutral bulk and the dark to the interfacial core.[151]
(Reprinted from J. Maier, "Thermodynamic aspects and
morphology of nano-structured ion conductors. Aspects of nano-
ionics. Part I." *Solid State Ionics* **154-155**, 291-301. Copyright ©
2002 with permission from Elsevier.)

charge potentials to describe it approximately by a superposition of
spatially separated regions of different transport parameters.

Now let us proceed to the description of a polycrystalline material
and the superposition of bulk and both perpendicular ($\perp$) and parallel
(‖) boundary effects in a bricklayer model (Fig. 38). The bricklayer
model replaces the complex microstructure[153] by a primitive periodic
arrangement of cubes. The superposition of bulk conductivity (σ_∞) and
boundary conductivities ($\sigma_{gb}^{\parallel}, \sigma_{gb}^{\perp}$) within such a model can be
approximately described by[120]

$$\sigma_m = [\sigma_\infty \sigma_{gb} + \beta_{gb}^{\parallel} \varphi_{gb} \sigma_{gb}^{\parallel} \sigma_{gb}^{\perp}] / [\sigma_{gb}^{\perp} + \beta_{gb}^{\perp} \varphi_{gb} \sigma_\infty] \qquad (90)$$

(φ_{gb} being the volume fraction of the boundary; $\beta_{gb}^{\parallel}, \beta_{gb}^{\perp}$ measuring the
fraction of pathways that percolate, ideally 2/3 and 1/3). These
transport parameters can also be conceived as complex quantities

allowing the interpretation of impedance spectra. The deconvolution into core (co) and space charge layers (sc) can be approximated by

$$\varphi_{gb} / \sigma_{gb}^{\perp} = \varphi_{sc} / \sigma_{sc}^{\perp} + \varphi_{co} / \sigma_{co}^{\perp} \qquad (91a)$$

$$\varphi_{gb} \sigma_{gb}^{\parallel} = \varphi_{sc} \sigma_{sc}^{\parallel} + \varphi_{co} \sigma_{co}^{\parallel} \qquad (91b)$$

The space charge values $\sigma_{sc}^{\parallel}$ and $\sigma_{sc}^{\perp}$ in Eq. (91) are mean values and calculated by integration of $c(x)$ or $1/c(x)$ (and normalization to the thickness of the space charge zone), respectively.

On this level we can state: Trivial size effects are size effects in which σ_m but not $\sigma_\infty, \sigma_{gb}^{\perp}, \sigma_{gb}^{\parallel}$ depend on φ, whereas the latter vary with φ in the case of the true size effects. (If edges and corners have to be considered, this distinction can be successively refined for different levels of resolutions.[152])

As the previous section showed, in a variety of examples severe enhancements of the ionic conductivity has been found and successfully attributed to space charge effects. Typical examples are silver halides or alkaline earth fluorides (see Section V.2.). How significantly these effects can be augmented by a particle size reduction, is demonstrated by the example of nano-crystalline CaF_2.[154] Epitaxial fluoride heterolayers prepared by molecular beam epitaxy not only show the thermodynamically demanded redistribution effect postulated above (see Section V.2.), they also highlight the mesoscopic situation in extremely thin films in which the electroneutral bulk has disappeared and an artificial ion conductor has been achieved (see Fig. 39).[119,155-157]

As already stated in Section V.2., materials such as acceptor doped ZrO_2, CeO_2 and $SrTiO_3$ show a space charge behavior that is characterized by a depletion of oxygen vacancies leading to a depression of the ionic conductivity.[158]

The positive space charge potential leads to peculiar effects in nano-crystalline CeO_2.[136,159] In weakly positively doped material in which the electronic concentration is quite high but still low enough for it to remain an ion conductor in the bulk, the space charge effect (depletion of positive carriers such as holes and oxygen vacancies but accumulation of negative carriers such as excess electrons) makes its overall conductivity change to n-type. The enhanced electron transport

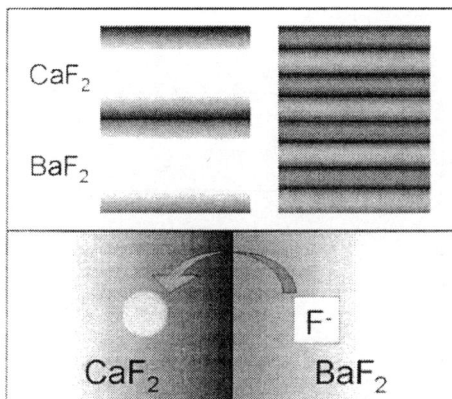

Figure 39. Ionic heterostructures composed of CaF_2 and BaF_2 leading to mesoscopic ion conduction.[36,119] (Reprinted from J. Maier, "Defect chemistry and ion transport in nanostructured materials. Part II. Aspects of nano-ionics." *Solid State Ionics*, **157**, 327-334. Copyright © 2003 with permission from Elsevier.)

along the boundaries as well as the depressed vacancy transport across them (which limits the bulk-to-bulk path) could be concluded from detailed electrochemical experiments (impedance spectroscopy, dc experiments using reversible and blocking electrodes).[136]

In the case of depletion layers the space charge overlap leads to extremely resistive situations. A relevant example are low angle grain boundaries in Fe-doped $SrTiO_3$. Such low angle grain boundaries consist of regularly arranged dislocations. Here the dislocations are positively charged, and the spacing between them is definitely smaller than the space charge width resulting in a severe and homogeneous blocking with respect to hole as well as ionic and atomic oxygen transport.[160]

Naturally, the space charge overlap is not the only possible true size effect.[37,161,162]

Others are structural effects that come into play when the effective size of a charge carrier (radius of influence) is of the order on the

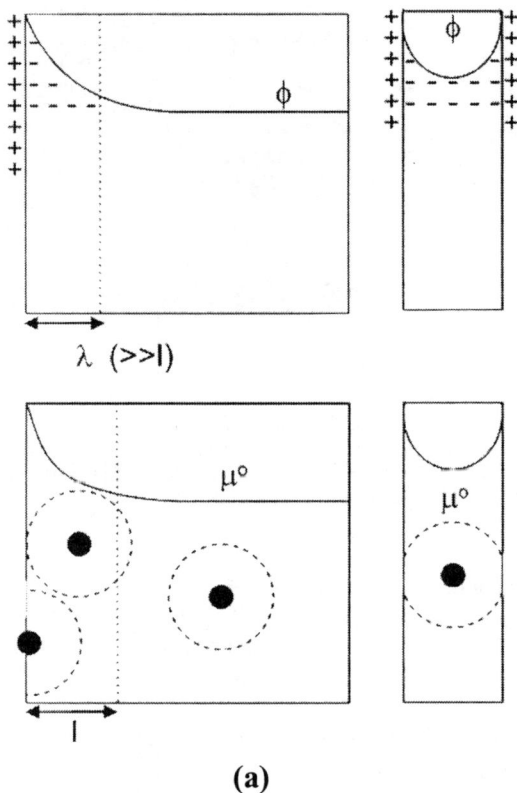

(a)

Figure 40. (a) Top: Space charge effects decaying with λ. Bottom: Structural effects that decay rapidly in the bulk. Within the indicated distance (ℓ) the structurally per-turbed core (not shown, extension s) is per-ceived by the defects as far as μ^o is concerning.[37] (b) Delocalized electrons perceive the boundaries much earlier ($\Delta\mu^o \propto L^{-2}$). (Reprinted from J. Maier, "Nano-Ionics: Trivial and Non-Trivial Size Effects on Ion Conduction in Solids." *Z. Physik. Chem.*, **217**, 415-436. Copyright © 2003 with permission from Oldenbourg Wissenschaftsverlag.)

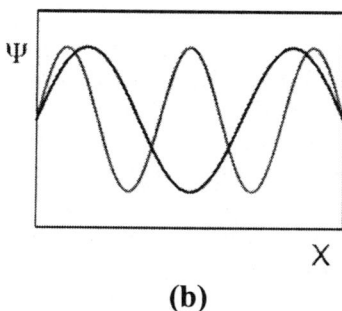

(b)

Figure 40. Continuation

interfacial spacing (or even earlier when e.g. elastic effects penetrate quite deeply). Then the standard chemical potential is affected (see Fig. 40). It is often found that the structure of crystals is well-maintained down to the range of a few nm. In NaCl, e.g. below 1 nm cluster forms represent the equilibrium structure, that are completely different from the rock-salt structure adopted at larger sizes.[163] Above 1 nm bulk and surface energies rapidly approach the values of the massive crystals.[164] In a very hand-waving way a parameter ℓ may be defined which is of such an order of magnitude. Since 1 nm is also typically the order of magnitude of interfacial core regions as well as typically the extension of the relaxation sphere around defects, it sets roughly the limit for the above core-space layer model.[36,37] Figure 40 refers to this point. Attention has to be paid to the fact that these two mesoscales are not necessarily de-convoluted, i.e., ℓ need not be small against λ nor need both parameters be constant concerning L. In the case of delocalised electrons (Fig. 40, bottom) this confinement (which affects μ^0) is already relevant for moderately small sizes and leads to the spectacular effects in the field of nano-electronics.[165] In reference[37] a variety of further geometrical and structural size effects are discussed. Only three shall be mentioned here: (i) One is the fact that the curvature introduces a Gibbs-Kelvin term into the chemical potential of two chemically different particles. This predicts a charging at the contact of two particles that are only different in terms of curvature. (Such a charging is even expected in the case of zero curvature, i.e., in the core of thin films, if we refer to the sub-Debye regime.[166]) (ii) In small crystallites defect pairs (excitons, Frenkel-pairs) may not be separated sufficiently resulting in severe interaction effects. The reader is referred to it for more details.[167] (iii) Of special importance are interfacial phase transformations.[168] One example are highly disordered stacking faults occurring at special interfaces of AgI (e.g. $AgI:Al_2O_3$). They can be conceived as heterolayers of $\gamma/\beta/\gamma$...-AgI being highly disordered (note the similarity to the above fluoride heterolayers).[169] The field of nanoionics is not only of fundamental interest but also of technological significance. Li-batteries may serve as an example. While in the rocking-chair secondary cells must reversibly insert Li into the homogeneous phase, heterogenous storage is possible in the nanocrystalline state.[170,171] Not only are the lengths of the transport paths (see following chapter) drastically diminished, but also the differences between double layer capacitance and insertion capacitance are getting blurred.[150,161] Already these few examples show that nanostructured matter, given sufficient stability, can combine advantages from both the solid and the fluid state and constitutes an exciting future area of solid state chemistry.

VI. KINETICS OF CHARGE TRANSFER AND TRANSPORT

1. Three Experimental Situations

Charge transport in the bulk has to obey electroneutrality. Figure 41 shows three simple experiments that comply with this restriction. For brevity let us call them the electrical (a), the tracer (b) and the chemical experiment (c); and, to be specific, let us consider an oxide. In the electrical experiment an electrical potential difference is applied, the electrons flowing in the outer circuit compensate the charge flow within the sample (Figure 41a shows this for the case of a pure ion conductor). If we apply reversible electrodes, in the steady state there is no compositional change involved. (At this point we are not interested in (electro-)chemical effects caused by non-reversible electrodes. This is considered in detail in Part II.[1]) The tracer transport (b) caused by the application of a chemical potential difference of the isotopes consists of a counter motion of the two isotope ions. Finally, experiment (c) presupposes a mixed conduction; her the outer wire is, as it were,

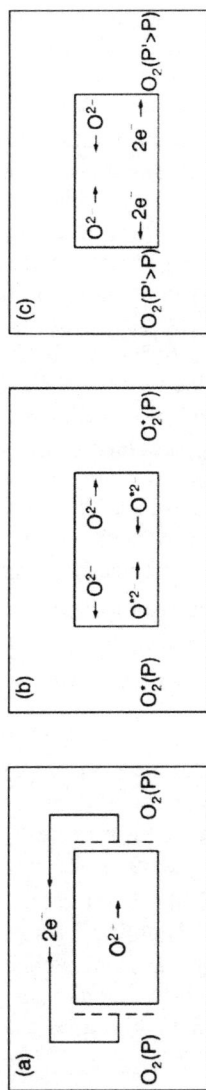

Figure 41. The three basic experiments discussed in the text: (a) electrical, (b) tracer, (c) chemical.[172] (Reprinted from J. Maier, "On the correlation of macroscopic and microscopic rate constants in solid state chemistry." *Solid State Ionics* **112**, 197-228. Copyright © 1998 with permission from Elsevier.)

internalized. The chemical transport is a consequence of different the chemical potentials of neutral oxygen and consists of an ambipolar motion of ions and electrons corresponding to a motion of "O". Unlike in the previous experiments now a compositional change occurs. Unlike experiment a, the experiments b and c represent true diffusion experiments. We will see that it is possible to decompose the diffusive transport coefficients into a resistive part and a chemical capacitance part. In the tracer case the chemical capacitance is simply given by the total ion concentration and hence a constant. The stationary electrical experiment only involves a resistance part (nevertheless diffusive transport coefficients, later labeled with upper index Q, are defined that allow a direct comparison); the transient behavior of the electrical experiment involves electrical capacitances.

If the bulk process is dominating, in the electrical experiment the total conductivity (ionic and electronic) is measured, the second gives information on the tracer diffusion coefficient (D^*) which is directly related to the ionic conductivity (or D^Q). In the third experiment one measures the chemical diffusion coefficient (D^δ), which is a measure of the propagation rate of stoichiometric changes (at given chemical gradient); it is evidently a combination of ionic and electronic conductivities and concentrations.[3,4,173-175]

In all three cases also surfaces or internal interfaces may be dominating which can be described in terms of effective rate constants $\bar{k}^Q, \bar{k}^*, \bar{k}^\delta$. As for the bulk processes $\bar{k}^Q$ is only formally obtained from the steady state parameters. The elucidation of $\bar{k}^Q$ is the basic problem of electrode kinetics extensively dealt with in liquid electrochemistry.

The three experiments do not only introduce decisive mass and charge transport parameters, they also permit their determination. Some points relevant in this context will be investigated in the following. (Note that electrochemical measurement techniques are covered by Part II.[1]) At the end of this section we will have seen that—close to equilibrium—not only all the D's and the $\bar{k}$'s can be expressed as the inverse of a product of generalized resistances and capacitances, but that these elements can be implemented into a generalized equivalent circuit with the help of which one can study the response of a material on electrical and/or chemical driving forces.

2. Rates, Fluxes and Driving Forces

(i) Transport, Transfer and Reaction

In spite of the complexity of real processes most cases of interest may be effectively described by a heterogeneous pseudo-monomolecular reaction of the type

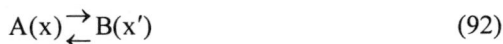

$$A(x) \underset{\leftarrow}{\overset{\rightarrow}{}} B(x') \tag{92}$$

i.e., the transport of A from x to $x' \equiv x + \Delta x$ while it simultaneously transforms into B. The special cases of $x = x'$ and $A = B$ correspond to a homogeneous chemical reaction

$$A \underset{\leftarrow}{\overset{\rightarrow}{}} B \tag{93}$$

and a pure transport $(A' \equiv A(x'))$

$$A \underset{\leftarrow}{\overset{\rightarrow}{}} A' \tag{94}$$

In both cases we are not specifically dealing with reorganization processes which are more complicated than in the case of liquids and specifically related with the crystal structure.[176,177] More precisely the first special case is met, if, for reactions, the difference in the space coordinate does not matter (i.e., if $z_A \phi(x) = z_B \phi(x')$). Hence pure gas phase reactions or neutral surface reactions can be described by Eq. (93), while Eq. (94) refers to pure transport steps (transport within the same structure). The description of typical electrochemical reactions such as charge transfer reactions require the analysis of Eq. (92). (We will see later that mechanistic equations are typically bimolecular, however, owing to the constancy of regular constituents, the consideration of Eq. (92) suffices in most cases of interest.)

Two approaches will be considered: one is to apply linear irreversible thermodynamics according to which the generalized rate $\mathfrak{R}$ depends linearly on the generalized driving force $\tilde{A}$ (corresponding to the first approximation of a Taylor expansion of $\widetilde{\mathfrak{R}}$ in $\widetilde{A}$, since $\widetilde{\mathfrak{R}}$ ($\widetilde{A} = 0$) disappears):

$$\widetilde{\mathfrak{R}} = L\widetilde{A} + (L'\widetilde{A}^2 + L''\widetilde{A}^3 + ...) \tag{95}$$

In a better approximation $\widetilde{\mathfrak{R}}$ (of process k) can also depend on secondary driving forces such that

$$\widetilde{\mathfrak{R}}_i = \Sigma_{i'} L_{ii'} \widetilde{A}_{i'} \tag{96}$$

where $L_{ii'} = L_{i'i}$ owing to microscopic reversibility.[178] This approach obviously holds only for small driving forces but is general otherwise.

An alternative approach that applies irrespective of the deviation from equilibrium but presupposes small concentrations, is chemical kinetics, according to which the rate of the elementary reaction is determined by

$$\widetilde{\mathfrak{R}} = \overrightarrow{\widetilde{k}}[A(x)] - \overleftarrow{\widetilde{k}}[B(x')] \tag{97}$$

(For the limits of Eq. 97 see e.g. Ref.[6]) The rate constants for forward and backward reaction contain a purely chemical part that is determined by the free activation energies $\Delta\overrightarrow{G}^{\neq}, \Delta\overleftarrow{G}^{\neq}$ and a factor that is determined by the portion of the electrical potential drop that adds to activation energy, i.e., $\overrightarrow{\widetilde{\alpha}}\Delta\phi$, $\overleftarrow{\widetilde{\alpha}}\Delta\phi$.[179] Hence $\Delta\overrightarrow{\widetilde{G}}^{\neq} = \Delta\overrightarrow{G}^{\neq} + \overrightarrow{\alpha}zF\Delta\phi$, $\Delta\overleftarrow{\widetilde{G}}^{\neq} = \Delta\overleftarrow{G}^{\neq} + \overleftarrow{\alpha}zF\Delta\phi$, and

$$\overrightarrow{\widetilde{k}} = k_0 \exp{-\frac{\Delta\overrightarrow{G}^{\neq} + \overrightarrow{\alpha}zF\Delta\phi}{RT}} \tag{98}$$

$$\overleftarrow{\widetilde{k}} = k_0 \exp{-\frac{\Delta\overleftarrow{G}^{\neq} + \overleftarrow{\alpha}zF\Delta\phi}{RT}} \tag{99}$$

$\vec{\alpha}$ and $\overleftarrow{\alpha}$ being symmetry factors.[180] The pre-factor k_0 is a measure of the transition attempts. Field corrections in Eqs. (98) and (99) are of equal magnitude if the activation threshold is halfway the reaction coordinate.

While the reaction rate is usually defined by the (normalized) concentration increase of species k, i.e., by $\partial c_k/\partial t$, the flux density j_k (concentration times velocity) can be introduced by the continuity equation

$$\partial c_k /\partial t = -\text{div } \mathbf{j}_k \qquad (100)$$

For a simple elementary reaction like Eq. (94) div j_k corresponds to $(\dot{j}_A - \dot{j}_A)/\Delta x$, and we can consider rates and fluxes almost synonymously in this case for pure transport. (Note that, unlike fluxes, usually rates are differently normalized with respect to stoichiometric number (ν_k) and geometry.) In many cases, in particular on a coarse grain-level it is meaningful to distinguish between transport effects (fluxes, transport rates) and generation/annihilation effects (local reaction rates). Then the continuity equation for k reads

$$\partial c_k /\partial t = -\text{div } j_k + \nu_k \mathfrak{R}^{(k)} \qquad (101)$$

where $\mathfrak{R}^{(k)}$ is the rate which produces (annihilates) the species k. In the steady state of pure transport the constancy of j_k follows, while a stationary conversion of $\mathfrak{R}^{(k)}$ into a flux j_k (consider e.g., electrode reaction at the electrode/electrolyte contact) implies $\mathfrak{R}^{(k)} \propto j_k$.

(ii) Transport in Terms of Linear Irreversible Thermodynamics and Chemical Kinetics

For Eq. (92) the driving force $\tilde{A}$ is obviously the difference in the electrochemical potentials

$$\tilde{A} = -[\tilde{\mu}_B(x') - \tilde{\mu}_A(x)] \qquad (102)$$

Table 3
Diagram for the non-equilibrium behavior in solids close to equilibrium. Conventional chemical reaction (x ≡ x') and particle transport (A ≡ B) are described in a general manner. Particle transport also includes the limiting cases of pure diffusion ($zF\Delta\phi = 0$) and pure electrical conduction ($\Delta\mu = 0$).[21] Note slight deviations to the notation in the text (e.g. β as pro-portionality factor between flux and driving force).

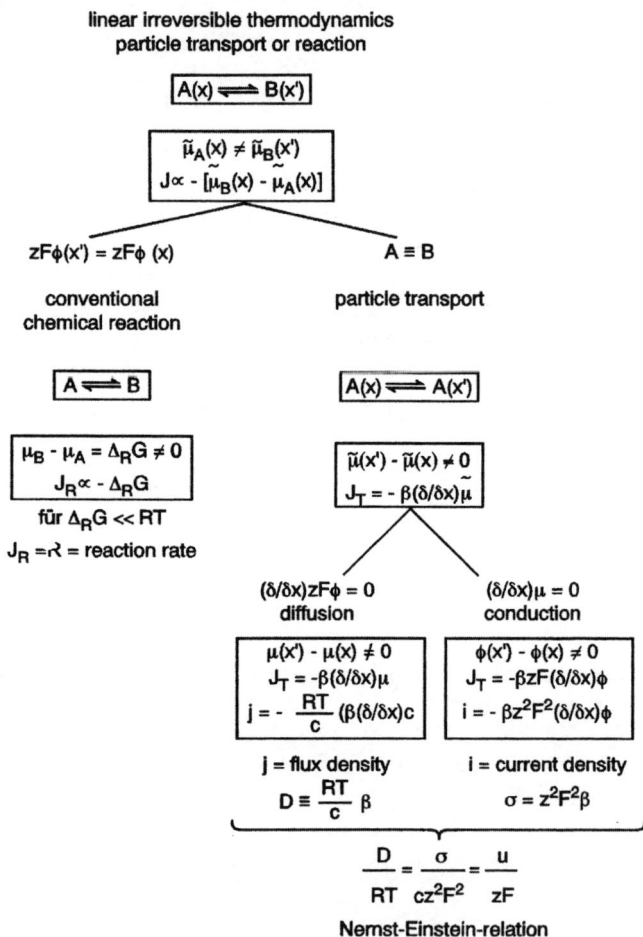

linear irreversible thermodynamics
particle transport or reaction

$$A(x) \rightleftharpoons B(x')$$

$$\tilde{\mu}_A(x) \neq \tilde{\mu}_B(x')$$
$$J \propto - [\tilde{\mu}_B(x) - \tilde{\mu}_A(x)]$$

$zF\phi(x') = zF\phi(x)$ $A \equiv B$

conventional   particle transport
chemical reaction

$$A \rightleftharpoons B$$ $$A(x) \rightleftharpoons A(x')$$

$\mu_B - \mu_A = \Delta_R G \neq 0$ $\tilde{\mu}(x') - \tilde{\mu}(x) \neq 0$
$J_R \propto - \Delta_R G$ $J_T = - \beta(\delta/\delta x)\tilde{\mu}$
für $\Delta_R G \ll RT$
$J_R = \mathcal{R}$ = reaction rate

$(\delta/\delta x)zF\phi = 0$ $(\delta/\delta x)\mu = 0$
diffusion   conduction

$\mu(x') - \mu(x) \neq 0$ $\phi(x') - \phi(x) \neq 0$
$J_T = -\beta(\delta/\delta x)\mu$ $J_T = -\beta zF(\delta/\delta x)\phi$
$j = - \dfrac{RT}{c}(\beta(\delta/\delta x)c)$ $i = - \beta z^2F^2(\delta/\delta x)\phi$

j = flux density   i = current density
$D \equiv \dfrac{RT}{c}\beta$ $\sigma = z^2F^2\beta$

$$\frac{D}{RT} = \frac{\sigma}{cz^2F^2} = \frac{u}{zF}$$

Nernst-Einstein-relation

which reduces to $-(\mu_B - \mu_A) = -\Delta_R G$ for a pure chemical reaction, and to $-[\tilde{\mu}_A(x') - \tilde{\mu}_A(x)] \propto -(\partial/\partial x)\tilde{\mu}_A$ for pure transport. If the electrical potential gradient is of no influence, e.g., in the case of neutral particles, the driving force is—according to $(\partial/\partial x)\mu_k = RT(w_k/c_k)(\partial/\partial x)c_k$ where w_k denotes the thermodynamic factor d ln a_k/d ln c_k and a_k the activity—finally determined by the concentration gradient.* If, however, the chemical potential is virtually constant, as it is the case for systems with a high carrier concentration (metals, superionic conductors), $z_k F(\partial/\partial x)\phi$ remains as driving force. For the linear approximation to be valid $\tilde{A}$ must be sufficiently small. Experimental experience confirms the validity of Fick's and Ohm's law (that immediately follow from Table 3 for diffusion* and electrical transport) in usual cases, but questions the validity of the linear relationship Eq. (96) in the case of chemical reactions. For a generalized transport we will use in the following the relation:[173,178,181]

$$\mathbf{j}_k = -\frac{\sigma_k}{z^2 F^2}\nabla\tilde{\mu}_k \qquad (103)$$

The interpretation of the pre-factor as a conductivity as well as the correlation between defect diffusion coefficient D_k and mobility u_k known as Nernst-Einstein equation follow directly from Table 3.

The chemical kinetics approach provides us with more insight with respect to the range of validity. We see from Table 4 that Fick's law indeed results without approximations (l.h.s.). For the pure electrical conduction (r.h.s.) we have to linearize the exponentials (Eqs. 98 and 99), i.e., to assume $|F\Delta\phi| \ll RT$ which is definitely a good approximation for transport in usual samples; it fails at boundaries or for ultrathin samples. Hence the application of Eq. (103) has to presuppose sufficiently thick samples and not too high fields. Table 4 also reveals the connection between D_k, u_k and the transport rate constant k_k and hence their microscopic meaning:

$$D_k \propto u_k \propto k_k \qquad (104)$$

---

* Close to equilibrium the pre-factors, even though containing concentrations terms, can be considered to be constant, yet the range of validity changes. The chemical kinetics treatment (Table 4) shows that for dilute conditions $\mathbf{j} \propto -\nabla c$ is valid at much larger derivations from equilibrium than $\mathbf{j} \propto -\nabla\mu$.

Table 4

Diagram of the kinetic treatment of the non-equilibrium behavior in the dilute states.[21] Reaction, diffusion and electrical conduction appear as special cases (ĉ: equilibrium concentration) of the electrochemical reaction $A(x) \rightleftarrows B(x')$.

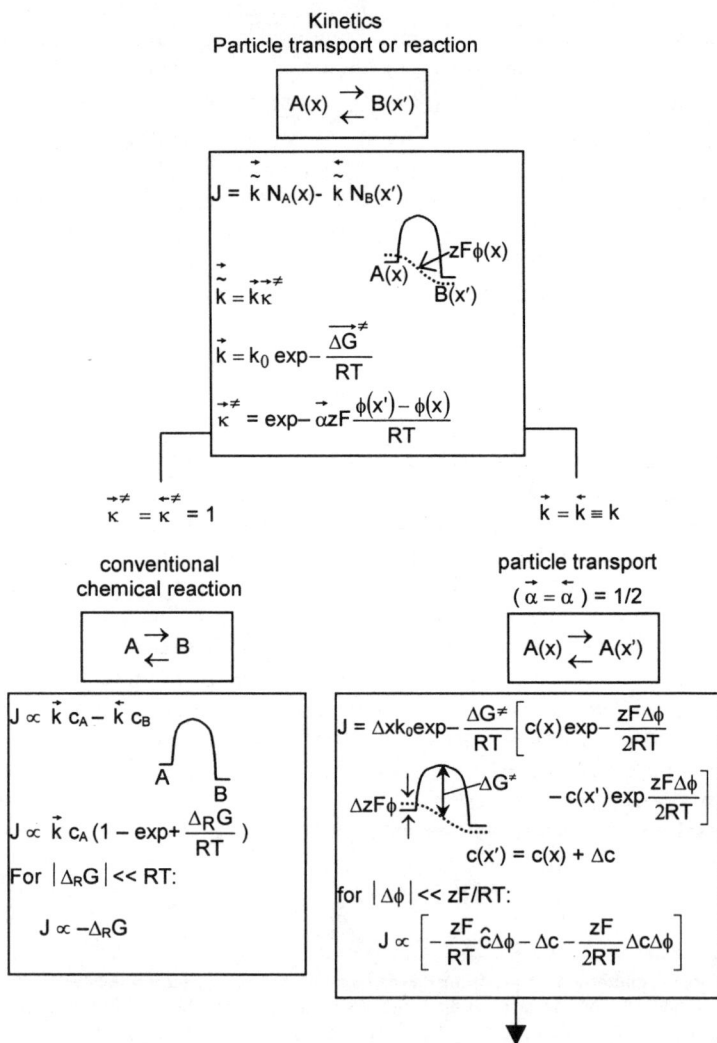

Kinetics
Particle transport or reaction

$A(x) \rightleftarrows B(x')$

$J = \overset{\rightarrow}{\tilde{k}} N_A(x) - \overset{\leftarrow}{\tilde{k}} N_B(x')$

$\overset{\rightarrow}{\tilde{k}} = \vec{k} \overset{\rightarrow\neq}{\kappa}$

$\vec{k} = k_0 \exp- \dfrac{\overset{\rightarrow\neq}{\Delta G}}{RT}$

$\overset{\rightarrow\neq}{\kappa} = \exp- \alpha zF \dfrac{\phi(x') - \phi(x)}{RT}$

$\overset{\rightarrow\neq}{\kappa} = \overset{\leftarrow\neq}{\kappa} = 1$

conventional chemical reaction

$A \rightleftarrows B$

$J \propto \vec{k} c_A - \overset{\leftarrow}{k} c_B$

$J \propto \vec{k} c_A \left(1 - \exp+ \dfrac{\Delta_R G}{RT}\right)$

For $|\Delta_R G| \ll RT$:

$J \propto -\Delta_R G$

$\vec{k} = \overset{\leftarrow}{k} \equiv k$

particle transport
$(\overset{\rightarrow}{\alpha} = \overset{\leftarrow}{\alpha}) = 1/2$

$A(x) \rightleftarrows A(x')$

$J = \Delta x k_0 \exp- \dfrac{\Delta G^{\neq}}{RT}\left[c(x)\exp- \dfrac{zF\Delta\phi}{2RT} - c(x')\exp \dfrac{zF\Delta\phi}{2RT}\right]$

$c(x') = c(x) + \Delta c$

for $|\Delta\phi| \ll zF/RT$:

$J \propto \left[-\dfrac{zF}{RT}\hat{c}\Delta\phi - \Delta c - \dfrac{zF}{2RT}\Delta c\Delta\phi\right]$

Table 4
Continuation.

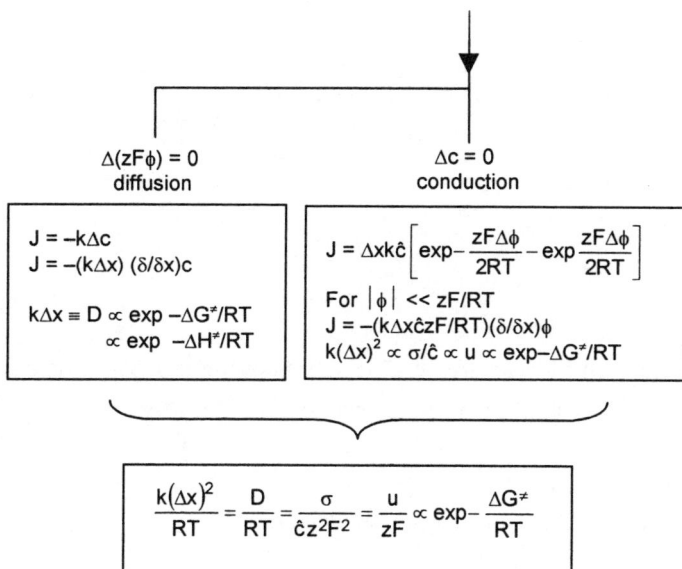

$$\Delta(zF\phi) = 0 \qquad\qquad \Delta c = 0$$
diffusion   conduction

$J = -k\Delta c$

$J = -(k\Delta x)(\delta/\delta x)c$

$k\Delta x \equiv D \propto \exp -\Delta G^{\neq}/RT$

$\propto \exp -\Delta H^{\neq}/RT$

$$J = \Delta x k\hat{c}\left[\exp-\frac{zF\Delta\phi}{2RT} - \exp\frac{zF\Delta\phi}{2RT}\right]$$

For $|\phi| \ll zF/RT$

$J = -(k\Delta x\hat{c}zF/RT)(\delta/\delta x)\phi$

$k(\Delta x)^2 \propto \sigma/\hat{c} \propto u \propto \exp-\Delta G^{\neq}/RT$

$$\frac{k(\Delta x)^2}{RT} = \frac{D}{RT} = \frac{\sigma}{\hat{c}z^2F^2} = \frac{u}{zF} \propto \exp-\frac{\Delta G^{\neq}}{RT}$$

Nernst-Einstein relation

(Eq. 104 ignores weak T-dependencies in the pre-factors.)

In the case of chemical processes, one also has to linearize exponentials, i.e., to assume $|\Delta_R G/RT \ll 1|$. This is, however, now a very severe approximation owing to the large values of reaction free energies (difference in μ^0-values now important), unless we slightly perturb an already existing equilibrium. Thus, for a pure chemical process but also for the general electrochemical reaction we are advised to use the full description.

This is of course also true if we need to consider the general electrochemical reaction Eq. (92). If the applied driving force (cf. electrical experiment) is an electrical potential gradient, Eq. (97) leads to the well-known non-linear Butler-Volmer equation.[79] We will become acquainted with equally important kinetic equations for the cases of the tracer and the chemical experiment.[172]

Let us briefly consider two further formulations of Eq. (97) that will prove helpful when dealing with reaction kinetics. The first is

$$\tilde{\mathfrak{R}} = \vec{\tilde{\mathfrak{R}}}\left(1 - \frac{\overleftarrow{\tilde{\mathfrak{R}}}}{\vec{\tilde{\mathfrak{R}}}}\right) = \vec{\tilde{\mathfrak{R}}}\left(1 - \exp{-\frac{\tilde{A}}{RT}}\right) \tag{105}$$

For $\tilde{A} \to 0$ (i.e., $\vec{\tilde{\mathfrak{R}}} \to \overset{\frown}{\tilde{\mathfrak{R}}} \equiv \mathfrak{R}°$) the result

$$\tilde{\mathfrak{R}} = \tilde{\mathfrak{R}}°\tilde{A} \tag{106}$$

complies with linear irreversible thermodynamics. The comparison with Eq. (95) shows that the exchange rate $\tilde{\mathfrak{R}}°$ plays the role of the decisive equilibrium permeability. (Note that in the case of transport $\tilde{A} \propto (\partial/\partial x)\tilde{\mu}$, and $\tilde{\mathfrak{R}}° \propto \sigma$.)

Equation (97) can also be rewritten as

$$\tilde{\mathfrak{R}} = \tilde{\mathfrak{R}}°\left[\frac{\delta \vec{\tilde{\mathfrak{R}}}}{\tilde{\mathfrak{R}}°} - \frac{\delta \overleftarrow{\tilde{\mathfrak{R}}}}{\tilde{\mathfrak{R}}°}\right] = \tilde{\mathfrak{R}}°\left[\frac{\delta\left(\vec{\tilde{k}}[A]\right)}{\overset{\frown}{\vec{\tilde{k}}}[\hat{A}]} - \frac{\delta\left(\overleftarrow{\tilde{k}}[B]\right)}{\overset{\frown}{\overleftarrow{\tilde{k}}}[\hat{B}]}\right] \tag{107}$$

where δ refers to the perturbation from the equilibrium value (denoted by an arc). We will exploit this formulation in Section VI.5.

It is well-known from thermodynamics that the entropy production or better the related quantity, the dissipation, $\Pi = T\delta S/\delta t$ is a positive-definite function being related with the rates and forces via[182,183]

$$\Pi \equiv \Sigma_i \tilde{\mathfrak{R}}_i \tilde{A}_i \geq 0 \tag{108}$$

If the linear relations are valid (Eq. 96), Π is a bilinear function in the $\tilde{A}$'s. One implication of the positive definiteness of Π is that all the pure transport terms L_{ii} cannot be negative. It can be shown that in steady states Π is at minimum for linear processes.[183] Moreover, processes near the steady state are characterized by a decrease of Π meaning that in the linear regime steady states are automatically stable, which is no longer the case for non-linear processes (see Section VI.9.).

3. Bulk Processes

(i) Electrical Conduction: Mobility, Conductivity and Random Walk

In the pure electrical experiment we measure the conductivity which can be broken down into contributions of ions and electrons and finally into the defect contributions (k: conduction electron, hole; vacancy, interstitial)

$$\sigma = \sigma_{ion} + \sigma_{eon} = \Sigma_k \sigma_k \qquad (109)$$

In Part II[1] we will become acquainted with methods to de-convolute ionic and electron conductivity. According to Table 4 σ_k can be broken down into charge number, concentration and mobility according to

$$\sigma_k = z_k Fu_k c_k \qquad (110)$$

the mobility being proportional to the defect diffusion coefficient D_k according to the Nernst-Einstein relation. Let us consider the case where the ion conductivity is dominated by a single defect k. Had we no idea about the defect concentration, then it would be reasonable to define an average ion mobility (corresponding to a self-diffusion coefficient D_{ion}) which could the be recast into an ionic diffusion D^Q coefficient:

$$D^Q = D_{ion} = D_k c_k / c_{ion} \qquad (111)$$

Equation (111) follows from the fact that the potential gradients and flux densities in Eq. (103) are invariant against the substitution ion $\leftrightarrow$ defect. Hence σ is invariant ($\sigma_k = \sigma_{ion}$) and so is the product Dc. (The generalization is given by Eqs. 150 and 151.)

Let us discuss in the following some fundamental aspects of the mobility of ions and electrons and consider specific conductors.

(ii) Mobility

The previous paragraphs (see Table 4) showed that the mobility of the defect is proportional to its (self-) diffusion coefficient. A realistic self-diffusion process involves jumps in different directions as well as a sequence of jumps leading to a long range transport. It is easy to show that Eq. (104) still applies in essence as long as the jump situation is isotropic and homogeneous and the long range transport follows a random walk process.[184,185] If the jump i refers to a spatial variation r_i, after n jumps the particle is displaced from the origin by $\sum_{i=0}^{n} r_i$

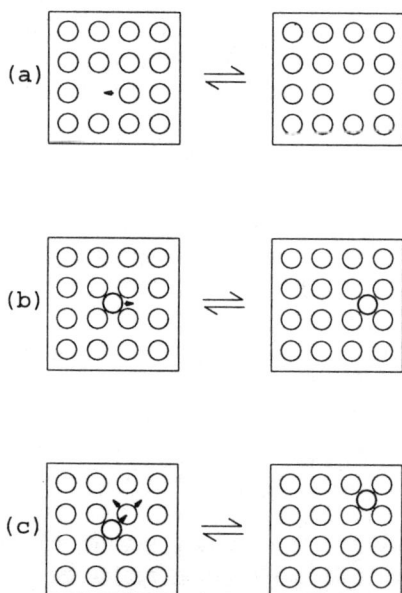

Figure 42. Elementary jump mechanisms in crystals: a) vacancy mechanism, b) direct interstitial mechanism, c) (collinear or non-collinear), indirect interstitial mechanism (interstitialcy mechanism).

corresponding to a square displacement $nr^2 + 2\sum\limits_{i=j+1}^{n}\sum\limits_{j=1}^{n-1} r^2 \cos_{ij}$ after n

steps. The mean square displacement for a random walk ($\overline{\cos\alpha_{ij}} = 0$) is then nr^2. On the other hand the probability function of reaching a given distance at n jumps is determined by a Gauss-function provided n is large enough. This yields the value 6Dt for the mean square displacement.[6] The combination of the two expressions leads to $n/t = 6D/r^2$ (where $r = \Delta x$ in the one-dimensional treatment). The quantity $\Gamma \equiv n/t$ is the jump frequency which is proportional to the rate constant k used for the simplistic elementary reaction in Eq. (94). Hence, D and thus the mobility (more precisely the product uT) of the defect is proportional to the jump frequency, the latter being activated via the migration threshold (Γ_o: attempt frequency)

$$uT = \text{const } \Gamma_o(\Delta x)^2 \exp\frac{\Delta S^{\neq}}{R}\exp-\frac{\Delta H^{\neq}}{RT} \tag{112}$$

Figure 42 shows the basic elementary ion migration processes in a low defective isotropic ion conductor with a mobility in the A-sublattice. The vacancy mechanism (Fig. 42 top) can be described by a transport process (z_v = effective charge of the A-vacancy) such as

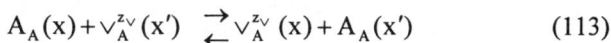

$$A_A(x) + v_A^{z_v}(x') \underset{\leftarrow}{\overset{\rightarrow}{\rightleftharpoons}} v_A^{z_v}(x) + A_A(x') \tag{113}$$

Interstitial ions ($z_i = -z_v$) are either transported directly (see Fig. 42 center) according to

$$A_i^{z_i}(x) + v_i(x') \underset{\leftarrow}{\overset{\rightarrow}{\rightleftharpoons}} v_i(x) + A_i^{z_i}(x') \tag{114}$$

or more frequently indirectly (see Fig. 42 bottom) by pushing a regular neighbor at x^* into the free interstitial site while occupying its position:

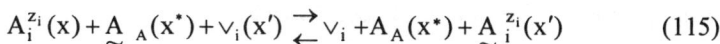

$$A_i^{z_i}(x) + \underset{\sim}{A}_A(x^*) + v_i(x') \underset{\leftarrow}{\overset{\rightarrow}{\rightleftharpoons}} v_i + A_A(x^*) + \underset{\sim}{A}_i^{z_i}(x') \tag{115}$$

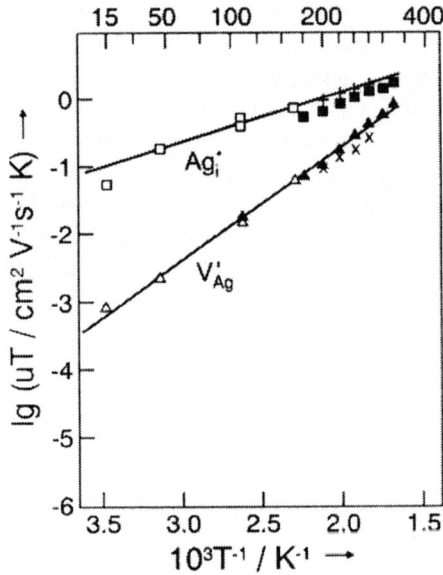

Figure 43. The mobility of the Frenkel defects in AgBr
as a function of the temperature (according to Eq. 113).
From Ref.[186]. (Reprinted from P. Müller,
"Ionenleitfähig-keit von reinen und dotierten AgBr- und
AgCl-Einkristallen." *Phys. Stat. Sol.* **12,** 775-794.
Copyright © 1965 with permission from WILEY-VCH
Verlag GmbH.)

It is clear that at high carrier concentrations Eq. (112) has to be
modified and at least a term that takes account of the (then varying)
concentrations of the regular partners have to be introduced ($u \propto 1-x$).
Figure 43 shows mobilities for vacancies and interstitials in AgBr and
Fig. 44 conductivities of Ag ion conductors.

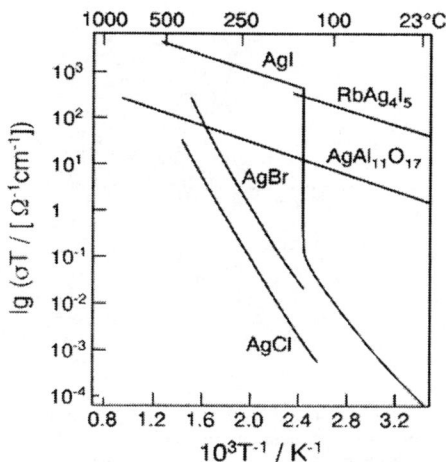

Figure 44. The temperature dependence of the specific conductivity for a series of selected silver ion conductors. ($AgAl_{11}O_{17}$ reads more correctly $Ag_{1+x}Al_{11}O_{17+x/2}$). The "superionic conductors" are characterized by flat slopes and high absolute values.[23]

In the so-called superionic conductors (see Fig. 44) virtually all constituents are carriers and usually the migration thresholds are very small (~ 0.1 eV in α-AgI) leading to very high ion conductivities (cf. Section IV.5.ii). The most popular example is α-AgI the crystal structure of which is displayed in Fig. 45.[187] Another example is β-alumina in which there is a ^{2}D superconduction in so-called conduction planes.[198] A detailed discussion of individual crystal structures would lead us too far from the scope of this contribution (see, e.g., reference[10]).

Figure 46 gives a selection of anion conductors in a log σ vs. 1/T representation. Proton transport deserves special treatment owing to the extremely polarizing nature of this elementary particle. Figure 47 shows a transition configuration of proton transport in $BaCeO_3$ (cf. Section IV.3.) according to molecular dynamics simulation.[192] The

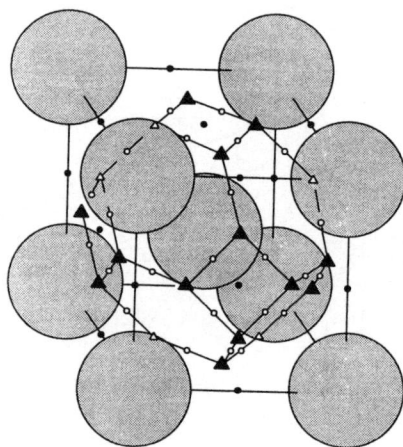

Figure 45. Crystal structure of α-AgI. The iodide ions are represented by large spheres. The multiplicity of (not precisely) energetically identical sites for Ag^+ ions and the low activation energy compared with the thermal energy when T > 146 °C lead to a "melting" of the Ag^+ partial lattice. According to Ref.[187].

transport is best described as a dynamic breaking and forming of hydrogen bonds in interaction with the lattice dynamics. The transport is liquid-like in ion exchange membranes in which protons of internal sulfonic acid groups are dissolved by water molecules and migrating within water channels in the polymer matrix. Heterocycles with proton-donor and acceptor function can replace water in this respect.[199]

For a discussion of conductivity aspects in disordered systems cf. Ref.[200] Polar groups of covalent polymers can also solvate ions (see Fig. 48) and lead to high conductivities, as PEO does with respect to LiX dissociation.[193,201]

Interesting in this context is the finding that oxides such as Al_2O_3, TiO_2 etc. can be used as fillers that enhance the ion conductivity.[116] Even though effects such as impact on crystallinity or on segmental motion are expected to have a great influence, the picture developed in Section V.2. should also be relevant: In a covalent matrix most of

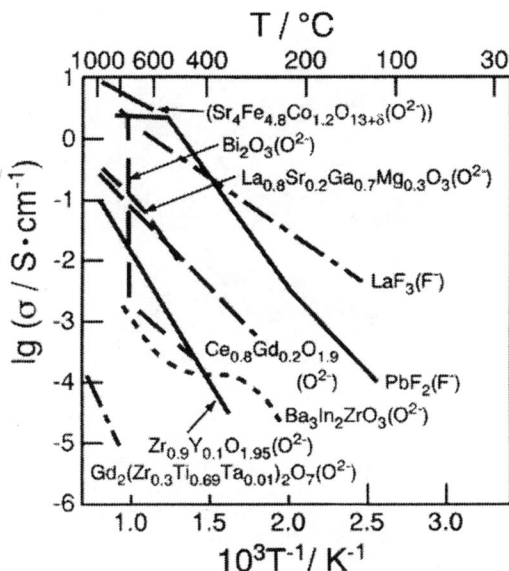

Figure 46. Specific conductivity of a few selected anion conductors as a function of the temperature (O^{2-}, F^-).[23] Very high O^{2-} conductivities are met in some perovskites, typically in doped ferrates.[188] These materials (cf. $Sr_4Fe_{4.8}Co_{1.2}O_{13+\delta}(O^{2-})$) are however mixed conductors and hence excellent materials for permeation membranes (cf. also Part II[1]). For more detailed data collections see Ref.[188-191].

the salt constituents should be associated (ion pairs). An acidic (basic) second phase should partly break these ion pairs by adsorbing the anion (cation) and releasing the counter ion. The experimental observation that in many cases acidic oxides enhance the Li^+ conductivity substantially, may be explained by this effect.[202]

The mobility of electrons will only be briefly touched upon as there is extensive solid physics literature on this (see e.g., Ref.[203]). In many cases electrons or holes behave as polarons, i.e., similar as the ionic defects. Then, they substantially polarize their environment,

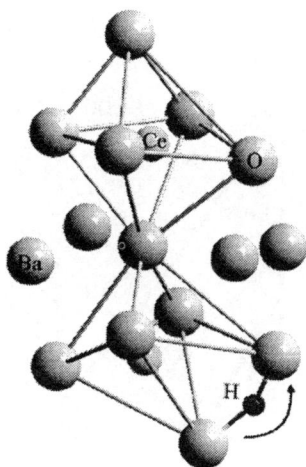

Figure 47. As oxygen ions move towards each other on account of lattice vibrations, the activation energy for proton jump is lowered, and the proton changes partner. According to Ref..[192]. (Reprinted from K. D. Kreuer, W. Münch, U. Traub and J. Maier, "On Proton Transport in Perovskite-Type Oxides and Plastic Hydroxides", *Ber. Bunsenges. Phys Chem.* **102,** 552-559. Copyright © 1998 with permission from WILEY-VCH Verlag GmbH.)

leading to perceptible migration thresholds on the order of 0.1 eV or more (small polarons). Large polarons can be approximately described by the band model. Whereas in the first case the T-dependence of the pre-factor may be neglected, in the second case it is the decisive contribution. For acoustic phonon scattering typically a $T^{3/2}$ law is obtained (Fig. 49), and the mobility decreases with increasing temperature (as assumed in Fig. 49 for high temperatures). Even though the phenomenon of nearly free electrons in metals is comparable with the superionic situation in terms of degree of disorder, the mobility is

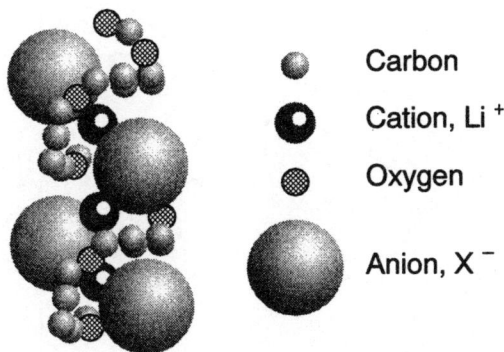

Figure 48. Schematic illustration of the structure of a PEO-LiX complex (Li$^+$X$^-$ dissolved in poly ethylene oxide[193]). (The relative magnitudes are not representative.) From Ref.[194]. (Reprinted from B. Scrosati, "Lithium Ion Plastic Batteries." in: *Lithium Ion Batteries* (M. Wakihara and O. Yamamoto eds.), VCH, Weinheim Copyright © 1998 with permission from WILEY –VCH, Verlag GmbH.)

very different and of course much higher owing to quantum mechanics. In superconductors[197] the mobility is formally infinite. Figure 50 shows the disappearance of resistivity in the high temperature superconductor "LSC" at T $\approx$ 40 K. Much literature is also concerned with organic metals based on conjugated polymers. Since one-dimensional delocalization cannot be very extended due to the Peierls-distortion,[204] an appropriate doping (redox effect) is necessary (shown in Fig. 51) to achieve high conductivities.

(iii) Tracer Diffusion

As already mentioned the tracer experiment (which can also be conceived as a counter motion of the isotopes) delivers the tracer diffusion coefficient. In the case of oxides ideally the natural oxygen gas phase is instantaneously replaced by a gas phase with the same oxygen partial pressure but exhibiting a different isotope ratio (or an oxide is contacted with the same oxide in which a cation isotope is

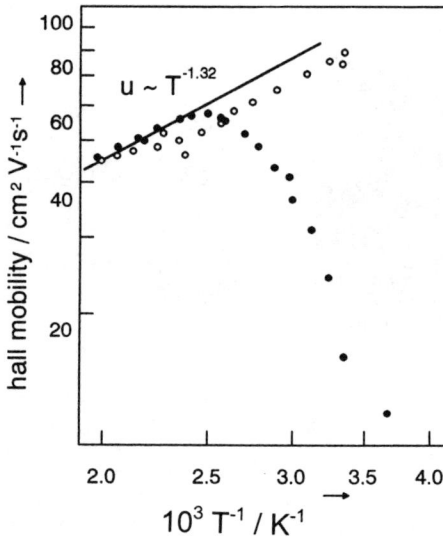

Figure 49. The mobility of the excess electrons in various SnO_2 samples determined by means of the Hall effect and conductivity. The high temperature behavior points to acoustic phonon scattering. Both samples differ in purity. According to Ref.[195]. (Reprinted from H. J. van Daal, "Polar Optical-Mode Scattering of Electrons in SnO_2.", *Solid State Commun.* **6**, 5-9. Copyright © 1968 with permission from Elsevier.)

enriched), then the diffusion is detected in-situ, e.g.,, by weight increase, or quenched non-equilibrium profiles are investigated, e.g., by SIMS (secondary ion mass spectroscopy). Owing to the ideality of the situation (asterisk denotes tracer)

$$j^* \propto \sigma^* \nabla \tilde{\mu}^* = \sigma^* \nabla \mu^* \propto \frac{\sigma^*}{c^*} \nabla c^* = \frac{\sigma}{c} \nabla c^* \tag{116}$$

(The transport quantities refer to the ions.) In this approximation $D^* = D^Q$. A closer inspection shows that correction factors f^* have to be introduced, that will be considered in Section VI.4.[205]

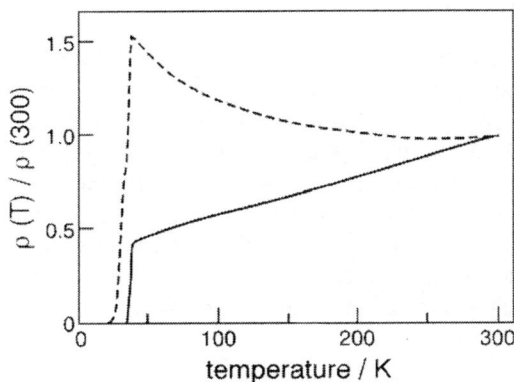

Figure 50. The resistivity of SrO-doped La_2CuO_4 disappears at temperatures less than $T_C \simeq 40$ K. The $(h^\bullet)_2$ "associates" (Cooper pairs) are responsible for this. The two curves are based on samples, that are annealed at differing partial pressures of oxygen.[196,197] Compare here Section IV..3. (Reprinted from J. C. Philips (ed.), *Physics of High Temperature Superconductors*, Academie Press, New York, Copyright © 1989 with permission from Elsevier.)

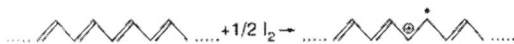

Figure 51. The doping of poly-acethylene with iodine leads to the formation of a hole. Iodide (or poly-iodide) is the counterion.

Irrespective of the relevance of tracer diffusion for the systematic discussion, it is a very important technique to detect the nature of conductive species and to identify transport mechanisms.[206,207]

(iv) Chemical Diffusion

The most important transport parameter for chemical processes is the chemical diffusion coefficient D^{δ}.[4,173-175,181] The relevance for electrochemistry is twofold. (i) Chemical diffusion is typically established by a counter flow of two charge carriers. (ii) It is the central bulk process occurring in response to an electrical polarization using non-reversible or selectively blocking electrodes (cf. Part II). Together with the geometry D^{δ} describes the time constant of the propagation of the chemical signal "stoichiometry" into the bulk. In the case that oxygen stoichiometry is concerned according to

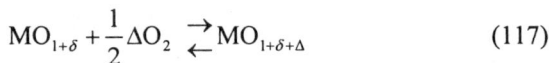

$$MO_{1+\delta} + \frac{1}{2}\Delta O_2 \underset{\leftarrow}{\overset{\rightarrow}{\rightleftharpoons}} MO_{1+\delta+\Delta} \tag{117}$$

the underlying process is an ambipolar motion of O^{2-} and $2e^-$.

If we consider O_i'' and $h^{\bullet}$ as carriers, we refer, e.g.,, to chemical diffusion of oxygen in La_2CuO_4. Writing down the one-dimensional transport equations (Eq. 103) for ionic and electronic carriers and considering flux coupling and electroneutrality, we immediately obtain for this case (see, e.g., Ref.[173])

$$j_{O_i'} = j_{O^{2-}} = j_O = D_O^{\delta}\frac{\partial c_O}{\partial x} = D_O^{\delta}\frac{\partial c_{O_i'}}{\partial x} = \frac{1}{2}D_O^{\delta}\frac{\partial c_{h^{\bullet}}}{\partial x} = \frac{1}{2}j_{h^{\bullet}}. \tag{118}$$

where

$$D_O^{\delta} = \frac{1}{4F^2}\sigma_O^{\delta}\frac{\partial \mu_O}{\partial c_O} = \frac{RT}{4F^2}\frac{\sigma_O^{\delta}}{c_O}\frac{\partial \ln a_O}{\partial \ln c_O} \equiv \frac{RT}{4F^2}\frac{\sigma_O^{\delta}}{c_O}w_O \equiv \frac{RT}{4F^2}\frac{\sigma_O^{\delta}}{c_O^{\delta}} \tag{119}$$

The ambipolar conductivity σ^{δ} represents the harmonic mean of $\sigma_{h^{\bullet}}$ and $\sigma_{O_i'}$ (i.e., $\sigma_{h^{\bullet}}\sigma_{O_i'}/\sigma$). The term $\partial \mu_O / \partial c_O$ can also be referred to as an inverse chemical capacitance (by analogy to the electrical capacitance ∂charge/$\partial\phi$) and is given by RTw_O/c_O, w_O being the thermodynamic factor.[15] The latter quantity can be expressed by the relevant parameters of the ionic and electronic constituent ($\partial\mu_O = \partial\mu_{O^{2-}} - 2\partial\mu_{e^-}$; $\partial c_O = \partial c_{O^{2-}} = -\frac{1}{2}\partial c_{e^-}$)

$$D_O^{\delta} = \frac{RT}{4F^2}\frac{\sigma_{O^{2-}}\sigma_{e^-}}{\sigma_{O^{2-}}+\sigma_{e^-}}\left(\frac{1}{c_{O^{2-}}}\frac{\partial\ln a_{O^{2-}}}{\partial\ln c_{O^{2-}}}+4\frac{1}{c_{e^-}}\frac{\partial\ln a_{e^-}}{\partial\ln c_{e^-}}\right) \qquad (120a)$$

A further discussion requires the translation into defect parameters ($\partial\mu_{O_i'} = \partial\mu_{O^{2-}}$, $\partial\mu_{e^-} = -\partial\mu_{h^{\bullet}}$; $\partial c_{e^-} = -\partial c_{h^{\bullet}}, \partial c_{O^{2-}} = \partial c_{O_i'}$)

$$D_O^{\delta} = \frac{RT}{4F^2}\frac{\sigma_{O_i'}\sigma_{h^{\bullet}}}{\sigma_{O_i'}+\sigma_{h^{\bullet}}}\left(\frac{1}{c_{O_i'}}\frac{\partial\ln a_{O_i'}}{\partial\ln c_{O_i'}}+4\frac{1}{c_{h^{\bullet}}}\frac{\partial\ln a_{h^{\bullet}}}{\partial\ln c_{h^{\bullet}}}\right) \qquad (120b)$$

Since (unlike $w_{O}, w_{O^{2-}}, w_{e^-}$) now $w_{O_i'}, w_{h^{\bullet}}$ are unity for a low defective material, it follows that

$$D^{\delta} = \frac{(D_{h^{\bullet}}c_{h^{\bullet}})(D_{O_i'}c_{O_i'})}{(D_{h^{\bullet}}c_{h^{\bullet}})+4(D_{O_i'}c_{O_i'})}\left(\frac{1}{c_{O_i'}}+\frac{4}{c_{h^{\bullet}}}\right)=t_{h^{\bullet}}D_{O_i'}+t_{O_i'}D_{h^{\bullet}} \qquad (120c)$$

(t: transference number).

This equation shows how defect concentrations and mobilities constitute D^{δ} and how D^{δ} depends on materials and control (P, T, C) parameters. D_O^{δ} obviously lies between $D_{O_i'}$ and $D_{h^{\bullet}}$, and is hence much larger than $D_{O^{2-}} = D_O^Q \approx D_O^*$. Since the mobility of the electronic carriers is usually much larger than that of the ionic carriers, a predominantly ionically conducting solid implies $c_{O_i'} \gg c_{h^{\bullet}}$ and hence $D_O^{\delta} \approx D_{h^{\bullet}}$; in the case of an electronic conductor, $c_{h^{\bullet}}$ may still be smaller than $c_{O_i'}$, if this is the case, then $D_O^{\delta} = 4(c_{O_i'}/c_{h^{\bullet}})D_{O_i'}$; only in the case of the electron-rich electronic conductor the other limiting case $D_O^{\delta} \approx D_{O_i'}$ is arrived at.

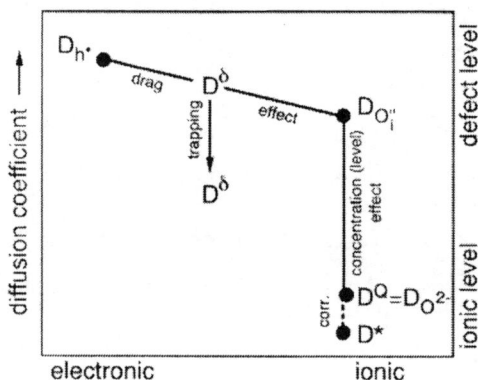

Figure 52. The relationships between the different diffusion coefficients discussed in the text. Holes and interstitial oxygen ions ars assumed as mobile defects.

Table 5
Comparison of ion conductivity tracer diffusion and chemical diffusion for electron-rich electron conductors, using

$$j \propto \Re = \vec{k}c_{O^{2-}}(x)c_{V_O^{\bullet\bullet}}(x') - \overleftarrow{k}c_{O^{2-}}(x')c_{V_O^{\bullet\bullet}}(x)$$

(Δk refers to the difference of the electrochemical rate constants $\vec{\tilde{k}}$ and $\overleftarrow{\tilde{k}}$. The tilde has been suppressed for simplicity.)

Experiment	Simplification ($\Re$)
Ionic conductivity	$-(\Delta k)\, c_{O^{2-}} c_{V_O^{\bullet\bullet}}$
Tracer exchange	$-k\,(\Delta c_{O^{2-}})\, c_{V_O^{\bullet\bullet}}$
Chemical diffusion	$-kc_{O^{2-}}\,(\Delta c_{V_O^{\bullet\bullet}})$

Figure 52 illustrates the situation. The more general case in which different defects contribute will be tackled below in the framework of Conservative Ensembles (Section VI.4.*ii.*) which also explains the displayed depression of D^δ by trapping effects (see Fig. 52).

It is instructive to compare the three transport processes (conduction, tracer diffusion and chemical diffusion) by using chemical kinetics and for simplicity concentrating on the electron-rich electron conductor, i.e., referring to the r.h.s. of Fig. 52. The results of applying Eq. (97) are summarized in Table 5 and directly verify the conclusions. Unlike in Section VI.2.*i.*, we now refer more precisely to bimolecular rate equations (according to Eqs. 113-115); nonetheless the pseudo-monomolecular description is still a good approximation, since only one parameter is actually varied. This is also the reason why we can use concentrations for the regular constituents in the case of chemical diffusion. In the case of tracer diffusion this is allowed because of the ideality of distribution.

The kinetic procedures also work in more general cases and also reproduce the detailed thermodynamic factor.[208] This approach is generally helpful when dealing with large driving forces.

There are cases in which compositional changes are only caused by ionic fluxes, e.g., by counter diffusion of cations during spinel formation (cf. Section VI.8) or diffusion of H_2O into perovskites. In such cases we do not deal with redox reactions. As regards the H_2O diffusion into oxides, the process is described by counter diffusion of H^+ (via $OH_O^\bullet$) and O^{2-} (via $v_O^{\bullet\bullet}$). The relations are completely analogous: A convenient representation of $D_{H_2O}^\delta$ is then[57,60]

$$D_{H_2O}^\delta = t_{OH_O^\bullet} D_{v_O^{\bullet\bullet}} + t_{v_O^{\bullet\bullet}} D_{OH_O^\bullet} \qquad (121)$$

D^δ can be measured by recording the time changes of a signal that is unambiguously correlated with stoichiometry or by spatial detection of diffusion profiles at a given time. While in Part II[1] we will encounter powerful electrochemical polarization methods, here we will just refer to chemical diffusion processes that are caused by changes in the chemical driving force. To be specific, we consider an oxide subject to a sudden change in the partial pressure of oxygen and follow an appropriate signal. Usually the signal integrates over the stoichiometry profile. If the stoichiometry change is substantial,[209] one may use weight as a measure. A sensitive and convenient measure is usually the electrical conductance which—according to Eq. (33)—depends in a power law on P_{O_2}. If the characteristic exponent N is appreciably large, this is a very recommended method (Fig. 53) (see, e.g.,, Ref.[210]). In the case of solid electrolytes (e.g., Y_2O_3-doped ZrO_2) $N \equiv 0$ and the method fails (unless one uses a blocking technique). In these cases spectroscopic methods relying on changes in EPR, NMR or optical absorption signals are appropriate.[211] A method that allows for spatial resolution, yet, only in the quenched state, is SIMS already mentioned in the context of tracer diffusion experiments;[207] in terms of chemical diffusion it is not sensitive enough to detect small relative changes in composition (like weight) but may be a good method to determine, e.g., H_2O incorporation in oxides.

In Ref.[212] an optical absorption technique was developed in order to follow chemical diffusion in Fe-doped $SrTiO_3$, which enables a spatially resolved in-situ detection, i.e., makes it possible to follow the stoichiometry as a function of space and time at high temperatures.

The solution to the diffusion equation are modulated Fourier series of the form[213]

$$\frac{S(x,t) - S_1}{S_2 - S_1} = \frac{c_O(x,t) - c_{O1}}{c_{O2} - c_{O1}} \tag{122}$$

$$= 1 - \sum_0^{\infty} \frac{4(-1)^i}{\pi(2i+1)} \cos\left[\pi(2i+1)\frac{x}{L}\right] \exp\left[-\pi^2(2i+1)^2 \frac{D^\delta t}{L^2}\right]$$

S can be any detection signal that is linearly correlated with the oxygen concentration c_O (here the absorption coefficient). The indices 1 and 2 denote the homogeneous equilibrium situations corresponding to P_1 and P_2. Switching from P_1 to P_2 occurs at $t = 0$.

Figure 54 displays the normalized profiles as a function of the normalized position coordinate for a sample half. The parameter at the curves is the normalized time $t/(L/2\sqrt{D^\delta})^2$. If this parameter reaches the value 1 (then $t \equiv \tau_{eq}$) the sample value equals the mean square displacement,

$$L^2 \approx 2Dt_{eq} \tag{123}$$

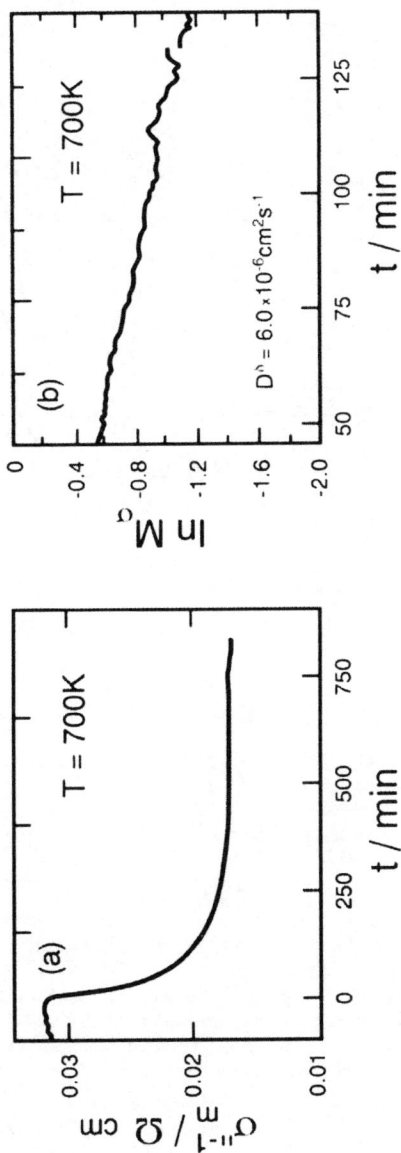

Figure 53. In the logarithmic plot (b) the measurement curve of the specific resistance of $YBa_2Cu_3O_{6-x}$ as a function of time (a) linearizes if the time is not too short; the slope is determined by D^δ and L.[215]

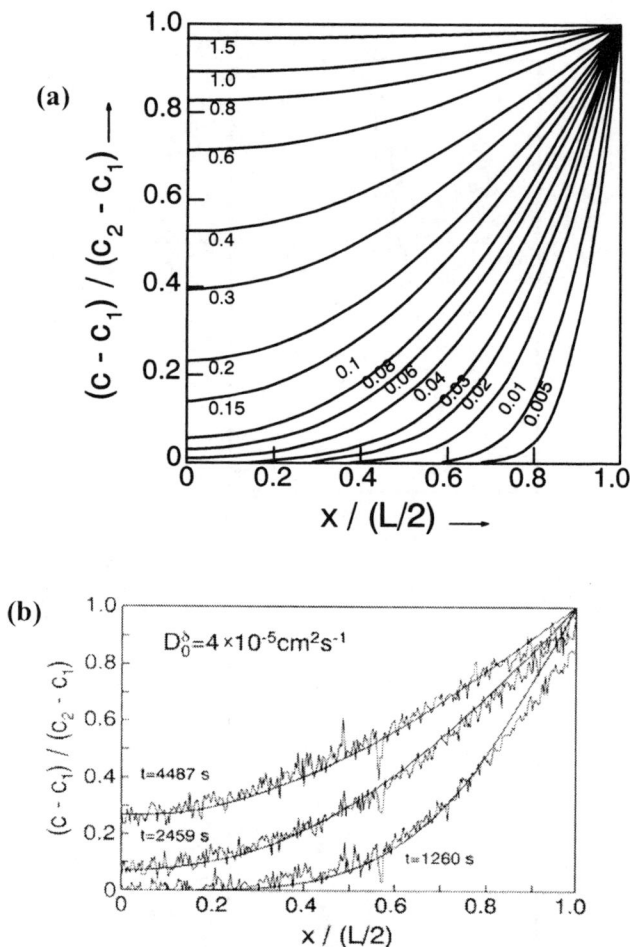

Figure 54. (a) Profiles for half the sample on rapid exchange of the surface concentration on both sides by $x = \pm L/2$. The value $x = 0$ refers to the center of the sample. The parameter on the curves is $D^\delta t/(L/2)^2$. (b) Concentration profiles during oxygen diffusion into $SrTiO_3$ (848 K), as a function of space and time. The smooth curves are obtained by fitting with a constant D^δ. The introduction of a concentration dependence only alters the result slightly.[214]

and the relative signal changes are less than 1% of the overall change (see Eq. 122). Figure 54 shows in-situ profiles for oxygen diffusion in the case of Fe-doped $SrTiO_3$. The absorption signal detects $[Fe^{4+}]$, which is linearly correlated with the oxygen content. The optical signal is followed by a CCD-camera.

If we follow the integrated signal $\overline{S}$ (conductance parallel or absorbance perpendicular to the interface), the position dependence disappears resulting in

$$\frac{\overline{S}(t) - S_1}{S_2 - S_1} = 1 - \frac{8}{\pi^2} \sum_0^\infty \frac{1}{(2i+1)^2} \exp\left(-\frac{t}{\tau^\delta}(2i+1)^2\right) \qquad (124)$$

with the time constant $\tau^\delta = L^2/\pi^2 D_0^\delta$. (It is worthy of note that τ^δ can be written as $R^\delta \cdot C^\delta$ with the chemical resistance $R^\delta \propto 1/\sigma^\delta$, and C^δ being the already mentioned chemical capacitance.) For long times an exponential law results (Fig. 53), which reads for the conductance experiment ($\sigma_m^{\parallel}$ being proportional to the parallel conductance)

$$M_\sigma \equiv \frac{\sigma_m^{\parallel}(t) - \sigma_2}{\sigma_1 - \sigma_2} = \frac{8}{\pi^2} \exp{-t/\tau^\delta} \qquad (125)$$

while for shorter times a $\sqrt{t}$ law is fulfilled.

Table 6 shows equilibration times (τ_{eq}^δ, see above) for one-dimensional diffusion into 1 mm thick samples for various diffusion coefficients, indicating the tremendous span of the rate of diffusion controlled reactions even for the same driving force. High diffusion coefficients are rather the exception in the solid state and, if they occur, they typically occur for one ion sort. This highlights the significance of transport steps in solid state science in general and the significance of frozen situations in particular.

Table 6

Equilibration times for one-dimensional diffusion in a 1 mm thick sample after a jump in external chemical component potential.

D^δ /cm^{2}s^{-1}	τ^δ_{eq} (L = 1 mm)	
10^{-20}	5×10^{17} s	~ 4 x Earth's age
10^{-10}	5×10^{7} s	~ duration of PhD work
10^{-8}	5×10^{5} s	~ 1 w
10^{-6}	5×10^{3} s	~ 1 h
10^{-5}	5×10^{2} s	~ 10 min
10^{-4}	50 s	~ 1 min
10^{-3}	5 s	$\left.\vphantom{\begin{matrix}a\\b\end{matrix}}\right\}$(fluid phases)
10^{-2}	0.5 s	

4. Correlations and Complications

(i) Conductivity: Static and Dynamic Correlation Factors

The above considerations relied on a site exchange describable by simple chemical kinetics and a random walk behavior over large distances in a homogeneous medium subject to a small driving force. In the following we will briefly consider some of the most important complications to this picture.

In general in each experiment individual deviations from random walk can occur. Noteworthy is the tracer correlation factor (f^*) (cf. Section VI.3.*iii.*) which is caused by the non-ubiquity of the jump partner (defect). If an ion has just changed its site from x to x' via hopping into a vacancy and so releasing a vacancy at x, a further vacancy is not immediately available and the dance of a particle between x and x' leads to a backward correlation ($\overline{\cos \alpha_{ij}} \neq 0$, cf. Section VI.3.*i.* and Ref.[205]); this is not the case for a vacancy migrating in an electrical experiment, nor for an isotope in a tracer experiment moving according to an interstitial mechanism (Fig. 42 bottom). The relation between D^* and D^Q then reads

$$D^* = f^* D^Q \qquad (126)$$

Whereas such correlation factors are typically between 0.5 and 1, severe discrepancies between D^* and D^Q can occur if the mechanisms differ. This happens in the case that defect contribute in different valence states, in particular if neutral transport is involved (see below). More generally then Eq. (126) has to be replaced by[3,4]

$$D^* = HD^Q \qquad (127)$$

In alkali hydroxides extended ring mechanism for proton motion can occur leading to a tracer propagation but not to any charge transport. The Haven ratio (H) amounts to several thousand in such cases.[216]

Dynamic correlation effects can occur in conductivity experiments if defect-defect interactions have to be considered. Then the relaxation

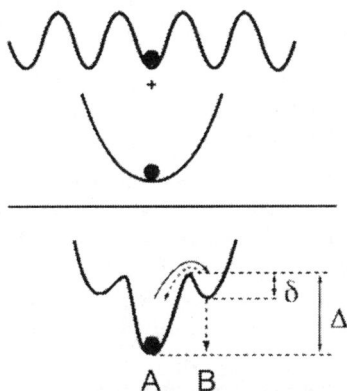

Figure 55. A jump within the true potential (lattice potential (top) plus defect potential (center)) requires the surmounting of a relatively high activation energy and a subsequent relaxation of the environment, before the site B takes on the original potential surroundings of A (see text). From Ref.[217]. (Reprinted from K. Funke, "Ion transport in fast ion conductors—spectra and models", Solid State Ionics, 94, 27-33. Copyright © 1997 with permission from Elsevier.)

of the surrounding is not very rapid compared to an ion hop such that backward jumps are favored as the environment at x' is not yet equilibrated, leading to a frequency dependent conductivity.

Figure 55 sketches the situation. The realistic potential includes lattice potential and interaction potential. Two functions are crucial in the description of the jump relaxation model, developed and refined by Funke et al.[217-219]: (i) the probability $W(t)$ that no correlated backward jump has occurred at time t (-W being the backward jump rate); and (ii) $g(t)$ describing the positional mismatch (-g measuring the stabilization rate). The basic assumption of the jump relaxation model is

$$\dot{W}(t) \propto W(t) \cdot \dot{g}(t) \tag{128}$$

implying that the tendency for backward hops and the tendency of rearrangement are proportional to each other. In Ref.[219] it was proposed by theoretical and empirical arguments that $W(t)$ and $g(t)$ can be obtained by combining Eq. (128) with the second differential equation

$$-\dot{g}(t) \propto g^k(t) W(t) \tag{129}$$

where $k = 2$ describes well the situation for high charge carrier concentrations. This conception which is based on dynamic inhomogeneities caused by the carrier transport, reproduces a manifold of experimental situations and particularly the frequency dependence of the complex conductivity. It is evident that the rigidity of the solid makes a general description more difficult than in fluids (cf. Debye-Hückel-Onsager theory[71]). It remains to be clarified in how far a quasi-Madelung approach for the dynamic relaxation can be helpful.[81,84]

In many cases—in particular in disordered media—static inhomogeneities (accompanied by a broad distribution of jump probabilities) have to be taken into account which obviously also result in a non-Debye relaxation and comparable frequency dependencies (see e.g., a recent review on this topic).[220]

Significant non-idealities can occur in the solids with two kinds of mobile cations (or anions). The so-called mixed alkali effect refers to the partly extremely strong depression of the alkali ion conductivity in crystals or glasses if substituted by another alkali ion. This is explained by the individually preferred environments and their interactions.[221-228]

Interestingly in the opposite case in which the counter ions are mixed, at least frequently, a conductivity enhancement has been observed (e.g., Ag(I, Br) or (Ca, Ba)F$_2$).[107,118,229]

Also in the case of chemical diffusion it may be questionable whether local equilibrium is always fulfilled. This is particularly the case at interfaces or in situations in which a very high electronic mobility enhances D^δ strongly compared with the ionic defect diffusivity.[177]

In the following we will study in more detail chemical diffusion and in particular the situation with various charge carriers. Let us concentrate on the case that, besides electrons and holes, oxygen ions in different valence states are mobile via interstitial and/or vacancy defects.

(ii) Chemical Diffusion and Conservative Ensembles

The chemical diffusion coefficient includes, as we know from the formal treatment in Section VI..3.*iv.*, both an effective ambipolar conductivity and an effective ambipolar concentration. The latter parameter is determined by the thermodynamic factor which is large for the components but close to unity for the defects.

If defects are not randomly distributed, in a better approximation $c^{1/2}$ or $c^{1/3}$ dependencies (see Section IV.5.*ii.*) may be introduced in the thermodynamic factors of the defects (Eq. 120b).

Formally, it will be even necessary to make corrections already in the starting flux equations. The detailed formulation of linear irreversible thermodynamics also includes coupling terms (cross terms) obeying the Onsager reciprocity relation. They take into account that the flux of a defect k may also depend on the gradient of the electrochemical potential of other defects. This concept has been worked out, in particular, for the case of the ambipolar transport of ions and electrons.[230]

A more explicit procedure is to introduce chemical interactions directly: There is a certain analogy with the equilibrium situation where it is possible to avoid the use of activity corrections over wide ranges by considering associates to correct for the interactions (see Section IV.5.*i.*). The relevance of this method of treatment is particularly evident if such associates can also be detected experimentally, e.g., by spectroscopic techniques in the case of ionic defects with differing charge states. This leads to a rescaling of defect concentrations and defect fluxes, which can now be taken to be ideal again to a good approximation (i.e., with negligible activity corrections and negligible coupling terms in the Onsager relationships). However, it is necessary to include source and sink terms from the start on account of the internal dissociation and association reactions (see Fig. 56). This leads us, in the case of local equilibrium, to the concept of "conservative ensembles",[231,232] which will now be described.

Figure 56 uses the example of associate formation between the ionic defect O_i'' and the electronic defect $h^\bullet$ to emphasize that the strict treatment requires the solution of coupled diffusion-reaction relationships, describing the general (electro-)chemical reaction scheme with individual diffusion or rate constants as parameters (cf. Section VI.2). Source terms (q) must be taken into account in the relevant continuity equations, e.g., for defect B that can be created by

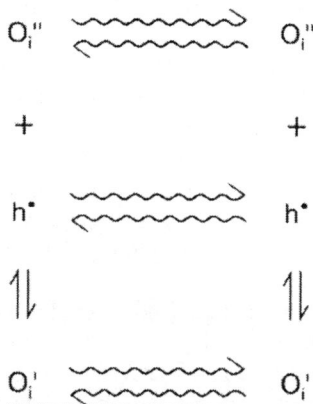

$$v_A A \underset{\leftarrow}{\overset{\rightarrow}{}} v_B B \qquad (130)$$

Figure 56. The internal mass and charge transport as a reaction-diffusion problem.

In the case that Eq. (130) describes the decisive elementary reaction, the continuity Eq. (130) for B is

$$\frac{\partial c_B}{\partial t} = -\text{div}\vec{j}_B + q_B \equiv -\text{div}\vec{j}_B + \nu_B \Re = -\text{div}\vec{j}_B + \nu_B(\vec{k}c_A^{\nu_A} - \overleftarrow{k}c_B^{\nu_B}) \quad (131)$$

Again we drop the tilde for simplicity. In the general case the problem depends on the rate constants of the association and dissociation processes. Since we assume local equilibrium (interaction much faster than transport) the situation simplifies considerably (only $K_{ass/diss}$ is included not the rate constants themselves). It would be wrong to believe that the second term in Eq. (131) can be neglected as a consequence of local equilibrium. It is true that the bracketed term in the formulation

$$\frac{\partial c_B}{\partial t} = -\text{div}\vec{j}_B + \nu_B \vec{k}c_A^{\nu_A}\left(1 - \frac{c_B^{\nu_B}}{c_A^{\nu_A}}\frac{\overleftarrow{k}}{\vec{k}}\right) \quad (132)$$

is close to zero because $c_B^{\nu_B}/c_A^{\nu_A} \simeq (\tilde{c}_B^{\nu_B}/\tilde{c}_A^{\nu_A}) = \vec{k}/\overleftarrow{k} = K$; however, the second part is not negligible in total in comparison with the flux divergence on account of the high values of the individual rate constants (see pre-factor).

It can now be shown that as far as the diffusion of certain ensembles, namely of the "conservative ensembles", is concerned, the source terms disappear. The resulting chemical diffusion coefficients then have to refer to this ensemble.

Let us consider an anti-Frenkel disordered material, taking into account both O_i'' and $v_O^{\bullet\bullet}$, but initially neglecting the occurrence of variable valence states. (In addition, we will assume a quasi one-dimensional situation.) Doing this we reduce the problem to a relatively trivial case. For it is immediately evident that the source terms disappear on consideration of the total ion flux density j (or current density i)

$$j_{O^{2-}} = j_{O_i'} - j_{v_O^{\bullet\bullet}} \quad \text{or} \quad i_{O^{2-}} = i_{O_i'} + i_{v_O^{\bullet\bullet}} \quad (133)$$

and the total electron flux

$$j_{e^-} = j_{e'} - j_{h^\bullet} \qquad \text{or} \qquad i_{e^-} = i_{e'} + i_{h^\bullet} \tag{134}$$

In this sense $(c_{O_i'} - c_{v_{\ddot{O}}})$ and $(c_{e'} - c_{h^\bullet})$ refer to "conservative ensembles". This is self-evident since all changes take place within these ensembles. Because

$$O_O + v_i \underset{\leftarrow}{\overset{\rightarrow}{\rightleftarrows}} v_{\ddot{O}} + O_i'' \tag{135}$$

and

$$\text{Nil} \underset{\leftarrow}{\overset{\rightarrow}{\rightleftarrows}} e' + h^\bullet \tag{136}$$

it follows that $q_{v_{\ddot{O}}} = q_{O_i'}$ and $q_{e'} = q_{h^\bullet}$ and, hence, $q_{O^{2-}} = q_{O_i'} - q_{v_{\ddot{O}}}$ $= 0$ and $q_{e^-} = q_{e'} - q_{h^\bullet} = 0$, if q represents the increase in concentration with time resulting from the defect chemical reactions. Since $\sigma_{O^{2-}} = \sigma_{O_i'} + \sigma_{v_{\ddot{O}}}$ and $\sigma_{e^-} = \sigma_{e'} + \sigma_{h^\bullet}$, the flux equations, formulated in terms of O^{2-} and e^-, are formally the same as those given in Section VI.3.*iv.* (cf. Eq. 120a). This is, however, not the case when we evaluate them in terms of defect parameters. Rather it follows because of $\partial \ln a_{O^{2-}} = \partial \ln a_{O_i'} = -\partial \ln a_{v_{\ddot{O}}}$, and $\partial \ln a_{e^-} = \partial \ln a_{e'} = -\partial \ln a_{h^\bullet}$ as well as $\partial c_{O^{2-}} = \partial c_{O_i'} - \partial c_{v_{\ddot{O}}}$ and $\partial c_{e^-} = \partial c_{e'} - \partial c_{h^\bullet}$ the more general equation

$$D_O^\delta = \frac{RT}{4F^2} \frac{(\sigma_{v_{\ddot{O}}} + \sigma_{O_i'})(\sigma_{e'} + \sigma_{h^\bullet})}{\sigma} \left(\frac{1}{c_{O_i'} + c_{v_{\ddot{O}}}} + \frac{4}{c_{e'} + c_{h^\bullet}} \right) \equiv \frac{RT}{4F^2} \frac{\sigma^\delta}{c^\delta} \tag{137}$$

If we also allow for interactions between ions and electrons, the situation becomes qualitatively different (see Fig. 56) even on the level of electronic and ionic constituents. In such cases we introduce variable valence states by permitting O^- and O^0 as valence states, i.e.,

O_i'', O_i', O_i^x, $V_O^{\bullet\bullet}, V_O^{\bullet}, V_O^x$ as defects. Then the electronic and ionic ensembles (O^2 and e^-) are no longer conservative but merely the combinations {O} and {e} as defined by Fig. 57. As can be derived from the reaction scheme, it follows that:

$$q_{\{O\}} \equiv q_{O^{2-}} + q_{O^-} + q_{O^x} = q_{O_i'} - q_{V_O^{\bullet\bullet}} + q_{O_i'} - q_{V_O^{\bullet}} + q_{O_i^x} - q_{V_O^x} = 0 \quad (138a)$$

and

$$q_{\{e\}} = q_{e'} - q_{h^{\bullet}} - q_{O_i'} + q_{V_O^{\bullet}} - 2q_{O_i^x} + 2q_{V_O^x} = 0 \quad (138b)$$

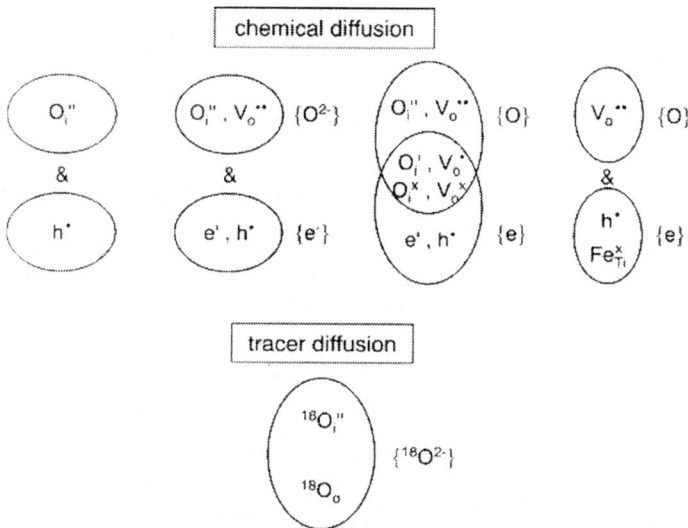

Figure 57. Conservative ensembles in the case of complex defect chemistry for various examples discussed in the text.[233,234] (Reprinted from J. Claus, I. Denk, M. Leonhardt, and J. Maier, "Influence of Internal Reactions on Chemical Diffusion: Application to Fe-doped SrTiO₃.", *Ber. Bunsenges. Phys. Chem.* **101**, 1386-1392. Copyright © 1997 with permission from WILEY-VCH Verlag GmbH.)

This result is immediately understandable because the concentration of the oxygen ensemble so-defined (see Fig. 57, third arrangement in the upper row)

$$c_{\{O\}} = c_{O_i'} + c_{O_i''} + c_{O_i^x} - c_{V_O^{\bullet\bullet}} - c_{V_O^{\bullet}} - c_{V_O^x} \qquad (139a)$$

$$c_{\{e\}} = c_{e'} - c_{h^\bullet} - c_{O_i'} + c_{V_O^\bullet} - 2c_{O_i^x} + 2c_{V_O^x} = -2c_{\{O\}} \qquad (139b)$$

obviously represents the deviation from the stoichiometric composition (δ), which is unchanged by the internal reaction. Eq. (138b) then follows as a consequence of the condition of electroneutrality ($-2c_{\{O\}} = c_{\{e\}}$).

For D^δ, now a structurally different expression is obtained after a more lengthy calculation:[231,232]

$$D_O^\delta = \frac{1}{4F^2}\left[2\sigma_{O^-} + 4s_{O^0} + \frac{(\sigma_{O^{2-}} + 2\sigma_{O^-})(\sigma_{e^-} - \sigma_{O^-})}{\sigma}\right]\frac{d\mu_O}{dc_O} \qquad (140)$$

with the abbreviations $\sigma_{O^-} = \sigma_{O_i'} + \sigma_{V_O^\bullet}$ and $s_{O^0} = \frac{F^2}{RT}(D_{V_O^x}c_{V_O^x} + D_{O_i^x}c_{O_i^x})$. In a purely formal manner, the conductivity factor in Eq. (140) now permits oxygen permeation even for a zero partial electronic conductivity (as a result of the migration of neutral defects but also as a result of an ambipolar migration of $2O^-$ and O^2). The term $d\mu_O/dc_O$ takes another form as well when reformulated in terms of defects, e.g., in terms of the fully ionized particles:

$$\frac{d\mu_O}{dc_O} = RT\left(\frac{\chi_{O_i'}}{c_{O_i'}} + 4\frac{\chi_{h^\bullet}}{c_{h^\bullet}}\right) = RT\left(\frac{\chi_{V_O^{\bullet\bullet}}}{c_{V_O^{\bullet\bullet}}} + 4\frac{\chi_{e'}}{c_{e'}}\right) \qquad (141)$$

The χ_k-terms refer to differential defect fractions and are defined via the corresponding conservative ensembles according to

$$\chi_k = \frac{\partial c_k}{\partial c_{\{k\}}} \qquad (142)$$

whereby $c_{\{k\}} = c_{\{O\}}$ or $c_{\{e\}}$ in the case of the excess particles (O_i'', e'), or $- c_{\{O\}}$ and $c_{\{e\}}$ in the case of missing ($v_O^{\bullet\bullet}$, $h^\bullet$) particles, respectively. These can be calculated from the defect chemistry.

The comparison of Eq. (141) with Eq. (120) shows that under simplified conditions the χ-terms can also be considered to be activity corrections.

If we can neglect associates, a quick calculation yields

$$\chi_{O_i''}^{-1} = \frac{\partial c_{O_i''} - \partial c_{v_O^{\bullet\bullet}}}{\partial c_{O_i''}} = \frac{c_{O_i''} + c_{v_O^{\bullet\bullet}}}{c_{O_i''}} \qquad (143)$$

and we return to our simple result, namely to Eq. (137).

An example for which different valance states are important is[53] $YBa_2Cu_3O_{6+x}$. A more recent example refers to the transport of different valence states of hydrogen in oxides.[235]

Another one is Fe-doped $SrTiO_3$, which has already been discussed above (see Fig. 53b). Hence, effects on σ_O^δ are not relevant, but $\partial\mu_O / \partial c_O$ and hence c_O^δ [236] are markedly influenced. The internal source and sink reaction is the conversion of Fe^{3+} to Fe^{4+}, which we formulate as

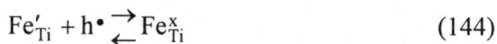

$$Fe_{Ti}' + h^\bullet \underset{\leftarrow}{\overset{\rightarrow}{\rightleftharpoons}} Fe_{Ti}^x \qquad (144)$$

in the high partial pressure regime. The mass action constant for Eq. (144) is well known as a function of temperature. The trapped hole states are not mobile but can supply mobile holes whenever necessary, hence strongly affecting the chemical capacitance $\partial c_O / \partial\mu_O$:

$$V_O^{\bullet\bullet} \rightsquigarrow V_O^{\bullet\bullet}$$

$$2h^{\bullet} \leftarrow\!\!\rightsquigarrow 2h^{\bullet}$$

$$\left\Vert\, {}_{-2Fe'_{Ti}}\right.$$

$$2Fe_{Ti}^{x} \tag{145}$$

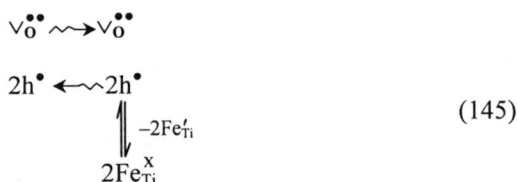

It is important to recognize that χ is a differential parameter, and the capacitance to be taken into account is a differential capacitance $(\partial c_O / \partial \mu_O)$. The electroneutrality relation yields $2[V_O^{\bullet\bullet}] \simeq [Fe'_{Ti}] = C - [Fe_{Ti}^{x}]$ and $2\partial[V_O^{\bullet\bullet}] = -\partial\left([h^{\bullet}] + [Fe_{Ti}^{x}]\right)$. Owing to conservation of mass and ionic equilibrium the correction follows as

$$\chi_{h^{\bullet}} = \frac{\partial[h^{\bullet}]}{\partial[h^{\bullet}] + \partial[Fe_{Ti}^{x}]} = \frac{(1 + K_{ass}[h^{\bullet}])^2}{(1 + K_{ass}[h^{\bullet}])^2 + CK_{ass}} \tag{146}$$

The concentration $[h^{\bullet}]$ follows from the set of defect chemical equations as a function of P, T, C; the quantity $\chi_{h^{\bullet}}$ itself is, thus, a function of these three parameters, too.

Instead of Eq. (120c) we now have to write

$$D^{\delta} = \frac{\sigma_{h^{\bullet}}}{\sigma} D_{V_O^{\bullet\bullet}} + \frac{\sigma_{V_O^{\bullet\bullet}}}{\sigma} \chi_{h^{\bullet}} D_{h^{\bullet}} \tag{147}$$

The values derived according to Eq. (147) (without any adjusted parameter!) agree extraordinarily well with the experiments (see Fig. 58).

The importance of internal redox reactions is perhaps even clearer for $ZrO_2(Y_2O_3)$. Without taking internal reactions into account, the result is simply $D^{\delta} = D_{e'}$ or $D_{h^{\bullet}}$ (cf. Section VI.3.iv.), depending on the oxygen partial pressure region. When ignoring the above complications it is not evident that in view of the high Y-doping further minority doping would affect the diffusion coefficient. It would also not be understandable that great differences from material to material are found for the same Y content as well as astonishingly strong dependences of D^{δ} on oxygen partial pressure and temperature. Again it is the role of redox-active impurities with respect to the chemical capacitance which explains these apparent anomalies according to

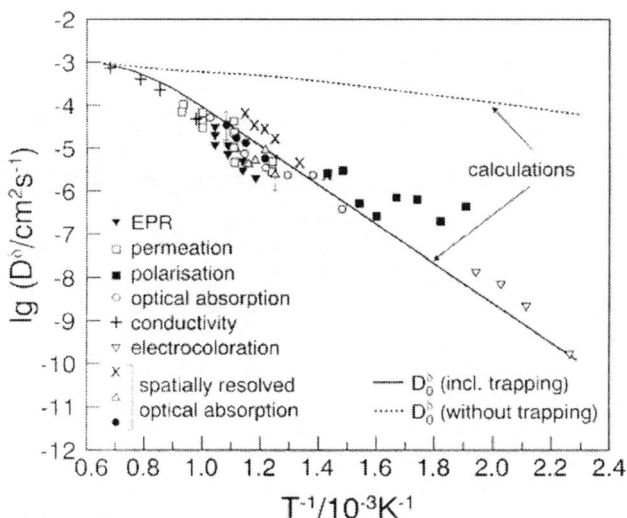

Figure 58. The chemical diffusion coefficient of oxygen in $SrTiO_3$ as a function of the temperature. The broken line includes the doping effect on σ^δ as determined by the ionization reaction (but in contrast to the continuous line ignores the effect on c^δ according to the χ-terms). The calculation applies to a doping content of 10^{19} cm^3 and an oxygen partial pressure of 10^5 Pa.[236] (Reprinted from J. Claus, I. Denk, M. Leonhardt, and J. Maier, "Influence of Internal Reactions on Chemical Diffusion: Application to Fe-doped $SrTiO_3$.", Ber. Bunsenges. Phys. Chem., **101**, 1386-1392. Copyright © 1997 with permission from WILEY-VCH Verlag GmbH.)

$$D^\delta(T,P,C) = \chi_{h^\bullet}(T,P,C)\,D_{h^\bullet} \qquad (148)$$

(whereby $\chi_{h^\bullet}$ is a well-defined function of the control parameters and $D_{h^\bullet}$ is a weak function of temperature).[237]

In Fig. 59 χ is considered as a function of the parameter $r \equiv K_{ass}^{-1}[h^\bullet]^{-1}$. In the case of the association of an acceptor A' and a hole to the oxidized form A^x, r is identical with the redox ratio $[A']/[A^x]$. According to Eq. (146) and Fig. 59, the maximum effect (minimum χ) does not lie at r = 0, but at r = 1 where the ratio[†] of oxidized and reduced species is unity (for minor redox-active doping). Obviously impurities are particularly relevant if their energy levels lie about in the center of the band gap (see Section IV.2) in the above cases, what reveals the similarity to static buffer effects in acid-base chemistry in aqueous solutions.[238]

The small maximum effect at r = 1 on the left-hand side of Fig. 59 is due to a small concentration of redox-active impurities. Plotting this maximum correction for r = 1, as a function of doping (r.h.s. of Fig. 59), shows that this "correction" can amount to several orders of magnitude.

Obviously, high D^δ values, e.g., as required for bulk conductivity sensors, demand materials that are free from redox centers, while the minimization of drift phenomena in boundary layer sensors demands just the opposite.[239]

The impact of internal equilibria on the evaluation of electrochemical measurement methods will be discussed in Part II.[1]

Let us now discuss the influence of variable valence states on the tracer (D^*) and charge diffusion coefficients (D^Q).

Here capacitance effects are absent and modifications are due to the different weight of the different charges. If we ignore correlation factors, the general result for the tracer diffusion coefficient is

$$D_O^* = \frac{RT}{4F^2} \frac{\sigma_{O^{2-}} + 4\sigma_{O^-} + 4s_{O^\circ}}{c_O} = \frac{1}{c_O} \Sigma_k [k] D_k \qquad (149)$$

whereby k indicates the individual ionic defects (O_i'', O_i', O_i^x, $V_O^{\bullet\bullet}$, $V_O^{\bullet}$, V_O^x). Equation (149) can be re-formulated as

$$D_O^* c_O = [O^{2-}] D_{O^{2-}}^* + [O^-] D_{O^-}^* + [O^\circ] D_{O^\circ}^* \qquad (150)$$

[†] If the redox-active doping is in the majority, [h$^\bullet$] depends on K_{ass}, and the minimum is somewhat displaced (see Ref.[238] for more precise details).

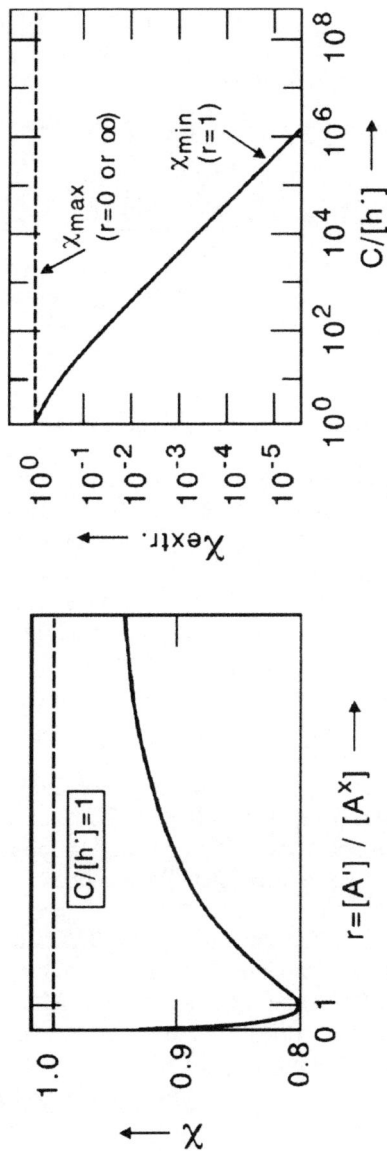

Figure 59. The χ factor of the holes as a function of the redox ratio r (left), if [h$^{\bullet}$] can be regarded as independent of r. Maximum ($\chi = 1$) and minimum ($\chi = \chi_{min}$) values as a function of doping (right).[238] (Reprinted from J. Maier, "Trapping Effects on the Diffusion in Mixed Conductors and Solid Electrolytes: Application to SrTiO$_3$ and ZrO$_2$(Y$_2$O$_3$)," in: *Ionic and Mixed Conducting Ceramics* Vol.94-12, Ed. by T. A. Ramanarayanan, W. L. Worrell, and H. L. Tuller, The Electrochemical Society, Pennington, NJ, pp 542-551, Copyright © 1994 with permission from The Electrochemical Society, Inc.)

The complicated ratio of D_O^δ to D_O^* (cf. Eqs. 139 and 149) reduces to the thermodynamic factor for a predominant electronic conductor.

The effects on D^Q are different: (i) The unionized defect does not appear in the conductivity experiment; (ii) according to Eqs. (110) and (111) a constant charge number is used to convert $\sigma_{ion} = \sigma_{O^{2-}} + \sigma_{O^-}$ formally to D_O^Q; by using $z = 2$ as the effective valence,

$$D_O^Q = \frac{RT\,(\sigma_{O^{2-}} + \sigma_{O^-})}{4F^2 c_O} \tag{151}$$

is obtained.

(iii) Diffusion Involving Internal Boundaries

According to Section V boundary layers exert a double influence. In the structurally altered core regions of the interface the mobilities and, hence, the defect diffusion coefficients are modified. In the space charge region the mobility, parallel to the interface, is assumed to be invariant while, perpendicular to the interface, electrical field effects are only negligible in the case of high migration thresholds. The anisotropy is more relevant for the conductivity (D^Q), on account of the strong anisotropy of the concentration (Section V). Hence in the boundary controlled cases D^Q, D^δ, D^* are all expected to be strongly anisotropic.

While D^Q directly reflects the ionic conductivity measured in a steady state experiment, the relationships are more complex for chemical diffusion and tracer diffusion. For simplicity we assume that the transport is so slow that local relaxation effects occur instantaneously.[240]

We first consider the chemical transport of oxygen in an oxide along a highly conducting interface (see Fig. 60 r.h.s.). Now the flux lines are not parallel to the interface for the whole experiment since also gradients in the orthogonal direction occur and the problem becomes complicated even if space charge effects are negligible. For further details and a full treatment the reader is referred to the relevant literature.[241] So long as ions and electrons are strictly coupled, tracer diffusion and chemical diffusion behave in an analogous way.

The treatment is greatly simplified, if the lateral diffusion into the bulk is absolutely negligible with respect to the boundary diffusion. The bulk transport then just takes place on a completely separate time scale; the effective diffusion length of the ceramic is the grain size (λ) rather than the sample size (L), but capacitive ($\propto 1/c^\delta$) and resistive terms ($\propto 1/\sigma^\delta$) are unchanged compared to the bulk values (cf. Section VI.3.*iv.*). In the case of the chemical experiment,

$$D_m^\delta = \frac{L^2}{\ell^2} D_{bulk}^\delta \propto \frac{L^2}{\ell^2} \frac{\sigma_{bulk}^\delta}{c_{bulk}^\delta} \tag{152}$$

Donor-doped BaTiO$_3$ and SrTiO$_3$ are relevant examples. There bulk diffusion is very slow because of the very low $v_O^{\bullet\bullet}$ concentration and the low mobility of the metal vacancies. If a perovskite ceramic,

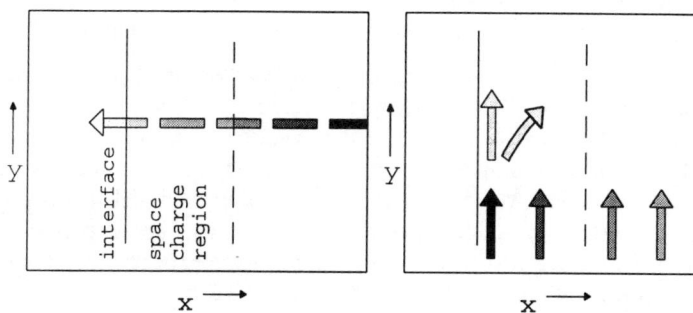

Figure 60. Diffusion across (left) and along (right) an interface (see text).

prepared under reducing conditions and exhibiting a high n-conductivity, is exposed to high partial pressures of oxygen, in-diffusion of oxygen takes place via the grain boundaries. As stated, bulk equilibration is very sluggish and only involves the upper layers of the grains. This leads to a ceramic material composed of highly conducting grains with insulating "skins", which exhibits very high effective (electrical) capacitances and finds application as a powerful capacitor material.[242]

In the case of chemical diffusion perpendicular to the interface, the complications are different ones (see Fig. 60, right). The problem is one-dimensional for reasons of symmetry, however, structural inhomogeneities (core versus bulk) and electrical field effects in the direction of transport are of importance here. Transfer processes through the interfacial core as well as through space charge zones have a profound influence on diffusion. In the instationary case the coupling of the partial fluxes according to Eq. (118) is violated and internal net fluxes occur also in the absence of external fluxes. Even though this complication disappears in the stationary state, the chemical diffusion coefficient is not a suitable transport coefficient any longer since ∇c_O is not a suitable driving force. It is important to note that–in addition to initially present equilibrium space charges–kinetic space charges build up during chemical diffusion in the general case (if oxygen diffusion into $SrTiO_3$ or SnO_2).[239,243]

Let us briefly highlight the effect of equilibrium space charges on the transport in our prototype oxide (with just O_i'' and $h^{\bullet}$ as defects).[244] The situation becomes particularly clear if we assume equal mobilities and equal bulk conductivities. Since in the case of chemical diffusion $\sigma_{O_i''}$ and $\sigma_{h^{\bullet}}$ are, as it were, in series, a space charge splitting according to the Gouy-Chapman situation always leads to a lowering of σ^{δ} and hence to a hindered chemical transport, while the sum of the conductivities increases and hence the electrical transport is facilitated (cf. Fig. 60). This slowing-down effect on D^{δ} explains the resistive impact of the Schottky boundary layer on the chemical diffusion in $SrTiO_3$; there, even both decisive charge carriers are depleted (see Section V.2). Figure 61 refers to such a bicrystal experiment measured by the optical technique described in Section VI.3.iv.

In a polycrystalline sample the situation is more complicated. Again, a simple result is only arrived at even if we assume extreme chemical resistance ratios for bulk and boundary values. Even if the grain boundary resistance dominates completely, the thermodynamic factor is still determined by the bulk, provided the density of boundaries is not too great. In other words, the largest part of the stoichiometric change $(d\mu_O/dc_O)$ takes place there. In this case[‡] we obtain, for hardly permeable boundaries of thickness d_{gb}, the surprisingly simple relationship

$$D_m^\delta \propto \frac{\ell}{d_{gb}} \frac{\sigma_{gb}^\delta}{c_{bulk}^\delta} \qquad \ell \qquad (153)$$

as a counterpart to Eq. (152).[245] Note that $d_{gb}/\ell \propto \varphi_{gb}^\perp$. The proportionality factor in Eq. (153) is $RT / 4F^2$.

The tracer diffusion is naturally influenced too by boundaries to be crossed. However, the situation is simpler here[§]. Since the tracer diffusion does not produce chemical or electrical effects, the tracer transport across a grain boundary, is characterized by the equilibrium O_i'' situation and similar to ionic conduction as described in Section V.2. It ought, in particular, to be added that, if the dominating charge carriers and $h^\bullet$ have the same mobility and the space charge potential is positive (negative), tracer diffusion is accelerated (retarded), while in chemical diffusion a retarding effect was to be expected for both signs. Just as for D^Q and D^δ, D^* is also subject to microstructural and bulk-crystallographic anisotropies, that endow the D values with tensor character (cf. again Fig. 60). In polycrystalline materials percolation processes via favorably orientated grains can be important.

[‡] This can be derived from the equivalent circuit shown in Section 6.7 (Fig. 69) for negligible C_{gb}^δ and dominating R_{gb}^δ.

[§] We neglect isotope effects on the mobility which is certainly incorrect in the case of the proton.

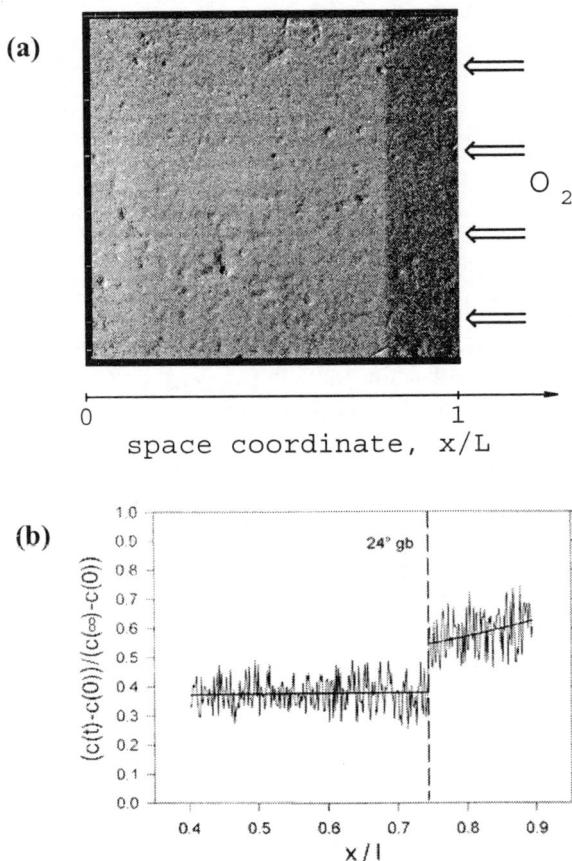

Figure 61. a) The figure shows the color change over a grain boundary
of low symmetry (of a bicrystal of Fe-doped $SrTiO_3$) during the in-
diffusion of oxygen through the right surface (partial pressure jump from
10^4 Pa to 10^5 Pa, T = 873 K, Fe concentration: 2.15 x 10^{18}cm^{-3}). The
others are not activated (24° tilt grain boundary, axis of rotation: [001],
near Σ 13). According to Ref.[142]. b)Normalized vacancy profiles for low
symmetry (~Σ 13) as obtained from in situ measurements shown in a).
According to Ref.[142]. Reprinted from M. Lionhardt, J. Jamnik and J.
Maier, *Electrochem. Solid State Letter* **333-335**, 2. (Copyright © 1999
with permission from The Electrochemical Society, Inc.)

As far as the one-dimensional dislocations are concerned, possible blocking effects retreat, in contrast to interfaces, into the background, while they very frequently provide rapid diffusion paths as a result of core and space charge effects.[246]

As in the case of the conductivity experiments, current-constriction effects can occur in the diffusion experiments,[247] if lateral inhomogeneities are present. In this way resistivities occur that can be easily misinterpreted in terms of sluggish surface steps. A concise treatment of proper surface kinetics will be given now.

5. Surface Kinetics

Figure 62 gives a sketch of oxygen incorporation from the gas phase into a mixed conductor for the case of chemical incorporation. One recognizes a variety of elementary steps: Gas diffusion followed by gas adsorption, dissociation, ionization, transfer into the crystal, transport through the subsurface (space charge zone) and bulk transport. Lateral inhomogeneities can lead to parallel processes which will be neglected here. In the case of the tracer experiment the oxygen refers to a given isotope while the counter isotope undergoes exactly the opposite process. Electrons are not of explicit importance but implicitly necessary as regards individual steps; nonetheless, it is, in principle, possible for the electrons to be directly transferred between tracer atoms (e.g., $^{18}O_{ad} + {}^{16}O_{ad}^- \underset{\leftarrow}{\overset{\rightarrow}{}} {}^{18}O_{ad}^- + {}^{16}O_{ad}$). Such a direct tracer mechanism may be of relevance in materials with extremely small concentrations of electronic carriers. In the electrical experiments the electrons stem from the electrode, and a fair comparison between e.g., tracer and electrical experiment needs to refer to a comparable phase distribution. Beyond these mechanistic differences manifested in the role of electrons, we also expect, similar as in the case of the diffusion coefficients, "conceptual" differences, i.e., differences that also occur even if we restrict ourselves to the case of the electron-rich electron conductor in which the mechanistic differences disappear.

A detailed treatment is given in Ref.[172] In the following we will assume two simplifying principles to be fulfilled. (1) The mechanism can be considered as a series mechanism in which one step has such small rate constants that it can be considered to be rate determining. (2) The storage of material in the surface during the diffusion experiments

$$\tfrac{1}{2}O_2 \; \rightleftharpoons \; \tfrac{1}{2}O_2 \; \rightleftharpoons \; \tfrac{1}{2}O_{2,\ ad}$$

$$T \qquad\qquad\qquad \downarrow\!\uparrow R$$

$$O_{ad} \qquad (oxide)$$

$$\downarrow\!\uparrow E \qquad E \qquad T \qquad T$$

$$\qquad \rightleftharpoons \; 2e^- \; \rightleftharpoons \; 2e^- \; \rightleftharpoons \; 2e^-$$

$$O^{2-}_{ad} \qquad \overset{V^{\cdot\cdot}_o}{\rightleftharpoons} \; O^{2-} \; \rightleftharpoons \; O^{2-} \; \rightleftharpoons \; O^{2-}$$

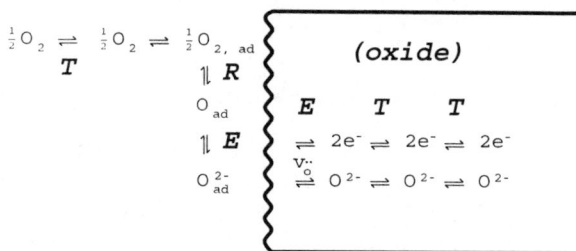

Figure 62. A selection of possible elementary steps for the incorporation of oxygen in an oxide. The surface reaction, in particular, is made up of complex individual steps. In reality the ionization degree of adsorbed atoms lies between zero and the bulk value (here −2). T: transport, R: chemical reaction, E: electrochemical reaction.

is assumed to be negligible compared to the material stored in the bulk (in other words: storage is small compared to flux divergence, and the flux is the same everywhere in the surface region). Then we neglect the chemical surface capacitances, and we only refer to surface resistances. Chemical capacitance effects will solely be determined by the bulk. In the case of the electrical experiment we refer to charge storage and hence to electrical capacitances that are, conversely, very substantial for interfaces. As the electrical experiment is not a diffusion experiment we will refer to it later (Section VI.6). Unlike electrode kinetics which is intensively considered in liquid electrochemistry, we will treat the surface part of the diffusion process—according to the great importance for the solid state—in more detail.

The inverse specific surface resistance, Λ_s, can be defined by the ratio of the flux and the affinity drop over the surface (A_s). Since the flux is the same everywhere and Λ_s by far takes its smallest value in the case of the rate determining step (*rds*), almost all the affinity drops over *rds*:

$$A_s = \Lambda_s^{-1} j_{O,s} = \Sigma_i \Lambda_{s,i}^{-1} j_{O,s} \simeq \Lambda_{rds}^{-1} j_{O,s} \simeq \Lambda_{rds}^{-1} j_{O,rds} \simeq A_{rds} \qquad (154)$$

(For simplicity we drop the tilde used in Section VI.2.) The comparison with Eq. (106) shows that Λ is identical with the exchange flux density j_{rds}^{o} of the *rds*. The affinity of the surface process A_s refers to the negative change in the respective chemical potential of oxygen between gas phase and first layer where the bulk process starts. If we ignore space charges, as we will do in the following for simplicity, this locus is x = 0 (see, e.g., Fig. 63). (In the case of space charges $x \simeq 2\lambda$ is an appropriate coordinate.)

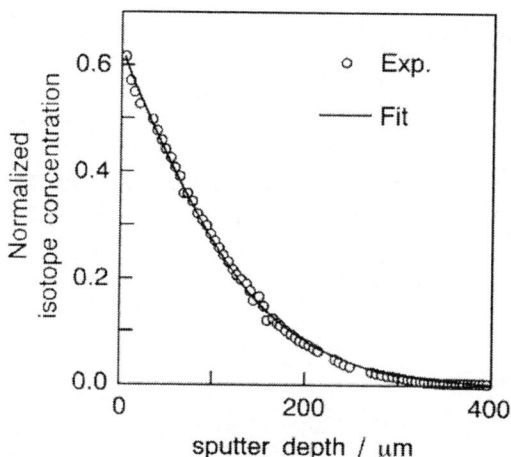

Figure 63. The kinetics in $La_{0.8}Sr_{0.2}CoO_{3-x}$, under the conditions given, is strongly influenced by the surface reaction. For pure diffusion control the normalized surface concentration would be unity.[207] (Reprinted from R. A. De Souza, J. A. Kilner, "Oxygen transport in $La_{1-x}Sr_xMn_{1-y}Co_yO_{3\pm\delta}$.", *Solid State Ionics*, **106**, 175-187. Copyright © 1998 wih permission from Elsevier.)

Since the chemical potential of oxygen (in the tracer case we refer to the chemical potential of the isotope) is spatially (Δ) constant in equilibrium, it holds that

$$j_{O,s} = -\Lambda_O \Delta\mu_O = -\Lambda_O \delta\mu_O(x = 0) \tag{155}$$

where, at the r.h.s., we refer to the deviation of μ from the final equilibrium state at $x = 0$. When investigating the time changes is of particular advantage to use concentration changes for the experimental detection. Then we rewrite Eq. (155) as

$$j_{O,s} = -\Lambda_O \frac{d\mu_O}{dc_O} \delta c_O(x = 0) \tag{156}$$

The thermodynamic factor $d\mu_O/dc_O$ refers to the bulk and reflects its (inverse) chemical capacitance. This term reduces to RTw_O/c_O for the chemical experiment and to RT/c_O for the tracer experiment. The prefactor $\Lambda_O d\mu_O/dc_O$ evidently defines what is called effective rate constant $\bar{k}_O^\delta$ or $\bar{k}_O^*$:

$$j_{O,s} \equiv -\bar{k}_O \delta c_O(x = 0) \tag{157}$$

Note that $\delta c_O(x = 0) \neq c_O(g) - c_O(0)$. The transition from Λ to $\bar{k}$ is analogous to the transition from σ (or σ^δ) to D (or D^δ). Obviously

$$\bar{k}_O^\delta = \Lambda_O^\delta w_O / c_O, \quad \bar{k}_O^* = \Lambda^* / c_O \tag{158}$$

and

$$\bar{k}_O^\delta / \bar{k}_O^* = (\Lambda_O^\delta / \Lambda_O^*) w_O \tag{159}$$

As we will see, these $\bar{k}$-values can be easily measured with the help of diffusion experiments. In the case that the same mechanism applies, $\Lambda_O^\delta = \Lambda_O^*$ (see also below) and $\bar{k}_O^\delta / \bar{k}_O^* = w_O$.[248] By analogy to D^Q we can also define a $\bar{k}$-value for the electrical experiment according to

$$\bar{k}_O^Q \equiv \Lambda_O^Q / c_O \qquad (160)$$

Let us consider the extreme cases and first lump together the surface reaction steps and also the individual bulk hopping steps and write the whole process in a two step form

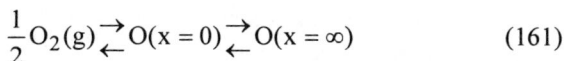

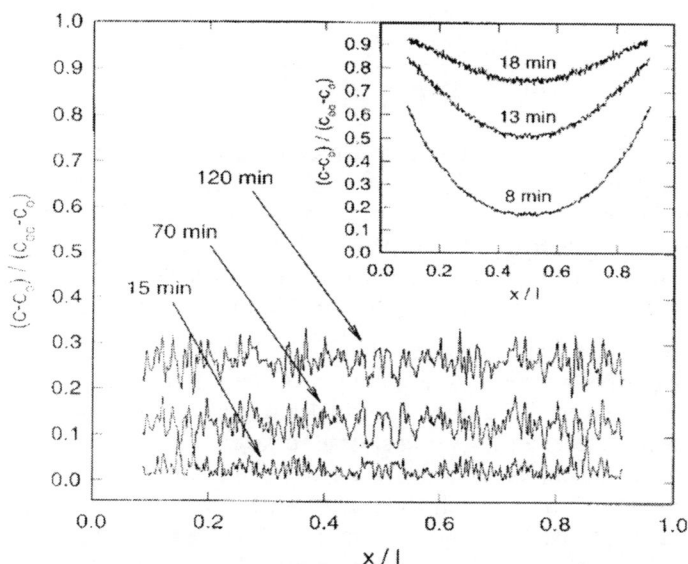

$$\frac{1}{2}O_2(g) \underset{\leftarrow}{\overset{\rightarrow}{}} O(x=0) \underset{\leftarrow}{\overset{\rightarrow}{}} O(x=\infty) \qquad (161)$$

Figure 64. Normalized concentration profiles obtained from the spatially resolved in-situ technique (see Section VI.3.*iv.*) for Fe-doped SrTiO₃ samples (Fe content: 4.6 x 10^{19}cm^{-3}) on rapid change of the gas atmosphere from 10^4 to 10^5 Pa at 923 K. The main diagram shows reaction control, the insert diffusion control. Both cases involve incorporation through a polished (100) surface. In the case of diffusion control a thin, porous Cr layer was used to activate the surface (according to Refs.[249,250]). (Reprinted from J. Maier, "Interfaces" in: *Oxygen Ion and Mixed Conductors and their Technological Applications*, Vol. 75-121, p. 368, Ed. by H. L. Tuller, J. Schoonman and I. Riess, NATO Science Series: E Applied Sciences, Kluwer Academic Publishers, Dordrecht, Copyright © 2000 with permission from Kluwer Academie Publishers B. V.)

If the bulk step (second step in Eq. 161) is rate determining, $\mu_O(g) - \mu_O(x = 0) = 0$, i.e., $\delta\mu_O(x = 0) = 0$ and $\delta c_O(x = 0) = 0$, and the surface reaction is in equilibrium. This is the situation considered in the previous sections. Conversely, if the surface step (first step in Eq. 161) is rate determining, there is transport equilibrium in the bulk $\mu_O(x = 0) = \mu_O(x = \infty)$ and $c_O(x = 0) = c_O(x = \infty)$. Then the diffusion profiles are horizontal with intercepts determined by $c_O(t,x = 0)$. The flux into the bulk is determined by $-\bar{k}\delta c$ while the flux out of it is zero, leading to $\partial\delta c/\partial t \propto -\delta c$ and

$$\frac{c_O(t) - c_O(t = \infty)}{c_O(t = 0) - c_O(t = \infty)} = \exp-\frac{\bar{k}t}{L} \qquad (162)$$

everywhere in the sample. Figure 64 shows almost horizontal profiles as predicted by Eq. (162) whose evaluation allows for an accurate determination of effective rate constants. Fig. 65 displays results of integral measurements. The exponential law for reaction control (Eq. 162) simplifies to a linear dependence for short times. As shown for the case of diffusion control in Section VI.3.iv. an exponential law is only valid for long times, while for short times a $\sqrt{t}$-behavior is fulfilled. The latter is seen in the l.h.s. of the figure. The switch-over from reaction to diffusion control typically occurs for thick samples, high temperatures and/or for catalyzing surfaces.

So far we considered the surface process in the limit of proximity to equilibrium which is, for the relaxation experiments discussed, often a sufficient approximation. In order to extend the validity range but also to highlight the situation from a different point of view, let us apply chemical kinetics to the surface process which we now decompose into the individual steps of the reaction sequence.[23,24,172,248] The rate determining step is supposed to be one of these surface reaction steps.

For simplicity let us assume the model mechanism

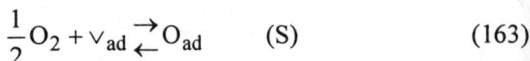

$$\frac{1}{2}O_2 + \vee_{ad} \underset{\leftarrow}{\overset{\rightarrow}{\rightleftarrows}} O_{ad} \qquad (S) \qquad (163)$$

$$O_{ad} + \vee_O^{\bullet\bullet} + 2e' \underset{\leftarrow}{\overset{\rightarrow}{\rightleftarrows}} O_O \qquad (T) \qquad (164)$$

with Eq. (164) being the *rds*, and—in order to emphasize the general principle—first consider an electron-rich electron conductor. Since the overall rate is equal to $\Re_{rds}$, we can generally state ($\varepsilon = \delta, *, Q$)

$$\Re^\varepsilon = \vec{k}_T[v_O^{\bullet\bullet}] - \overleftarrow{k}_T[O_O] \qquad (165)$$

Again, in order to simplify the notation, we suppress the tilde, even though the k's can contain electrical fields and do definitely so for $\varepsilon = Q$. The rate constant $\vec{k}_T$ contains also the constant electron concentration as well as the equilibrium concentration of the adsorbed species. (The appearance of the concentration of regular species in Eq. (165), even though not dilute, is justified, since it is constant in the cases $\varepsilon = \delta$ and $\varepsilon = Q$, while it refers to the ideal tracer concentration for $\varepsilon = *$.)

In the tracer case $[v_O^{\bullet\bullet}]$ is invariant and $[O_O]$ is perturbed, while exactly the inverse is true for chemical diffusion (as long as $[v_O^{\bullet\bullet}] \ll [O_O]$). The rate constants can also be considered to be invariant. However, in the electrical experiment, the latter are the variables of interest as they contain the driving force. Simplifying the notation by setting $v \equiv [v_O^{\bullet\bullet}]$, $O \equiv [O_O]$, $k' \equiv \vec{k}$, $k'' \equiv \overleftarrow{k}$, and using the arc to designate the state at $t \to \infty$, we can write with the help of Eq. (107)

$$\Re^\varepsilon = \begin{cases} -\Re^\circ \dfrac{\delta O}{\overset{\frown}{O}} & \text{tracer experiment} & (\varepsilon = *) \\[2ex] \Re^\circ \dfrac{\delta v}{\overset{\frown}{v}} & \text{chemical experiment} & (\varepsilon = \delta) \quad (166) \\[2ex] -\Re^\circ \left(\dfrac{\delta k'}{\overset{\frown}{k'}} - \dfrac{\delta k'}{\overset{\frown}{k''}} \right) & \text{electrical experiment} & (\varepsilon = Q) \end{cases}$$

with the exchange rate $\Re^\circ$ being $\overset{\frown}{k'v} = \overset{\frown}{k''O} = \sqrt{\overset{\frown}{k'v}\,\overset{\frown}{k''O}}$. Evaluating the last relation of Eq. (166), the Butler-Volmer equation originates. Linearization leads to Eq. (160). Apparently the other two relations represent similarly important rate equations for the surface processes of

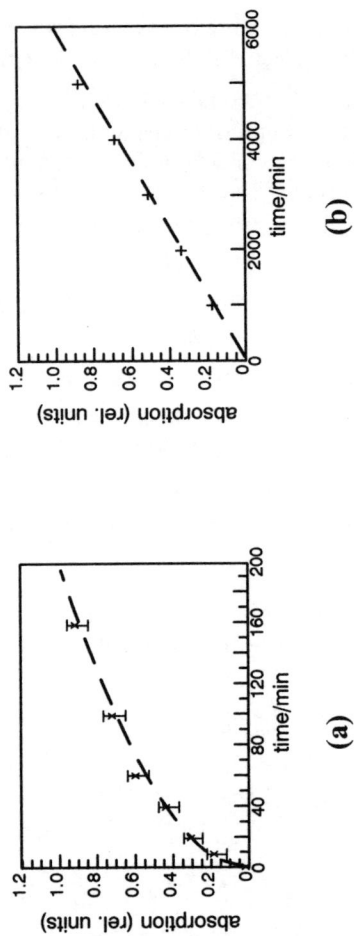

Figure 65. Results of the optical analysis of the incorporation of oxygen in $SrTiO_3$. Time course of the intergral optical absorption in the diffusion direction. a) diffusion control, b) reaction control.[251]

tracer and stoichiometry experiments from the linearization of which we derive Eq. (158).

If the compounds are no longer electron-rich, two difficulties are met. (i) The electron concentrations in the above equations may no longer be invariant in the chemical equations, and (ii) the experiments may be characterized by different mechanisms.

Let us assume identical mechanisms (cf. Eqs. 163 and 164). The changed electron budget only matters for $\varepsilon = \delta$ (for $\varepsilon = *$, Q the electron concentration is invariant). For the chemical process we find instead of Eq. (166) more generally

$$\mathfrak{R}^\delta = \mathfrak{R}^\circ \frac{\delta\left\{[e']^2[v_O^{\bullet\bullet}]\right\}}{[\bar{e'}]^2[\bar{v}_O^{\bullet\bullet}]} \tag{167}$$

which yields after linearization

$$\mathfrak{R}^\delta = \mathfrak{R}^\circ\left(\frac{\delta[v_O^{\bullet\bullet}]}{[\bar{v}_O^{\bullet\bullet}]} + 2\frac{\delta[e']}{[\bar{e'}]}\right) = \mathfrak{R}^\circ\left\{\frac{1}{[\bar{v}_O^{\bullet\bullet}]} + 2\frac{1}{[\bar{e'}]}\frac{\delta[e']}{\delta[v_O^{\bullet\bullet}]}\right\}\delta[v_O^{\bullet\bullet}] \tag{168}$$

$$= -\mathfrak{R}^\circ(w_O/c_O)\,\delta c_O(x = 0)$$

For a trap free situation the term $\delta[e']/\delta[v_O^{\bullet\bullet}]$ simplifies to 2, otherwise to $2\chi_n$ (cf. Section VI.4.*ii.*).[175] The r.h.s. term in {} is obviously exactly the term w_O/c_O discussed for D^δ. Other mechanistic cases are more difficult to handle, especially if the *rds* does not refer to $x = 0$ and coupled equilibria have to be taken account of; however, the results are comparable.

This is shown in Refs.[24,172,248] in great detail. In agreement with the thermodynamic considerations we find in all cases close to equilibrium that

$$\bar{k}^\delta = \mathfrak{R}^{\circ\delta}(w_{O^{2-}}/c_{O^{2-}})_{bulk}$$

$$\bar{k}^* = \mathfrak{R}^{\circ*}/c_{O^{2-}_{bulk}} \tag{169}$$

$$\bar{k}^Q = \mathfrak{R}^{\circ Q}/c_{O^{2-}_{bulk}}$$

where we now also distinguish between different exchange rates in view of the possibility that different mechanisms $\varepsilon = \delta$, *,Q occur. The molecularity of the *rds* can give rise to additional constant factors (ν) in Eq. (169). (This may occur, e.g., if the adsorption is rate determining. In the simplest case P_{O_2} appears as a factor with the exponent 1 in the kinetic ansatz leading to $\nu = 4$, whereas $\nu = 1$ results if $P_{O_2}^{1/2}$ enters the rate equation which is the case for a rapid predissociation.

The advantage of the kinetic treatment lies in the fact that (i) also solutions far from equilibrium can be handled and (ii) the range of validity of Eq. (169) can be given (similarly as in the diffusion case, cf. Section VI.2.*i*). Since in the above derivations bulk defect chemistry was assumed to be established at x = 0, the index "bulk" was used in Eq. (169) to allow for more general situations. Note that these explicit formulae predict defined dependencies on the control parameters which can be checked provided defect chemistry is known. For simple situations (see Refs.[252,253]) a power law relationship results (is a constant)

$$\bar{k}^{\varepsilon} = \alpha^{\varepsilon} P^{N^{\varepsilon}} \Pi_r K_r^{\gamma_r^{\varepsilon}} C^{M^{\varepsilon}} \tag{170}$$

describing the dependence on the control parameters temperature, component activity (partial pressure P) and doping content (C). The measurements of the dependencies provide extremely useful mechanistic information. Such dependencies and also the relationships between the $\bar{k}$'s and the D's are discussed for specific cases in Refs.[252,253]

Let us turn to our $SrTiO_3$ (Fe-doped) example and to Fig. 66. The parameter $\bar{k}^{\delta}$ is indeed found to be markedly higher than $\bar{k}^{*}$ but less high than expected for $\mathfrak{R}^{o\delta} = \mathfrak{R}^{o*}$. In other words, $\bar{k}^{\delta} / w_O \ll \bar{k}^{*}$. Moreover, if we coat the surface with Pt we enhance $\bar{k}^{\delta}$ drastically while $\bar{k}^{*}$ remains unchanged. If this is indeed caused by different mechanisms is still under debate.

While a direct tracer exchange would not be substantially influenced by Pt, the Pt-coating makes the chemical measurement more similar to the electrical measurement as in both cases the electrons are

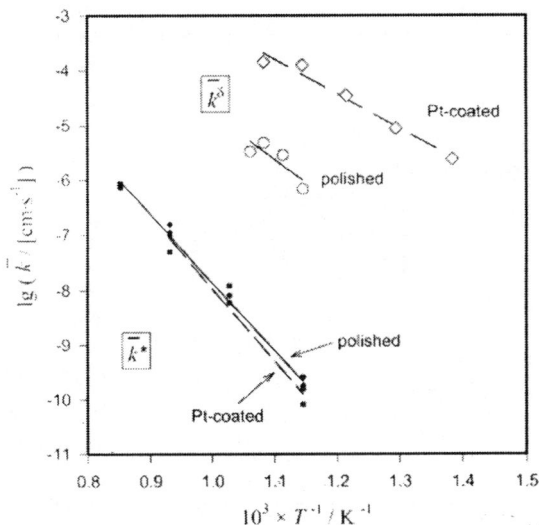

Figure 66. Chemical $\bar{k}'s\,(\bar{k}^{\delta})$ and tracer $\bar{k}'s\,(\bar{k}^{*})$ as a function of 1/T for differently prepared (100) surfaces of Fe-doped SrTiO$_3$.[250,254]

taken up from this metal. In this case $\bar{k}^{\delta}/w_O$ would approach $\bar{k}^Q \neq \bar{k}^{*}$. Such "surface Haven ratios" (cf. Eq. 127) are still to be verified. The respective mutual inter-dependencies of the $\bar{k}'s$ are displayed in Fig. 67 which may be compared to Fig. 52. There are almost no cases for which a clearcut mechanistic picture could be given, what is not surprising, in view of the manifold of possible reaction mechanisms. A comparatively well investigated process is chemical oxygen incorporation into SrTiO$_3$. By combining the knowledge of $\bar{k}^{\delta}(P,T,C)$ with experiments far from equilibrium and with UV-experiments that allow one to tune the concentration of electronic carriers, a mechanism could be made probable that involves (i) a first ionization of the adsorbed molecule by consuming a regular electron, (ii) a second ionization by consuming an energetically higher lying conduction electron; (iii) then the peroxide decomposes, finally

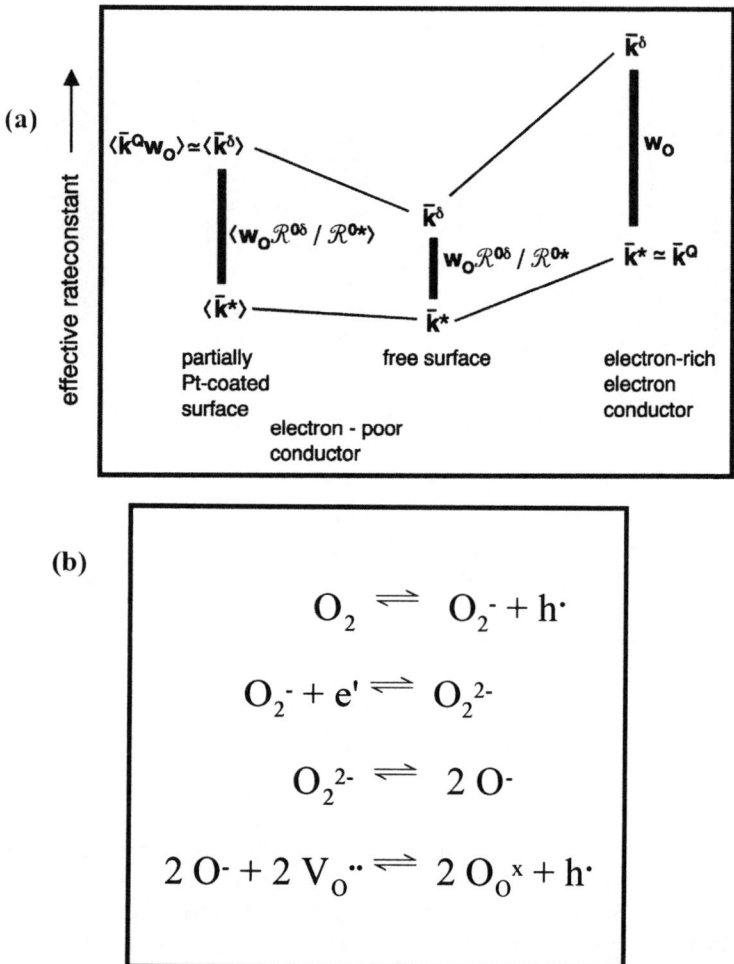

Figure 67. (a) Relation between the $\bar{k}$'s for high electron concentration (r.h.s. column) and for low electron concentrations of the oxide. Free surface: center column, partial Pt coverage: left column. The brackets indicate averaging over the heterogeneous surfaces. There is a close correspondence to Fig. 52 (according to Ref.[248]) (b) Probable mechanism of oxygen incorporation in $SrTiO_3$ ($\bar{k}^\delta$). (Reprinted from J. Maier, "Interaction of oxygen with oxides: How to interpret measured effective rate constants?", *Solid State Ionics*, **135**, 575-588. Copyright © 2000 with permission from Elsevier.)

(iv) O⁻enters the vacancy while being simultaneously fully ionized (see Fig. 67b).[255]

6. Electrode Kinetics

Now let us consider briefly electrode kinetics, i.e., $\overline{k}^Q$ or—since we have not involved any chemical capacitance effects—the exchange rate $\mathfrak{R}^{oQ}$. Let us follow the work by Wang and Nowick[256] on electrode kinetics of Pt, O_2/CeO_2 (doped) which was indeed invoking the model mechanism given by Eqs. (163) and (164). If the transfer step is assumed to be rate determining and the symmetry factors are set to 1/2, the exchange current density reads ($i^o \equiv zFj^o \propto \mathfrak{R}^{oQ}$)

$$i^o = zF\sqrt{\widehat{\Theta}(1-\widehat{\Theta})}\sqrt{\widehat{k}_T\widehat{k}_T} \tag{171}$$

(Θ is the degree of coverage and refers, as the rate constants, to equilibrium values, i.e., bias free). The index T refers to the transfer reaction, while the index S refers to the sorption pre-equilibrium (cf. Eq. 163) characterized by the mass action constant K_s. Under Brouwer-like conditions (that is conditions necessary for Eq. 170 to apply) either $\widehat{\Theta}$ or $(1 - \widehat{\Theta})$ is unity. Since the adsorption enthalpy is smaller than zero for high temperatures and/or low P_{O_2}, for which $K_s P_{O_2}^{1/2} \ll 1$, it follows that $(\widehat{\Theta} \ll 1)$

$$\widehat{\Theta} = K_s P_{O_2}^{1/2} \tag{172}$$

The P_{O_2} - and T-dependencies of the exchange current density are then given by

$$\frac{\partial \ln i^o}{\partial \ln P} = 1/4$$

$$-R\frac{\partial \ln i^o}{\partial 1/T} = \frac{1}{2}\Delta H_s^o + \frac{1}{2}(\Delta \overrightarrow{H}_T^{\neq} + \Delta \overleftarrow{H}_T^{\neq}) \tag{173}$$

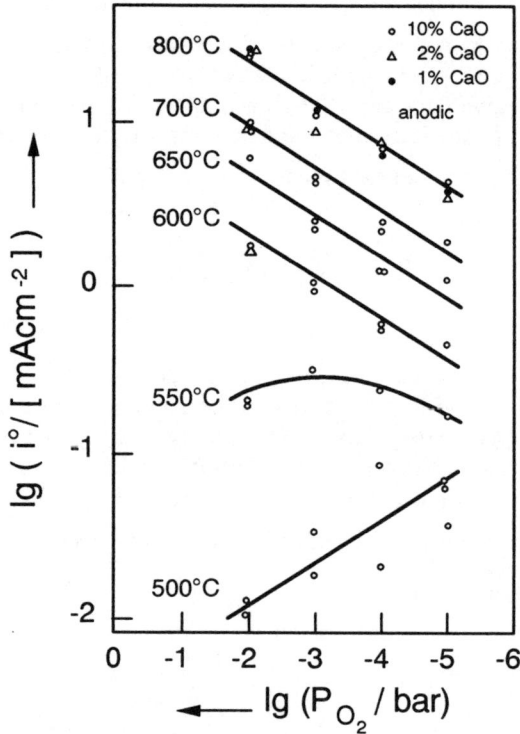

Figure 68. The exchange current density as a function of oxygen partial pressure for different temperatures confirming the electrode kinetical model given in the text.[256] (Reprinted from D. Y. Wang, A. S. Nowick, "Cathodic and Anodic Polarization Phenomena at Platinum Electrodes with Doped CeO_2 as Electrolyte. I. Steady-State Overpotential.", *J. Electrochem. Soc.*, **126**, 1155-1165. Copyright © 1979 with permission from The Electrochemical Society, Inc.)

In the other limit $K_s P^{1/2} \gg 1$ we obtain analogously

$$\frac{\partial \ln i^\circ}{\partial \ln P} = -1/4$$

$$-R \frac{\partial \ln i^\circ}{\partial 1/T} = -\frac{1}{2} \Delta H_s^\circ + \frac{1}{2} (\overrightarrow{\Delta H_T^\neq} + \overleftarrow{\Delta H_T^\equiv})$$

(174)

As σ in the bulk case, the exchange rate includes reaction and activation energies. The results shown in Fig. 68 as a function of P_{O_2} at different temperatures are consistent with these predictions. Not very many studies of electrode kinetics involving solid/solid contacts are known. What makes the treatment particularly complex are lateral inhomogeneities. A recent example which refers to the cathodic situation in high temperature fuel cells is given in Ref.[257].

7. Generalized Equivalent Circuits

In the framework of linear irreversible thermodynamics it is possible to handle responses to electrical and/or chemical driving forces, as could be seen from the previous sections, by what is called Nernst-Planck-Poisson set of equations plus rate equations of local reactions.[258-262] The transport and reaction coefficients are determined by inverse resistances and inverse capacitances. Hence it is, after all, not surprising that the situation can be mapped by generalized equivalent circuits. A recent general approach used network thermodynamics and is given in Ref.[263]. The treatment is partly based on earlier literature.[258,260,264] For the description three types of elements are needed. The first (i) is the electrochemical resistor ($\propto \sigma^{-1}$ or more generally $\propto \Lambda^{-1}$). As far as the resistor is concerned, we do not need to distinguish between chemical and electrical effects, since they refer to the same mechanism (cf. Nernst-Einstein-equation). This is different in the case of the capacitances. The chemical capacitance (ii) refers to a stoichiometry polarization, whereas the electrical capacitance as the third element (iii) refers to dielectric displacement. The equivalent circuits shown in Fig. 69 map the underlying equations 1:1. The horizontal axis refers to the positional coordinate while the orthogonal axis refers to the deviation from equilibrium. In this way complex

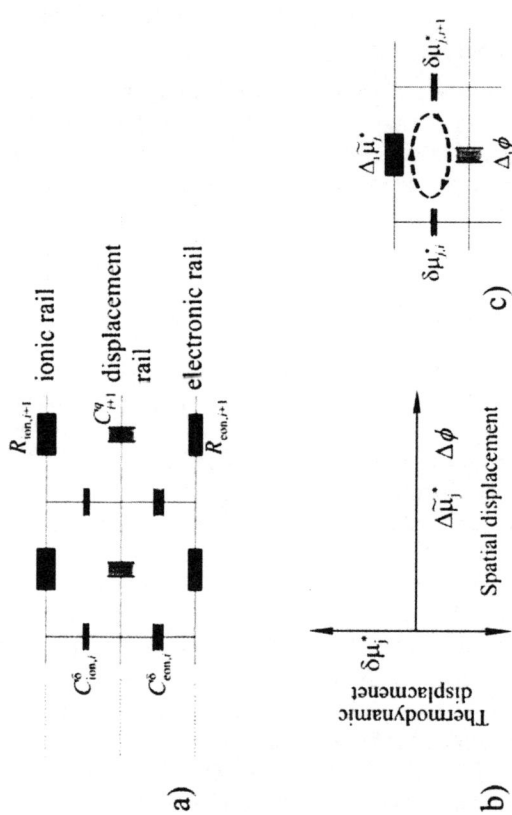

Figure 69. Equivalent circuit for the description of generalized electrochemical processes in linear systems (see text).[263] (Reprinted from J. Jamnik, "Impedance spectroscopy of mixed conductors with semi-blocking boundaries" *Solid State Ionics*, **157**, 19-28. Copyright © 2003 with permission from Elsevier.)

electrochemical phenomena can be handled, including the pure electric transport and pure chemical diffusion. An example of the latter, viz. chemical diffusion in a polycrystalline ceramic was addressed in Section VI.4.*iii.*, while electrochemical polarization effects are to be discussed in Part II.[1]

8. Solid State Reactions

Since the constituents are usually charged, the field of inorganic state reactions exhibits a substantial intersection with electrochemistry. The incorporation of a neutral component (e.g., O into an oxide) has already been considered as a most simple reaction. Here we concentrate on typical reactions that involve the formation of a new phase. Owing to the manifold of individual problems, a detailed treatment cannot be given here and only a few points relevant in our context shall be highlighted.

Let us consider the oxidation of a metal, e.g., Zn to ZnO.[4,75,265] The reaction starts with nucleation and early growth phenomena (lateral mass transport, tunneling, space charge phenomena). Usually at larger thicknesses the interface smoothens and at sufficiently large thicknesses the transport becomes controlled by chemical diffusion. Even though it is not clear whether the ion transport is due to oxygen vacancies or zinc interstitials, we will assume the second case ($\sigma_{ion} = \sigma_v$) without great restriction of generality. Then in the diffusion controlled regime the flux and thus the reaction rate is determined by $\sigma_O^\delta \nabla \mu_O$ (with $\sigma^\delta = \sigma_{eon} \sigma_{ion}/\sigma$ being the ambipolar conductivity, see Section VI.4.*ii*). Owing to its constancy the flux can be recast as $(1/L) \int \sigma_O^\delta d\mu_O$. Thus the growth rate measured by the thickness change $\dot{L}$ is inversely proportional to L and and the well-known square-root rate law ($L \propto t^{1/2}$) follows (see Fig. 70).

Since for ZnO, $\sigma_{eon} \gg \sigma_{ion}$, and thus $\sigma^\delta \approx \sigma_{ion}$, the experimentally observed steep temperature dependence of the oxidation rate is explained by the activation enthalpy of σ_{ion}. Furthermore, the well-established doping effect[266] by Al^{3+} or Li^+ in depressing or enhancing the corrosion rate is immediately derived from the defect chemistry: $Al_{Zn}^\bullet$ defects decrease σ_{ion} by decreasing $[V_O^{\bullet\bullet}]$ (or $Zn_i^{\bullet\bullet}$) while the

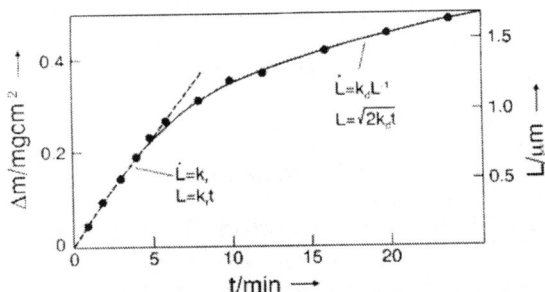

Figure 70. Change from reaction to diffusion control in a spinel formation reaction[4] indicated by the mass change. (Reprinted from H. Schmalz-ried, *Solid State Reactions*, Verlag Chemie, Weinheim. Copyright © 1981 with permission from WILEY-VCH Verlag GmbH.)

Li'_{Zn} defects increase σ_{ion}. The detailed partial pressure dependence follows through integration.

Figure 70 shows that the square-root law is preceded by a linear law for short times, which is indicative of a surface control. In the steady state of the surface control, $\dot{L}$ does not depend on t, neither explicitly or implicitly. (If $\Re$ is the rate of the surface reaction we may write $\dot{L} = f[L(t), \Re(t)]$, hence $\ddot{L} = \dfrac{\partial \dot{L}}{\partial L}\dfrac{dL}{dt} + \dfrac{\partial \dot{L}}{\partial \Re}\dfrac{d\Re}{dt} = 0$; the first term vanishes because $\partial \dot{L}/\partial L = 0$ and the second because of the steady state condition $d\Re / dt = 0$.) In the stage of early growth $\dot{L}$ depends also on L, and a more complicated law follows. Generally, the description of real solid state reactions can be awfully complicated owing to the role of the space coordinate in general, and the interfacial problems in particular.[4] Qualitatively speaking, it is clear from the above considerations that good contact, high conductivities (high T, appropriate doping), high driving forces and in particular small diffusion lengths are necessary to provide fast reaction.

Interrupting reading for a moment and looking around in the environment, tells us that there are essentially two categories of structures. Firstly, simple mostly planar geometries, particularly due to artificial architecture. There we rely on non-equilibrium processes such as cutting and polishing, the structure being maintained thanks to the extremely low diffusion constants. Secondly, complex interface morphologies, partly self-similar or self-affine, created under extreme conditions in the earth's history (e.g., rock-formation) and now frozen-in, or formed under mild in-situ conditions such as natural growth but still far from equilibrium. Such typical non-linear processes will be considered below. At this stage let us consider Fig. 71. It highlights that a planar interface may become unstable for kinetic reasons: If the diffusion of the metal M through the oxide is rate limiting for the corrosion of an M-N alloy (N: noble metal), any perturbation of the M, N/MO interface will be washed out (as shown in Fig. 71), but it will be augmented, possibly leading to dendrites, if the diffusion of M through the alloy is rate limiting.[265] In this case the noble metal will remain embedded in the oxide matrix. Local stability criteria like this may be extremely helpful even if they cannot predict the growth process.[177] (In the stable case the perturbation δL is positive, while the time change of it, $(\partial/\partial t)(\delta L)$, is negative; such a function guaranteeing stability is called a Ljapunow-function.)

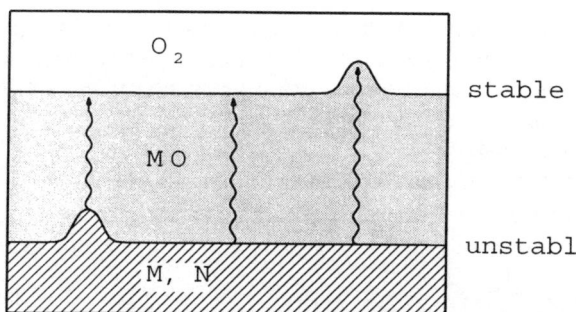

Figure 71. Morphological stability/ instability of the interfaces in a corrosion experiment.[265]

9. Non-linear processes

The vividness of our world does not rely on processes that are characterized by linear force-flux relations, rather they rely on the non-linearity of chemical processes. Let us recapitulate some results for proximity to equilibrium (see also Section VI.2.ii.): In equilibrium the entropy production (Π) is zero. Out of equilibrium, $\Pi \equiv T\delta_i S/\delta_i t > 0$ according to the second law of thermodynamics. In a perturbed system the entropy production decreases while we reestablish equilibrium ($\dot{\Pi} < 0$), (Fig. 72). For the cases of interest, the entropy production can be written as a product of fluxes and corresponding forces (see Eq. 108). If some of the external forces are kept constant, equilibrium cannot be achieved, only a steady state occurs. In the linear regime this steady state corresponds to a minimum of entropy production (but non-zero). Again this steady state is stable, since any perturbation corresponds to a higher Π-value ($\delta\Pi > 0$) and $\dot{\Pi} < 0$.[183] The linear concentration profile in a steady state of a diffusion experiment (described in previous sections) may serve as an example. With $\Pi \propto \int J\nabla c dV \propto \int (\nabla c)^2 dV$. We argue as follows: The function $c(x)$ satisfying this condition will be the same if we change the integrand of the variational problem to $(\nabla c)^2 + 1$ or even to $\sqrt{(\nabla c)^2 + 1}$. While the first point is trivial, the second one can be justified by Euler's variational theorem.[23] Since $\int \sqrt{1 + (dc/dx)^2}\, dx = \int \sqrt{(dx)^2 + (dc)^2}$ $= \int$ (length element), the variational problem is obviously analogous to finding the curve of minimum length between two terminals, which is the line in the x-c-plane. Hence the linear profile of the steady state refers to a minimum entropy production. It is also straightforward to show that this is stable.

All this does no longer apply to non-linear systems,[183,267-270] there (if at all) only the part of Π that is caused by varying the forces, reaches a minimum. For those states, however, a stability is not guaranteed ($\delta_X \Pi \gtrless 0$) (Fig. 72). Close to a steady state this partial variation of $\Pi = \Sigma_i \Re_i A_i$ is given by $\delta_X \Pi = \Sigma_i \Re_i^{ss} \delta A_i + \Sigma_i \delta\Re_i \delta A_i$ which reduces to

$$\delta_X \Pi = \Sigma_i \delta\Re_i \delta A_i \qquad (175)$$

ENTROPY PRODUCTION CLOSE TO AND FAR FROM EQUILIBRIUM

EQUILIBRIUM

$\Pi \propto$ entropy - production

$\Pi = Td_iS/dt \geq 0$

$\dot{\Pi} \leq 0$

$\delta\Pi^{eq} \geq 0$

(prehistory unimportant, stable)

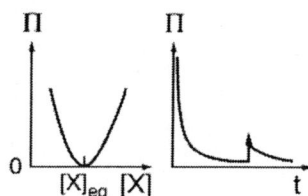

STEADY STATE CLOSE TO EQUILIBRIUM

$\Pi \geq \Pi^{ss} > 0$

$\dot{\Pi} \leq 0$

$\delta\Pi^{ss} \geq 0$

(prehistory unimportant, stable steady states)

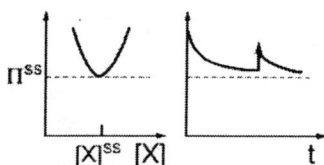

STEADY STATE FAR FROM EQUILIBRIUM

$\Pi \geq 0$

$_x\dot{\Pi} \leq 0$

$\delta_x\Pi^{ss} \lessgtr 0$

(prehistory important, unstable steady states possible)

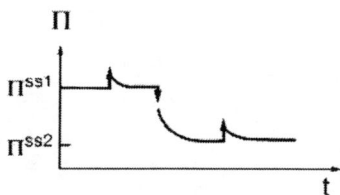

Figure 72. Entropy production close to and far from equilibrium

Table 7
Typical non-equilibrium phenomena far from equilibrium. [20,21]

autocatalytic reaction	$A + X \longrightarrow 2X$	[X] vs t	growth, positive feedback
autocatalytic + decay reaction	$A + X \underset{}{\overset{k_1}{\rightleftharpoons}} 2X$ $X \xrightarrow{k_2} Z$	[X] vs t, (ss), (ss)	growth or death
no selection pressure	$[A] = \text{const}$ $W_c = k_1[A] - k_2$	$[X]_{ss}$ vs [A], 0, k_2/k_1	nonequilibrium phase transformation
autocatalytic + decay + selection pressure (+ perturbation)	$A + X \longrightarrow 2X$ $X \longrightarrow Z$ $[A] \neq \text{const}$ $W_c = f(t)$	[X] vs t	competition selection mutation

Table 7
Continuation

Reaction scheme			Behaviour
Lotka-Volterra reaction scheme	$A + X \longrightarrow 2X$ $X + Y \longrightarrow 2Y$ $Y \longrightarrow Z$	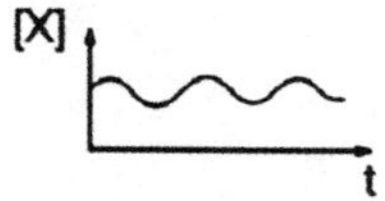	structurally unstable oscillation
Brusselator reaction scheme or	$A \longrightarrow X$ $2X + Y \longrightarrow 3X$ $B + X \longrightarrow Y + D$ $X \longrightarrow Y$		limit cycle oscillation
catalytic CO oxidation	$(CO + O \rightleftharpoons CO_2;$ $Pt_{hex} \rightleftharpoons Pt_{sq})$	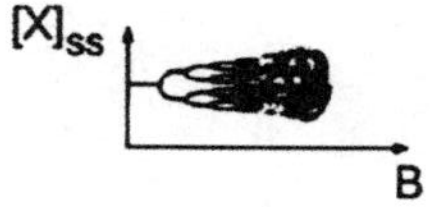	symmetry breaking bifurcation deterministic chaos
above reaction scheme + diffusion	(e. g. $X \rightleftharpoons X$)	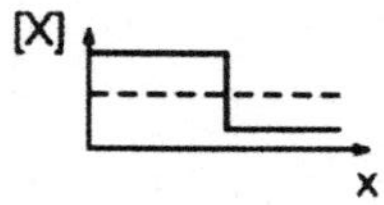	compartmentation, dissipative structure in space

since the first order term must vanish for the steady state (ss).

A key ingredient of exciting reaction-schemes that drives reactions away from harmless states is autocatalysis. Let us consider the simplest autocatalytic reaction[270]

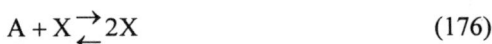

$$A + X \underset{\leftarrow}{\overset{\rightarrow}{\rightleftharpoons}} 2X \tag{176}$$

and first neglect the back reaction. Owing to

$$\Re \propto d[X]/dt \propto [A][X] \tag{177}$$

the stationary (initial) state $[X] = 0$ is unstable, and any finite concentration of X leads to an exponential growth. Indeed the variation $\delta_X \Pi$ is negative ($\delta\Re \propto \delta[X]$, $\delta A \propto -\delta[X]/[X]$ for ideal conditions; cf. Eq. 175). If the back reaction is taken account of, the system is driven towards equilibrium, as because of $\overleftarrow{\Re} \propto [X]^2$ the negative feedback becomes dominating. The final steady state is then stable because $\delta\Re$ is approximately $-[X]\delta[X]$ and $\delta_X\Pi > 0$.[269] If we return to the initial situation and add a decay reaction

$$X \rightarrow Z \tag{178}$$

to the growth reaction, i.e.,

$$\Re = (\overline{k}[A] - \overrightarrow{k'}) \, [X] = W[X] \tag{179}$$

the situation is clearcut if the "food", i.e., [A] is constant. Then the bracketed term, i.e., W, determines whether X grows or decays. The situation becomes more interesting if [A] is exhaustible and one allows for similar parallel reactions that also involve A, to occur. This leads automatically to competition between the species and only one survives.[270] Tuning W leads to the possibility to decide about "dead or alive" (non-equilibrium phase transformation, cf. second row in Table 7). Introducing the possibility of fluctuations in k, automatically mutation and selection processes are obtained.[267,270] Marginally more complicated reaction mechanisms lead to oscillations or chaos, which can also affect the position coordinate if the processes are coupled to

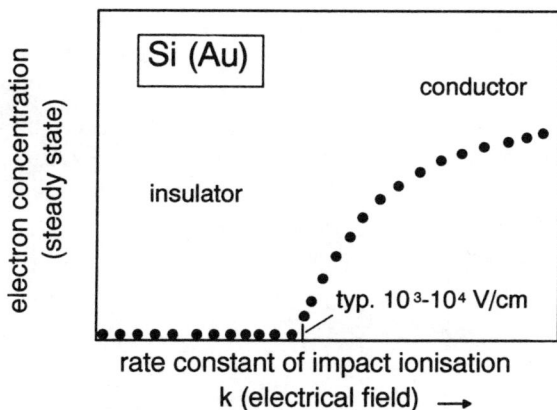

Figure 73. Non-equilibrium phase transition (insulator-conductor transition) in Au-doped silicon as a result of voltage variation in the nonlinear range.[271,272]

transport. Table 7 gives an overview. It is certainly not the place here to treat these phenomena in more detail. It is, however, clear that similar processes occur in solids, in particular if the interaction with electric field and/or electromagnetic waves are included.[271] An important autocatalytic charge carrier process in solids is impact ionization, crudely written as

$$e' \underset{\leftarrow}{\overset{\rightarrow}{}} 2e' + h^{\bullet} \qquad (180)$$

(The electron-hole annihilation serves as decay process.) Figure 73 refers to Au-doped silicon. If the rate constant is tuned by a bias one can switch from an insulating to a conductive behavior (cf. non-equilibrium phase transformation).

Figures 74 and 75 show oscillations, chaotic behavior and pattern formation observed in Ge under appropriate conditions.

Related phenomena are voltage oscillations at Ag/AgI contact observed in Ref.[275] periodic bands in precipitation reactions (Liesegang phenomenon)[177] or oscillations involving proton conductors.[276]

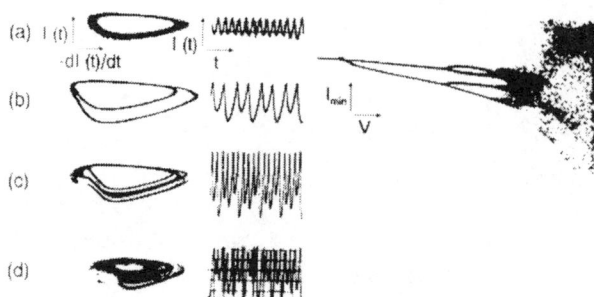

Figure 74. Periodic structures and chaotic behavior in Ge as a result of irradiation and voltage variation in the non-linear range (I: current, V: potential).[273] (Reprinted from S. W. Teitsworth, R. M. Westervelt, E. E. Haller, "Nonlinear Oscillations and Chaos in Electrical Breakdown in Ge." *Phys. Rev. Lett.*, **51**, 825-828. Copyright © 1983 with permission from the American Physical Society.)

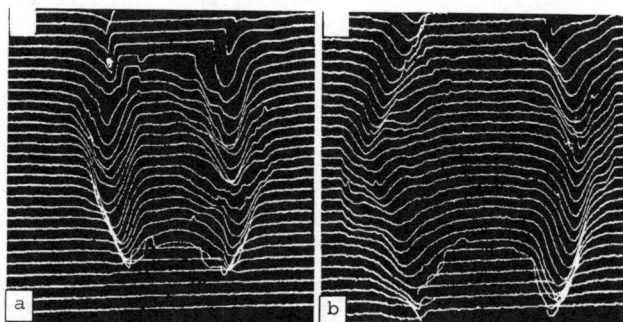

Figure 75. Current filaments in p-Ge at various voltages, made visible by the EBIC technique (electron beam induced currents). [274] (Reprinted from K. M. Mayer, R. Gross, J. Parisi, J. Peinke, R. P. Huebener, "Spatially Resolved Observation of Current Filament Dynamics in Semiconductors.", *Solid State Commun.*, **63**, 55-59. Copyright © 1987 with permission from Elsevier.)

The importance of the transport steps in the solid state has severe consequences: (i) Morphological questions are extremely relevant and can often be described by coupled diffusion reaction equations[183,277] (also the anisotropy of the surface energies is crucial here[278]). (ii) The "mean-field" character of the above reactions restricts the range of validity; more accurately inhomogeneities at the reaction centers[279] but also long range percolation phenomena have to be taken into account.[280]

A useful (also extreme) counterpart to the also idealized linear geometry is fractal geometry which plays a key role in many non-linear processes.[280,281] If one measures the length of a fractal interface with different scales, it can be seen that it increases with decreasing scale since more and more details are included. The number which counts how often the scale ε is to be applied to measure the fractal object, is not inversely proportional to ε but to a power law function of ε with the exponent d being characteristic for the self-similarity of the structure; d is called the Hausdorff-dimension. Diffusion limited aggregation is a process that typically leads to fractal structures.[283] That this is a non-linear process follows from the complete neglect of the back-reaction. The impedance of the tree-like metal in Fig. 76 synthesized by electrolysis does not only look like a fractal, it also shows the impedance behavior expected for a fractal electrode.[284]

Beyond that, fractal geometry is of direct relevance for the transport in inhomogeneous systems, since the percolation cluster (immediately at the percolation threshold, φ_c, c.f. Fig. 77) assumes a fractal shape. In two-phase mixtures of an insulating and a conducting phase this predicts a power law dependence of the conductivity on $\varphi - \varphi_c$ (φ: volume fraction).[281] A relevant example in the context of our considerations is the conduction behavior of $AgCl:\alpha\text{-}AgI$ composites.[118]

The present text could only address a few relevant topics within the rich field of solid state electrochemistry. It will have fulfilled its purpose if it not only convinces the reader of the diversity and colorfulness of the field, but also reveals to him or her the possibility of purposefully tuning electrochemical properties of solids.

Figure 76. Electrode tree (Wood´s metal). Its impedance corresponds to its fractal geometry.[282] (Reprinted from G. Daccord, R. Lenormand, "Fractal patterns from chemical dissolution.", *Nature*, **325,** 41-43. Copyright © 1987 with permission from Nature Publishing Group.)

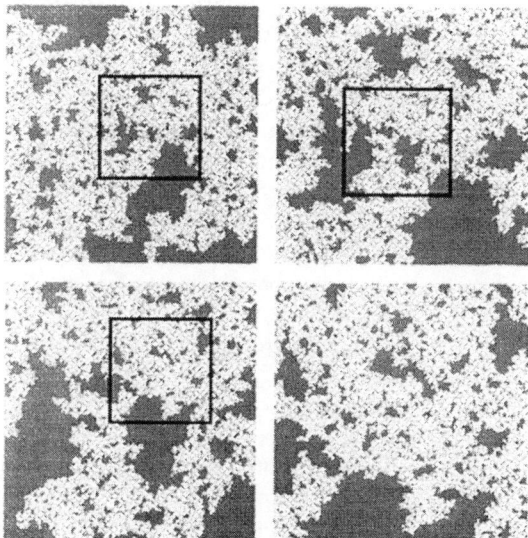

Figure 77. Selfsimilarity of a large percolation cluster at the critical concentration. The windows indicate the section which is enlarged in the respective succeeding figure.[281] (Reprinted from A. Bunde and S. Havlin, "Percolation I", in: *Fractals and Disordered Systems*. Ed. by A. Bunde and S. Havlin, Springer-Verlag, Berlin. Copyright © 1996 with permission from Springer-Verlag GmbH & Co. KG.)

ACKNOWLEDGMENT

The author is very much indebted to Dr. Janez Jamnik for helpful discussions and critically reading the manuscript.

SYMBOLS

C	doping content, capacitance
D	density of states, diffusion coefficient
E	electrical field
G	free enthalpy
ΔG^o	standard free energy of reaction
$\Delta G^{\neq}$	activation free enthalpy of reaction
H	Haven ratio, enthalpy
J	interaction term in the quasi-Madelung approach
K	mass action constant
L	cross coefficient, sample thickness
N	particle number
N_m	Avogadro's number
P	partial pressure
S	entropy
T	temperature
U	voltage
X	generalized force
a	activity, area
c	concentration (by volume)
f	activity coefficient, correlation factor
f^c	configurational activity coefficient
f^{nc}	non-configurational activity coefficient
g	free enthalpy per particle, nano-size factor, mismatch function
i	current density
j	flux density
k	rate constant
$\overline{k}$	effective rate coefficient
m	mass
n	mole number
$\overline{n}$	effective mole number of states
q	source term
t	transference number
u	mobility
v	stoichiometric number
x	mole fraction, position coordinate
z	charge number

Γ	jump frequency
Θ	degree of coverage
Λ	chemical permeability
Π	dissipation function
Σ	surface charge
α	symmetry factor
β	percolation parameter
ε	dielectric constant
ζ	$\equiv c / c_{\infty}$
ϑ	degree of influence
λ	Debye length
λ^*	Mott-Schottky length
μ	chemical potential
$\tilde{\mu}$	electrochemical potential
ξ	$\equiv x / \lambda$
σ	conductivity
σ_m	mean conductivity
τ	time constant
ϕ	electrical potential
φ	Madelung number, volume fraction
χ	differential defect fraction
ψ	wave function
$\vee$	vacancy
A	affinity
$\Re$	reaction rate
$\Re^{\circ}$	exchange rate

INDICES

$\sim$	electrochemical
$\frown$	equilibrium
$*$	tracer
$\perp$	perpendicular

‖	parallel
∞	bulk
→	forward
←	backward
Q	charge
i	interstitial site
δ	stoichiometry

REFERENCES

[1] J. Maier, in *Modern Aspects of Electrochemistry*, Ed. by B. Conway, C. G. Vayenas, R. E. White, Kluwer Academic/Plenum Press, to be published.

[2] F. A. Kröger, *Chemistry of Imperfect Crystals*, North Holland Publ. Comp., Amsterdam, 1964.

[3] H. Rickert, *Einführung in die Elektrochemie fester Stoffe*, Springer-Verlag, Berlin, 1973.

[4] H. Schmalzried, *Solid State Reactions*, Verlag Chemie, Weinheim, 1981.

[5] H. Schmalzried and A. Navrotsky, *Festkörperthermodynamik*, VCH, Weinheim, 1975.

[6] A. R. Allnatt, A. B. Lidiard, *Atomic Transport in Solids*, Cambridge University Press, Cambridge, 1993.

[7] A. B. Lidiard, in *Handbuch der Physik*, Ed. by S. Flügge, Springer-Verlag, Berlin, 1957, Vol. 20, pp. 246.

[8] P. Kofstad, *Nonstoichiometry, Diffusion and Electrical Conductivity in Binary Oxides*, John Wiley & Sons, New York, 1972.

[9] *Solid State Electrochemistry*, Ed. by P. Bruce, Cambridge University Press, Cambridge, 1994.

[10] T. Kudo and K. Fueki, *Solid State Ionics*, VCH, Tokyo-Kodansha, 1990.

[11] L. W. Barr, A. B. Lidiard in lit.[16]

[12] I. Riess, Solid State Ionics **157** (2000) 1.

[13] I. Riess, in *Encyclopedia of Electrochemistry*, Ed. by A. J. Bard and M. Stratmann, WILEY-VCH, Weinheim, Vol. 1 (E. Gileadi and M. Urbakh (eds.)) to appear.

[14] *CRC Handbook of Solid State Electrochemistry*, Ed. by P. J. Gellings and H. J. M. Bouwmeester, CRC Press, Boca Raton, 1997.

[15] J. Maier, *Solid State Phenomena* **39-40** (1994) 35.

[16] *Physical Chemistry, An Advanced Treatise*, Ed. by H. Eyring, D. Henderson, and W. Jost, Academic Press, New York, 1970, Vol. 10,

[17] J. Corish, and P. W. M. Jacobs, *Surf. Def. Prop. Solids* **2** (1973) 160; ibid. **6** (1976) 219.

[18] F. Aguillo-Lopez, C. R. A. Catlow and P. D. Townsend, *Point Defects in Materials*, Academic Press, New York, 1988.

[19] W. D. Kingery, H. K. Bowen, D. R. Uhlmann, *An Introduction to Ceramics*, John Wiley & Sons, New York, 1976.

[20] J. Maier and Angew. *Chem. Int. Ed. Engl.* **32** (3) (1993) 313.

[21] J. Maier and Angew. *Chem. Int. Ed. Engl.* **32** (4) (1993) 528.

[22] P. Knauth and H. L. Tuller, *J. Am. Ceram. Soc.* **85** (2002) 1654.

[23] J. Maier, *Physical Chemistry of Ionic Materials: Ions and Electrons in Solids*, John Wiley & Sons, Chichester, 2004.

[24] J. Maier, *Festkörper—Fehler und Funktion: Prinzipien der Physikalischen Festkörperchemie*, B. G. Teubner Verlag, Stuttgart, 2000.

[25] N. F. Mott, *Metal-Insulator Transitions*, Taylor & Francis, London, 1974.

[26] See textbooks of solid state physics and chemistry.

[27] G. Wulff and Z. Krist. **34** (1901) 449.

[28] R. Defay, I. Prigogine, A. Bellemans and H. Everett, *Surface Tension and Adsorption*, John Wiley & Sons, New York, 1960.

[29] F. A. Kröger and H. J. Vink, J. van den Boomgaard, *Z. phys. Chem.* **203** (1954) 1.

[30] W. Schottky, *Halbleiterprobleme I*, Ed. by W. Schottky, Vieweg, Braunschweig, 1954.

[31] R. Kirchheim, *Prog. Mater. Sci.* **32** (1988) 261.

[32] N. W. Ashcroft and N. D. Mermin, *Solid State Physics*, HRW Int. Ed., Philadelphia, 1976.

[33] H. Gerischer, in *Physical Chemistry, An Advanced Treatise*, Ed. by H. Eyring, Academic Press, New York, 1970.

[34] R. Memming, in *Comprehensive Treatise of Electrochemistry*, Ed. by B. E. Conway et al., Plenum Press, New York, 1983, Vol. 7, pp. 529.

[35] J. Maier, *Solid State Ionics* **143** (2001) 17.

[36] J. Maier, *Solid State Ionics* **157** (1-4) (2003) 327.

[37] J. Maier, *Z. Physik. Chem.* **217** (2003) 415.

[38] M. Born, *Z. Physik* **1** (1920) 45.

[39] W. Jost, *J. Chem. Phys.* **1** (1933) 466.

[40] N. F. Mott and M. J. Littleton, *Trans. Faraday Soc.* **34** (1938) 485.

[41] E. A. Guggenheim, *J. Phys. Chem.* **33** (1929) 842.

[42] G. Brouwer, *Philips Res. Repts.* **9** (1954) 366.

[43] C. G. Fonstad and R. H. Rediker, *J. Appl. Phys.* **42** (1971) 2911.

[44] J. Maier and W. Göpel, *J. Solid St. Chem.* **72** (1988) 293.

[45] J. Maier and Radiat. *Eff. Defects Solids* **158** (2003) 1.

[46] J. C. Philips, *Physics of High Temperature Superconductors*, Academic Press, New York, 1989.

[47] L. Heyne, M. Beekmans and A. de Beer, *J. Electrochem. Soc.* **119** (1972) 77.

[48] J. Maier and G. Schwitzgebel, *Mater. Res. Bull.* **18** (1983) 601.

[49] J. Teltow, *Ann. Phys.* **5** (1950) 63; *Z. phys. Chem.* **195** (1950) 213.

[50] J. Corish and P. W. M. Jacobs, *J. Phys. Chem. Solids* **33** (1972) 1799.

[51] C. G. Fonstad and R. H. Rêdiker, *J. Appl. Phys.* **42** (1971) 2911.

[52] J. G. Bednorz and K. A. Müller, *Z. Physik B* **64** (1986) 189.

[53] J. Maier and G. Pfundtner, *Adv. Mater.* **30** (1991) 292.

[54] G. M. Choi and H. L. Tuller, *J. Am. Ceram. Soc.* **71** (4) (1988) 201.

[55] S. Stotz and C. Wagner, *Ber. Bunsenges. Phys. Chem.* **70** (1966) 781.

[56] H. Iwahara, T. Esaka, H. Uchida and N. Maeda, *Solid State Ionics* **314** (1981) 259.

[57] R. Waser, *Ber. Bunsenges. Phys. Chem.* **90** (1980) 1223.

[58] Y. Larring and T. Norby, *Solid State Ionics* **77** (1995) 147.

[59] K. D. Kreuer, *Chem. Mater.* **8** (1996) 610.

[60] K.-D. Kreuer, E. Schönherr and J. Maier, *Solid State Ionics* **70/71** (1994) 278.

[61] W. Münch, K.-D. Kreuer, G. Seifert and J. Maier, *Solid State Ionics* **136-137** (2000) 183; M. Tuckermann, L. Laasonen, M. Sprik and M. Parrinello, *J. Phys.: Condens. Matter.* **6** (Supl. 23A) (1994) 99.

[62] A. S. Nowick, *Solid State Ionics* **77** (1995) 137.

166   Joachim Maier

[63]R. Waser, *J. Am. Ceram. Soc.* **74** (1991) 1934.
[64]D. M. Smyth, M. P. Harmer and P. Peng, *J. Am. Ceram. Soc.* **72** (1989) 2276.
[65]K. Sasaki and J. Maier, *J. Appl. Phys.* **86** (10) (1999) 5422.
[66]K. Sasaki and J. Maier, *J. Appl. Phys.* **86** (10) (1999) 5434.
[67]J. Maier, *Phys. Chem. Chem. Phys.* **5** (2003) 2164.
[68]I. Denk, W. Münch and J. Maier, *J. Am. Ceram. Soc.* **78**(12) (1995) 3265.
[69]N. Bjerrum, *Kgl. Danske Videnskab. Selskab. Mat. Fys. Medd.* **7** (1926) 3.
[70]P. Debye and E. Hückel, *Phys. Z.* **24** (1923) 185; 305.
[71]See textbooks on electrochemistry.
[72]J. B. Wachtman and W. C. Corvin, *J. Res. Nat. Bur. Std.* **69 A** (1965) 457.
[73]M. B. Armand, *Solid State Ionics* **9/10** (1983) 745; P. G. Bruce, J. Evans and C. A. Vincent, *Solid State Ionics* **28-30** (1988) 918.
[74]J. J. Markham, *F-Centres in Alkali Halides*, Academic Press, New York, 1966.
[75]S. M. Sze, *Physics of Semiconductor Devices*, Wiley-Interscience, New York, 1981.
[76]T. Bieger, J. Maier and R. Waser, *Ber. Bunsenges. Phys. Chem.* **97** (1993) 1098.
[77]J. Maier, *Chem. Eur. J.* **7** (22) (2001) 4762.
[78]W. Buckel, *Supraleitung*, VCH, Weinheim, 1990.
[79]J. O'M. Bockris and A. K. V. Reddy, *Modern Electrochemistry*, Plenum Press, New York, 1970.
[80]J. K. Aboague and R. S. Friauf, *Phys. Rev. B* **11**(4) (1975) 1654; R. S. Friauf, *J. Phys. (Paris)* **38** (1977) 1077, *J. Phys. Paris Colloq.* **41** (1980) C6-97.
[81]N. Hainovsky and J. Maier, *Phys. Rev. B* **51**(22) (1995) 15789.
[82]J. Maier and W. Münch, *Z. Anorg. Allg. Chem.* **626** (2000) 264.
[83]A. R. Allnatt and M. H. Cohen, *J. Chem. Phys.* **40** (1964) 1860, 1871; R. A. Sevenich and K. L. Kliewer, *J. Chem. Phys.* **48** (1964) 3045; A. R. Allnatt, E. Loftus and L. A. Rowley, *Crystal Lattice Defects* **3** (1972) 77.
[84]J. G. Ghosh, *J. Chem. Soc.* **113** (1918) 707.
[85]N. H. March and M. P. Tosi, *J. Phys. Chem. Solids* **46** (1985) 757.
[86]F. Zimmer, P. Ballone, J. Maier and M. Parrinello, *J. Chem. Phys.* **112**(14) (2000) 6416.
[87]F. Zimmer, P. Ballone, M. Parrinello and J. Maier, *Solid State Ionics* **127** (2000) 277.
[88]R. Merkle and J. Maier, *Phys. Chem. Chem. Phys.*, **5**(11) (2003) 2297.
[89]The fact that attractive interaction can lead to phase transition is a well-established result of the theory of phase transition and has been referred to in the case of superionic free situations by a variety of authors.[285]
[90]E. F. Schubert, *Doping in III-V-Semiconductors*, Cambridge University Press, Cambridge, 1993.
[91]See e.g., H. Schilling, *Festkörperphysik*, Verlag Harri Deutsch, Thun, 1977.
[92]A direct proof of Eq. (67) for the case of ionic crystals is given in Ref.[286]. Mass action laws are locally still valid, since then the $\Delta\phi$-term in $\Sigma \nu_j \tilde{\mu}_j$ disappears because of
$$\Sigma_j \nu_j z_j = 0 .$$
[93]In the literature sometimes the statement is made that the Poisson-Boltzmann equation is only compatible with electrostatics if linearized, which is not correct. The argument refers to the superposition principle which relies on the presupposed linearity of Poisson's equation. Note, however, that Poisson's equation is not linear if the charge density depends on ϕ itself in a non-linear way as it is the case here.
[94]J. Maier, *Prog. Solid St. Chem.* **23**(3) (1995) 171.
[95]G. Gouy, *J. Physique* **9** (1910) 457; D. L. Chapman, *Phil. Mag.* **25** (1913) 475.
[96]D. C. Grahame, *Chem. Rev.* **41** (1947) 441.

[97] S. M. Sze, *Semiconductor Devices*, John Wiley & Sons, New York, 1985.

[98] G. E. Pike, *Phys. Rev. B* **30** (1984) 795; G. E. Pike and C. H. Seager, *J. Appl. Phys.* **50** (1979) 3414.

[99] K. Bohnenkamp and H.-J. Engell, *Z. Elektrochem.* **61** (1957) 1184, VCH, Weinheim.

[100] I. Denk, J. Claus and J. Maier, *J. Electrochem. Soc.* **144**(10) (1997) 3526.

[101] J. Jamnik, *Appl. Phys.* **A55** (1992) 518.

[102] R. D. Armstrong and R. Mason, *J. Electroanal. Chem.* **41** (1973) 231.

[103] R. D. Armstrong and B. R. Horrocks, *Solid State Ionics* **94** (1997) 181.

[104] J. Jamnik, H.-U. Habermeier and J. Maier, *Physica B* **204** (1995) 57.

[105] S. Kim, J. Fleig and J. Maier, *Phys. Chem. Chem. Phys.* **5** (11) (2003) 2268.

[106] C. C. Liang, *J. Electrochem. Soc.* **120** (1973) 1298.

[107] K. Shahi, J. B. Wagner. *Appl. Phys. Lett.* **37** (1980) 757.

[108] J. Maier, *J. Electrochem. Soc.* **134** (1987) 1524.

[109] J. Maier, *J. Phys. Chem. Solids* **46** (1985) 309.

[110] Y. Saito, K. Hariharan and J. Maier, *Solid State Phenomena* **39-40** (1994) 235.

[111] A. Bunde, W. Dieterich and E. Roman, *Solid State Ionics* **18/19** (1986) 147; E. Roman, A. Bunde and W. Dieterich, *Phys. Rev. B* **34** (1986) 331; P Knauth, G. Albinet and J. M. Debierre, *Ber. Bunsenges. Phys. Chem.* **98**(7) (1998) 945; J. C. Wang, N. J. Dudney, *Solid State Ionics* **18/19** (1986) 112.

[112] J. Maier and B. Reichert, *Ber. Bunsenges. Phys. Chem.* **90** (1986) 666.

[113] J. Maier, *Mater. Chem. Phys.* **17** (1987) 485.

[114] U. Lauer and J. Maier, *J. Electrochem. Soc.* **139** (5) (1992) 1472.

[115] J. Maier, *Ber. Bunsenges. Phys. Chem.* **93** (1989) 1468; 1474.

[116] F. Croce, G. B. Appetecchi, L. Persi and B. Scrosati, *Nature* **394** (1998) 456; W. Wieczorek, Z. Florjanczyk and J. R. Stevens, *Electrochim. Acta* **40**(13-14) (1995) 2251. The effect can be even seen when the polymer is replaced by a liquid solvent ("soggy sand electrolytes") cf. Ref.[202].

[117] J. Maier, *Ber. Bunsenges. Phys. Chem.* **89** (1985) 355.

[118] U. Lauer and J. Maier, *Ber. Bunsenges. Phys. Chem.* **96** (1992) 111.

[119] N. Sata, K. Eberman, K. Eberl and J. Maier, *Nature* **408** (2000) 946.

[120] J. Maier, *Ber. Bunsenges. Phys. Chem.* **90** (1986) 26.

[121] J. Fleig and J. Maier, *Solid State Ionics* **86-88** (1996) 1351.

[122] A. S. Skapin, J. Jamnik and S. Pejovnik, *Solid State Ionics* **133** (2000) 129.

[123] Y. Saito and J. Maier, *J. Electrochem. Soc.* **142**(9) (1995) 3078.

[124] J. Maier, U. Lauer, *Ber. Bunsenges. Phys. Chem.* **94** (1990) 973.

[125] M. Holzinger, J. Fleig, J. Maier and W. Sitte, *Ber. Bunsenges. Phys. Chem.* **99**(11) (1995) 1427.

[126] *Sensors, A Comprehensive Study*, Ed. by W. Göpel, J. Hesse, J. N. Zemel, VCH, Weinheim, 1987.

[127] G. Simkovich and C. Wagner, *J. Catal.* **1** (1967) 340.

[128] J. Maier and P. Murugaraj, *Solid State Ionics* **40/41** (1990) 1017.

[129] F. Besenbacher, I. Chorkendorff, B. S. Clausen, B. Hammer, A. Molenbrock, J. K. Norskøv and I. Stensgaard, *Science* **279** (1998) 1913.

[130] G. C. Vayenas, S. Bebelis and S. Neophytides, *J. Phys. Chem.* **92** (1988) 5085.

[131] R. Hagenbeck and R. Waser, *Acta Mater.* **48** (2000) 797.

[132] Y.-M. Chiang and T. Tagaki, *J. Am. Ceram. Soc.* **73** (1990) 3278.

[133] R. Moos and K. H. Härdtl, *J. Appl. Phys.* **80** (1996) 393.

[134] X. Guo, J. Fleig and J. Maier, *J. Electrochem. Soc.* **148**(9) (2001) J50.

[135] X. Guo, W. Sigle and J. Maier, *J. Am Ceram. Soc.* **86**(1) (2003) 77.

[136] S. Kim and J. Maier, *J. Electrochem. Soc.* **149**(10) (2002) J73.

[137]N. M. Beekmans and L. Heyne, *Electrochim. Acta* **21** (1976) 303.

[138]S. H. Chu and M. A. Seitz, *J. Solid State Chem.* **23** (1978) 297.

[139]D. Bingham, P. W. Tasker and A. N. Cormack, *Philos. Mag. A* **60** (1989) 1.

[140]X. Guo, *Solid State Ionics* **96** (1997) 247.

[141]M. J. Verkerk, B. J. Middelhuis and A. J. Burggraaf, *Solid State Ionics* **6** (1982) 159.

[142]M. Leonhardt, J. Jamnik and J. Maier, *Electrochemical and Solid-State Letters*, **2** (1999) 333.

[143]J. Jamnik, X. Guo, J. Maier, *Appl. Phys. Lett.* **82**(17) (2003) 282.

[144]R. Hagenbeck and R. Waser, *J. Appl. Phys.* **83** (1998) 2083.

[145]J. Jamnik, J. Maier, S. Pejovnik, *Solid State Ionics* **75** (1995) 51.

[146]Structural changes, i.e., changes in the bond lengths can be of longer range in the case of hard materials and in the case of pronounced misfit, in particular for epitaxial films. Effects of longer range are also expected for polar surfaces.[147] Higher dimensional defects can largely contribute to the absorption of "mechanical" stress.

[147]E. Heifets, E. A. Kotomin, *J. Maier, Surf. Sci.* **462** (2000) 19; E. Heifets, R. I. Eglitis, E. A. Kotomin, J. Maier and G. Borstel, *Phys. Rev. B* **64** (2001) 23417.

[148]J. Maier, *Solid State Ionics* **23** (1987) 59.

[149]J. Maier, *Phys. Stat. Sol.* (a) **112** (1989) 115.

[150]J. Maier, *Solid State Ionics* **148** (2002) 367.

[151]J. Maier, *Solid State Ionics* **154-155** (2002) 291.

[152]J. Maier, *Proc. Electrochem. Soc.*, in press.

[153]Cubic clusters do usually not represent the equilibrium situation even for large crystals, nor is it represented by spheres. Owing to different free energies of crystallographically different interfaces or edges (more exactly interface and edge tensions γ_s, χ_t) the minimization of the total thermodynamic potential requires:[287]

$$\frac{1}{h_s}\left(\gamma_s + \Sigma_t \frac{\partial L_t}{\partial a_s}\chi_t\right) = \text{const.}$$

If we ignore edge tensions, the well-known Wulff-shape results for constant interfacial tensions.[27] This presumed constancy and the neglect of edges limits the validity of the Wulff-concept to small dimensions, while, for large crystals, a limit to its realization is given by the high kinetic demands. So it is not surprising that the Wulff-shape is not often found (not to mention the fact that contacts by condensed phases additionally change the thermodynamics).[288] In a polycrystal global morphological equilibrium is not established anyway, and we can only rely on local equilibration effects. If the anisotropy of the interfacial tensions is negligible, the equilibrium shape of a single crystal is a sphere, while in a "mean-field" ceramic the grain-shape is close to tetrakaidecahedron.[19] Its shape fulfills best the space requirements and the conditions of local equilibrium (e.g., contact angle of 120° at a three-grain contact). Since obviously more specific situations do not very significantly change the picture or lead to very individual corrections we go back to the simple bricklayer model and to Eq. (90). As far as the overall conductance effect is concerned, the bricklayer model also works surprisingly well for more complicated microstructures. For a consideration of its limits cf. Ref.[289,290].

[154]W. Puin, S. Rodewald, R. Ramlau, P. Heitjans and J. Maier, *Solid State Ionics* **131**(1, 2) (2000) 159.

[155]N. Sata, N. Y. Jin-Phillipp, K. Eberl and J. Maier, *Solid State Ionics* **154-155** (2002) 497.

Solid State Electrochemistry I   169

Solid State Electrochemistry I   169

—

[184]G. H. Weiss, in *Fractals in Science*, Ed. by A. Bunde and S. Havlin, Springer-Verlag, Berlin, 1994; G. Daccord, R. Lenormand, *Nature* **325** (1987) 41.

[185]A. D. Le Claire and A. B. Lidiard, *Phil. Mag.* **1** (1956) 1; A. D. LeClaire, *Brit. J. Appl. Phys.* **14** (1963) 351.

[186]P. Müller, *phys. stat. sol.* **12** (1965) 775.

[187]L. W. Stock, *Z. phys. Chem.* **B25** (1934) 441; R. Cava, F. Reidinger, B. J. Wuensch, *Solid State Commun.* **24** (1977) 411.

[188]B. Ma, J. P. Hodges, J. D. Jorgensen, D. J. Miller, J. W. Richardson Jr and U. Balachandran, *J. Solid State Chem.* **141** (1998) 576.

[189]Data complied by K. Sasaki after: T. Kudo and K. Fueki, *Solid State Ionics*, Kodansha, 1990, and references therein; A. Rabenau, *Solid State Ionics* **6** (1982) 277; K. D. Kreuer, *Chem. Mater.* **8** (1996) 610 and references therein; K. D. Kreuer, Th. Dippel, J. Maier, in Proc. Electrochem. Soc., Pennington (NJ), 1995, Vol. PV 95-23, pp. 241; K. Schmidt-Rohr, J. Clauss, B. Blümich and H. W. Spiess, *Magn. Reson. in Chem.* **28** (1990) 3; K. D. Kreuer, *Solid State Ionics* **97** (1997) 1; H. Iwahara, T. Esaka, H. Uchida and N. Maeda, *Solid State Ionics* **3/4** (1981) 539; B. C. H. Steele, *Solid State Ionics* **75** (1995) 157; T. Takahashi, H. Iwahara and T. Esaka, *J. Electrochem. Soc.* **124** (1977) 1563; T. Ishihara, H. Furutani, H. Nishiguchi and Y. Takita, in *Ionic and Mixed Conducting Ceramics III*, Ed. by T.A. Ramanarayanan, W.L. Worrell, H.L. Tuller, M. Mogensen, and A.C. Khandkar, The Electrochemical Society, Pennington (NJ), 1994, Vol. PV 94-22, pp. 834; T. Ishihara, H. Matsuda and Y. Takita, *J. Am. Ceram. Soc.* **116** (1994) 3801; J. B. Goodenough, J. E. Ruiz-Diaz and Y. S. Zhen, *Solid State Ionics* **44** (1990) 21; I. Kontoulis, Ch. P. Ftikos and B. C. H. Steele, *Mater. Sci. Eng.* **B22** (1994) 313; H. L. Tuller, *Solid State Ionics* **94** (1997) 63; J. T. Kummer, *Prog. Solid St. Chem.* **7** (1972) 141; G. Farrington, B. Dunn, *Solid State Ionics* **7** (1982) 287.

[190]S. Chandra, *Superionic Solids*, North-Holland, Amsterdam, 1981; *Superionic Solids and Solid Electrolytes*, Ed. by S. Chandra and A. S. Laskar, Academic Press, New York, 1989.

[191]O. Yamamoto, Y. Takedo and R. Kanno, *Kagaku* **38** (1983) 387, Kagaku-Dojin Publishing Company, Kyoto.

[192]K. D. Kreuer, W. Münch, U. Traub and J. Maier, *Ber. Bunsenges. Phys. Chem.* **102** (1998) 552.

[193]C. A. Vincent, *Chemistry in Britain*, April, 1981, pp. 391.

[194]B. Scrosati, in *Lithium Ion Batteries*, Ed. by M. Wakihara and O. Yamamoto, VCH, Weinheim, 1998.

[195]H. J. van Daal, *Solid State Commun.* **6** (1968) 5, Elsevier, Amsterdam.

[196]R. J. Cava, R. B. van Dover, B. Batlogg and E. A. Rietman, *Phys. Rev. Lett.* **58** (1987) 408, The American Physical Society, College Park (MD).

[197]J. C. Philips, *Physics of High Temperature Superconductors*, Academic Press, New York, 1989.

[198]C. A. Beevers and M. A. R. Ross, *Z. Krist.* **95** (1937) 59; J. T. Kummer, *Prog. Solid St. Chem.* **7** (1972) 141.

[199]K.-D. Kreuer, A. Fuchs, M. Ise, M. Spaeth and J. Maier, *Electrochim. Acta* **43**(10-11) (1998) 1281.

[200]P. Maass, M. Meyer, A. Bunde and W. Dieterich, *Phys. Rev. Lett.* **77** (1996) 1528; P. Pendzig, W. Dieterich and A. Nitzan, *J. Non-Cryst. Solids* **235-237** (1998) 748.

[201]M. B. Armand, *Ann. Rev. Mat. Sci.* **6** (1986) 245; M. A. Ratner and D. F. Shriver, *Mater. Res. Bull.* **14** (1989) 39.

[202]A. Bhattacharyya and J. Maier, *Advanced Materials*, in press.

[203]O. Madelung, *Introduction to Solid State Theory*, Springer-Verlag, Berlin, 1978.

[204]R. E. Peierls, *Quantum Theory of Solids*, Clarendon Press, Oxford, 1955.

[205]J. R. Manning, *Diffusion Kinetics for Atoms in Crystals*, Van Nostrand, Princeton, 1968; A. D. Le Claire and A. B. Lidiard, *Phil. Mag.* **1** (1956) 1.

[206]R. Dieckmann and H. Schmalzried, *Ber. Bunsenges. Phys. Chem.* **81** (1977) 344.

[207]R. A. De Souza and J. A. Kilner, *Solid State Ionics* **106** (1998) 175.

[208]J. Maier, in preparation.

[209]Since we only allow for small driving forces $\Delta\mu_0$, this implies large chemical capacitances $\partial c_0/\partial\mu_0$; a relevant example is the high temperature superconductor $YBa_2Cu_3O_{7-\delta}$.

[210]R. Metselaar and P. K. Larsen, *J. Phys. Chem. Solids* **37** (1976) 599.

[211]K. Sasaki and J. Maier, *Phys. Chem. Chem. Phys.* **2** (2000) 3055; K. Sasaki and J. Maier, *Solid State Ionics* **134** (2000) 303.

[212]I. Denk, U. Traub, F. Noll and J. Maier, *Ber. Bunsenges. Phys. Chem.* **99** (1995) 798; M. Leonhardt, Ph D thesis, University of Stuttgart, 1999.

[213]H. S. Carslaw and J. C. Jäger, *Conduction of Heat in Solids*, Clarendon Press, Oxford, 1959.

[214]I. Denk, Ph D thesis, University of Stuttgart, 1995; I. Denk, F. Noll and J. Maier, *J. Am. Ceram. Soc.* **80** (2) (1997) 279.

[215]G. Pfundtner, Ph D thesis, University of Tübingen, 1993.

[216]M. Spaeth, K. D. Kreuer, C. Cramer and J. Maier, *J. Solid St. Chem.* **148** (1999) 169.

[217]K. Funke and I. Riess, *Z. phys. Chem. Neue Folge* **140** (1984) 217; K. Funke, *Solid State Ionics* **94** (1997) 27.

[218]K. Funke, Prog. *Solid St. Chem.* **22** (1993) 11.

[219]K. Funke, R. D. Banhatti, S. Brückner, C. Cramer, C. Krieger, A. Mandanici, C. Martiny and I. Ross, *Phys. Chem. Chem. Phys.* **4** (14) (2002) 3155.

[220]W. Dieterich and P. Maass, *Chem. Phys.* **284** (2002) 439.

[221]J. A. Bruce and M. D. Ingram, *Solid State Ionics* **9/10** (1983) 717.

[222]G. V. Chandrashekar and L. M. Foster, *Solid State Commun.* **27** (1978) 269.

[223]P. K. Davies, G. I. Pfeiffer and S. Canfield, *Solid State Ionics* **18/19** (1996) 704.

[224]M. Tatsumisago, Y. Akamatsu, T. Minami, *Mater. Sci. Forum* **32-33** (1988) 617.

[225]H. Hrugchka, E. E. Lissel and M. Jansen, *Solid State Ionics* **28-30** (1988) 159

[226]J. A. Bruce, R. A. Howic and M. D. Ingram, *Solid State Ionics* **18/19** (1986) 1129, Elsevier, Amsterdam.

[227]M. Meyer, V. Jaenisch, P. Maass and A. Bunde, *Phys. Rev. Lett.* **76** (1996) 2338.

[228]A. Bunde, *Solid State Ionics* **105** (1981) 1; A. Bunde, M. D. Ingram, P. Maass, *J. Non-Cryst. Solids* **172/174** (1994) 1222.

[229]S. Rodewald and J. Maier, unpublished.

[230]M. Martin and H. Schmalzried, *Solid State Ionics* **20** (1986) 75; H. I. Yoo, H. Schmalzried, M. Martin and J. Janek, *Z. phys. Chem. Neue Folge* **168** (1990) 129.

[231]J. Maier, *J. Am. Ceram. Soc.* **76** (1993) 1223.

[232]J. Maier, G. Schwitzgebel, *phys. stat. sol. (b)* **113** (1982) 535.

[233]This representation of "Conservative Ensembles" was used by M. H. R. Lankhorst, H. J. M. Bouwmeester and H. Verweij, in *Electroceramics IV*, Ed. by R. Waser, S. Hoffmann, D. Bonnenberg, Ch. Hoffmann, Augustinus Buchhandlung, Aachen, 1994, Vol. II, pp. 697.

[234]J. Claus, I. Denk, M. Leonhardt and J. Maier, *Ber. Bunsenges. Phys. Chem.* **101**(9) (1997) 1386.

[235]M. Widere, W. Münch, Y. Larring and T. Norby, *Solid State Ionics* **154-155** (2002) 669.

[236] J. Claus, I. Denk, M. Leonhardt and J. Maier, *Ber. Bunsenges. Phys. Chem.* **101** (1997) 1386.

[237] R. I. Merino, N. Nicoloso, J. Maier, in *Ceramic Oxygen Ion Conductors and Their Technological Applications*, Ed. B. C. H. Steele, British Ceram. Proc., The Institute of Materials, Cambridge, 1996, pp. 43; K. Sasaki and J. Maier, *Solid State Ionics* **134** (2000) 303.

[238] J. Maier, in *Ionic and Mixed Conducting Ceramics*, Ed. by T. A. Ramanarayanan, W. L. Worrell and H. L. Tuller, The Electrochemical Society, Pennington (NJ), 1994, Vol. PV 94-12, pp. 542; J. Maier and W. Münch, *J. Chem. Soc. Faraday Trans.* **92**(12) (1996) 2143.

[239] J. Jamnik, B. Kamp, R. Merkle and J. Maier, *Solid State Ionics*, **150** (2002) 157.

[240] H. Schmalzried and J. Janek, *Ber. Bunsenges. Phys. Chem.* **102** (1998) 127.

[241] R. T. Whipple, *Phil. Mag.* **45** (1954) 1225; A. D. LeClaire, *Brit. J. Appl. Phys.* **14** (1963) 351; Y.-Ch. Chung and B. J. Wuensch, *J. Appl. Phys.* **79** (1996) 8323; O. Preis and W. Sitte, *J. Appl. Phys.* **79** (1996) 2986.

[242] J. Daniels, K. H. Hardtl, D. Hennings and R. Wernicke, *Philips Res. Repts.* **31** (1978) 489.

[243] R. Meyer and R. Waser, *J. Eur. Ceram. Soc.* **21**, (2001)1743.

[244] J. Jamnik and J. Maier, *Ber. Bunsenges. Phys. Chem.* **101**(1) (1997) 23; J. Jamnik and J. Maier, *J. Phys. Chem. Solids* **59**(9) (1998) 1555.

[245] J. Jamnik, in *Solid State Ionics: Science & Technology*, Ed. by B. V. R. Chowdari, K. Lal, S. A. Agnihotry, N. Khare, S. S. Sekhon, P. C. Srivastava and S. Chandra, World Scientific Publishing Co., Singapore, 1998, pp. 13.

[246] J. D. Eshelby, C. W. A. Newey, P. L. Pratt and A. B. Lidiard, *Phil. Mag.* **3**(25) (1958) 75; A. Atkinson, *Adv. Ceram.* **23** (1987) 3; *Solid State Ionics* **28-30** (1988) 1377.

[247] J. Jamnik, J. Fleig, M. Leonhardt and J. Maier, *J. Electrochem. Soc.* **147** (8) (2000) 3029.

[248] J. Maier, *Solid State Ionics* **135**(1-4) (2000) 575.

[249] J. Maier, in *Oxygen Ion and Mixed Conductors and their Technological Applications*, Ed. by H. L. Tuller, J. Schoonman and I. Riess, NATO Science Series: E Applied Sciences, Kluwer Academic Publishers, Dordrecht, Vol. 368, (2000) pp. 75.

[250] M. Leonhardt, Ph D thesis, Stuttgart, 1999.

[251] T. Bieger, J. Maier and R. Waser, *Ber. Bunsenges. Phys. Chem.* **97** (1993) 1098.

[252] J. Maier, *Solid State Ionics* **112** (1998) 197.

[253] J. Maier, *Mat. Res. Soc. Proc.* **548** (1999) 415.

[254] M. Leonhardt, R. A. De Souza, J. Claus and J. Maier, *J. Electrochem. Soc.* **149**(2) (2002) J19; R. A. De Souza, J. Fleig, R. Merkle, and J. Maier, *Z. Metallkd.* **94**(3) (2003) 218.

[255] R. Merkle and J. Maier, *Phys. Chem. Chem. Phys.* **4**(17) (2002) 4140.

[256] D. Y. Wang and A. S. Nowick, *J. Electrochem. Soc.* **126** (1979) 1155; D.Y. Wang and A. S. Nowick, *J. Electrochem. Soc.* **126** (1979) 1166.

[257] V. Brichzin, J. Fleig, H.-U. Habermeier and J. Maier, *Electrochemical and Solid-State Letters* **3** (9) (2000) 403.

[258] R. P. Buck and C. Mundt, *Electrochim. Acta* **44** (1999) 1999.

[259] D. R. Franceschetti, *Solid State Ionics* **70/71** (1994) 542.

[260] G. C. Barker, *J. Electroanal. Chem.* **41** (1973) 201; T. R. Brumleve and R. P. Buck, *J. Electroanal. Chem.* **126** (1981) 73.

[261] J. R. Macdonald, *J. Chem. Phys.* **58** (1973) 4982.

[262] J. R. Macdonald and D. R. Franceschetti, *J. Chem. Phys.* **68** (1978) 1614.

[263] J. Jamnik and J. Maier, *Phys. Chem. Chem. Phys.* **3** (2001) 1668; J. Jamnik, *Solid State Ionics* **157** (2003) 19.

[264] J. Maier, *Z. Physik. Chemie N. F.* **140** (1984) 191.

[265] C. Wagner, *J. Electrochem. Soc.* **103** (1956) 571.

[266] C. Gensch and K. Hauffe, *Z. phys. Chem.* **196** (1951) 427.

[267] W. Ebeling, *Strukturbildung bei irreversiblen Prozessen*, B. G. Teubner, Leipzig, 1976.

[268] J. Meixner, *Z. Naturforschung* **4A** (1949) 594; G. Nicolis and I. Prigogine, *Self-Organisation in Non-Equilibrium Systems*, John Wiley & Sons, New York, 1971.

[269] M. Eigen, P. Schuster, *The Hypercycle*, Springer-Verlag, Heidelberg, 1979.

[270] M. Eigen, *Naturwiss.* **58** (1971) 465.

[271] E. Schöll, *Phase Transitions in Semiconductors*, Springer-Verlag, Berlin, 1987.

[272] S. H. Koenig, R. D. Brown and W. Schillinger, *Phys. Rev.* **128** (1962) 1668; M. E. Cohen and P. T. Landsberg, *Phys. Rev.* **154** (1967) 683; A. E. McCombs and A. G. Milnes, *Int. J. Electron.* **32** (1972) 361.

[273] S. W. Teitsworth, R. M. Westervelt and E. E. Haller, *Phys. Rev. Lett.* **51** (1983) 825.

[274] K. M. Mayer, R. Gross, J. Parisi, J. Peinke and R. P. Huebener, *Solid State Commun.* **63** (1987) 55.

[275] J. Janek and S. Majoni, *Ber. Bunsenges. Phys. Chem.* **99** (1995) 14.

[276] E. D. Wachsman, *J. Electrochem. Soc.* **149** (2002) A242.

[277] H. Meinhard and A. Gierer, *J. Cell. Sci.* **15** (1974) 312; H. Haken, *Synergetik*, Springer-Verlag, Berlin, 1990.

[278] T. Ihle and H. Müller-Krumbhaar, *Phys. Rev. E* **49** (1994) 2972.

[279] E. Kotomin and V. Kuzovkov, *Modern aspects of diffusion controlled reactions*, Elsevier, Amsterdam, 1996.

[280] B. Mandelbrot, *The Fractal Geometry of Nature*, Freeman, San Francisco, 1982.

[281] A. Bunde and S. Havlin, *Fractals and Disordered Systems*, Springer-Verlag, Berlin, 1996.

[282] G. Daccord and L. Lenormand, *Nature* **325** (1987) 41.

[283] J. Nittmann and H. E. Stanley, *Nature* **321** (1986) 663.

[284] E. Chassaing and B. Sapoval, *J. Electrochem. Soc.* **141** (1994) 2711; G. Daccord and R. Lenormand, *Nature* **325** (1987) 41.

[285] R. A. Huberman, *Phys. Rev. Lett.* **32** (1974) 1000; A. Bunde, *Z. Physik B* **36** (1980) 251; H. Schmalzried, *Z. phys. Chem. NF* **22** (1959) 199 and others.

[286] K. L. Kliewer and J. S. Koehler, *Phys. Rev. A* **140** (1965) 1226.

[287] A. I. Rusanov, *Phasengleichgewichte und Grenzflächenerscheinungen*, Akademie-Verlag, Berlin, 1978.

[288] R. Kaischew, *Bull. Acad. Bulg. Sci. Phys.* **1** (1950) 100; **2** (1951) 191.

[289] J. Fleig and J. Maier, *J. Electrochem. Soc.* **145** (6) (1998) 2081.

[290] T. Mason, *Mat. Res. Soc. Proc.*, in press.

[291] I. Kosacki, B. Gorman and H. U. Anderson, in *Ionic and Mixed Conductors*, Ed. by T. A. Ramanarayanan, W.L. Worrell, H. L. Tuller, M. Mogensen, and A. C. Khandkar, The Electrochemical Society, Pennington, NJ, 1998, pp. 631; I. Kosacki, talk given at EURODIM 2002, Wrocaw, Poland.

2

Electron Transfer Reactions With and Without Ion Transfer

A. J. Appleby

238 Wisenbaker Engineering Research Center
Texas A&M University, College Station, TX 77843-3402

I. INTRODUCTION

The most frequent molecular model used for electrolytic electron transfer since the 1930s is similar to the Franck-Condon (FC) principle[1] for spectral electronic transitions.   Kinetic activation occurs until a rapid radiationless electronic transition becomes possible.   Following FC and the equivalent Born-Oppenheimer approximation,[2] it is assumed that classical nuclear motion during the electronic transition is slow enough to be negligible.   In condensed media, the potential energy of a reactant involves an extended number of nuclei and many degrees of freedom.   To reconcile the energy requirements in condensed media with electron transfer has resulted in many ingenious mechanistic proposals.   Charged molecules surrounded by rather tightly-bound solvent dipoles have potential energies different from vacuum values because of the presence of the surrounding dielectric solvent.   These tightly-bound "Inner Sphere" solvent dipoles may or may not be free to move before rapid electron transfer of FC type.

Modern Aspects of Electrochemistry, Number 38, edited by B. E. Conway *et al.* Kluwer Academic/Plenum Publishers, New York, 2005.

Since the 1950s, the surrounding dielectric solvent has usually been considered to determine kinetics because of its assumed immobility during rapid electron transfer. Efforts to summarize reaction rate evidence in the 1970s[3,4] and more recently[5,6] have failed to leave the theory in a satisfactory state.

This chapter reviews electron transfer models, develops molecular models of the solvent surrounding ions of different types to provide a more complete picture of the orientational changes taking place, especially in cases where electron transfer is combined with atom transfer, where assembly of a solvation sphere is required during the process. Finally, the activation energies of some charge-transfer processes are given.

II. THE FRANCK-CONDON PRINCIPLE AND ELECTRON TRANSFER

1. Historical Development

The electrochemical FC principle can be traced to 1931 papers by Franck and Haber on photochemical electron transfer[7] and by Gurney on electrochemical hydrogen evolution.[8] Landau's publication on gaseous electron transfer followed in 1932.[9] The electron makes a single, rapid transition from a donor to an acceptor state in which all heavy particle motions are frozen in time. The electronic transition may be adiabatic, i.e., with a transition probability of unity with no tunneling through an energy barrier, or non-adiabatic, i.e., with tunneling. Transitions may take place with or without radiation or photon absorption. Gurney's model[8] for proton dissociation from the H_3O^+ ion[10] with simultaneous electron acceptance from a metal energy level introduced the concept, accounting for the overpotential[11], the Tafel equation[12] and its Butler-Volmer extension,[13,14] then recently confirmed by Bowden.[15] The H_3O^+ ion[10] was formalized as the oxonium or hydronium ion by Bernal and Fowler.[16]

Gurney[8] used Hund's molecular orbital model[10] for H_3O^+ to deduce that its dissociation energy was 8.3 eV, close to the 7.9 eV value estimated using a thermochemical cycle.[17] A thermally activated H_3O^+ ion[10] with one bond randomly stretched to a higher potential energy state was postulated to accept an electron from a metal electrode level and spontaneously "fly apart" (sic) to $H_2O + H$, since the force between a neutral hydrogen atom and a water molecule would always be repulsive. The final state products would be in the same nuclear FC configuration as that of the activated initial state. For radiationless electron transfer, the potential energy difference between the initial and final states must be equal to that of the electron in the metal level. The probability of a given transition is given by the product of the (inverted) Fermi-Dirac distribution for electrons in the electrode multiplied by both the Boltzmann distribution for activated H_3O^+ ions and the electron tunneling probability through any energy barrier which may be present.[18,19] Gurney linked the increased potential energy of the activated state of H_3O^+ (ΔU) to the energy difference between this excited state and that of the products in the same configuration (ΔE) by supposing that the potential energy-distance terms for the initial and final states were broadly linear, with similar positive and negative slopes for reactant bond-stretching and for product repulsion. Then $\Delta U \approx 0.5(\Delta E - E_o)$, where E_o is equal to the potential energy difference between the ground states of the reactants and products in the same nuclear configuration, giving an expression containing only the metal electron energy level. Integration over all metal energy levels gives a rate $\propto \exp -0.5E/RT$, where E is the reactant ground state energy. Gurney thereby accounted for what was later called the symmetry factor, approximately 0.5. The activation energy is therefore represented by the energy to reach the crossing point between the energy-distance curves for reactants and the products, after the difference between their absolute energies is removed by subtracting the potential energy of reaction. Bowden's earlier theory[15] supposed that adsorbed dipoles of a certain critical energy carrying a partial charge were laid down on the electrode surface. The Boltzmann distribution then gave an exponential expression between rate and voltage or overpotential, with a fractional term because of the partial charge. The first excited level for the proton vibration lies at +16.2kT[17] above the +24.3kT ground state for the symmetrical three-dimensional oscillator at 298 K, so in the absence of other interactions, $H_2O–H^+$ bond-stretching is very improbable. The ground state of the H_3O^+ ion must therefore carried on a classical thermally-excited translational-vibrational subsystem of neighboring water molecules, so the classical energy of the entire solvated H_3O^+ ion must be the excited state.

Later, Gurney pointed out the irreversibility of his model, since the transfer of electrons from hydrogen atoms to metal levels is improbable.[20] He also considered metal deposition[21] using Morse-function energy-distance diagrams for the reactants and products, so that the reaction energy pathway may be represented by "joining together the curves" on their repulsive sides, giving a low barrier between reactants and products. This is analogous to the early Eyring and Polanyi reaction energy barriers derived from Morse functions,[22] which were followed by the 1935 Eyring absolute rate expression,[23] following Herzfeld.[24] In 1935, Horiuti and Polanyi considered the activation energy barrier for adsorbed atomic hydrogen to be the crossing point of the reactant and product energy states.[25] Butler[26] extended Gurney's 1931 theory to adsorbed hydrogen, implying that the electron makes an instantaneous FC transition, i.e., no time is allowed to form a true transition state.

Bates and Massey[27] extended Landau's work on gaseous systems in 1943. In 1952, Platzmann and Franck[28] applied similar concepts to the spectra of halide ions in solution. Randles[29] and Libby[30] respectively applied the FC principle to electrochemical reactions and homogeneous electron transfer between isotopic ions. Libby used the hydrogen molecule ion as a model, because its vacuum energy levels were known. He stressed that the FC principle would apply, since the velocity ratio for water molecule and electron motion would approximate the square root of their mass ratio, i.e., 200. Libby also introduced a "catalysis" concept for FC electron transfer by "complexing the exchanging ions in such a way that the complexes are symmetrical providing their geometries are identical to within the vibration amplitude involved in zero point motion." In other words, electron transfer is more likely if the initial and final state structures are identical. He considered that the electron wave function could only effectively penetrate the coordination shell of the ion if two reacting ions were bridged by an ion of opposite charge on closest approach. Assuming that the classical Born changing equation[31] could be applied to the ion energy, then the difference in free energy between the z and z + 1 states would be:

$$\Delta u = -[e^2 z^2 / 2r\varepsilon_0 - e^2 (z + 1)^2 / 2r\varepsilon_0] = e^2 (2z + 1) / 2r\varepsilon_0 \qquad (1)$$

where e is the electronic charge, ε_0 is the static dielectric constant of the medium and r is the ion radius in both the initial and final states, following the FC principle. This expression would lead to a small energy barrier because of the high value of ε_0 for water, so he postulated a much smaller ε_0 value close to an ion. He also suggested that the effective dielectric constant to be used for high-frequency electron motion would be the optical infrared value, i.e., the square of the IR refractive index n, or about 1.8.

2. Inner and Outer Sphere Concepts

In 1954 Weiss[32] used Bernal and Fowler's simplified solvation model,[16] with an "Inner Sphere" of ionic coordination, i.e., a small spherical double layer around the ion of charge ze, followed by a sharp discontinuity at radius r_i, the edge of the "Outer Sphere" or "Dielectric Continuum." He used a simple electrostatic argument to determine the energy to remove an electron at optical frequency from the Inner Sphere:

$$\Delta u_i = cze\mu/\epsilon_{opt}r_i^2 \tag{2}$$

where c is the coordination number of inner sphere water dipoles of dipole moment μ, and ϵ_{opt} is an optical frequency dielectric constant for the inner sphere, if this is physically meaningful. He used the Born charging equation was to estimate the energy change on both low frequency and optical frequency transfer of charge ze from the inner sphere to a vacuum at infinity:

$$\Delta u = -(z^2e^2/2r_i)(1 - 1/\epsilon_o) \tag{3}$$

$$\Delta u_o = -(z^2e^2/2r_i)(1 - 1/n^2) \tag{4}$$

The difference between the two expressions would be the residual energy left in the dielectric continuum on transferring the charge at optical frequency under FC conditions, i.e., permanent dipoles remain in their original positions during charge transfer. Weiss noted that the expressions only apply in the region of bulk dielectric constant, beyond the "Debye sphere" immediately surrounding the solvated ion.[33] Debye estimated its radius as 11, 31, 57, and 88 Å for mono-, di-, tri- and tetravalent ions respectively.[16,33] As is discussed in Section III, the distances beyond which water reaches its bulk dielectric properties are in fact much less than these.

Weiss considered that the total energy on removing an electron from the central ion (of initial charge z + 1) at optical frequency to infinity should be on the order of:

$$\Delta u_a = -[cze\mu/\epsilon_{opt}r_i^2 + (2z + 1)e^2/2r_i)(1/n^2 - 1/\epsilon_o)] \tag{5}$$

The "Outer Sphere" part of this expression is similar to the functions used by Landau[34] and Mott and Gurney[35] for the polarization due to displacement of a charge in a polar medium (c.f., Platzmann and Franck[28]). Weiss did not make use of this expression to estimate the height of the homogeneous electron exchange energy barrier which must be overcome by thermal vibrational-rotational energy. He assumed that the energy charge on discharging one ion and charging another in a homogeneous exchange reaction is equal to the algebraic sum of the optical frequency continuum terms for charge and discharge from the standard states (a vacuum at infinity), plus that of the Inner Sphere energy changes, plus the energy of assembly of the ions in the initial state. The latter is $e_i e_j / \varepsilon_o (r_i + r_j)$, where the e and r terms are the individual charges and radii. Weiss also said nothing about the crossing points of energy terms for reactants and products in or out of equilibrium, although he did point out that if weak interaction occurs, the probability of transition will be low and must be determined using the Landau-Zener adiabatic transition expression (see below). He also discussed electron tunneling, and was careful to distinguish between the transition of the reactant through the adiabatic "barrier" to the products, and electron tunneling through a Gamow energy barrier.[18,19]

Electron tunneling of Gamow type was discussed by (R. J.) Marcus, Zwolinski, and Eyring in 1954-55.[36] The rate was given by a transition state model with an activation energy term containing coulombic repulsion and reorganizational energy differences between reactants and products, multiplied by a frequency factor containing the tunneling coefficient. Somewhat similar approaches were taken by Laidler and coworkers.[37-39] Refs. 38 and 39 considered the possibility of some inner sphere rearrangement before electron transfer, with a change in solvation energy corresponding to a change in ion size. The concept of Inner Sphere and Outer Sphere reactions, depending on the type of activated state, was introduced by Taube.[40]

3. Dielectric Continuum Theory

The next developments of the FC approach were in papers by (R. A.) Marcus,[41-49] and a later series from the Soviet Union. About the same time Hush[50] introduced other concepts, to be discussed below. The early work of Marcus[41] considered the Inner Sphere to be invariant with frozen bonds and vibrational coordinates up to the time of electron transfer. The "classical subsystem" for ion activation has its ground state floating on a continuum of classical levels, i.e., vibrational-librational-hindered translational motions of solvent molecules in thermal equilibrium with the ground state of the "frozen" solvated ion.

The movement of a permanent or induced dipole changes the energy of the ion by Coulomb's law. The Inner Sphere dipoles are frozen, so only those in the outer sphere can move to change the potential energy of the central ion, reducing the system to the external electrolyte and the electron to be transferred. To avoid the difficulty of summing the energy changes imparted to the ground state of the ion due to dipole movements at progressively greater distances, the medium may be regarded as a dielectric continuum, and that the electron may be regarded as a charge distributed over the surface of a conducting spherical condenser of radius r, a non-physical adjustable parameter accounting for the ion solvation energy. The application of the FC principle results in a polarization of the environment of the charge on rapid electron transfer.

The early papers of Marcus[41, 42] describe an activated state X* with the electronic configuration of the reactant(s) and the Outer Sphere atomic configuration of the activated product X. A collective displacement vector describes outer motions in a classical Hooke's law elastic medium. Microscopically, this would have Gibbs free energies of $e_{ind}^2/2\alpha_e$ for induced dipole charges (e_{ind}), where the electronic polarizability[51] α_e is $e^2/m_e(2\pi\nu_e)^2$ and m_e, ν_e are the electronic mass and frequency. The maximum energy of rotating permanent dipoles of moment μ at distance r from charge ze is $-ze\mu/\varepsilon_0 r^2$. Marcus first looked at the medium macroscopically, regarding it as a dielectric with complete dielectric saturation in the Inner Sphere surrounding the ion.[41] He then considered partial saturation of the dielectric, with averaged dielectric constants.[42] A permanent dipole vector **u** cannot keep up with the sudden change of field under FC conditions, so it leaves a residual Gibbs energy in the medium $u^2/2\alpha_u$[42,49] where α_u is the polarizability per unit volume. The classical induction equalities are (**P.E**)dV = $(u^2/2\alpha_u)dV = (\alpha_u E^2/2)dV$ integrated over the dielectric volume not occupied by the reactants, where **P** is the polarization vector dV or dipole moment **u** per unit volume, and **E** is the corresponding field. Two **P** and **E** terms are distinguished, according to their frequency response, P_u, E_u for nuclear and molecular motions, and P_e, E_e for electronic motions.[41,42] Later, a particulate charge, rather than dipole description of the electrolyte was given, to account for a multipolar medium.[43]

The most probable X* and X were obtained by setting the differentials of the Gibbs energy of X* (δF^*) and the difference between those of X* and X ($\delta F^* - \delta F$) equal to zero. The sum δF^* +

$m(\delta F^* - \delta F)$ is set equal to zero to define the crossing point for the X^* and X electron states, where m is an undetermined Lagrangian multiplier. The (Gibbs) energy of activation (G_{act}) is the residual Gibbs energy barrier in the dielectric on FC charge transfer, after allowance for the Gibbs energy of reaction. The latter is $F^* - w^*$ at $\delta F^* = 0$, where w^* is the Gibbs energy of the initial state, and is given by the volume integral:

$$G_{act} = F^* - w^* \qquad (6)$$
$$= (1/8\pi) \int [(E^{*2}/\varepsilon_o) + m^2(E^* - E)^2(1/n^2 - 1/\varepsilon_o)]dV$$
$$- e_i^2/2\varepsilon_o r_i - e_j^2/2\varepsilon_o r_j$$

where the final two terms are the Born charging equation expressions for the w^*, the Gibbs energy of the components, with two initial components with charges e_i, e_j, and radii r_i, r_j. E^* is the volume-element dependent field vector due to the activated initial electronic state, and E is the value simultaneously present (in FC terms) in the final electronic state. It will be recognized that the integral of the first term inside the bracket is an energy of assembly under static dielectric conditions, whereas that of the second is the energy charge on going from reactants to products at optical frequency.

By conservation of energy, $F^* - F$ must be equal to $\Delta G^\circ - T\Delta S_e + w^{**}$, where ΔG° is the standard Gibbs energy of reaction, ΔS_e is any electronic entropy change on electron transfer, and w^{**} is the work required to bring the activated reactants and activated products from infinity. The second minimizing condition, i.e., $\delta F^* - \delta F = 0$ gives:

$$F^* - F = (1/8\pi) \int \{ ([E^{*2} - E^2]/\varepsilon_o) - (2m - 1)(E^* - E)^2(1/n^2 - 1/\varepsilon_o)\}dV$$
$$(7)$$

The volume integrals were evaluated by substituting $dV = 4\pi r^2 dr$, using the Coulomb's law assumption that the E terms are given by the vector sum of the negative gradients of the potential due to each charge e_i at distance r_i, i.e, $-\Sigma\nabla(e_i/r_i)$ in the space outside of the reacting particles. Within the particles, the usual Born charging assumption for conducting spheres makes the E terms zero. The E^2 terms contain $[e_i\nabla(1/r_i)]^2$ and $2e_ie_j\nabla(1/r_i)\nabla(1/r_j)$. The integral of the former from r_i to infinity is $4\pi e_i^2/r_i$, and that of the latter is $8\pi e_ie_j/R$, where $R = r_i + r_j$ at closest approach. The first term inside the bracket in Eq. (6), after subtraction of the last two terms on the right, becomes $e_i e_j / \varepsilon_o R$, and the integral of $(\mathbf{E}^* - \mathbf{E})^2$ simplifies to $8\pi(\Delta e)^2(1/2e_i + 1/2e_j - 1/R)$. Multiplied by $(1/8\pi)m^2(1/n^2 - 1/\varepsilon_o)$, this is λ, the reorganizational energy. The multiplier m is calculated from Eq. (7), and substituted in Eq. (6). Libby's concept (c.f. Eq. 1) considered the energy charge on changing charge to be proportional to e.g., $z^2 - (z - 1)^2$. However, all the z terms cancel in the Marcus equations, leaving only Δe, the amount of charge transferred in the forward and backward processes, i.e., by one electron unit so that the activation energy remains at the lowest level. The system is symmetrical, the electronic energy curves for the reactants and products being exactly similar and parabolic. This is because of the use of the expressions containing $m\mathbf{E}.m\mathbf{P} = m^2\mathbf{E}^2\alpha_u$ to give the continuum energy changes, i.e., m acts on the differential of the energy of a state, i.e., on its field, and not on its Gibbs energy. To quote Marcus, "in the absence of specific interactions, the (separate Born charging formulas for each reactant will hold at all separation values), since in the equation (for charging) each ion would merely see another charge, $-m\Delta e$, and the surrounding medium, in both the homogeneous and electrode cases".[44]

The reorganizational energy of the system, λ, was defined as the residual or inertial energy in the dielectric when the atomic configuration of the reactants (i.e, that in the Outer Sphere) is equal to that of the products. If the length of the many-dimensional reaction coordinate from the ground state of the reactants to that of the products is put equal to x, then $\lambda = kx^2$, where k is a constant. Ignoring the small $-T\Delta S_e + w^{**}$ terms, the similar parabolic terms for the reactants and products are vertically separated by ΔG°. The position of the crossing point, mx, along the reaction coordinate is given by a simple geometrical argument as $(\Delta G^\circ + \lambda)/2\lambda$. This gives a geometrical identity to the Marcus m. The Gibbs energy of activation ΔG^* is then equal to the energy at this crossing point, i.e.,

$$\Delta G^* = k(mx)^2 = (m^2)\lambda = (\Delta G^\circ + \lambda)^2/4\lambda \qquad (8)$$

The reorganizational energy λ was calculated using the Born charging formula energy differences for the ion with its primary solvation sheath under static and optical conditions, using the difference between the squares of the charges $(\Delta e)^2$. Thus, λ_{hom} for the homogeneous electron transfer reaction between two ions of radii r_i and r_j at the closest approach distance $R = (r_i + r_j)$:

$$\lambda_{hom} = (\Delta e^2)(1/2r_i + 1/2r_j - 1/R)(1/n^2 - 1/\varepsilon_o) \qquad (9)$$

The electrode case may also be thought of as involving two ions, one having infinite radius. However, Marcus considered that there was also an image charge situated at R', twice the distance of the reacting ion from the electrode.[44-46] Thus, λ_{elec} was given by:

$$\lambda_{elec} = (\Delta e^2)(1/2r_i - 1/2R')(1/n^2 - 1/\varepsilon_o) \qquad (10)$$

We should note that for water, $1/\varepsilon_o$ is negligible, and the infrared value of $1/n^2$ is about 0.55. The overall rate is $\exp-m^2\lambda/kT$ multiplied by a frequency equal to the collision number and by the concentrations of the reactants.

Other Marcus papers correlated theory and experiment.[48,49] He discussed adiabatic and non-adiabatic homogeneous transfer, pointing out that the degree of broadening and splitting between the electron energy terms of reactants and products is related to the lifetime τ of each excited state via the uncertainty principle, i.e., $\Delta E\tau = h/4\pi$, where ΔE is the broadening, and $2\Delta E$ is the corresponding splitting, i.e., the overlap or interaction energy between reactants and products. The splitting $2\Delta E$ is about 0.6, 6.0, and 60 kJ/mole for electron transition times of 10^{-13}, 10^{-14}, and 10^{-15} s, so rapid electron transitions will greatly reduce the Gibbs energy of activation. The FC, i.e., Born-Oppenheimer) approximation for water dipoles will hold at least to 10^{-12} s, where the amount of splitting is negligible. Marcus indicated[47] that the calculated electron tunneling probability determined by modeling the solvent energy barrier between the reacting species[56] could be used to determine the amount of splitting, since the transition time is $1/\rho v$, where ρ is the tunneling probability and v is the electron frequency in the reactant. Thus, $2\Delta E$ is approximately $h v \rho/2\pi$. This can be used in the Landau-Zener formula[9,52] or in more exact and general procedures[53] to determine the probability of non-adiabatic reactants to product transitions. Thus, the use of ρ as a prexponential term in the rate equation[36,39] is incorrect. The Marcus papers generally assume adiabaticity, unless proved otherwise. He pointed out[48] that no theory existed (indeed exists) for the large- to medium-overlap Landau-Zener transition at a metal electrode, because the multiplicity of levels allows electrons to make unsuccessful transfers to one level, and successful ones to another during the cause of small fluctuations in space, energy, and time. Electrons would emerge at energies of ±2-3kT of the Fermi level, as Gurney had approximately demonstrated by integration.[8] A similar procedure to that of Gurney integrated to the band edge energy[54] was used by Gerischer for semiconductor electrodes.

4. Activation via the Dielectric Continuum

The Marcus theory used thermal activation via collisional interactions to overcome the electrostatic (or other) energy barriers. The Soviet school[55-59] used a similar dielectric continuum concept derived from work by Platzman and Franck[28] and Pekar[60] for radiationless transitions in polar crystals, and by Lax[61] in polyatomic molecules (see also Frohlich, Ref. 62). The work is most accessible in later review articles.[57-59] It used Pekar's Hamiltonian description of the energy of the medium derived from nonequilibrium thermodynamics, rather than the classical electrodynamics used by Marcus. Unlike Marcus, they regarded the dielectric continuum as capable of activating the discharging ion via coupled harmonic electrostatic motions. The Born-Oppenheimer approximation was used to separate the wave functions of low-frequency dipole vibrations from high-frequency electronic vibrations. Dipoles performing small oscillations around the discharging ion may each contribute one vibrational energy quantum (about 0.016kT at 298 K for a frequency of 10^{11} s^{-1}, or about 4 x 10^{-4} eV on a molar basis). Thus, 1300 dipoles are required for an activation energy of 0.5 eV, in a radius of 20 Å, corresponding to a large polaron.[58,59] The charge is transferred over a distance of about 6 Å, which is on the order of, and probably less than, the distance through which dielectric saturation occurs close to the discharging ion (the correlation radius of the dipole moments). The theory was considered valid at 20 Å, but was considered less so at 6 Å, because of vibrational frequency dispersion due to dielectric saturation at small distances[59] (c.f., Marcus[42]). However, the theory of small polarons[63,64] allowed estimates to be taken to smaller ions and charge and energy transfer distances.

The transition probability was calculated using the Landau-Zener equation for homogeneous adiabatic electron transfer, and for the electrode case with very weak coupling, for which an exact solution exists. In many cases, the rate was expressed with a non-calculable preexponential interaction Hamiltonian for the overlap terms. The polaron approach differs from normal thermal activation. The latter may be regarded as via acoustic waves that result in collisions. The activation frequency is a collision number for Marcus, and the frequency of long-wave coupled dipole vibrations for the Soviet school. However, both are of the same order of magnitude at ordinary temperatures (about 10^{11} s^{-1}). One difference between the Marcus continuum theory and that of the Soviet school is the nature of the energy corresponding to the electron terms, which Marcus regarded as a Gibbs energy, because it uses the Born charging equation. However, $(1/n^2 - 1/\varepsilon_o)$ results from the difference between two Born charging equations and is largely temperature-independent,[59,65] so λ has the properties of potential energy.

5. Inner Sphere Rearrangement With "Flow of Charge"

Hush, followed by others[40,66] assumed that all rearrangement did not necessarily occur in the Dielectric Continuum. His early papers[50] appeared to take exception to the assumption of FC restrictions. He considered that adiabatic and non-adiabatic FC electron-transition processes of gas-phase type were not appropriate for electron-exchange reactions in solution at metal electrodes. The eigenfunction of the transferring electron might flow between reactants and products over a relatively long time, the behavior of the dielectric medium governing the course of the reaction. This implies a transition state of "normal" type with an energy col, where the electron density differs from that of the initial or final state by an amount equal to the symmetry factor. The slow change in charge density as the reaction proceeds implies an adiabatic process. The energy along the reaction coordinate would then only contain ion-solvent interactions, not electronic terms. These interactions were given simple forms, i.e., an energy of cavity formation in the solvent, which is largely independent of charge, the ion-dipole interaction depending linearly on charge, and a Born charging energy depending on the square of charge. No induced dipole effects (also depending on the square of charge) were considered. A further simplification assumed that energy changes resulting from the overall change of charge from reactants to products would overwhelm the differences resulting from the small changes in ligand-ion bond-length. Thus, the changing energy along the reaction coordinate due to ion-dipole interactions will be given by $e\mu q/r_d^2$, where μ is the dipole moment of water, q is the fraction of electronic charge at a given point on the reaction coordinate, and r_d is the approximately constant ion-dipole distance. As charge flows in, the energy of the system at equilibrium must be corrected by an amount proportional to the change of charge.

The charge-transfer barrier due to ion-permanent dipole interactions was rather small, and since ion-induced dipole terms ($\propto z^2/r_d^4$)[16] were ignored, the only place to seek a charge-transfer barrier was in the Born charging energy. This varies as z^2/r_i, where r_i is again the Bernal and Fowler[16] Born ionic radius, approximately the radius of the solvated ion. To a first approximation, r_i, like r_d, may be considered constant, so that the (positive) change in energy on going from charge z to $z - q$ is $(2qz - q^2)B$, where B is the Born continuum energy multiplier, equal to $e^2(1 - 1/\varepsilon_o)/2r_i$ (Eq. 3). To correct the energy after charge q has been transferred to the equilibrium value for the overall process $M^{z+} \rightarrow M^{(z-1)+}$, $[z^2 - (z-1)^2]B$, the absolute (positive) energy difference between the initial and final states, must be multiplied by q and subtracted from $(2qz - q^2)B$, giving:

$$\Delta G^* = q(1-q)e^2(1 - 1/\varepsilon_o)/2r_i \qquad (11)$$

which differentiation shows has a maximum at $q^* = 0.5$, so that the Gibbs energy of activation under equilibrium conditions is $e^2(1 - 1/\varepsilon_o)/8r_i$. For $r_i = 3.5$ Å, this is about 49.6 kJ/mole, which is reasonable. Replacing $[z^2 - (z-1)^2]B = C$ by $C + f\eta$, where f is F/RT adds the overpotential, η, to the expression. Differentiation now shows a maximum at $q^* = (C + f\eta)/2C$, so that $\Delta G^* = (C + f\eta)^2/4C$, which except for the multiplier $z^2 - (z-1)^2$ is identical in form to the Marcus Eq. (8), with C equivalent to Marcus λ and $\Delta G^o = f\eta$. The electrochemical symmetry factor $(1/f)d\Delta G^*/d\eta$ is therefore equal to $(f/2)(C + \eta)/C$.

In a later paper,[67] Hush further developed his theory and compared it to that of Marcus. He took into account a change in ion-ligand distance in the inner sphere by adding a Mie repulsive energy cAr_d^{-n} to the ion-dipole electrostatic term $-cze\mu r_d^{-2}$, where A is constant, n is large (9-12), and c is the coordination number. By setting the derivative equal to zero and eliminating A (assumed to be constant along the reaction coordinate) from the equilibrium expressions for the states with charges z and $z - q$, we obtain:

$$r_{d,e,z-q} = r_{d,e,z}[z/(z-q)]^{1/(n-2)} \qquad (12)$$

where the $r_{d,e}$ values refer to equilibrium states, and q replaces Hush's λ to avoid confusion with the Marcus λ. Mie's equations of $Ar_d^{-n} - Br_d^{-m}$ type have an equilibrium energy equal to $-B(r_{d,e})^{-n}[1 - (m/n)]$. Again ignoring ion-induced dipole effects, the energy associated with the inner sphere ΔE_i is approximately given by the ion-dipole term, with m = 2 and B proportional to z. Using (12):

$$\Delta E_{i,z-q} = -[n - 2/n]B(z - q)(r_{d,e,z})^{-2}(1 - q/z)^{2/(n-2)} = \Delta E_{i,z}(1 - q/z)^{n/(n-2)} \tag{13}$$

As before, $-(\Delta E_{i,z} - \Delta E_{i,z-q}) + q(\Delta E_{i,z} - \Delta E_{i,z-1})$ can be expanded as far as the quadratic terms the energy change along the reaction coordinate above the ground states of the reactants and products in equilibrium. The first two binomial terms vanish, so:

$$\Delta E_i^* = -\Delta E_{i,z}[q(1 - q)][n/(n - 2)^2 z^2] \tag{14}$$

Again $q^* = 1/2$, so $\Delta E_i^* = -\Delta E_{i,z}[n/4(n - 2)^2 z^2]$ and is rather small (7-11 kJ/mole). Hush considered ΔE_i^* to be $\approx 50\%$ of the total activation energy, the remainder being Outer Continuum energy. Since it again contains $q(1 - q)$, introduction of the overpotential as before again gives a symmetry factor $(C' + f\eta)/2C'$, where $C' = -\Delta H_{i,z}[m/4(m - 2)^2 z^2]$.

A complete model must include the induced dipole term. If it is initially assumed that $r_{d,e}$ is invariant, this is $-cz^2 e^2(\alpha_e/2)(r_{d,e})^{-4}[1 - (4/n)]$. This has the same form and properties of the Born charging term, again giving a symmetry factor proportional to overpotential. Without the aid of further simplifying assumptions, it is not possible to obtain any exact expressions to show the behavior of the terms when $r_{d,e}$ is allowed to vary. Hush's expansion to the binomial quadratic term leads to a harmonic approximation. This suggests that the equilibrium force constants may be used to estimate the energy changes, since these will include the effect of the ion-dipole and ion-induced dipole terms in harmonic approximation. However, the way in which force constant change as z − q changes is unknown, although if z − q is identified with bond order,[68] the relationship between the force constant and this term may be linear. Another possibility is the use of a relationship between z and the enthalpy of solvation, $\Delta H_{i,z}$. Assuming the straight line log-log relationship $\Delta H_{i,z} = -Kz^x$ for the Inner Sphere where K, x are constants, expansion of $z^x(1 - q/z)^x$ gives $q* = 1/2 + f\eta/x(x - 1)Kz^{x-2}$, and:

$$\Delta H_i* = C'' [q(1 - q)] + qf\eta = C''/4 + (f\eta)^2/4C'' + f\eta/2 \qquad (15)$$

where $C'' = [x(x - 1)Kz^{x-2}]/2$. For ions of valence 1-4 in the same range of atomic weight, a good case may be made for the enthalpy of solvation being approximately proportional to z^2. If the inner sphere part follows the same relationship, the approximation in (14) becomes exact, with $C'' = K$. The symmetry factor will again be of the form $(f/2)(C'' + f\eta)/C''$.

In his 1961 paper,[67] Hush modified his outer sphere energy equation by changing the $(1 - 1/\varepsilon_o)$ term to the Marcus expression $(1/n^2 - 1/\varepsilon_o)$ because the electron transition time, though longer than that under FC conditions, would still be short compared to solvent molecule motion. This proposition will be discussed later. The q value at the highest point of the energy surface ($q*$, Hush's $\lambda^{\neq}$) is then apparently formally identical to the Marcus m. The comparison of the Marcus and Hush crossing terms is indicated in Figure 1. Hush used similar expressions to those of Marcus for the continuum parts of the reorganizational energy in the homogeneous and heterogeneous electron transfer cases, but he correctly pointed out that R' in Eq. (10) for the heterogeneous electrode reaction at infinite dilution should be left out in the real solutions which are normally studied.[67] Equation (9) would therefore somewhat understate the activation energy. Hush used the collision number $(kT/2\pi\mu*)^{1/2}$ for the mean frequency for crossing the barrier, where $\mu*$ is the reduced mass of the activated complex. He considered that reaction would be adiabatic, since the energy gap ΔE between the upper and lower electronic states at the crossing point should be about 0.03 eV or 2.75 kJ/mole, the order of magnitude of crystal interactions for transition metal ions. This is sufficient to give a transition probability close to unity.[67] This corresponds to a transition time $h/2\Delta E$ for the electron states on the order of 2×10^{-14} s, allowing a continuum energy expression containing $(1/n^2 - 1/\varepsilon_o)$. In a later paper,[69] Hush attempted to generalize his model, introducing some of the Soviet school ideas into the continuum description using the Kubo model for the small polaron.[70] He also discussed adiabatic cases. He treated all vibrations, in both the inner and outer spheres, as quadratic functions of frequency tensors operating on the vector differences between the equilibrium positions of the initial and final states.[69]

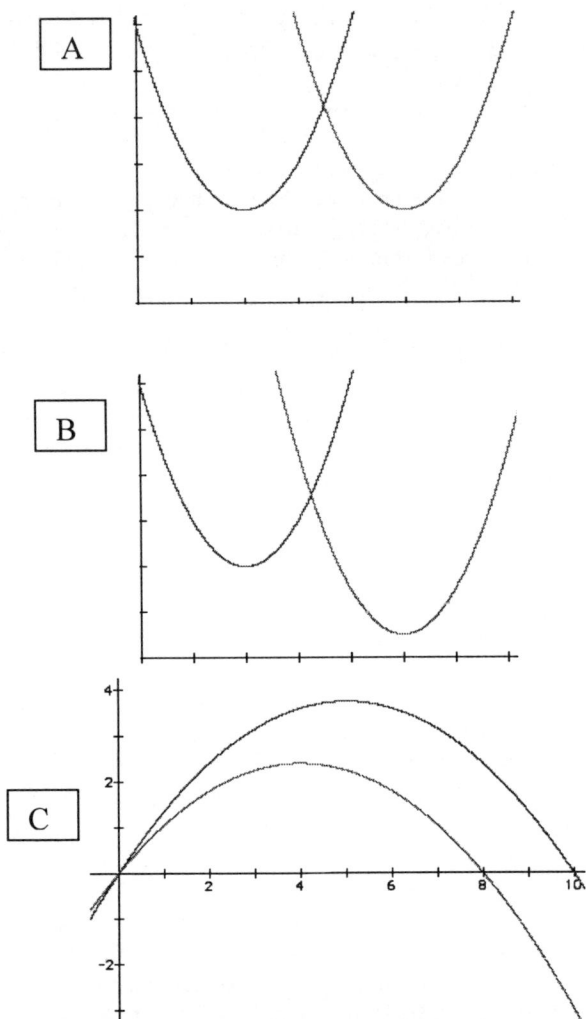

Figure 1. Two-dimensional potential energy surfaces in equilibrium (A) and out of equilibrium (B) for Marcus[41-49] Outer and Inner Sphere and Hush Outer Sphere;[67,69,176] (C) Hush Inner Sphere[50,67,176] in and out of equilibrium. y-axes: Nominal energy. x-axes: Nuclear configuration as a function of change of charge. C shows activation energies of 3.75 units at equilibrium (transition state charge 0.5), 2.4 units at –3.0 units energy displacement (transition state charge 0.4, mean symmetry factor 0.45).

Whether the electron transition time is $\approx h/2\Delta E$ is debatable. It corresponds to a single FC transition time, whereas the postulate of "flow of charge" to the inner sphere suggests a change in electron probability over time, i.e., a multiplicity of electron transfers between the donor and the acceptor, with the electron spending progressively larger amounts of time on the acceptor as the transitions proceed. Each of these may be regarded as a single radiationless FC electron transition. Between transitions, a small amount of bond-stretching occurs. Thus, a finite number of molecular vibrations of the reactant occurs as the charge changes from z to z − q and the ion-ligand bond lengthens. The symmetrical stretching frequencies[71] for typical 3+ and 2+ transition metal ions are about 390 and 490 cm^{-1}, corresponding to periods of 8.6 x 10^{-14} s and 6.8 x 10^{-14} s respectively. Thus, a reasonable minimum time for a flow of charge might be about 4 x 10^{-13} s, corresponding to about 5 vibrational periods. In the absence of a molecular energy model for the continuum (as distinct from the Born charging concept), it is difficult to give an opinion as to whether this time is sufficiently short for the polarization energy term containing $(1/n^2 - 1/\varepsilon_o)$ to apply. The relaxation time for distant continuum dipoles may be about 10^{-11} s.[57] The classical expression for $cos\theta$, the limiting value of the Langevin function for small θ in fields of energy less than kT at some distance from an ion is $ze\mu/3\varepsilon_o r^2 kT$.[72,73] If ε_o has its bulk value at r = 10 Å (see Section III) the $cos\theta$ values will be 0.03 and 0.02 for the transition z = 3 to z = 2, corresponding to a change in the angle of a water dipole to the field of 89° to 88°. To go from the torsional equilbrium position for a state z to a state z − 1 will require only a small fraction of a period. If this scenario is correct, the more distant continuum may always be in equilibrium with the changing field under Hush's "flow of charge" concept.

6. Inner Sphere Rearrangement and Force Constants

The first Inner Sphere treatment to use force constants to estimate energy changes between valency states in isotopic homogeneous electron exchange reactions under FC conditions was given by George and Griffith,[74] following Orgel.[75] The reactants (e.g. M^+ and M^{2+}) are considered to be at the distance of closest approach with excited inner spheres. If the difference in distance between the ion and the c identical ligands in the ground state is d, M^+ in its excited state has the ligand-ion bonds in compression through a portion of this distance, whereas M^{2+}

has its bonds stretched. For radiationless FC electron transfer, the sum of the potential energies of the two reactants above their ground states should be equal to that of the two (identical) products. The degree of harmonic compression of each of the n inner sphere bonds of force constant f in the M^+ reactant is x, and the degree of stretching of the n bonds of force constant f' in M^{2+} is x'. On instantaneous conversion to the products, M^+ becomes M^{2+} with its bond stretched by the amount $d - x'$, and M^{2+} becomes M^+ compressed by $d - x$. The condition for radiationless transfer is therefore $1/2fx^2 + 1/2f'x'^2 = 1/2f'(d - x)^2 + 1/2f(d - x')^2$, which can only be true if $d = x + x'$, i.e., the excited system is symmetrical. Possible activation energies are each equal to the sum of the energies of the excited levels of the two reactants above their ground states. Using the symmetry condition, each must be equal to $1/2fx^2 + 1/2f'(d - x)^2$. The most probable is the minimum value at x_{min} where $x_{min}(f + f') = f'd$. Thus, the minimum value of the activation energy for a homogeneous redox exchange reaction (i.e., at a Gibbs energy of reaction equal to zero) will be given by the reduced force constant expression:

$$U^* = cd^2ff'/2(f + f') \qquad (16)$$

Marcus generalized this reduced force constant expression to account for other cases, e.g., homogeneous reactions between different ions.[44,46] This harmonic approximation under FC conditions implies the reaction energy surface represented by the intersection of two parabolas, which are similar in the Marcus generalization. Thus, the relationship between Gibbs energy of inner sphere activation and the Gibbs energy of reaction is given by an equation of the same form as Eq. (16). As in the Hush approach, the inner and outer sphere energies at all stages of reaction are additive.

For the electrode case only one ion is involved, and it has been stated that U^* would be half of the value given by Eq. (16).[76] When the electrode is a reactant, the electron energy is the Fermi Energy,[8] and there is only one activated reactant. In thermodynamic equilibrium, the Gibbs energy of the grounds states of the reduced and oxidized species are equal. Consider the harmonic portions of the energy-distance curves for the reactants and products as in Figure 2, in which the equilibrium ligand to ion distances are d_1 and d_2 respectively. Writing $d = d_1 - d_2$, as in the homogeneous case, solving for the point of intersection gives the activation energy under equilibrium conditions, i.e.,

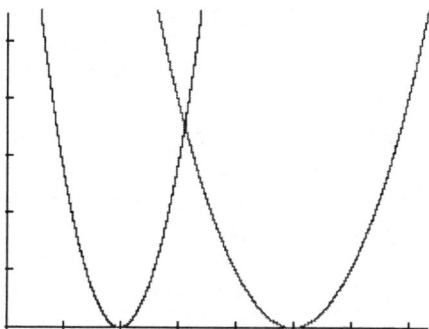

Figure 2. Two-dimensional potential energy surfaces in equilibrium for the heterogeneous (electrochemical) equivalent of the George and Griffith[74] homogeneous case involving Inner Sphere bond-stretching. y-axis: Nominal potential energy. x-axis: Nominal ion-water molecule distance. $d = d_1 - d_2$, where d_1, d_2 are the equilibrium distances of the initial and final states, giving a crossing point energy U^* under equilibrium conditions of $cd^2ff'/2(f^{1/2}+ f'^{1/2})^2$, where c is the coordination number and f, f' are the force constants of the initial and final states.

$$U^* = cd^2ff'/2(f^{1/2} + f'^{1/2})^2 \qquad (17)$$

For probable f, f' values (f = 2.8f', Sections IV-4, IV-9, VI-1), Eq. (17) is indeed about half of Eq. (16).

7. The Franck-Condon Approximation

The FC or Born-Oppenheimer approximation is physically clear if the activation energy barrier is in the Dielectric Continuum. The reacting ion is activated by some collisional or vibrational-librational means from the classical Boltzmann thermal pool, so that the rate of activation is equal to the rate of arrival of energy, which is equal to a characteristic classical electrolyte frequency. The electron transfers when its energy exceeds that of the barrier due to the inertia of the solvent permanent dipoles. Marcus[41-49] consistently supposed that the medium may be regarded as a dense gas phase with a collision frequency, which in its simplest form for the heterogeneous case is given by $(kT/2\pi\mu^*)^{1/2}$ per unit area of electrode (c.f., Hush). For a typical transition metal ion coordinated by six water molecules, this is about 4.9×10^3 cm s^{-1}, or 10^{11} s^{-1} when divided by the width of the reaction zone (≈ 5 Å). It may be equally logical to regard a liquid as a mobile solid, with a range of classical frequencies to a Debye limit v_m of *restrahlen* type, such that $v_m = c(9N_o/4\pi V)^{1/3}$, where c is the velocity of sound, N_o is Avogadro's number, and V is the molar volume.[77] This is equal to 4.3×10^{12} s^{-1} for water, which is close to kT/h (6.2×10^{12} s^{-1} at 298 K). After multiplication by a 5 Å reaction zone width, this corresponds to 3.1×10^5 cm-s^{-1}. Levich[57] stated that there are two types of polarons in polar liquids, namely acoustic polarons and optical polarons, with molecular vibrations in the same and opposite directions, respectively. He only considered optical polarons as important for energy transfer. However, an equally good case could be made for acoustic polarons.

Another evident mechanism for energy transfer to activated ions may be by bimolecular collisions between water molecules and solvated ion reactants, for which the collision number is $n(r_1 + r_2)^2(8\pi kT/\mu')^{1/2}$, where n is the water molecule concentration, r_1 and r_2 are the radii of the solvated ion and water molecule of reduced mass μ'. With r_1, r_2 = 3.4 and 1.4 Å, this is 1.5×10^{13} s^{-1}. The Soviet theoreticians believed that the appropriate frequency should be for water dipole librations, which they took to be equal 10^{11} s^{-1}. This in fact corresponds to a frequency much lower than that of the classical continuum in water.[78] Under FC conditions, the net rate of formation of activated molecules (the rate of formation minus rate of deactivation) multiplied by the electron transmission coefficient under nonadiabatic transfer conditions, will determine the preexponential factor. If a one-electron redox reaction has an exchange current of 10^{-3} A/cm^2 at 1.0 M concentration, the extreme values of the frequency factors (10^6 and 4.9×10^3 cm^{-2} s^{-1}) correspond to activation energies of 62.6 and 49.4 kJ/mole respectively under equilibrium conditions for adiabatic FC electron transfer.

8. The Tafel Slope

Electrochemical kinetic measurements show that many reactions have a rather constant Tafel slope over a wide overpotential range. This is true for both redox[79] and combined electron- and atom-transfer reactions, particularly proton transfer.[3,80] None of the approaches discussed above account for the experimental facts. References 41-50, 55-59, and 67 all give a curved Tafel slope in which the symmetry factor depends linearly on overpotential. In the Marcus and Soviet FC approaches, this results from the intersection of two similar parabolas as electron energy terms in molecular configuration space. For the (apparently) non-FC Hush inner sphere expressions, it may result from the binomial square-term approximations used. Introduction of $f\eta$ into the harmonic force constant expression given in Eq. (17) gives a rather complex expression corresponding to the intersection of two dissimilar parabolas, giving a curved Tafel plot.

Further problems occur with combined electron- and atom-transfer. Marcus[81] modified his electrostatic theory for bond-breaking and forming cases by using Johnston's semi-empirical bond-energy-bond-order (BEBO) model.[68] This does not significantly straighten the log current density-overpotential dependence.[3] The first Soviet evaluation of bond-breaking was applied to proton transfer[82]. The first difficulty is the energy gap between the $n = 0$ and $n = 1$ vibration states for protons in the $O-H^{\delta+}$ bond in solvated H_3O^+ of 16.2kT at 298 K, so the probability of the proton occupying higher vibrational states and the proton vibrational partition function are small. Thus, Ref. 82 considered the proton to undergo FC reaction in its ground state, which floats on the classical levels of the electrolyte thermal bath, which are continuous from 60 cm^{-1} to ca. 3444 cm^{-1} (2×10^{12} to 10^{14} s^{-1}).[78] As discussed earlier, thermal activation was considered to be via coupled librations, not modified gas-phase collisions. The proton could move from its initial to final state over a short distance (0.5 Å) by tunneling. The exact mechanism releasing it from its "cage" of water molecules was not addressed. In heavy particle transfers, e.g., chlorine evolution from chloride ion, a tunneling transition could not occur, so the corresponding theory required ingenious modifications.[83] The use of the FC expressions resulted in the same curved log i – η plot as that for redox reactions. These approaches were criticized on the grounds that rotational-librational-vibrational exchange may be enough to give a sufficient number of excited proton states to sustain the reaction, and because the energy fluctuations resulting from coordinated dipole reorientation in the continuum would be too small to allow activation[4a].

Later publications of the Soviet group attempted to straighten the log i – η plot, starting by postulating straight electron energy terms.[59] Later proposals included transitions from excited proton states, and a dynamic ionic atmosphere which could modulate the charge on the proton, abandoning strict FC conditions.[80,84] A semiclassical treatment of the inner sphere was introduced for redox processes.[85] The hydrogen evolution process was regarded as a supermolecule reaction, in which the introduction of anharmonicity could result in linear Tafel behavior.[86] Later papers[80] and reviews of proton[87] and heavy particle discharge[88,89] regarded the system in a similar way. A recent review[90] regards redox processes (simple electrochemical electron transfer or ECET reactions) as FC Outer Sphere transfers, whereas for reactions involving ion-atom electron transfer (electrochemical ion transfer, ECIT), e.g., iodide/iodine,[91] it proposes that the FC condition should be relaxed.

The theory of charge transfer is apparently still in an unsatisfactory state. Few developments have occurred over the past twenty years, and theories are still not in good accord with experiment, particularly for the Tafel slope. It is claimed that introduction of anharmonicity into the intersecting terms under FC conditions can result in straight Tafel lines,[79,86] but no molecular calculations substantiate this. A problem in developing an improved theory are the fact that the energy of interaction of ions with the Dielectric Continuum is restricted to Born charging, which has no molecular basis. The second problem is the fact that simple models of the Inner Sphere, even if they include ion-dipole, ion-induced dipole, and dipole-dipole repulsion terms,[16,92,93] still result in bond lengths and force constants which do not agree with experimental values.[71] EXAFS results show an inner sphere coordination number of 6 for typical transition metal ions, and the inner sphere ion-ligand bond is rather short, i.e., 2.0-2.1 Å.[94] Addition of quadrupole terms[95] is not enough to correct the simple theoretical models (see Appendix). Although solvation energy can be represented as a nominal power series to include the change of dielectric constant with field strength,[96] this is not helpful in understanding the physics of solvation.

III. INTERACTION OF IONS WITH POLAR MEDIA TO DIELECTRIC SATURATION

1. The "Electrostatic Continuum"

The model used for ion solvation energies since the work of Bernal and Fowler[16] has considered simple electrostatic interactions between the ion and the permanent and induced charges on solvent molecules, plus a term for estimating the work involved in importing the ionic charge from its standard state at infinity to the neighborhood of the ion situated in the Electrostatic or Dielectric Continuum. The Born charging equation[31] for the Continuum treats the ion as a Gaussian sphere with a solid metallized surface (i.e., a spherical condenser) of nominal radius r_o to which small elements of electronic charge are slowly and reversibly transported from infinity. The sphere has a minute initial charge, and work is done in bringing up each succeeding element. The resulting integration in a medium of static dielectric constant ε_o gives a Gibbs energy of $-z^2e^2/2\varepsilon_o r_o$. If the Gibbs energy of the ion *in vacuo* at infinity is arbitrarily zero, then the value in the medium of static dielectric constant ε_o is $-(1 - 1/\varepsilon_o)z^2e^2/2r_o$. The assumption of infinitesimally slow, reversible charging is hardly appropriate for a change in unit electronic charge in a time period of ca. 10^{-14} seconds. The application of Born charging to electron transfer has been recently reviewed.[97] Objections have been raised to the validity of the calculation, particularly in regard to the indivisibility of the electronic charge.* [93,98] If the same argument were applied to an electron, the charging energy and the corresponding relativistic mass would be impossibly large.[97] The equation also does not take into account partial or complete dielectric saturation near the ion, whose effective Born radius cannot in any case cannot be considered as a perfectly conducting sphere. It also does not take into account the differences in experimental Gibbs energies between positive and negative ions of similar radius.[98,99] It has been claimed that the expression for the "self-energy" of an ion interacting with the Continuum is free the above objections.[42] This is $(\varepsilon_o E_A.E_A/8\pi)dV$ integrated from nominal ionic radius r_o to ∞, where E_A is the field in volume element dV of state A. This integral gives $z^2e^2/2\varepsilon_o r_o$, which when subtracted from the *in vacuo* value gives the same expression as Eq. (3).[100] The Born charging energy and the self-energy expressions are formally identical, and both derive from the work of charging a molecular spherical condenser with fractional multiples of electronic charge. Thus the use of "self-energy" instead of Born charging[42] still depends on the validity of the Born charging assumptions. Born charging will be replaced here by a molecular model for the continuum interaction. In it, the standard state

---

* "If an ion can be treated as a conducting or non-conducting sphere, of radius r_i , and the solvent as a uniform medium of unvarying dielectric constant D, electrostatic theory can provide an explanation of the energy and entropy of the ions. The first supposition is possibly, and the second certainly, false..." (Ref. 17b, p. 881).

of the solvent dipoles is for the liquid state at infinity and for the ion it is its vacuum value. The model may be regarded as one in which dipoles are transported from infinity and assembled around the ion.

2. The Electrostatic Gibbs Energy in a Continuous Polar Medium

The equilibrium electrostatic Gibbs energy ΔG_A of an isolated volume element dV of a state A *in vacuo* containing a polarization vector $\mathbf{P}_A$dV is $(\mathbf{P}_A.\mathbf{A}_A)$dV, where $\mathbf{A}_A$ is the electric displacement vector, assumed constant within dV, and $\mathbf{P}_A$ is the dipole moment per unit volume induced by the local field. In a real medium, a multiplicity of polarization elements are present, all of which contribute to a polarization field $\mathbf{E}_{p,A}$. In addition, the imaginary cavity containing a reference dipole contributes a cavity field, $\mathbf{E}_c$, which locally decreases $\mathbf{E}_{p,A}$. Hence, there is an internal field $\mathbf{F}_A$ in dV, whose value is $\mathbf{A}_A + (\mathbf{E}_{p,A} + \mathbf{E}_c)$. The polarization element $\mathbf{P}_A$ interacts with this field, which results from the vector sums of the polarizations induced in other volume elements. The energy $\mathbf{P}_A.(\mathbf{E}_{p,A} + \mathbf{E}_c)$ summed over all volume elements is equal to the sums of the individual interactions of the polarization in one unit with that of the polarizations in all other elements. Hence, the $\mathbf{P}_A.(\mathbf{E}_{p,A} + \mathbf{E}_c)$ energies are associated with each pair of units, and therefore each must be divided by 2 to avoid double counting.

The local internal field $\mathbf{F}_A$ is produced by an external field $\mathbf{E}_A$. To account for the Onsager cavity[101] in the medium (see below), the ratio of $\mathbf{F}_A$ to $\mathbf{E}_A$ is q_A for the sake of generality. By definition, $\mathbf{A}_A/\mathbf{E}_A = \varepsilon_A$, the *microscopic* or *local* static dielectric constant in element dV, which may be a function of the (external or internal) field. Since $\mathbf{E}_{p,A} + \mathbf{E}_c = \mathbf{F}_A - \mathbf{A}_A$, the total electrostatic energy in the volume element dV is given by:

$$
\begin{aligned}
d\Delta G_A &= (-)[\mathbf{P}_A.\mathbf{A}_A + \mathbf{P}_A.(\mathbf{E}_{p,A} + \mathbf{E}_c)/2 - W_A]dV \\
&= (-)(\mathbf{P}_A.\mathbf{A}_A/2 + \mathbf{P}_A.\mathbf{F}_A/2 - W_A)dV \\
&= (-)(\mathbf{P}_A.\mathbf{A}_A/2 + q_A\mathbf{P}_A.\mathbf{E}_A/2 - W_A)dV \\
&= (-)(\mathbf{P}_A.\mathbf{A}_A/2 + q_A\mathbf{P}_A.\mathbf{A}_A/2\varepsilon_A - W_A)dV
\end{aligned} \tag{18}
$$

where (−) represents the overall sign of the term, W_A is the (+) work required to create the polarization $\mathbf{P}_A$ when the external field $\mathbf{E}_A$ is applied, i.e., when the electric displacement $\mathbf{A}_A$ is switched on. This work is usually expressed as the integral of the charge e associated with the dipole P_A multiplied by the Onsager cavity field F_A and by the element of distance dl through which the charge e is extracted by the field, i.e., the expression $-\int_0^d e F_A\, dl$.[102] While P_A is produced by and reacts with the *internal* field, its value is conventionally given in terms of the measurable external field. If the molecular polarization P_A is proportional to the internal field F_A, so that $P_A = el = n_o \alpha_{T,A} E_A = n_o \alpha_{T,A} F_A/q_A$, where n_o is the number of solvent dipoles per unit volume and $\alpha_{T,A}$ is the molecular polarizability of the medium in dV, this integral becomes $+q_A(el)^2/2n_o\alpha_{T,A} = +q_A(P_A)^2/2n_o\alpha_{T,A} = +P_A.F_A/2 = (+)q_A P_A.E_A/2$. So, provided P_A is proportional to the external field, i.e., $\alpha_{T,A}$ is constant, W_A will be exactly equal to $+P_A.F_A/2$. This will certainly be true for solids, in which permanent dipoles in fixed positions realign themselves along the field. In liquids, essentially free rotation of dipoles is assumed,[72,73] but the creation of polarization in each volume element by an applied field to give a time-averaged realignment still requires extraction of charge along the field vector, i.e., work to create the polarization. The total electrostatic energy in dV with constant $\alpha_{T,A}$ is therefore $(-)(P_A.A_A/2)dV$, i.e., $(-)[n_o\alpha_{T,A}(A_A)^2/2\varepsilon_A]dV$. This shows that the induction is reduced by ε_A from the vacuum value, not by ε_A^2, as implied in derivations putting the interaction energy equal to $(-)[n_o\alpha_{T,A}(E_A)^2/2]dV$.[99,100]

The local field consists of the difference between A_A and the vector sum of the displacement vectors A_A' from all charges in the system acting in the same direction as A_A. The definition of the *macroscopic* dielectric constant ε_A is A_A/E_A. This corresponds to the definition from Gauss's theorem for parallel fields $E_A = 4\pi\rho/\varepsilon_A$, where ρ is the charge density on a surface. Comparison of the capacities of a generalized parallel-plate condenser *in vacuo* and filled with a medium of dielectric constant ε_A shows that $A_A = 4\pi\rho$; $E_{p,A} = -4\pi P_A$; and $E_A = 4\pi(\rho - P_A)$,[100] where $4\pi P_A$ is the counterfield induced by the applied external field E_A; $P_A = n_o(\alpha_e E_A + \mu\cos\theta)$ is the induced moment per unit volume; n_o is the number of molecular dipoles of individual moment μ per unit volume; α_e is their electronic polarizability; $\cos\theta$ is the Langevin function; and $dP_A/dE_A = \alpha_{T,A}$, the total optical frequency and molecular (inertial) polarization in element A. Because $\varepsilon_A = A_A/E_A = A_A/(A_A - 4\pi P_A)$ and $P_A = n_o\alpha_{T,A}E_A = n_o\alpha_{T,A}A_A/\varepsilon_A$, it follows that ε_A must be generally equal to $1 + 4\pi P_A/E_A$, i.e., $1 + 4\pi n_o\alpha_{T,A}$, since the derivation does not depend on any particular model for the condenser. If $\alpha_{T,A}$ is constant in dV, then ε_A must also be constant. This expression for ε_A is valid for a uniform parallel field and for a spherically uniform charge density and field distribution. It is at least approximately correct for a spherical distribution in which the charges (dipoles around an ion) are not continuous (see Appendix). We shall now use three simple assumptions: (i) The center of the sphere is occupied by an ion of charge ze, e being the electronic charge; (ii) the electrostatic energy in dV, $(-)n_o\alpha_{T,A}(\mathbf{A}_A.\mathbf{A}_A)/2\varepsilon_A]dV = (-)[(\varepsilon_A - 1)(\mathbf{A}_A.\mathbf{A}_A)/8\pi\varepsilon_A]dV$, may be summed from a_o to infinity to give the total energy present in the medium outside a sphere of radius a_o; (iii) $\alpha_{T,A} = \alpha_T$ and $\varepsilon_A = \varepsilon_o$ are constant throughout the medium. The validity of (ii) and (iii) will be examined below.

Since $\mathbf{A}_A = ze/a^2$ for both a point charge and for a charge uniformly distributed over a surface, and dV is $4\pi a^2 da$, where a is the ion-dipole center distance, the electrostatic energy due to the presence of an ion in the surrounding dielectric continuum of dielectric constant ε_o may be approximated by the integral:

$$\Delta G'_{cont} = \int (\mathbf{P}_A \cdot \mathbf{A}_A / 2) dV$$

$$= -\int_{a_o}^{\infty} \alpha_T \, n_0 (\mathbf{A}_A.\mathbf{A}_A) / 2\,\varepsilon_0]dV$$

$$= -(1-1/\varepsilon_o)\int_{a_o}^{\infty} (z^2 e^2/2a^2)\,da \qquad (19)$$

$$= -(1-1/\varepsilon_o)(z^2 e^2/2a_1)$$

i.e., it is numerically equal to the Born charging energy of the ion regarded as a spherical condenser of radius a_o compared with the corresponding value *in vacuo*. It is therefore either a Gibbs or Helmholtz free energy, since liquids have very small isothermal compressibilities. However, the "Continuum" is no longer continuous, but molecular, and the distance a_1, defining the energy of interaction of the charge ze with the Dielectric Continuum is now clearly defined. It is not the radius of a "hard" solvation sphere treated as a hypothetical metallic spherical condenser, but the distance from the ion to the dipole centers of the first layer of solvent molecules for which the solvent has bulk values of $\alpha_{T,A}$ ($= \alpha_T$), and ε_A ($= \varepsilon_o$). Integration will be inaccurate when a_o and da are of similar order of magnitude and the total energy decreases in large finite steps with increasing a. Other inaccuracies may result from the approximation $dV = 4\pi a^2 da$ for the volume element, the effect of multipole terms, and from the approach to dielectric saturation. By substituting the exact expression $4\pi(a_{n+1}^3 - a_n^3)/3$ for dV in Eq. (19), putting the radii a_n, a_{n+1} of the inner edges of the nth and (n + 1)th shells equal to $2r(s + n - 1/2)$ and $2r(s + n + 1/2)$ respectively, where r is the effective solvent molecule radius, and s is constant, we obtain:

$$\Delta G_{cont} = -\left(1 - \frac{1}{\varepsilon_0}\right)\left(\frac{z^2 e^2}{2r}\right)\sum_{n=1}^{n=\infty}\left[\frac{1}{(s+n)^2} + \frac{1}{12(s+n)^4}\right] \qquad (20)$$

The theory of the dielectric constant of liquid water considers it to consist of rather freely rotating approximately tetrahedral superdipole groups.[72,73] Its radial distribution function[103] shows closest neighbors at the hydrogen-bond distance of 2.9 Å, and second nearest neighbors at about the tetrahedral diagonal distance, i.e., 4.74 Å, with no defined structure beyond. If we assume the superdipoles to be close-packed, the volume of each five-molecule group from the bulk density of water will be 1.11×10^{-22} cm^3, i.e., their effective radius will be 2.98 Å. Allowing for, e.g., 8% defects suggests that 2.9 Å is a reasonable effective radius, so that the dipole center of a time-averaged water molecule may be considered to be located at the center of the shell of thickness $2r = 5.8$ Å.

The series was summed to 20 terms, then extrapolated to infinity using the integral, which is a good approximation for the small residual energy. Values of s in the range 0.08 to 0.3 were selected, corresponding to superdipole center distances varying from 6.30 Å to 7.55 Å. As discussed in Section IV-8, these values lie around a reasonable range for encountering free water superdipoles from a central ion, the shorter ones being distance from a small tetrahedral 1+ ion, and the longer one for 3+ ions. The calculations showed that putting a_0 in Eq. (19) equal to a_1 in Eq. (20) understates the interaction calculated by the integral compared with the summation, whereas putting a_0 equal to $2r(s + 1 - 1/2)$, i.e., the first shell edge, overstates it. The relation a_0 (Å) = $0.94a_1 - 1.96$ reconciles the two. At s = 0.09, 61% of the interaction is in the first shell, and 77% is in the first and second shell. At s = 0.18 and 0.3, the values are 58%, 75%, and 55%,72%.

Since Bernal and Fowler,[16] the charging radius r_o in the Born equation has been put equal to the Inner Sphere radius, or approximately the ion to water molecule center distance plus 1.4 Å. At least for 1+ ions, this gives a fairly good approximation to the Gibbs energy of interaction of the ion with the outer Dielectric Continuum if α_T and ε_o are constant throughout the medium. High-valency ions are discussed in Section IV.

We note that ΔH_{cont}, ΔS_{cont} contain initial multipliers $-[1 - (1 - LT)/\varepsilon_o]$, and $-L/\varepsilon_o$, where $-L$ is $(\partial \ln\varepsilon_o/\partial T)_P$.[98] For water at 298.2 K, the ratio $\Delta H_{cont}/\Delta G_{cont}$ is 1.023.

3. The Approach To Dielectric Saturation

When $\alpha_{T,A}$ varies with the ion field as dielectric saturation is approached, A_A/E_A and P_A/E_A become nonlinear:

$$\varepsilon_A = 1 + 4\pi n_o \alpha_{T,A} = 1 + 4\pi n_o(P_A/E_A); \quad \varepsilon_A E_A = A_A = E_A + 4\pi n_o P_A \quad (21)$$

Here, ε_A is the *integral* dielectric constant, i.e., the value corresponding to the measured values of A_A and E_A. When ε_A is not constant, the work to create P_A may not be equal to $(P_A.F_A/2).dV$. When $\partial P_A/\partial E_A$ is a function of E_A, we require a means of calculating both P_A and W_A. We use the integral $(-)\int_0^{F_A} P_A \, dF_A = XA$.[42] Testing with various $P_A.F_A$ functions verifies that this gives the *net* energy of interaction of P_A with E_A, i.e.,

$$W_A = X_A + |P_A.F_A| = (-)\int_0^{F_A} P_A \, dF_A + |P_A.F_A| \quad (22)$$

where W_A, X_A are always positive and negative, respectively. Thus, in general:

$$d\Delta G_A = (-)(P_A.A_A/2 - P_A.F_A/2 - X_A)dV \quad (23)$$

4. Polarization

In the Marcus nomenclature,[41,42] the polarization in polar liquids is u- or e-type. The u-type is due to reorientation of molecular dipoles by an applied field. Its time constant is comparatively long ($\approx 10^{-12} - 10^{-11}$ s, depending on the type of motion and the degree of molecular association). In water, maximum absorption occurs in the broad Debye bands of frequencies $5 \times 10^9 - 5 \times 10^{11}$ s^{-1}, followed by thermal infrared resonance absorption corresponding to restricted dipole rotations or librations at $10^{12}-10^{12}$ s^{-1}. This is followed by small infra-red bands for intermolecular vibrations to $\approx 10^{14}$ s^{-1}. The e-type results from high-frequency distortion of electronic orbitals by the applied field in the range 5×10^{14} to beyond 10^{15} s^{-1} (Dogonadze and Kornyshev[104]). The assumption that all nuclear motions are a single u-type is an oversimplification, but the e-type has special properties. Its polarization vector is always considered proportional to the field at accessible field strengths, and it always lies along the field vector. In general:

$$\mathbf{P}_A = \mathbf{P}_{u,A} + \mathbf{P}_{e,A} = n_o \alpha_{T,A} \mathbf{F}_A / q_A = n_o \alpha_{u,A} \mathbf{E}_A + n_o \alpha_e \mathbf{E}_A \qquad (24)$$

where $\alpha_{u,A}$ is a function of $\mathbf{F}_A$, but the electronic or optical polarizability $\alpha_e = 1 + 4\pi n_o n^2$ is not, with limitations to be discussed later. $\mathbf{P}_{u,A}$ may be subdivided into two types of molecular displacement polarizations, namely free or hindered rotations, $\mathbf{P}_{ul,A}$, and translational or electrostrictive motions, $\mathbf{P}_{ut,A}$. In a uniform applied field, only the first is significant. However, in the non-uniform field near an ion, both types will occur, since the dipole will move up-field to maximize its interaction energy until the coulombic force is opposed by non-electrostatic contact forces. In a uniform field, non-polar molecules show no rotation since α_e is always parallel to the field. In a non-uniform field, electronic induction will cause translation of non-polar molecules, giving ut-type polarization.

Only $\mathbf{P}_{ul,A}$ has been generally considered in the literature. It may be written $\mu_{eff} \cos \theta_A$, where μ_{eff} is the effective value of the permanent dipole moment of a molecule in the polar medium, lying at a time-averaged angle θ_A to the internal field $\mathbf{F}_A$. Because of electronic polarization due to the fields of neighboring permanent dipoles, μ_{eff} differs from the vacuum value, μ_v. In the absence of a field, the permanent dipoles perform thermal motions around positions determined by their attractive and repulsive forces. The *net* electrostatic forces averaged over all polar solvent dipoles are always attractive, whereas the non-electrostatic contact forces are always repulsive.[105] The classical statistical mechanical distribution of $\cos\theta_A$ values is given by the Langevin function, $L(x_A)$,[106] i.e.,

$$L(x_A) = \cos\theta_A = (\coth F_A\mu'/kT) - kT/F_A\mu' \qquad (25)$$

where $x_A = F_A\mu'/kT$, and μ' is a further effective dipole moment value which depends on the structure of the polar solvent, i.e., the attractive and repulsive forces between individual dipoles. It is the moment of the smallest average pseudo-freely-rotating spherical specimen of the polar medium, containing μ_{eff} at its center.[72] Equation (25) should be rigorous provided that the structure of the polar medium is everywhere uniform, there are sufficient dipoles in each population for Stirling's factorial formula to reasonably apply, and the energy levels of the dipoles are classical and continuous, permitting an integration to obtain the partition function.

The molecular ul-polarization induced in dV by the external field E_A in a polar solvent is given by:

$$P_{ul,A}dV = \alpha_{ul,A}E_AdV = \mu_{eff}\cos\theta_AdV = \mu_{eff}L(x_A)dV \qquad (26)$$

From Eq. (25), for small x, $F_A\mu' < kT$, and $L(x_A) \approx x/3$. Under these conditions:

$$\alpha_{ul,A} = P_{ul,A}/E_A = (n_o :_{eff}/E_A)\cdot(F_A \cdot :'/3kT) = n_oq_A:_{eff}\cdot :'/3kT \qquad (27)$$

Thus, from Eqs. (24) and (27):

$$\varepsilon_o - 1 = 4\pi n_oq(\alpha_e + :_{eff}\cdot :'/3kT)$$
$$n^2 - 1 = 4\pi n_oq\alpha_e$$
$$\varepsilon_o = n^2 + 4\pi n_oq(:_{eff}\cdot :')/3kT \qquad (28)$$

where q is a generalized value of q_A. The last expression is the Kirkwood equation[72] for the static dielectric constant of associated polar liquids. From Eq. (26), in general:

$$\varepsilon_A = n^2 + 4\pi n_o\mu_{eff}L(x_A)/E_A \qquad (29)$$

These equations were first derived by Debye,[33,107] with the assumption that the relationship between F_A and E_A was for a Lorentz cavity containing a medium with the macroscopic dielectric constant and no permanent dipole, when $F_A = E_A + 4\pi P_A/3$. By elimination of P_A with $E_A = A_A - 4\pi P_A = \varepsilon_A E_A - 4\pi P_A$, q_A for this cavity is $(\varepsilon_A + 2)/3$. Moving the $(\varepsilon_A + 2)$ term to the denominator of the left side of the first part of Eq. (28) gives the Clausius-Mossotti equation. Onsager[101] showed that this theory was incorrect for media containing permanent dipoles, and that q_A was $3\varepsilon_A/(2\varepsilon_A + 1)$ if the imaginary cavity containing the dipole has a dielectric constant of unity. Before the hypothetical introduction of its dipole, but with the field switched on, opposite walls of the cavity were already polarized by the opposing charges of neighboring liquid dipoles, partly aligned along the direction of the field. Thus, the cavity already contained a virtual dipole before introduction of real dipole, increasing both its net value and that of the cavity field. The corresponding factor[101] for a dipole bathed in a medium of dielectric constant n^2, i.e., the electron cloud surrounding the permanent dipole) is $3\varepsilon_A/(2\varepsilon_A + n^2)$. Frölich[108] considered that if μ_{eff} is a *point* dipole in the center of a relatively large (in molecular terms) Lorentz cavity with a dielectric constant n^2, then its field within the Onsager cavity will be $(n^2 + 2)/3$ times the vacuum value, i.e., $\mu_{eff} = (n^2 + 2)\mu_A/3$. This expression is limited by the dimensional constraints of the Onsager and Lorentz cavities,[108] and will fail in non-uniform radial fields near an ion. For water $(n^2 + 2)/3 \approx 1.25$, and the practical value of μ_{eff} can be adjusted to agree with that of ε_0.

The total local field F_A' is equal to the vector sum of the Onsager cavity field F_A, and the Onsager reaction field.[101] The latter is independent of E_A, and results from the dipoles' own fields.[101] In the medium of dielectric constant n^2, F_A' is:

$$F_A' = [3\varepsilon_A/(2\varepsilon_A + n^2)]E_A + [2(\varepsilon_A - 1)/(n^2 r^3)(2\varepsilon_A + n^2)]_{:eff} \quad (30)$$

where the terms are respectively the cavity and reaction fields, where r is the dipolar molecule radius.[101,108a] Frölich[108a] showed that only the first term is important for permanent dipoles, since the second only changes the absolute value of the dipole moment, which may be regarded as an adjustable parameter.

5. The Static Dielectric Constant of Water

The effective value of ':eff:' in Eq. (28) may be calculated using various assumptions. Kirkwood[72] proposed that ':eff:' for associated liquids is $\mu_A^2(1 + n_o f_V)$, where f_V is the electrostatic radial distribution function summed over the selected molecule and its partners in space. He substituted $z\langle\cos\gamma\rangle_{av}$ for $n_o f_V$, where z is the number of closest neighbors whose moments are an angle γ, and $\langle\cos\gamma\rangle_{av}$ is the cosγ value averaged over all possible orientations. Neglecting second-nearest neighbors and multipole effects, he assumed a quasi-tetrahedral structure for water with 4 nearest neighbors,[16] each dipole being at the O–H bond angle, taken as $\approx 50°$ (in reality, $52.25°$[*]). Each hydrogen bond on the central water dipole is directed towards the π-orbitals of the oxygen on the neighboring molecule, which are also at an angle to the dipole, and perpendicular to the O-H bond plane. Assuming that free rotation[109] of the hydrogen bond, the effective moment of the neighboring water dipole along the axis of rotation is $\mu_{eff}\cos 52.25$, which is in turn at $52.25°$ to the dipole axis of the central water molecule. Hence, the effective value of $\mu_{eff}\cdot\mu'$ in Eq. (28) for torsion of the tetrahedral four-molecule water cluster is $\mu_{eff}^2(1 + 4\cos^2 52.25)$. Simple electrostatics[99] shows that the averaged vector sum of the fields of the 4 nearest neighbors along the axis of μ_{eff} is $(8\cos^2 52.25)\mu'/a^3$, where a is the distance between dipole centers. This increases the value of the vacuum moment μ_V due to the dipole field electronic polarization, i.e., $\mu_{eff} = \mu_v + (8\cos^2 52.25)\alpha_e\mu'/a^2$, hence:

$$:_{eff}:' = \mu_v^2(1 + 4\cos^2 52.25)/[1 - (8\cos^2 52.25)\alpha_e/a^3)]^2 = Q\mu_v^2 \quad (31)$$

where a is the effective hydrogen bond length of 2.9 Å at 298 K,[103] which Kirkwood took as 2.75 Å. From Eqs. (28) and (31), we obtain:

$$(\varepsilon_o - 1)(2\varepsilon_o + 1)/3\varepsilon_o = 4\pi n_o(\alpha_e + Q\mu_v^2/3kT) \quad (32)$$

For water, with $\alpha_e = 1.444 \times 10^{-24}$ cm^3/molecule, Kirkwood's original assumptions ($Q = 4.47$) give an excellent ε_o value (78.0 at 298 K). However, the use of the more realistic $52.25°$ and 2.9 Å ($Q = 3.44$) gives 63.9.

The radial distribution function was obtained by Pople,[103] c.f., Harris and Alder,[110] Haggis, Hasted, and Buchanan.[111] Pople showed that Kirkwood's assumption of complete hydrogen bonding in the first shell with none in the second was oversimplified. The first shell dipoles are bonded to the second via bent hydrogen bonds of bending

---

[*] Discussed in Ref. 99, pp. 10-12.

force constant g, so that the Kirkwood expression $\mu' = \mu_{eff}(1 + 4\cos^2 52.25)$ should be summed over all shells and replaced by $\mu' = \mu_{eff}[1 + \Sigma n_i \langle\cos\theta_i\rangle_{av}]$, where n_i is the number of dipoles in the ith shell, which lie at angle $\cos^{-1}\theta_i$ to the central dipole. For the first shell, $\langle\cos\theta_i\rangle_{av}$ is $(\cos 52.25 \text{Lexp} + g/kT)^2$. The second shell has one repulsive and two attractive orientations with $\cos^{-1}\theta_2 = (1/3)(\cos 52.25 \text{Lexp} + g/kT)$,[4] the general attractive orientation being $\cos^{-1}\theta_1 = 3^{3\,i}(\cos 52.25 \text{Lexp} + g/kT)^{2i}$. Pople found the effect of each shell on the static dielectric constant using the n_i values from the radial distribution function. With $g/kT = 10$ at 273 K, the relative contributions of the first, second, and third shells were found to be 1.20, 0.33, and 0.07, and the temperature dependence of the dielectric constant showed good agreement between theory and experiment.[103]

6. The Dielectric Constant at High Field Strengths

Approaching dielectric saturation, $P_{ul,A}$ is no longer proportional to E_A, and we require a usable expression for $\alpha_{ul,A}$ as a function of E_A. Malsch[112] showed that ε_A appeared to be proportional to $-E_A^2$ for $E_A < 3 \times 10^5$ V/cm (1000 esu-cm^2). The interaction of such fields with dipole moments of ≈ 2.0 D will result in interaction energies of less than 0.1kT, far from the saturation requirement $F_A\mu' > kT$. Several equations have been suggested to relate ε_A and E_A[114-117] (see Ref. 113). Theories of the dielectric constant in high fields by Debye,[33,107] Sack,[118] and Webb[119] using the obsolete Lorentz cavity approach are only of historical interest. They show a much slower recovery from dielectric saturation with distance from an ion[120] than those based on the Onsager cavity.[*] The first attempt to extend the Onsager-Kirkwood theory to high fields was by Ritson and Hasted,[122] who used empirical

---

[*] Grahame[117] and Conway, Bockris and Ammar[120] used the normal Poisson distribution between potential ϕ_A and charge density ρ_A, i.e., $\nabla^2\phi_A = -4\pi\rho/\varepsilon_A$, under conditions of a field-dependent dielectric constant. Buckingham[121] showed that this is incomplete if there is a gradient of E_A and ε_A. The Maxwell relation $\text{div}A_A = 4\pi\rho/\varepsilon_A$ is $\varepsilon_A \text{div}E_A + E_A.\text{grad }\varepsilon_A$, the second term arising from the differential. Hence, the complete Poisson equation becomes $\nabla^2\phi_A = -4\pi\rho/\varepsilon_A + E_A.\text{grad}|E_A|(d\ln\varepsilon_A/dE_A)$.

expressions similar to Eq. $(29)^{**}$ to calculate ε_A parallel and perpendicular to the radius vector of a (univalent) ion as a function of the field $\mathbf{E}_A$ and distance. They did not discuss the validity of their approach, but they used $4\pi\partial\mathbf{P}_A/\partial\mathbf{E}_A$ to obtain the dielectric constant. Since $\mathbf{E}_A\varepsilon_A = \mathbf{A}_A = n^2\mathbf{E}_A + 4\pi\mathbf{P}_A$ by definition, their expressions are identical to the *differential dielectric constant* $\partial\mathbf{A}_A/\partial\mathbf{E}_A{}^{108b}$. The results show a very rapid fall in $\partial\mathbf{A}_A/\partial\mathbf{E}_A$ at distances less than ≈ 3.5 Å from a univalent ion, the fall being more rapid along the radius than perpendicular to it.

Booth[73] used Frölich's[108] modification of the Onsager expressions[101] for the cavity field in non-associated polar liquids, and corresponding modifications of Kirkwood's equation for associated polar media. Booth's assumptions in deriving Eq. (28) for the Kirkwood case are important to determine the validity of his final expressions. He used the Onsager-Frölich cavity field ratio $3\varepsilon_A/(2\varepsilon_A + n^2)$ as the value for $q_A = \mathbf{F}_A/\mathbf{E}_A$ in the Langevin function, pointing out that the cavity field expression would be exact only if $\varepsilon_A = \varepsilon_0$, i.e., $\mathbf{E}_A$ is everywhere small and constant. To make the calculation tractable, he assumed the validity of Eq. (30), and that $\varepsilon_A \gg n^2$, so that Eq. (30) became:

$$\mathbf{F}_A' = 3/2(\mathbf{E}_A + \mu_{eff}/n^2r^3) \tag{33}$$

The reaction field is small and independent of $\mathbf{E}_A$ and $\mathbf{F}_A'$, so he put $\mathbf{F}_A' \approx \mathbf{F}_A = q_A\mathbf{E}_A = (3/2)\mathbf{F}_A$ for simplicity. Using Frölich's Lorentz cavity with the dipole in an n^2 medium[108]:

$$\mu_{eff} \approx A\mu_v(n^2 + 2)/3; \quad \mu' \approx B\mu_v(n^2 + 2)/3 \tag{34}$$

where A and B are structure-dependent constants, both unity for a non-associated (Onsager) liquid. For water, Booth[73] used a tetrahedral model, first indicating that $B = (5 + 8 \cos\gamma_{av} + 12 \cos\gamma'_{av})^{0.5}$, and $A = (1 + 4 \cos\gamma_{av})/B = X/B$, where γ is the angle between the axes of nearest neighbor moments, and γ' is the angle between the axes of the nearest neighbors of a given molecule. He used 1/3 for $\langle\cos\gamma\rangle$, i.e., he

---

** Their equations for the Onsager theory put μ^2 in the Langevin function, instead of one μ inside and one outside. However, the error is not serious since the calculations and graphs are correct.

took γ to be $180\text{-}109.47°$, and $\langle\cos\gamma'\rangle$ $1/27$, the approximate value for $[90°-(109.47°-104.5°)]$, where $104.5°$ is the angle subtended by the protons in the water molecule. Hence B $= (\sqrt{73})/3 = 2.848$ and A $= 7/(\sqrt{73}) = 0.819$. Booth's equations[73] for ε_A (high fields) and ε_o (low fields) are:

$$\varepsilon_A = n^2 + 4\pi n_o A[\mu_v(n^2 + 2)/3E_A]L\{1.5BE_A[\mu_v(n^2 + 2)/3]/kT\} \quad (35)$$

$$\varepsilon_o = n^2 + 4\pi n_o(1.5)X[\mu_v(n^2 + 2)/3]^2/3kT \quad (36)$$

where X $=$ AB $= (1 + 4 \cos\gamma\rangle_{av}) = 7/3$. Making the same assumptions as Booth's, Kirkwood's expression for $:_{eff.}:' = 3.443\mu_v^2[(n^2 + 2)/3]^2$ would give:

$$\varepsilon_o = n^2 + 4\pi n_o(1.5)(3.443)[\mu_v(n^2 + 2)/3]^2/3kT \quad (37)$$

From the Clausius-Mosotti equation, with $n_o = 3.33$ x 10^{22} molecules-cm^{-3} and with $\alpha_e = 1.444$ x 10^{-24} cm^3-molecule at 298 K, n^2 can be taken as 1.76. Using $\varepsilon_o = 78.5$, Eq. (36) gives 2.03 x 10^{-18} esu-cm (2.03 D, Debye units) for μ_v, or 2.04 D if the correct (1.484), rather than the approximate (1.5), value of q is used. In contrast, with the correct value of q_A, Eq. (37) gives 1.92 D. The value of μ_{eff} using Booth's preliminary model[73] is $A\mu_v(n^2 + 2)/3$, i.e., 2.09 D, assuming that $\varepsilon_o - n^2$ $= 76.7$ is used to calculate the value of μ_v from Eq. (36).

Calculations by Coulson and Eisenberg[123] from the charge distribution of water molecules suggest a value of about 2.42 D at 298 K due to induction by the fields of surrounding dipoles. In an erratum[124] to Ref. 73, Booth used a statistical analysis to derive more generalized values of A and B, with a value of X $=$ AB $= (1 + 4 \cos\gamma\rangle_{av})$ $= 7/3$ to give the correct result for ε_o in Eq. (36). This reduced the value of B from $(\sqrt{73})/3$ to $7/3$, giving A $= 1$, which increases μ_{eff} to 2.54 D. Kirkwood suggested that μ_{eff} should be that for the moment of the central water molecule, whereas the torsional term in the Langevin function should be that for the group of associated molecules preventing the rotation of the central dipole. Hence Booth's result for B $= (1 + 4 \cos\gamma\rangle_{av}) \approx 7/3$ seems reasonable, at least for low fields, provided that longer-range effects are ignored, and the cavity field assumptions hold. Booth's final calculation[125] increased the value of B by 10% to 1.1(7/3), by reducing the value of A from 1.0 to 0.909 to account for the effect of distortion of the Onsager cavity field when the field is non-uniform close to an ion, therefore no longer parallel within the cavity. This results in a value of μ_{eff} equal to 2.31 D. Booth also thought that the reaction field (Eq. 31) might also influence the value of ε_A in intense fields, and proposed a small correction in the opposite direction to account for this. A further effect maintaining the value of the dielectric constant as the field increases may result from electrostriction, which should locally increase n_o.[125] In view of the uncertainties in Booth's assumptions and calculations, a nominal value of 2.02 D is initially used here. This may be regarded as the local value of the dipole moment when the effect of the Lorentz-Frölich expression cavity expression containing n^2 has broken down due to dimensional constraints. The dipole moment value will be modified as necessary to enable a best fit to experimental bond-lengths for the inner sphere.

If the Langevin function in Eq. (35) is expanded[*] to the second term, we obtain:

$$\varepsilon_A = n^2 + 4\pi n_o q_A X[\mu_v(n^2 + 2)/3]^2 \{1 - [q_A BE_A \mu_v(n^2 + 2)/3kT]^2/15\}/3kT \tag{38}$$

This is only valid for $BE_A \mu_v(n^2 + 2)/2kT < 1$, but owing to the form of the series, it is a good approximation for values < 1.2. The change in ε_A in field E_A from that at limitingly low fields, i.e., $\Delta\varepsilon_A$, may be written by reintroducing the Onsager-Frölich cavity field, instead of the factor $q_A = 1.5$, using Eq. (38) and the general relation:[*]

$$f(\varepsilon_A) = f(\varepsilon_o) + \Delta\varepsilon_A[\partial f(\varepsilon_A)/\partial\varepsilon_A]\varepsilon_A = \varepsilon_o \tag{39}$$

where $f(\varepsilon_A)$ from rearranged Eq. (38) is $(\varepsilon_A - n^2)(2\varepsilon_A + n^2)/3\varepsilon_A$. After differentiation and solving for $\Delta\varepsilon_A$, we obtain:

$$\Delta\varepsilon_A/\varepsilon_o = -3(4\pi n_o\varepsilon_o XB^2)[\mu_v(n^2 + 2)/3]^4 A_A^2/[5(kT)^3(2\varepsilon_o^2 + n^4)(2\varepsilon_o + n^2)^2] \tag{40}$$

---

[*]The expansion of $L(x)$ is $x/3[1 - x^2/15 + 2x^4/315 - x^6/1575....+ (-1)^{n+1}(3)(2^{2n})[B_{2n}/(2n)!]x^{2n-2}]$, where B_{2n} is the $(2n)$th Bernouilli number. The three terms to x^4 give quite accurate results to $x = 1.5$ (for $x = 1.5$ or $L(x) = 0.4381$, its value is 0.4411). For large x ($x > 1.5$), $1 + 2e^{-2x} + 2e^{-4x} - 1/x$ gives accurate results (0.4379 at $x = 1.5$). Thus, from Equations 35 and 36, $\varepsilon_A = n^2 + (\varepsilon_o - n^2)p$, where p is the even term power series in the square brackets.

$$= -3(\varepsilon_o - n^2)\mu'^2 A_A{}^2/5(kT)^2(2\varepsilon_o{}^2 + n^2) \tag{41}$$

which is a generalized form of the Van Vleck,[114] Böttcher,[115] and Schellman[116] equations, which are applicable to weak fields ($E_A\mu_v/kT <$ 1). The g_2 term in Schellman's Eq. $(19)^{116}$ is equal to XB^2.

7. Inadequacies of the Booth Theory

Although Booth separated the electronic and permanent dipole polarization, Buckingham[121] contended that omitting polarization energy from the potential energy function may affect ε_A. The interactions between pseudo-freely-rotating superdipole groups of effective moment $\mu' = 7\mu_{eff}/3$ [124] may be estimated using Pople's geometry[103] for the first to third, and second to fourth neighbors at minus and plus the tetrahedral angle to each other, which results in a third nearest neighbor distance of a = 7.35 Å with a nearest neighbor distance of 2.92 Å. The maximum displacement at this distance due to a rotating superdipole is 24,000 esu-cm^{-2}, which will result in only a small change in ε_A from the vacuum value (Eq. 35). Ignoring the Lorentz ($n^2 + 2)/3$ cavity term for closest-neighbor groups, and taking the mean interaction between each pair[103] as $-2\alpha_T\mu'^2/3\varepsilon_A a^6$, with $\alpha_T = (\varepsilon_A - n^2)/4\pi n_o$, we find it to be $-0.008kT$, far less than the free rotation requirement. For this, the thermal energy per rotational axis should exceed twice the electrostatic interaction energy per axis. Essentially free rotation coupled to partial hydrogen bonding of the larger groups therefore can occur, whereas the individual dipoles are not free. The specific heats at constant volume for liquids are explained by supposing that hindered rotations or librations with an energy barrier $\approx$ kT occur.[126] Similarly, the large change in the static dielectric constant on melting in polar media, going from a small temperature-independent value to a large value proportional to the Langevin polarizability 1/T suggests very different of solid and liquid dipoles (c.f., nitromethane[127]).

Buckingham objected to calculating ε_A via the averaging (i.e., Langevin) function, since this assumes a simple solution before the problem is fully stated. However, this seems to be the only realistic method of handling the problem. Pople's modified treatment of Kirkwood's dipole clusters[103] in which the internal field distorts hydrogen bonding between shells outside of the first cluster, has been discussed in Section III-4. Pople did not consider that work must be done to bend the hydrogen bonds when the field is applied. This is because hydrogen bonds are continuously making and breaking, so that the bent bond is the statistical average, and no net work is required to go from one configuration to the next. This will be true until the field becomes sufficiently great to orient, then twist, and finally to completely line up the individual dipoles in a superdipole group, resulting in a breakdown of the local water structure.

A further objection by Buckingham was Booth's use of the rigid dipole in a Lorentz cavity containing a medium of refractive index n^2, which will certainly break down at short distances. Booth also assumed that n^2 is field-independent, which will not be true in intense fields. Buckingham and Pople[128] showed that the induced dipole was in fact a power series in the odd terms of E_A, and could not contain the even terms, which would have no directionality. The permanent dipole value is also a series, with weak field and strong field terms.[128] Following Debye,[129] Buckingham[121] showed that the electronic polarizability may be anisotropic. Hence, the dielectric constant $\varepsilon_A = A_A/E_A$ is also a power series, this time with even terms in E_A for an isotropic material to avoid directionality (c.f., the expansion of the Langevin function, footnote under Eq. 38). However, Booth had pointed out the even power series would not be valid in intense fields, which would themselves introduce anisotropic effects.[125] Thus, the expression for the dielectric constant as a function of the field becomes completely nominal.   Buckingham also showed that $\partial A_A/\partial E_A$ (c.f., Ritson and Hasted[122]), which he called the *incremental dielectric constant,* was also a power series of even E_A^{2n} terms, each in the series for ε_A being multiplied by $n + 1$. This is easily verified by differentiation.

8. The Dielectric Constant as a Function of Displacement

The best-known empirical expression for ε_A as a function of E_A^2 is Grahame's.[117,130] Its differential form with vector notation dropped, an with n^2 isotropic and constant is:

$$\partial A_A/\partial E_A = n^2 + \{(\varepsilon_o - n^2)/[1 + (b/m)E_A^2]^m\} \tag{42}$$

where b and m are constants, the latter considered by Grahame to lie between 0.5 and 2.[117] Here $\partial A_A/\partial E_A$ was defined by Grahame as the *differential dielectric constant* or *dielectric coefficient,*[44,108b,122] i.e.,

Buckingham's *incremental dielectric constant*. The corresponding *integral* dielectric constant $\varepsilon_A = A_A/E_A$ given by Booth's equation[124] is the integral of Eq. (42) between the limits 0 and E_A, divided by E_A. Its value depends on the value of m, and is:

$$\varepsilon_A = n^2 + (\varepsilon_0 - n^2)(bE_A^2)^{-1/2} \tan^{-1}(b^{1/2}E_A) \qquad (43)$$

$$\varepsilon_A = n^2 + (\varepsilon_0 - n^2)(2bE_A^2)^{-1/2} \ln\{(2b)^{1/2}E_A + (1 + 2bE_A^2]^{1/2}\} \qquad (44)$$

for m = 1 and m = 0.5, respectively. It has been stated[39,99] that Eq. (42) was based on the theoretical ideas of Booth,[73,124,125] but this does not appear to be the case, since it predates Booth's work. Grahame[130] compared these expressions with Booth's after his first publications,[73,124] using $b^* = 1.08$ x 10^{-8} esu^{-2}–cm^4 derived from Malsch's results.[112] He showed a rather good correspondence between the two expressions, particularly for m = 1. For m = 0.5, Grahame's integral predicts slightly lower ε_A values for moderate fields (e.g., 73.3 compared with 73.5 for Booth's expression for $E_A = 4,600$ esu-cm^{-2}, corresponding to 3.75 Å from a univalent ion). At very high fields, Eq. (44) predicts higher values than Booth's expression (e.g., 19.4 compared with Booth's 11.9 for $E_A = 10^5$ esu-cm^{-2}). However, Booth's value corresponds to a distance of only 2.0 Å from a univalent ion. Values at such high displacements are irrelevant, since Booth's assumptions concerning water structure break down as x_A approaches 3, i.e., for fields greater than about 4,600 esu-cm^{-2}. This is discussed later in this Section. Equation (43) (m = 1) was used by Laidler and coworkers[37,131] with b = 1.08 x 10^{-8} esu^{-2}–cm^4 to examine the dielectric behavior of water near ions and their electrostatic repulsions.

The relationship of Eq. (42) to Booth's is of interest. For $E_A^2[B\mu_v(n^2 + 2)/2kT]^2/15 \ll 1$, Eq. (38) may be written:

$$\varepsilon_A = n^2 + (\varepsilon_0 - n^2)/\{1 + E_A^2[B\mu_v(n^2 + 2)/2kT]^2/15\} \qquad (45)$$

Equation (45) has the same form as Eq. (42) with m = 1, but it represents the integral dielectric constant. The value of $[B\mu_V(n^2 + 2)/2kT]^2/15$ is 3.110 x 10^{-9} esu^{-2}–cm^4 with $\mu_v(n^2 + 2)/3 = 2.540$ x 10^{-18}, calculated from Eq. (36) using the experimental value of

---

* Units of b are E_A^{-2}. For units of E_A, 1.0 V–m^{-1} = 0.01 V–cm^{-1} = 3.33 x 10^{-5} esu-cm^{-2}.

$(\varepsilon_o - n^2)$. This we may put equal to b'/m, where m is unknown. The quantity b' is analogous to Grahame's b, but is an integral quantity derived from the experimental $\varepsilon_o - n^2$, rather than a differential quantity derived from Malsch's experiments with fields up to 833 esu-cm^{-2} (2.5 x 10^5 V-cm^{-2}).[112] The correctness of Malsch's data has in any case been questioned, since he did not allow for adiabatic heating.[132] As Grahame pointed out,[117] m does not enter into the E_A^2 term in the binomial expansion of Eq. (42). We can infer the value of m by comparing the coefficients of these and the E_A^4 term in expansions of Eqs. (35) (see Eq. 38 footnote) and 42, i.e., $x^2/15 = bE_A^2$, and $2x^4/315 = (b/m)^2$, giving m = 7/13 = 0.538. This approximate fitting using the first two terms provides a starting point to give a simple and convenient function mimicking Booth's. As might be expected, the error is least at small to relatively large fields (at 10^4 esu-cm^{-2}, ε_A = 61.17 vs. 61.18 for Booth, a difference of –0.8%). It becomes greater as fields increase, e.g., at 10^5 esu-cm^{-2}, it is 10.3, vs. 11.9 for Booth, a difference of – 15%. All things considered, representing Eq. (42) as an expression for the *integral,* rather than the *differential* dielectric constant, using a slightly adjusted b = 3.15 x 10^{-9} esu^{-2}–cm^4, with m = 0.5 in Eq. (42), gives a very good fit with Eq. (35), as Figure 3 shows. At 10^4 esu-cm^{-2}, the error is –0.05% compared with Booth, with –5.8% at 10^5 esu-cm^{-2}. This use of smaller m values can further improve accuracy in very high fields (e.g., m = 0.485, –1.0% at 10^5 esu-cm^{-2}) but small changes then occur at lower fields (+0.09% at 10^4 esu-cm^{-2}).

Why this equation fits so well is clear from the expansion of Eqs. (42) and (35). In weak fields $(b/m)E_A^2 < 1$, and Eq. (42) becomes $\varepsilon_A = n^2 + \{(\varepsilon_o - n^2)(1 - bE_A^2)$, which corresponds to Eq. (41), whereas Eq. (35) may be written:

$$\varepsilon_A = n^2 + (\varepsilon_o - n^2)(5bE_A^2/3)^{-1/2}L(x_A) = n^2 + (\varepsilon_o - n^2)(3/x)L(x_A) \qquad (46)$$

In intense fields, $L(x_A)$ approaches $(1 - 1/x_A)$, and it is easily verified by calculation that $(3/x_A)(1 - 1/x_A)$ in Eq. (46) coincides closely with $(1 + 2x_A^2/15) = (1 + 2bE_A^2)^{-1/2}$ from Eq. (42) for x_A > 2.5-3.0. Thus Grahame's Eq. (42) with m = 0.5 may act as an excellent replacement for Booth's expression (Eq. 35) for the *integral* dielectric constant to simplify calculation, if necessary. It also suggests a function which may be used to mimic $L(x_A)$:

$$L(x_A) = (5b/3)^{1/2}E_A/(1 + 2bE_A^2)^{1/2} \qquad (47)$$

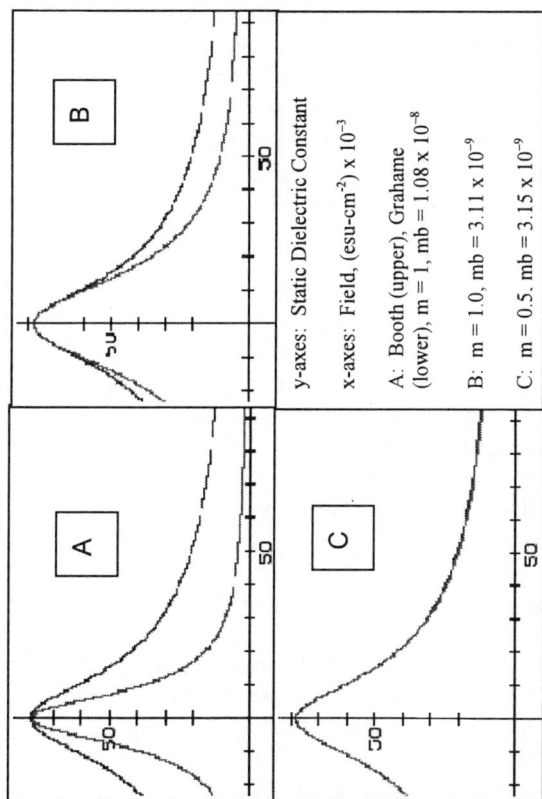

Figure 3. Grahame's equation[117, 130] $\partial A_A/\partial E_A = n^2 + \{(\varepsilon_o - n^2)/[1 + (b/m)E_A{}^2]^m\}$ (Eq. 42) for the *differential* dielectric constant, plotted as an *integral* dielectric constant ε_A with m = 0.5, b = 3.15 x 10^{-9} esu^{-2}-cm^4 along with Booth's equation (Eq. 35) for ε_A as a function of field E_A with A = 1, B = 7/3[124], μ_v = 2.027 D. Ordinate: ε_A (dimensionless units). Abcissa: E_A, esu-cm^{-2} x 10^{-3}.

y-axes: Static Dielectric Constant

x-axes: Field, (esu-cm^{-2}) x 10^{-3}

A: Booth (upper), Grahame (lower), m = 1, mb = 1.08 x 10^{-8}

B: m = 1.0, mb = 3.11 x 10^{-9}

C: m = 0.5. mb = 3.15 x 10^{-9}

Further progress requires an expression for ε_A as a function of $\mathbf{A}_A$, i.e., as a function of distance from an ion. From Eqs. (35) and (36), the Booth expression for ε_A in terms of ε_0 is:

$$\varepsilon_A = n^2 + (\varepsilon_0 - n^2)(3/x_A)\mathbf{L}(x_A) \qquad (48)$$

This expression may determine the relationship between ε_A as a function of $\mathbf{A}_A$ in intense fields, e.g., for $x > 3$, when the approximation $\mathbf{L}(x)_A = 1 - 1/x_A$ applies. However, it assumes classical statistics, so it will require a quantum correction for x values in this range, assuming that the Langevin function is still applicable. For smaller values of x_A, Eq. (48) may be solved graphically. A more transparent expression for small x_A values would be useful. Equation (42) with $m = 1/2$ leads to a quartic equation in ε_A which may be simplified by expansion to a cubic expression. A solution is the use of Booth's approximation ε_A and $\varepsilon_0 \gg n^2$, with substitution of Eq. (47) into Eq. (42). This gives a simple and useful approximation:

$$\varepsilon_A = (\varepsilon_0^2 - 2b\mathbf{A}_A^2)^{1/2} \qquad (49)$$

with $b = 3.15 \times 10^{-9}$ esu^{-2}-cm^4, which is rather accurate for $x_A < 2.5$, above which the classical theory may in any case not be applicable. The value of both ε_0 and b will depend on temperature and solution composition and concentration.

How the polarization opposes the displacement in intense fields can be shown by using the approximate expression for $\mathbf{L}(x_A)$ derived from Eqs. (45) and (49):

$$\mathbf{L}(x_A) = (5b/3)^{1/2}\mathbf{A}_A/\varepsilon_0 = a_{ul,o}\mathbf{A}_A/\mu_{eff}\varepsilon_0 = (1 - n^2/\varepsilon_0)\mathbf{A}_A/4\pi\mu_{eff}n_0 \qquad (50)$$

Since $(1 - n^2/\varepsilon_0)$ is close to unity for associated polar liquids, $\mathbf{L}(x_A)$ only depends strongly on the value of the electric displacement. From Eq. (50), to a good approximation up to $x = 2.5$:

$$\mathbf{P}_{ul,A} \cdot \mathbf{A}_A/2 = -\mu_{eff}(\mathbf{A}_A \cdot \mathbf{A}_A)8\pi n_0 \qquad (51)$$

So long as $\mathbf{P}_{ul,A}$ and $\mathbf{L}(x_A)$ are proportional to $\mathbf{A}_A$, the polarization produced can oppose the displacement. When $\mathbf{L}(x_A)$ tends to its high-field value of $1 - 1/x$ starting at about $x = 3$ [$L(x_A) = 0.67$)], the polarization can no longer oppose the displacement, so ε_A falls rapidly. For $x_A = 3$, Eq. (35) gives $\varepsilon_A = 53.2$ at $A_A = 7.39 \times 10^5$ esu-cm^{-2}, whereas Eq. (42) with m = 0.5, and Eq. (49), both with b = 3.15 x 10^{-9}, give $\varepsilon_A = 52.0$ and 53.2 at the same displacement, corresponding to distances of 2.55 Å, 3.60 Å, and 4.41 Å from ions with z = 1, 2, and 3 respectively. In this range of x_A and A_A, the Booth expression may be in error as the structure of water begins to break down in high fields. At smaller values of A_A the error would much less, e.g., for x = 1, the values are 73.6 and 73.5, respectively at $A_A \approx 3.41$ x 10^5, i.e., distances of 3.75, 5.31, and 6.50 Å for z = 1, 2, 3. For x = 2 the corresponding values would be 63.5 and 64.0, with $A_A \approx 5.88$ x 10^5, and 2.86, 4.04, and 4.95 Å. The extent of structure-breaking at x = 3 (i.e., when the interaction energy between the librating or rotating dipole and the cavity field is 3kT) is indicated by the fact that the average angle of the superdipole to the field is then $\cos^{-1}L(3) = \cos^{-1}0.672 = 47.8°$. This limiting value of A_A should cover all requirements for calculations involving the outer solvation sphere and corresponds to a upper limiting value of x and $L(x)_A$ which may be reasonably consistent with Booth's simplifying assumptions. This is further discussed in Section III-12.

So far the mean Gibbs energy of interaction between an ion and a single dipole (Eq. 18) has been ignored. This should not be confused with the effective potential energy to describe the dynamics of superdipoles of effective moment $\mu' = 1.5B\mu_v(n^2 + 2)/3$. The moment μ_{eff} of the central dipole of each superdipole group is $A\mu_v(n^2 + 2)/3$. The latter, multiplied by the field and $L(x_A) = L\{1.5BE_A[\mu_v(n^2 + 2)/3]/kT\} = \cos\theta$ is the potential energy required to torque the central dipole to a thermally-averaged angle θ to the internal field. The Gibbs energy of interaction between an ion and a single dipole (from Eq. 18) at low to moderate fields is $-P_A \cdot A_A/2n_o = -\mu_{eff}L(x_A)A_A/2$, while the Langevin potential energy is $-\mu'A_A/\varepsilon_A$, i.e., their ratio is $(A/3B)L(x_A)\varepsilon_A$, where A and B are from Eq. (35). From Eq. (49), this becomes $(A/3B)L(x_A)(\varepsilon_o^2 - 2bA_A^2)^{1/2}$. At $x_A = 0.3$, when the Booth theory predicts $\varepsilon_A = 77.9$ at an A_A value of 1.08 x 10^5 (6.7, 9.4, and 11.5 Å from ions with z = 1, 2, and 3), the Gibbs energy of interaction is already 0.33kT. At $x_A = 1$, it has become 3.3kT, and at $x_A = 2$, 9.6kT. This approaches the strength of hydrogen bonds, so water molecules will become successively stripped from the tetrahedral super-dipole, and the associated Kirkwood-Booth structure will finally approach that of a non-associated Onsager liquid. The tetrahedral group will first line up with the three nearest dipoles and the central dipole towards the ion, giving a B value $> 4^*$. The initial increase in B, i.e., increase in x, will result in an approximately 30% decrease in ε_A, but both of these are more than compensated by the increase in B in the expression for energy of interaction. As B decreases to the Onsager value of 1.0, the interaction energy becomes slightly more negative, but the dielectric constant will show an increase of ca 25% over the original value, since x becomes smaller. Whether these changes are significant is difficult to tell, since they should be approximately compensated by the fact that Booth's assumptions regarding the Onsager-Frölich cavity field for F_A and the Lorentz cavity for n^2 will begin to fail. As stated earlier, the Langevin derivation assumes free rotation. The transition from free rotation to vibration occurs at $x_A > 0.25$ per degree of freedom. At energies above the limit, the fraction of molecules still capable of rotation will drop rapidly. Most molecules will be vibrating in alignment with the field. Thus, the mean angle to the field will drop more rapidly than predicted by the Langevin function, which in any case will require a quantum correction. Hence, under intense-field conditions, Booth's "bottom-up" approach to the dielectric constant would be best replaced by a "top-down" analysis.

For dipoles close to an ion, it may be assumed that the interaction between the permanent moment and the ion displacement may be given by rewriting Eq. (23) as follows for a single dipole:

$$\Delta g_{ul,A} = (-)(:_v.A_A/2 - :_v.E_A/2 - X_A) \qquad (52)$$

where X_A is defined in Eq. (22). This assumes that the Lorentz and Onsager cavities have disappeared, so $q_A = 1$ and the permanent moment has reverted to its vacuum value. The angle between the moment and the displacement can be assumed to be 0. The induced dipole will be calculated separately. Since $E_A = A_A/\varepsilon_A = A_A + E_{p,A}$, we obtain from Eq. (22):

$$\Delta g_{ul,A} = (-)(:_v.A_A + :_v.E_{p,A}/2 - W_A) \qquad (53)$$

---

* Booth[124] suggests 5, but this seems unlikely. The tetrahedral group will be expected to take up an orientation with the base of the pyramid towards the ion. Two the base water molecules can rotate to align themselves in the direction of the ion, while the other must be bent through about 50°. The other two dipoles will be less affected by the field.

which is equivalent to the familiar expression used by Bernal and Fowler,[16] in which the negative $\mu_V.E_{p,A}/2$ term is the repulsive energy per dipole due to the presence of all other dipoles in the system, and W_A is the work to detach the selected dipole. This work may be calculated by integration:

$$X_{ul,A} = - (P_A.E_A + W_A) = - \int_{F_{A,1}}^{F_A} P_A \, dF_A = -\mu_{eff} \int_{F_{A,1}}^{F_A} L(x_A) \, dF_A$$

$$= -(AkT/B)ln[x_{A,1}(\sinh x_A)/x_A(\sinh x_{A,1})] \tag{54}$$

where $F_{A,1}$, $x_{A,1}$ are the values corresponding to the lower limit of structure-breaking, and F_A, x_A are the values close to the ion. However, in view of all the uncertainties given above, $X_{ul,A}$ may be equated to the work of breaking one hydrogen bond (although some degree of hydrogen bonding may be possible at the side of the dipole away from the ion).

We now note that in Eq. (35), the local dielectric constant ε_A is equal to $A_A/(A_A + E_{p,A})$, so that it may be calculated in a "bottom-up" manner if the average geometries of successive shells of dipoles are known. The dipole field is proportional to $1/a^3$, but the number of dipoles per shell increases as a^2, so the summation should be carried out until sufficient accuracy is achieved. Section IV attempts to do this for the inner, second, and third shells.

9. The Ion-"Continuum" Interaction in Polar Liquids

From Eqs. (18), (23), and (24):

$$d\Delta G_A = (-)(P_{ul,A}.A_A/2 + P_{e,A}.A_A/2)dV \tag{55}$$

From $P_{ul,A} = \alpha_{ul}n_oE_A = (A_A/\varepsilon_A)(\varepsilon_A - n^2)/4\pi n_o$; $P_{e,A} = \alpha_e n_o E_A = (A_A/\varepsilon_A)(n^2 - 1)/4\pi n_o$, we obtain:

$$d\Delta G_A = -(A_A^2/8\pi)[(1 - 1/\varepsilon_A)]dV \tag{56}$$

Using Eq. (49), which is valid to $L(x_A) \approx 3$, Eq. (44) may put in the form:

$$d\Delta G_A = -(A_A^2/8\pi)[1 - (\varepsilon_o^2 - 2bA_A^2)^{-1/2}]dV \tag{57}$$

After putting $dV = 4\pi a^2 da$, expanding, and integrating from a_0 to ∞, the result is:

$$\Delta G_A = -(z^2 e^2/2a_0)[(1 - (1/\varepsilon_0)(1 + t/5a_0^4 + t^2/6a_0^8......)] \qquad (58)$$

where $1/\varepsilon_0 \approx 0.013$, and $t = bz^2 e^2/\varepsilon_0^2 \approx 11.7z^2$ (with a in Å). We are interested in a_0 values for $z = 3$ of about 5.2 Å, where the terms corresponding to the advent of dielectric saturation make no more that about an additional 0.04% contribution to the total Gibbs energy of interaction of the ion with the continuum.

Buckingham[96] reached a similar conclusion using a Born "self energy" argument. Instead of using $(\varepsilon_0 \mathbf{E}_A.\mathbf{E}_A/8\pi)dV$, with $dV = 4\pi a^2 da$, integrated from a_0 to ∞, he used its equivalent differential for the electrostatic energy density in dV, $\mathbf{E}_A d\mathbf{A}_A/4\pi$[133], integrated from 0 to $\mathbf{A}_A$, multiplied by $dV = 4\pi a^2 da$ integrated from a_0 to ∞. He expanded ε_A in an even power series,[121] concluding that the distances in second and higher water dipole shells made the higher terms negligible, thus dielectric saturation there could be ignored.

As in Eqs. (19) and (20), the integration here is to ∞, and so represents the infinite dilution case. In the relatively concentrated (0.1-1.0 M) solutions used in electrochemical kinetic experiments, the summation or integration should be taken over the appropriate number of shells to electroneutrality. Since the Debye-Hückel approximation* will not apply to such cases, this will not be identical with the Debye reciprocal length, but may be given by a pseudo-lattice approximation.[134]

It follows that an equation of Born type, but based on different physical principles (Eq. 56) is a good approximation for the continuum energy in dipolar liquids up to the onset of dielectric saturation at $x = 3$, provided it is integrated from an appropriate distance somewhat less than that of the superdipole center of the innermost solvation shell from the central ion. This corrected radius will differ from the distance from the ion to the dipole centers of the solvation shell under consideration by about 50% more than the radius of a water molecule.

---

* The Poisson equation should in any case not be applied where there is a gradient of E, e.g., in radial geometry[121], see footnote on p. 207.

10. Induction Effects

Moelwyn-Hughes[93] examined the ion-solvent interaction energy outside of the first coordination shell or Inner Sphere by a non-Born charging method using the same Inner Sphere induction term as Bernal and Fowler[16] and Eley and Evans,[92] i.e., $-\alpha_e(\mathbf{E}_A)^2/2 = -\alpha_e(\mathbf{A}_A)^2/2$ with $\varepsilon_A = 1$ (c.f., discussion under Eq. 18). He also used the Inner Sphere expression for the nuclear part of the polarization, i.e., $-ze\mu_v/a^2$ without no dielectric constant or orientational term $\cos\theta$. The use of this (incomplete)** expression (see Eq. 18) for the nuclear part of the polarization led him to believe that the interaction between an ion and permanent dipoles outside the first coordination shell could be disregarded, since it results in a physically meaningless integral for the Gibbs energy, i.e.,

$$- \int_{a_0}^{\infty} 4\pi n_0 a^2 (ze\mu_V / a^2) da = (-\infty) \qquad (59)$$

The correct expression for this interaction to $L(x<3)$, i.e., $[\alpha_{u,A}\varepsilon_A(\mathbf{E}_A)^2/2]dV$ (Eq. 19) has the same form as that for the induced dipole energy $[\alpha_e\varepsilon_A(\mathbf{E}_A)^2/2]dV$. The use of the combined expression with $\alpha_A = (\varepsilon_A - 1)/4\pi n_0$ (Eq. 19) ensures that it contains the induced electronic component, equal to about 2.3% of the total at 298 K.

11. Polarization of ut Type

The possible presence of polarization of ut type ($\mathbf{P}_{ut,A}$) induced in the non-uniform field rather close to an ion, but still within the validity of the assumptions of Eq.s (48)-(51), (57), and (58) is a possible source of error. Physically, ut polarization consists of dipole translation along field towards the ion, increasing the local value of $\mathbf{A}_A$, $\cos^{-1}L(x)_A$, and $\mathbf{P}_{ul,A}$. To examine this requires knowledge of the binding forces and energies between water molecules. The potential energy Mie function for each pair is[137]

$$\Phi = Ad^{-12} - (1 + s_c)Cd^{-3} - (1 + s_b)Bd^{-6} \qquad (60)$$

** Marcus[43a], following Mandel and Mazur[135], and Brown[136] stated that many earlier expressions for the polar term for intermolecular energy were incomplete or wrong.

where A, B, and C are constants, d is the distance between the dipole centers of nearest neighbors, and s_c, s_b are structural factors covering interactions with next-nearest and subsequent neighbors. B is the sum of two terms, the first being the dispersion energy, $0.75Z^{1/2}hv_e\alpha_e^2$ (= 6.81×10^{-59} erg-cm^{-6}), where Z is the number of electrons, of frequency v_e, the second being the electronic induction energy,[137] $2\alpha_e\mu^2$, where μ may be put equal to an effective liquid water dipole moment, e.g., 2.42 D,[123] giving 1.69×10^{-59} erg-cm^{-6}. Ref. 137 assumed no dielectric constant in the summation of s_b, s_c to infinity, which cannot be correct. No dielectric constant may be required between nearest neighbors (see later discussion), but an ε_A term is required for next-nearest and subsequent neighbors. Since this will be large, longer-range interactions and s_b, s_c may be neglected.

The minimum energy for two pairs of nearest neighbors, is 49.18 kJ-mole^{-1} in liquid water.[138] The hydrogen bond distance d is 2.76 Å at the melting point and 2.9 Å at 293 K.[103,138] C is $f\mu^2$, where f is a function of the nearest-neighbor angles, with a maximum value of 2.[137] Putting the differential of Eq. (60) equal to zero at d = 2.9 Å, and using Eq. (60) with Φ equal to 4.082×10^{-13} ergs per dipole pair gives A = 6.50×10^5 erg-Å^{12} and C = 1.095×10^{-11} erg-Å^3, i.e., C = $1.87\mu^2$. The number of nearest neighbors in liquid water[138] is actually about 4.4-4.6. With this correction, C falls to $1.57\mu^2$–$1.66\mu^2$, but Pople[103] indicated that using more than 4 nearest neighbors is unjustified, since the larger numbers result from the partial collapse of the second shell. The expression for the interaction of two dipoles[139] inclined at half of the H–O–H bond angle of 104.45°[140] gives f = 1.23. The simplest model for the charge distribution[95] is derived from Rowlinson,[141] but it places all of the negative charge on the oxygen, rather than located symmetrically out of the H-O-H plane.[142] Better models will include the lone pair geometry[143,144] or use the central force potential[145] using repulsive terms for H...H and O O, and an attractive-repulsive potential well for O H.

Duncan and Pople (DP)[143] placed the lone pairs at ±60.1° to the dipole axis. With one end of each molecule aligned at this angle, and the other aligned at half of the bond angle gives f = 1.30. These values assume that the field corresponds to that of a simple dipole of vanishingly small length. An accurate calculation using the DP charge distribution (Section V) shows much higher local fields than predicted by the assumption of simple dipoles at distances of 2.9 Å from the oxygen atom. The fields in line with the H–O axes are ca. 40% higher than those for an oriented dipole with the Coulson-Eisenberg moment,[123] and those along the sp hybrid orbital–O axes are 10% higher. This explains why the calculated dielectric constant is somewhat low.[103]

The above results give us the possibility of calculating the distortion of the water structure in the shells surrounding the coordination shell or inner solvation sphere of an ion, and an estimate of magnitude of the electrostrictive or translational polarization energy in the continuum to the point of partial dielectric saturation near an ion. The inner shell for 2+ and 3+ ions is a puckered trikisoctahedron,[39,146,147] with dipoles in two overlapping layers at different distances. The dipoles in the (distorted) shell next to the eight outer trikisoctahedron dipoles for ions with $z = 3$ have centers averaging approximately 5.0-5.2 Å from the central ion. Looking at the system from a simple one-dimensional viewpoint (i.e., along the direction of the electric displacement), it is possible to set up a series of simultaneous equations involving the differentials of Eq. (60) and the differential of $-P_{ul,A} \cdot A_A/2$ to give the forces between the various layers. These equations can be solved graphically after being all set equal to zero to allow the amount of electrostriction between the layers to be estimated. The results show that $z = 3$, the distance between a water dipole with its center initially situated at 5.2 Å from the ion before its charge is "switched on" will translate down the field. However, the one-dimensional translation will be hindered by lateral compression, so that the effective translation will be only about 0.025 Å. A dipole initially situated at 5.5 Å will correspondingly move by 0.02 Å. A dipole situated in the next shell, at about 9 Å from the central ion will move by ≈ 0.005 Å. The movement in subsequent shells will be negligible. The corresponding increases in interaction energy (proportional to r^{-4}) is about 2% at 5.2 Å for $z = 3$, and 1.8% at 5.5 Å. The cumulative motion in the shell at 8.0-8.3 Å (about 0.03 Å) results in increase of about 1.5%, a percentage which falls slowly in subsequent layers (e.g., 3% at 10.9 Å). Both the small displacement, and the small change in interaction energy are approximately proportional to z^2, so these figures must be multiplied by about 0.44 for $z = 2$, and by 0.11 for $z = 1$. A calculation to determine the energy in individual shells suggests that the predominant $(-P_{ul,A} \cdot A_A/2).dV$ terms adds about 1.5% to the total electrostatic energy in the continuum measured from $r = 5.2 - 5.5$ Å for an ion with $z = 3$, about 0.7% for $z = 2$, and less than 0.2% for $z = 1$. The electronic polarization energy terms $(-P_{e,A} \cdot A_A/2).dV$ will be similarly affected. For $z = 3$ and $r = 5.2$

Å, the estimated dielectric constant is 63.5 from Eq. (35), and $E_A = 8.3$ x 10^3 esu-cm^{-2}. The estimated electrostrictional molar volume change (0.144 x 18) is 0.26 cm^3-mole^{-1}. This is in good agreement with the estimates of Desnoyers, Verrall, and Conway[148] for this field strength. The effect of the term $(-P_{ut,A} \cdot A_A/2).dV$ on the interaction energy of interaction up to dielectric saturation is small, but not negligible.

12. Quadrupole or Multipole Effects

A trigonometrical calculation[96] shows that an water quadrupole-ion displacement interaction energy at ionic (charge ze) distance a from the dipole center is:

$$d\Delta\Phi_A = -ze(\mu/l)[(l\cos\theta/a^2 \pm y^2(1 - 3\cos^2\sigma\sin^2\theta)/2a^3] \qquad (61)$$

where partial charges $\pm\mu/2l$ are on the protons and oxygen, l, y are the dipole length and half of the distance between the partial positive charges, σ is the angle of y to the plane containing a and the dipole axis, and θ is the mean angle $\cos^{-1}L(x)_A$ of the dipole axis to the field. The ± signs before the angle independent quadrupole terms refer to positive and negative ions respectively, i.e., if a is measured from the dipole center, the interaction energy to a positive ion is more negative than that of an equivalent simple dipole. In most texts, ion distance is measured from either the geometric center or center of gravity of the water molecule. The latter is situated about 0.066 Å from the oxygen atom, so a further term $\pm[(0.066)^2 - (l - 0.066)^2]/a^3$ must be added within the square brackets. The effect of the differently defined quadrupole moment may be to reverse the difference between the ions at constant a (see Appendix).

Equation (61) is $-P_{ul,A} \cdot A_A$ for a simple molecular quadrupole. Assuming free rotation, the average value of $\cos^2\sigma \approx 1/3$, giving a ratio of the mean quadrupole to dipole interaction equal to $[1 \pm (y^2\cos\theta)/2al)$, or from Eq. (50) and the definition of A_A, $[1 \pm (1 - n^2/\varepsilon_o)y^2ze/8\pi\mu_{eff}n_oa^3l]$. After inserting numerical values,* this is $(1 \pm 1.085z/a^3)$, with a in Å. Multiplying Eq. (57) by this amount and integration from a_o to ∞ gives:

---

* Dipole length l = 0.5871 Å, bond angle 104.45°, bond length 0.9584 Å[140], hence y = 0.7575 Å.

$$\Delta G_A = -(z^2 e^2/2a_o)[(1 - 1/\epsilon_o) \pm 1.085z(1 - 1/\epsilon_o)/4a_o^3 - t/5\epsilon_o a_o^4$$
$$\pm 1.085z(-1)t/8\epsilon_o a_o^7 - t^2/6\epsilon_o a_o^8..)] \tag{62}$$

where t is defined under Eq. (57). For $z = 3$ and $a_o = 5.2$Å, the first quadrupole term contributes only about 0.1% to ΔG_A, with negligible contributions for higher terms. The same argument should be applied to DP multipoles in Kirkwood tetrahedral groups, but there is no reason to believe that the result would be very different.

13. At Dielectric Saturation

The local dielectric constant $\epsilon_A = A_A/(A_A + E_{p,A})$ is $n^2 + 4\pi P_A/E_A$ only for a constant, parallel field. The factor 4 in $4\pi P_A/E_A$ will become smaller at small distances, when individual dipoles are no longer parts of quasi-continuous shells of charge (Appendix 1). Within the limits of his assumptions, Booth rather rigorously obtained an expression for ϵ_A containing the Langevin function, whose derivation uses the Boltzmann distribution and classical statistical mechanics. These may both become invalid for a small cohort of dipoles whose interaction with the field approach kT. In a parallel field, a large number of dipoles of all energies may be readily found, so that Stirling's formula for factorials of large numbers used in deriving the Boltzmann distribution depends should be satisfied.

In radial fields near an ion, the population of Langevin dipoles in a given interaction range becomes rapidly less with decreasing ion-dipole distance, e.g., in a shell of thickness 5.8 Å with a center at 6.3 Å there are about 20 superdipoles, with about 5 times as many in the next shell. A less approximate Stirling expression

$$\ln N! = 1/2\ln 2\pi N + N\ln N - N \tag{63}$$

gives an acceptable value for N! even for small N (−7.8%, − 4.0%, −2.7%, −1.4% for $N = 1, 2, 3, 6$ respectively). The modified Boltzmann distribution then is approximately:

$$N_i/N = (\exp - e_i/kT)(\exp - 1/2N_i)/\Sigma(\exp - e_i/kT)(\exp - 1/2N_i) \tag{64}$$

where N_i is the number of molecules in the ith state of energy e_i, in a total population of N molecules. The numerator and the partition function f in the denominator contain the weighting factor $\exp - 1/2N_i$

for each N_i. As a minimum (see above), we are interested in N_i values corresponding to a fraction of the subpopulation of molecules in the first shell for 1+ ions, whose motions (see below) have transitioned from free rotations to large amplitude librations under the influence of the ion field. The distortion in the energy distribution function is to give smaller weighting factors for large e_i and small N_i. The smaller f values will result in smaller mean thermal energies than the classical Boltzmann values, which will generally result in smaller mean angles of the dipole to the field, i.e., somewhat smaller effective values of ε_A near the ion than Booth's theory predicts. This error will become progressively less so for successive shells.

The Boltzmann distribution assumes no interaction between molecules, i.e., an ideal system. When dipoles are oriented by a strong field, they interfere with each other. In relatively weak fields, free rotation of solvent dipole groups occurs, as the Kirkwood-Booth model of the low-field value of the dielectric constant requires. The Schrödinger equation for a rigid rotating dipole cluster of effective moment undergoing planar rotation in an internal electrostatic field $F_A = zeq_A/\varepsilon_A a_A^2$ is:

$$\partial^2\psi/\partial\phi^2 + (8\pi^2 I/h^2)(E - V\cos\theta_A)\psi = 0 \qquad (65)$$

where θ_A is the mean angle between the dipole and the field, ϕ is the angular displacement from θ_A, E is kinetic energy, and $V = zeq\mu'/\varepsilon_A a_A^2$, with μ' as in Eq. (34) and a_A equal to the distance from an ion. The other symbols have their usual meanings. This equation has one total energy minimum at $\theta = 0$. Similar wave equations with two minima (at $\theta = 0$ and π, for hindered rotation in a crystal) and for three-dimensional rotation with n minima (at $q = 0$, $2\pi/n$, $2\pi/2n$, etc., for hindered internal rotation of, e.g., CH_3 groups within a molecule rotating in space) have been evaluated.[149,150] The eigensolutions for such equations transition from those for a non-degenerate free rotation (when $V \ll 0.25kT$) to approximately those for a harmonic vibration (when $V \gg 0.25kT$). From Eqs. (35) and (36), this transition will occur at practically the bulk ε_0 value at 7.3, 10.3, and 12.6 Å from ions with $z = 1$, 2, and 3 respectively. For $z = 1$, this is a little beyond the center of the first Continuum shell, while for $z = 3$, it is closer than the center of the second shell. At the center of the first Continuum shell at 6.3 Å for $z = 1$ ions (Sections III-2 and IV), Eqs. (49) and (35) show $x = 0.34$ and $\varepsilon_A = 77.8$. For $z = 3$ ions at 7.6 Å, the corresponding values are $x = 0.72$

and $\varepsilon_A = 75.8$. Under these circumstances, Booth's assumptions (including the use of the classical Boltzmann distribution may still approximately apply), and the superdipoles with maximum energy of kT per axis will be performing large-amplitude librations. The frequency of free rotation (v_o) of superdipoles under energy equipartition conditions at 298 K is estimated as 1.25×10^{11} s^{-1} in Section VI-1. The libration frequency[151] will be slightly less at $v_o(1 + 1/16x)^{-1}$. Both are well within the classical range, so quantum corrections to the energy function are not required. Thus, we conclude that while physically, Booth's assumptions and the approximations considered here may still have value to x = 3, in reality, they can be abandoned at x = 0.3-0.75. At higher field strengths, the dipoles become fully oriented and dielectric saturation rather suddenly occurs.

We therefore arrive at Eq. (53), the classical expression used by Bernal and Fowler[16] for the inner sphere, with the addition of a small additional term, which may be regarded as a $+T\Delta S$ term for the break-up of the solvent structure, which can be replaced by an estimate of the energy to break the structure of water.

14. Summary of Ion-Solvent Interactions

The above discussion shows that the electrostatic free energy of solvation can be divided into an coordination shell or inner solvation sphere in which ε_A is close to 1, where the $\mathbf{P}_u$ interaction depends only on $-\mu A_A$, and an outer solvation sphere where the $\mathbf{P}_A$ interaction depends to a good approximation on Eqs. (55)-(57), but in which the electrostatic Gibbs energy may be approximated by the integral in Eq. (58), which resembles the Born charging equation, but it is obtained in a different way with a more definite physical meaning.

An accurate estimate of the electrostatic Gibbs energy of interaction between an ion and the polar molecular "continuum" should be summed to the point of electroneutrality over all time-averaged solvent molecule charge positions and configurations, but a sufficiently accurate expression may be obtained by the above integration from a carefully chosen minimum distance. This distance is the weighted mean of the dipole center distances in the first shell of the continuum surrounding the solvated ion, minus a correction term equal to about 2.00 Å. Almost 60% of the continuum energy is typically in the first "outer" shell, which is characterized by a somewhat lower dielectric constant than the bulk value estimated by the Kirkwood-Booth theory.[72,73,124,125]   However, since the local value of the dielectric constant, ε_A enters into the final energy expression via a factor $[1 - 1/\varepsilon_A]$, the overall change in Gibbs energy due to the change in dielectric constant is small. Grahame[117] reached a similar conclusion using Born charging in conjunction with the differential dielectric constant, Eq. (42).

To a good approximation, the quadrupole or multipole model of the water molecule may be considered to be a dipole to estimate the electrostatic Gibbs energy of interaction with the continuum. Booth's expressions[73,124,125] for the dielectric constant break down more rapidly in large radial fields than in parallel fields. The dielectric constant to be used under these conditions can only be determined by evaluating the opposing local counterfields. The limit corresponds to $x \approx 3$, where Booth's model would begin to be increasingly inaccurate if it were still physically relevant. Because of the abruptness of the change from high values of the local dielectric constant to essentially vacuum values, Bernal and Fowler's division of the energy of interaction of an ion and a dipolar solvent into Inner and Outer Spheres[16] is justified.

IV. THE INNER SPHERE(S)

1. Multivalent Cations as Trikisoctahedra

A model is given here for multivalent cations with large Gibbs energies of hydration with "puckered" trikisoctahedral[39,146,147] primary or inner solvation shells. These dipoles are in direct line-of-sight of the ion and have an intense interaction with the displacement A_A in a given volume element dV. It has been suggested that they may be surrounded by monomeric water, depending on the ionic charge[152], but this improbable given the discussion in Sections III-12 and -13. Multivalent single-atom anions should behave like cations, but large multiatomic anions should have a single-shell structure.

It has been generally assumed[16,92,93] that the Langevin function $L(x)_A$ for these dipoles will be close to unity. Because of the close line-of-sight interactions, the assumptions required for the presence of Onsager and Lorenz cavities fail under these conditions. Frölich's requirement[108] that the cavity must be a spherical volume large enough to have the dielectric properties of a macroscopic specimen, and contain a sufficient number of charges to be treated by classical statistical mechanics cannot be so next to an ion, where the internal to external field ratio q_A will be close to unity, and the effective moment μ_{eff} should be close to the vacuum value (discussed under Eq. 37).

Booth's theory[73,124,125] predicts that ε_A will fall to n^2 in intense fields, however the space involved in ion to nearest-neighbor-dipole interactions is not bathed in an electronic medium. The ion itself is not polarized, since there is no net field in spherical symmetry. Thus, n^2 should be unity for the primary shell and other nearest-neighbor interactions, c.f., Bernal and Fowler,[16] Eley and Evans.[92] The dipole fields $\mathbf{E}_{p,A}$ opposing the ion dielectric displacement $\mathbf{A}_A$ may be calculated from the fields of nearest-neighbor and more distant dipoles to obtain the local dielectric constant $\varepsilon_A = \mathbf{A}_A/(\mathbf{A}_A + \mathbf{E}_{p,A})$. Components of $\mathbf{E}_{p,A}$ determine the repulsive energies between pairs of dipoles, and produce the net external field $\mathbf{A}_A/\varepsilon_A = \mathbf{A}_A + \mathbf{E}_{p,A}$, whose internal field creates induced dipoles. Previous work[16,92,93,152] has generally ignored the effect of the dipole fields on induction, which can lead to considerable error for high-z ions with up to 14 trikisoctahedral dipoles. At dielectric saturation, the Gibbs energy of ion-nearest-neighbor-dipole interaction is given by Eqs. (52) and (53), with q_A, n^2 and $\cos\theta$ close to unity, with μ_{eff} close to μ_v. With $\mathbf{E}_A = \mathbf{A}_A/\varepsilon_A = \mathbf{A}_A + \mathbf{E}_{p,A}$, we obtain the equivalent of the Bernal and Fowler expression[16] for the inner sphere ion-permanent dipole interaction:

$$\Delta g_{ul,A} = (-)(:_v.\mathbf{A}_A + :_v.\mathbf{E}_{p,A}/2 - W_A) \qquad (66)$$

in which $:_v.\mathbf{E}_{p,A}/2$ is the energy per dipole due to the presence of all other dipoles in the system, and W_A is the work to detach the dipole from its environment in the presence of $\mathbf{A}_A$. In principle, W_A may be calculated (Eq. 54). Work is first required to orient the axes of freely-rotating transient groups of superdipoles, then more is needed to detach water molecules from these. Per dipole, this may be equated to the work of breaking a maximum of one hydrogen bond, since some hydrogen bonding should always be possible in the direction away from the ion.

The inner sphere of relatively small ions is generally accepted to have 4 or 6 dipoles in symmetrical tetrahedral or octahedral coordination, depending on the ion size and charge. In principle, these might respectively accommodate 4 and 8 other oriented intercalated dipoles as a second shell. In the tetrahedral case, the second dipole group would be at $(180 - \phi)°$ from the first, where ϕ is the tetrahedral angle, i.e., at 70.53°. In the octahedral arrangement, the second set are at $\phi/2$, i.e., at $\sin^{-1}(2/3)^{0.5} = 54.74°$ from the first. The octahedral arrangement is most frequent with high-z ions with octahedral d-orbitals, which make at least some contribution to their solvation energies.[153]

If the ion-dipole center distance is 2.75 Å and the overall packing corresponds to the closest approach distance for water (taken as 2.76 Å), then the second set of dipoles are 15% further away than the first, whereas if it is 2.5 Å, the second group are 31% further away. The concept of a tightly bound inner shell and a more loosely bound second shell offers a good explanation for the coordination number (c) and hydration number (n).[154,155]

Inner sphere equilibrium solvation energy calculations have generally used rather simple models, rather than attempting to determine the energy well parameters. Bernal and Fowler[16] included ion-permanent dipole and ion-induced dipole terms with Van der Waals attractions and Mie ion-water repulsions, with no exact calculations. Eley and Evans[92] replaced the permanent dipole term by point charges, and Moelwyn-Hughes[93,98] used Mie repulsive terms. Since the quadrupole concept was introduced,[96,141] a quadrupole term has generally been added to the permanent dipole term.[154] Often[154,156] no Mie repulsion term has been used, the ion-dipole distance being taken as the ion crystal- plus water molecule radii, where a reactive wall occurs. Such a model cannot give a correct description of the energy well and the force constant. The pair potentials[144,157] used in molecular dynamic simulations in recent years are also rather simple, and do not use the feedback of electrostatic fields.[145,158]

2. Inner and Second Sphere Energies

From Eq. (18), the potential energy of interaction Φ_{inner} of permanent $:_p$ and induced dipoles $:_i$ close to an ion may be written:

$$\Phi_{inner} = -:_p \cdot \mathbf{A}_A - \frac{1}{2}:_p \cdot \mathbf{E}_{p,2} + \mathbf{W}_{A,p} - :_i \cdot \mathbf{A}_A - \frac{1}{2}:_i \cdot \mathbf{E}_{p,1} + \mathbf{W}_{A,i} + \Sigma \phi_i$$

(67)

where $\mathbf{W}_{A,p}$, $\mathbf{W}_{A,i}$ are the work to create the dipoles, $\mathbf{E}_{p,1}$, $\mathbf{E}_{p,2}$ are the dipole field vectors along the ion displacement and μ_p axes, and the ϕ_i terms are the Mie repulsions between the dipole, its closest neighbors, and the ion. The factor 1/2 in the $\mathbf{E}_p$ terms avoids double counting. Following Eq. (18), $\mathbf{W}_{A,i} = +:_i \cdot \mathbf{F}_A/2 = +:_i \cdot (\mathbf{A}_A + \mathbf{E}_{p,1})/2$, giving:

$$\Phi_{inner} = -:_p \cdot A_A - \frac{1}{2}:_p \cdot E_{p,2} + W_{A,p} - :_i \cdot A_A/2 + \Sigma \phi_i \qquad (68)$$

The work $W_{A,p}$ to break up the superdipole structure in associated polar liquids is small compared with $+:_p \cdot A_A$. We see that the repulsive interaction between the induced electronic dipole and the total dipole field along the displacement axis has disappeared, i.e., all repulsive energy terms for each $:_i \cdot :_i$ and $:_i \cdot :_i$ pair are included in their work of formation, and repulsive terms containing $:_p \cdot E_{p,2}$ only involve those between permanent dipoles, giving an important simplification. If required, the appropriate quadrupole corrections[*] can be added to the $E_{p,2}$ values along the axis of each $:_p$, but this greatly complicates the calculations. By definition:

$$:_i = \alpha_e(A_A + E_{p,1}) \qquad (69)$$

The dipole field creating $:_{i,1}$ along the ion displacement axis is a geometrical function of all permanent dipole $:_p$ and induced dipole $:_i$ terms for a system of n dipoles:

$$:_{i,1} = \alpha_e A_{ion} + \Sigma g_{p,j} :_{p,j} + \Sigma g_{i,j} :_{i,j} \qquad (70)$$

where the $g_{p,j}$, $g_{i,j}$ terms are geometrical factors associated with the displacements for jth dipoles of each type. If appropriate, these may contain the optical refractive index n^2. A similar equation is written for each $\mu_{i,j}$ and the simultaneous equations are solved for each $:_{i,j}$ for the given system geometry. The net field $E_A = A_A + E_{p,1}$ along the ion displacement axis is thus obtained from Eq. (68), giving the local dielectric constant $\varepsilon_A = A_A/(A_A+E_{p,1}) = \alpha_e A_A/:_i$. A simple illustration for two dipoles of induced moment μ_i with centers at distance x from an ion, both permanent moments μ_p inclined at θ to the ion displacement has $\mu_i = \alpha_e(A_{ion} - g\mu_p \cos\theta/x^3 - g\mu_i/x^3)$. Rearranging:

---

[*] The first water molecule model used here assumed for simplicity that the negative dipole charges are at 0.15 Å from the oxygen atom center, in the H-O-H plane[92, 159]. This was modified as necessary as the fitting proceeded. This model results in a dipole length of a 0.436 Å. At the distances considered (with x at about 5 times the dipole length), the repulsive terms for a dipole differ from those of an aligned water quadrupole by less than 1%. However, both the axial field term, and the potential energy of interaction estimated from the point charges are 12-13% higher than those for the corresponding x^3 and x^2 terms for the corresponding quadrupole approximation.

$$\mu_i = (zex - g\mu_p\cos\theta)/(x^3/\alpha_e + g) = \mu_p(2.377zx - g\cos\theta)/(0.693x^3 + g)$$
(71)

where g is the appropriate geometrical factor, the numerical factor 2.377zx (with x in Å) is the ratio zex/μ_p, with μ_p nominally equal to 2.02 D, and $0.693x^3$ is x^3/a_e ($a_e = 1.444 \times 10^{-24}$ cm^3). At typical values of x for z = 1, A_{ion} is such that μ_i is rather small (e.g., $0.3\mu_p$). It becomes much more important for z = 3. This is performed for all dipoles in the system until the necessary accuracy is obtained. Often it is only necessary to consider dipoles of one type in a single shell, as above. Some properties of such an assembly of permanent dipoles may be illustrated by considering a simple model of a line of permanent dipoles with an effective number of closest neighbors equal to n (a Madelung constant for the line[160]), at a closest neighbor distance x'. The dipole displacement at an angle α to a line intersecting the center of a dipole μ_p at an angle β at a distance x', and where γ, γ', are the angles subtended by the planes containing the angles α and β and a reference plane containing the line (Figure 4)* is given by[139]:

$$E_p = -(\mu_p/x'^3)[2\cos\alpha\cos\beta - \sin\alpha\sin\beta\cos(\gamma - \gamma')]$$
(72)

For the dipole array in question, $\gamma = \gamma' = 90°$, and putting $\alpha = \beta = (1 - \theta)$, the dipole field $E_{p,2}$ along the axis of each dipole will be $+n\mu_p(1 - 3\sin^2\theta)/x'^3$, so the interaction energy of one dipole with the others is $+\mu_pE_{p,2}/2$, where the 1/2 avoids double counting. The dipole field $E_{p,1}$ perpendicular to the line opposing the displacement $-A_{ion}$ will be equal to $+n\mu_p\cos\theta/x'^3$. The field orienting the dipole array is $-A_{ion} + (n\mu_p\cos\theta)/x'^3$, so $\varepsilon_A = A_{ion}/[A_{ion} - (n\mu_p\sin\theta)/x'^3] = \alpha_eA_{ion}/:_i$, allowing $:_i$ to be calculated. For simplicity, $\alpha_eA_{ion}/:_i$ is assumed independent of $\cos\theta$, but the electron distribution will be anisotropic due to the directional nature of the chemical bonds.[99,121] The electrical potential energy of each permanent dipole is:

---

* The description of the angles is unclear in Ref. 139.

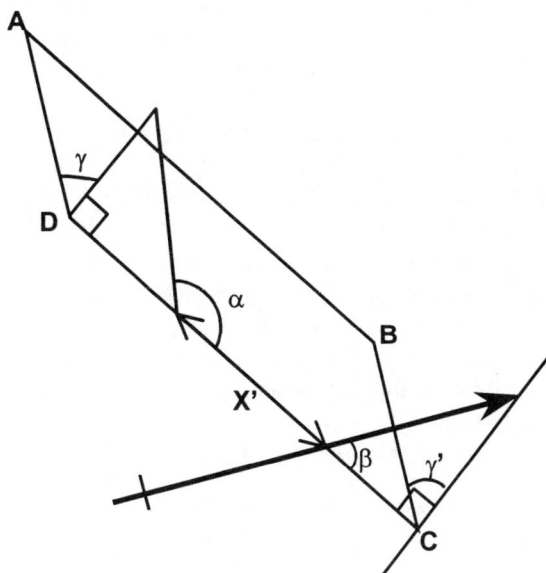

Figure 4. Dipole field angles. ABCD is a reference plane.

$$\Phi = -A\mu_p\cos\theta + n\mu_p^2(1 - 3\sin^2\theta)/2x'^3$$
$$= -A_{ion}\mu_p\cos\theta + 1.5n\mu_p^2\cos^2\theta/x'^3 - n\mu_p^2/x'^3 \tag{73}$$

where the $-n\mu_p^2/x'^3$ term is the potential energy of a dipole in the undisturbed dielectric fluid. Equation (73) minimizes at $\cos\theta = 1$ for $A_{ion} > 3n\mu_p^2/x'^3$, and at $\cos\theta = A_{ion}/(3n\mu_p/x'^3)$ at $A_{ion} < 3n\mu_p/x'^3$.

To illustrate the properties of a spherically symmetrical dipole assembly around an ion, we first consider only permanent dipoles to see the effects of counterfields and repulsion terms. If the examination of second and third shells shows little effective energy for permanent dipoles, there is no point in examining induced dipoles, since the induction fields will be negligible. In spherically symmetrical systems are more complex than the above examples, but show similar characteristics. The $\mathbf{E}_{p,1}$ axial field terms due to each neighbor type depend on the angles via a multiplier equal to $(2 - \cos\gamma)\cos^2\alpha_i\cos\theta_i + \cos\gamma_i\cos\theta_i - (2 - \cos\gamma_i)(\sin\alpha_i\cos\alpha_i)\sin\theta_i$, where α_i is the angle between the displacement axis and the line joining the centers of neighboring dipoles of the ith type, and γ_i is the angle to the dipole to the plane of these neighbors. Molecular models shows that rotation away from the plane to a point mid-way between nearest neighbor sites reduces overall interactions to a minimum. The angular rotations, δ_i, are then $\pm 60°$ for tetrahedral, and $\pm 45°$ for octahedral symmetry.

The $E_{p,2}$ repulsive fields between symmetrically arranged dipoles in the same shell along their axes are $+(\mu_p/x_i^3)(n_i/y_i^3)[2(\cos^2\theta_i - \sin^2\alpha_i) + (\cos^2\theta_i - \cos^2\alpha_i)(\cos 2\gamma_i)] = g_{p,i}(\mu_p/x_i^3)$, where $g_{p,i}$ is the geometrical factor (n_i/y_i^3) of the ith type. From geometrical considerations, $\sin\gamma_i = \sin\theta_i\sin\delta_i$, so $\cos 2\gamma_i = 1 - 2\sin^2\theta_i\sin^2\delta_i$, and $\cos\gamma_i = (1 - \sin^2\theta_i\sin^2\delta_i)^{0.5}$.

The potential energy of interaction of an ion with a single solvent permanent dipole in a spherical shell is given by:

$$\Phi_{inner} = -A\mu_p\cos\theta + \sum g_{p,i}\mu_p^2/2x^3 + \sum\phi_i \qquad (74)$$

If required, the attractive interaction can be represented by the quadrupole expressions (Eqs. 61 and 62), and the $E_{p,A,1}$ field term can be estimated assuming point charges. This results in the appearance of additional terms in $\cos^3\theta$ and higher powers. These refinements will be ignored for the present.

Equation (74) minimizes for $\cos\theta$ for any given value of x, and for x at any given value of $\cos\theta$. We can find the distance at which $\cos\theta$ is equal to unity in simple systems (tetrahedral, octahedral) by differentiating Eq. (74) with respect to $\cos\theta$, and putting $\cos\theta$ equal to unity in the equilibrium condition. For tetrahedral groups $\cos\theta = 1$ for x > 1.594 Å for univalent ions, assuming an effective value of $\mu_p = \mu_v$ in water = 2.02 D, to be refined as required. However, the minimum energy with $\cos\theta = 1$ would be at 0.362 Å i.e., at $x = 3(n/y^3)(1 - \sin^2\alpha/2)\mu_p/2ze$, which is physically meaningless. In tetrahedral symmetry, the corresponding figures are x > 2.676 Å for $\cos\theta = 1$, with minimum potentials are x = 1.394 Å for z = 1, and 0.697 Å for z = 2. Graphical calculation* shows that the function is unstable, with no potential well, if x and θ are allowed to vary simultaneously where this is physically meaningful. A symmetrical group of dipoles will then move down the field towards the central ion, and $\cos\theta$ will go from

---

* Graphical simulations were performed using the graphing calculator function on an Apple Macintosh PowerPC personal computer.

unity to successively smaller values as the repulsive energy term increases, but the energy of attraction still continues to increase as x becomes less, until the Mie repulsive energy terms between the ion and the dipoles, and between the dipoles themselves, take effect.

Because of the instability of the functions, we are justified in making the simplifying assumption that the dipole axes are aligned along the displacement axis in the high fields close to the ion, when the basic assumptions for the theory of the Langevin function have broken down. It has been suggested that this may not occur, to allow for the possibility of more favorable hydrogen bonding to associated water molecules.[161] This is discussed in the Appendix, which shows that the *effective* dipole axis for multipoles may not coincide with the normal symmetrical dipole axis.

3. Dipole Nearest Neighbors

In tetrahedral symmetry, there is only one first shell interaction (nearest neighbor, n = 3), whereas in octahedral symmetry, there are both nearest-neighbors (n = 4) and second-nearest neighbors on the other side of the ion (n = 1). The interactions between the first shell and the second may be handled similarly. In this case, the geometry of the axes is not an isosceles triangle and the equations for calculating the angle expressions in the g_i terms are more complex, but the principles are the same. Since the problem is a many-bodied one, reasonable simplifying assumptions are needed to make it tractable. The first is that $\cos\theta = 1$ for the inner shell. Hence, the weak interactions of second-shell dipoles with the central ion, along with the various repulsions, may give a $\cos\theta$ term in the second shell. For the second shell of a trikisoctahedron, there are three nearest-neighbor interactions between a second shell dipole and first shell dipoles, and three second-nearest-neighbor interactions. There are three nearest-neighbor interactions between second shell dipoles, three second-nearest neighbor (diagonal) interactions, and one diagonal interaction with the dipole on opposite side of the ion.

So far, all interactions have assumed that μ_p can be considered to be a dipole. This is a good approximation for repulsions, but not for attractive interactions at shorter range, which require a multipole approach to improve accuracy. Since all calculations were numerical, it was easy to regard the water molecules as having point charges. For simplicity, in initial simulations the center of each positive charge was at each proton, separated by 2 x 0.757 Å, and the center of negative charge was considered to be at the molecular center of water[16,159] or 0.15 Å from the oxygen nucleus, i.e., a dipole length of 0.436 Å with an effective charge of 2.317 x 10^{-10} esu for a moment of 2.02 D. Changes were made as needed to improve fitting. The interaction energy (in ergs) of a cation with a fully-aligned permanent quadrupole at an ion-dipole center distance a (in Å) will then be $-ze^2(0.964x10^8)\{(a-0.218)^{-1}-[(0.757)^2+(a+0.218)^2]^{-0.5}\}$.

A test was conducted on the eight identical dipoles of the second shell of a trikisoctahedron. The inner six dipoles in octahedral symmetry with $\cos\theta = 1$ were located nominally at 2.5 Å, to be adjusted to account for experimental energy values and force-constants. In setting up the preliminary equations, electrostatic repulsions were expected to predominate in the loosely-bound second shell, so Mie repulsions between nearest-neighbor water dipoles were first of all ignored, as were attractive London d^{-6} dispersion forces, i.e., crystal field effects, since these are overwhelmed by r^{-4} ion-induced dipole terms for 2+ and 3+ ions. This procedure is the same as eliminating the Mie repulsive terms by differentiating, setting to zero, and minimizing the energy expression.[98] For the two trikisoctahedron shells and the "Outer Sphere," three simultaneous equations were used, with water-water repulsions and dipole-dipole repulsions for all possible dipole pairs included between the first two shells, and the system was iterated until it was self-consistent.

The nearest neighbor distance between a first shell dipole and one in the second shell was initially assumed to be the diameter of a water molecule (2.76 Å). Under these conditions, the distance between the central ion and the center of a second shell dipole is 3.30 Å when the primary shell ion-dipole center distance 2.5 Å. After various iterations described below, the view that adjacent water molecules had their geometric centers at 2.76 Å was abandoned.

The first problem for trikisoctahedral symmetry was to find the minimum energy configuration for the second shell. One assumption made was that rotation of the dipole in the second shell was the same as that in octahedral symmetry, i.e., towards diagonal neighbors, or towards one nearest neighbor in the first shell. Molecular models show that there appear to be nine second-shell nearest-neighbor pairs in which one neighbor lies on the plane containing the dipole centers and the central ion. In these cases, the second nearest neighbor lies on a plane of rotation inclined at 54.74° to this plane. Of the remaining three pairs of nearest neighbors, one has both neighbors lying on the plane joining their centers and the central ion, one has one neighbor at +54.74° to the plane with the other at +90°, and the other is at +54.74° and −90°. Similarly, of the twelve diagonal pairs, nine are at 60° and 0° to their plane, one has both dipoles on the plane, and two pairs are at +60° and ±54.74°. If ϕ, ϕ' are the angles between the planes of rotation and the plane which includes the dipole centers and the central ion, and θ is the (common) angle on each plane of rotation between the corresponding dipole and the displacement axis on the plane of rotation, the required functions in Eq. (72) may be calculated. We find that $\cos\alpha$, β are equal to $\cos\theta(1 - \sin)^2\theta\sin^2\phi, \phi')^{-0.5}$, $\sin\alpha$, β are $\sin\theta\cos\phi$, $\phi'(1 - \sin^2\theta\sin^2\phi, \phi')^{-0.5}$, $\cos\gamma$, γ' are $\cos\theta(1 - \sin^2\theta\sin^2\phi, \phi')^{-0.5}$, and $\sin\gamma$, γ' are $\sin\theta\cos\phi$, $\phi'(1 - \sin^2\theta\sin^2\phi, \phi')^{-0.5}$. When these are inserted into Eq. (74), complex expressions result, which contain sums of $\cos^2\theta$, $\sin\theta\cos\theta$, $\cos^2\theta$, $\sin\theta\cos\theta$, and $\sin\theta\cos^2\theta$ terms. The latter have alternate ± signs whose values depend on the method of counting. They certainly cancel for $\theta = 0$, and they appear to cancel (or nearly cancel) for all values of θ. They were therefore ignored. The repulsive terms between one second shell dipole and the seven other second shell dipoles were then calculated. The interaction between the three first shell nearest neighbors assumed a rotation towards one neighbor, and angles of 120° with the remaining pair, with opposite rotations in respect to the positions of the inner shell second neighbors. A simpler expression was obtained for the interactions between the first and second shell dipoles, since the first shell permanent dipoles were assumed to have $\theta = 0$. It is clear from Eqs. (67) and (68) that :$_p$ interacts repulsively with the E_p due to the permanent dipoles, whereas :$_i$ does not, so only the effects of the permanent dipoles in the first shell on those of the second need be considered. Having determined the orientations and fields of the permanent dipoles, the values of the induced dipoles can then be determined by calculating the relative field multipliers in Eq. (70), and solving the simultaneous equations for each shell.

The potential energy of each of the eight second shell permanent dipoles is given by $[-245.8(2.377z(\cos\theta)/a_2^2 + f(\theta,\phi,\phi')/a_2^3]$ kJ-mole^{-1}, where a_2 is the second shell ion-dipole center distance (3.30 Å in this example), and the factor 245.8 is $N\mu_p^2$ in (Å^{-3}-erg-mole^{-1}) x 10^{-10}. The potential thus obtained corresponds to the potential energy change in solvating a gas-phase ion with a gas-phase dipole. To obtain the approximate Gibbs energy of solvation of a gaseous ion by liquid water in the second shell, a quantity equal to the energy of the number of hydrogen bonds broken per water molecule surrounding the ion should be removed. This average number is uncertain. However, at least one hydrogen-bonding position (i.e., two half-bonds) is blocked by the ion. In tetrahedral symmetry, it is possible to argue, from geometrical considerations, that the effective value is one hydrogen bond per water molecule[95], but in octahedral symmetry the geometry makes this less certain. However, the equivalent of one hydrogen bond per solvating water molecule, or about 20.9 kJ-mole^{-1}, is plausible and was adopted. In addition, there is the work of forming a cavity in the liquid large enough to contain the ion and its oriented solvation shell(s). The radius of the cavity may be put equal to the distance b to the centers (strictly, the dipole centers) of the first shell of bulk water molecules (i.e., of "continuum" water). The energy to form such a cavity can be estimated electrostatically, or more simply from the surface energy, γ, of water. The energy of formation of the cavity is $+4\pi b^2\gamma$, but it will include other terms such as those for the formation of straight or bent[103] hydrogen bonds, which are accounted for separately. The compressive forces in the cavity are opposed by an unknown repulsive Mie term acting over part of b, which is ultimately borne by a repulsive terms between the ion and the first shell of water molecules. However, we can estimate the value of this term by putting the differential of the electrostatic and Mie energy terms equal to zero to determine the equilibrium condition when the opposing forces are equal.[93,98] If the Mie energy term depends on $(qb)^{-12}$, where q is a fraction, it can be put equal to $+A'b^{-12}$, where A' can be eliminated from the equilibrium condition at b_o, which is $+12A'b_o^{-13} = 8\pi b_o\gamma$. Thus, $A'b^{-12}$ is $+4\pi b_o^2\gamma/6$, and the total energy of cavity formation is $+4.67\pi b_o^2\gamma$. We will see below that for an ion of z = 3, b_o is about 6.2 Å. Thus, the energy of cavity formation is about 4.5 x 10^{-12} ergs for a trikisoctahedron of 14 water molecules, or 280 kJ/14 per mole of water molecules. This corresponds almost exactly to the energy of one hydrogen bond.

4. Initial Simulation Results for First and Second Trikisoctahedron Shells

Some EXAFS data on metal ion-oxygen distances are available for hexaaquo transition metal ions.[94] These are generally about 2.1 Å for 2+, and 2.0 Å for 3+ ions, and correspond to those determined by X-ray diffraction in crystals.[94] Some symmetrical breathing mode vibrational frequencies are also available, allowing force constants for a reduced mass equal to that of one water molecule to be calculated.[71,162] They are about 390 and 490 cm^{-1} for 2+ and 3+ states.

The ion-dipole interaction may be approximately expressed as $f = +Aa^{-n} - \Sigma Ba^{-m}$, where the first term is for Mie repulsions and the second is the effective sum of the ion-permanent dipole ($ze\mu a^{-2}$) and ion-induced dipole ($z^2 e^2 \alpha_i a^{-4}/2\varepsilon_A$) interactions and electrostatic repulsions. The latter are dominated by the repulsions in the first and second shells (see later discussion), which are generally proportional to a^{-3}. From the equilibrium condition, $nAa_o^{-n} = \Sigma m Ba_o^{-m}$, where a_o is the equilibrium ion-dipole center distance at ϕ_O, the bottom of the energy well. Writing $a = a_o(1 + x/a_o)$ and expanding the powers of $(1 + x/a_o)$ to the quadratic terms, the first binomial terms vanish and we obtain:

$$f = 2(\phi - \phi_o)/x^2 = \Sigma m(n - m)Ba_o^{-(m+2)} \tag{75}$$

where f is the force constant. Both the ion-permanent dipole and ion-induced dipole terms are significant. The force constants vary between approximately $1.5r^{-4}$ and $2.25r^{-6}$ for 2+ and 3+ ions where r is the ratio of the bond lengths in the 3+ and 2+ states, depending on the relative importance of the two terms. The relative frequencies ($\propto \sqrt{f}$) will therefore vary by factors of between 1.65 and 2.6, the most likely value being about 2, which is much more than the experimental result.

Initial simulation results showed that second shell water molecules regarded as dipoles aligned along the axis of the displacement, i.e., $\theta = 0$ for all z values, thus simplifying the problem. The equilibrium position of the second shell could not be determined by assuming that nearest-neighbor water molecules were located at between 2.76 and 2.9 Å apart. Repulsive Mie energy terms between water molecules were added to the electrostatic attractive and repulsive forces added to reproduce experimental solvation energies, the best fit being n = 12. Other powers in the range 9-14 and exponential expressions[163] were examined, but these were less satisfactory for bond-length reasons. The value of the dipole moment was chosen to give the best fit, starting with 2.02 D (Eq. 36).

Bernal and Fowler[16] and Eley and Evans[92] suggested that positive ions should align along the dipole axis since they considered the negative charges to be on the centerline, on or close to the oxygen, negative ions aligning more or less along one O–H bond axis. Verwey[161a] went further in considering that positive ions should also be off axis, following his earlier suggestion[161b] that the negative charges and bonds are arranged tetrahedrally. These "Verwey Positions" would be stabilized by allowing three hydrogen bonds towards the bulk solvent, rather than only two[161a].*

Early calculations in the present work for higher-valency ions assumed water dipoles of Rowlinson type in axial positions, using both dipole and point-charge calculations. Any refinement with DP multipoles in the Verwey and axial positions (see Appendix for $z = 1$) requires considerable time and effort, and is unlikely to give radical changes, since it results in only slightly different bond-lengths. For positive ions, the Verwey positions may be favored because the dipole-dipole interactions appear to be more attractive in the most favorable rotational orientations for four tetrahedral or six octahedral dipoles (average about $+1.7kT$ per dipole at 298 K for the octahedral case at a dipole center-positive charge distance of 2.1 Å, compared with $+10.8kT$ per dipole for aligned dipoles. For tetrahedral solvating dipoles, see Appendix). Whether this is enough to offset the lower positive ion attraction in the Verwey position compared with that for oriented dipoles and to swing the balance for 2+ ions and particularly, 3+ ions, for which a much larger part of the interaction is via induced dipoles, remains to be seen. Since the electronic polarizability is not isotropic,[121,129] it may maximize along the dipole axis, giving a further reason for steering the water molecules in this direction (see Appendix 2).

The Mie term used for second-shell water molecules was the same as that used following Eq. (60) (6.50 x 10^{-8} erg–Å^{12} per pair). However, there is no net potential energy in the combined first and second shells due to water molecule repulsions. The second shell molecules press the first shell inwards, and are repelled by the ion Mie terms, so the repulsive energies cancel.

The two simultaneous equations for the first and second shell energies were solved to obtain the induced dipoles in the second shell, which gave $0.624\mu_p$, $0.408\mu_p$, and $0.213\mu_p$ for $z = 4+, 3+$, and 2+ respectively. For $z = 1$, a large value of θ was obtained, giving counterfields of opposite sign, with $\mu_i = 0.338\mu_p$, giving -9.1 kJ-mole^{-1} (3.6kT) for the induced dipole interaction. This effect is an artifact, which would disappear if a third shell is considered. The corresponding ε_A values for the second shell, ignoring the effects of

---

* From Pople's work[103], these would be bent, since a solvated ion has a moment of inertia at least as great as that of a water superdipole.

counterfields due to the third or subsequent shells (see below), were 2.02, 2.32, 2.96, and 0.93 for z = 4+, 3+, 2+ respectively. Thus, the overall interactions for the eight members of the second shell of the trikisoctahedron situated at a distance $a_{o,2}$ = 3.30 Å from a central ion with a primary shell ion-dipole center distance $a_{o,1}$ of 2.50 Å with the above assumptions represent about 20%, 18%, and 18% of the total solvation energies of about –8,300, –4,600, and –1,890 kJ-mole^{-1} for z = 4+, 3+, and 2+ ions.

Similar functions were examined for $a_{o,1}$ = 2.3 Å and 2.8 Å. Somewhat less negative values of the interaction were found at both distances. At 2.8 Å, the packing of the first shell is more open and x is slightly shorter than at 2.5 Å (3.164 Å vs. 3.30Å), giving higher repulsions. At 2.3 Å, the first shell is more densely packed, squeezing the second shell dipole out to 3.35 Å, reducing the net interaction somewhat. Further examination showed that at x = 2.5 Å, a marginally higher net interaction was possible if the nearest-neighbor distance was increased to give $a_{o,2}$ = 3.4 Å, slightly reducing the repulsive terms.

5. The Third and Fourth Shells

Assuming for the moment complete water molecule separation from superdipoles, the third trikisoctahedron shell can have 24 or 12 dipoles, occupying the space between one in the first shell and two in the second. A reasonable range of first shell ion to water molecule center distance $a_{o,1}$ for high-z and low-z ions is from 1.9 Å to 2.6 Å. The oriented second shell dipoles then have ion-to-center distances $a_{o,2}$ of 3.428, 3.429, 3.426, and 3.327 Å for $a_{o,1}$ values of 1.9, 2.0, 2.1, and 2.6 Å respectively, illustrating the squeezing effect on the second shell as $a_{o,1}$ becomes shorter. The 24 third shell dipoles lie at $a_{o,3}$ distances of 4.454, 4.504, 4.571, and 4.882 Å at the same respective $a_{o,1}$ distances. The z = 4 case at a bond-length of 1.9 Å slightly squeezes the first shell water molecules to a distance of 1.344 Å, less than the normal radius of 1.38 Å. The second shell water molecules do not interfere with those in the third shell for z = 1, 2 and 3, but do for z = 4, giving a center-to-center distance of 3.01Å, rather than 2.9 Å.

The alternative second shell of 12 dipoles lie adjacent to two second shell dipoles, equidistant from, and touching, two first shell dipoles. However, due to interference from second shell dipoles, their closest approach is the same as that for the 24 dipoles shell discussed above. The third or fourth shells are more complex to model than the

Table 1
Calculated Third-Shell Interactions (24 Dipoles)

z	ε_A	$L(x_A)$	$-\phi_{ul,3}$ kT units	$-\phi_e$ kT units
1	15.01	0.212	1.11	0.05
2	13.37	0.475	5.71	0.28
3	10.39	0.702	13.29	0.85
4	6.83	0.855	23.07	2.45

second sphere, so approximations were made as required. The most important terms were the symmetrical repulsions between the second and third shell dipoles, rather than the more neutral terms resulting from repulsions in the first and third shells. The latter orient added dipoles to a greater θ value, reducing repulsive interactions. We may rewrite Eq. (35) as

$$\varepsilon_A = n^2 + 4\pi n_o \mu_v (\varepsilon_A/A_A) L(x_A) \qquad (76)$$

for intense fields, where $L(x_A)$ is assumed to equal $A_A\mu_v/\varepsilon_A kT$. Approximate interactions were obtained by solving Eq. (76) graphically for ε_A at the appropriate distances for 24 dipoles, assuming the above $a_{o,1}$ values for z = 4+, 3+, 2+, and 1+, respectively, and nominally putting μ_v equal to 2.0 D. The results are shown Table 1 for overall interactions $-\phi_{ul,3}$ per permanent dipole for third shells of 24 dipoles in kT units at 298 K. The final column contains the corresponding electronic induced dipole interaction $-\phi_e$ from Eq. (55) with $q_A = 1$, i.e., using $-P_{ul,A}.A_A(1 + q_A/\varepsilon_A)/2 = -\mu_v L(x_A)A_A(1 + 1/\varepsilon_A)/2$, and $-P_{e,A}.A_A/2 = -\alpha_e A_A^2/2\varepsilon_A$.

The dielectric constants have reasonable values, considering that the packing of 24 dipoles around the trikisoctahedron preserves the four-nearest-neighbor water structure. Induced electronic dipole effects at this distance are small for z < 4. Whether a 24 or 12 dipole model is reasonable is obtained by estimating the volume of the shell extending from the inner to the outer edge of the 24 dipole distance (i.e., $a_{o,3} \pm r_w$, where r_w is the water molecule radius of 1.38 Å) for z = 4. This volume is 7.04 x 10^{-22} cm^3, which would contain 23.5 (i.e., 24) water molecules at the bulk water density. For z = 4, the total interaction per dipole is about 3 times the hydrogen bond energy, so the water structure will be significantly broken. After subtracting the hydrogen bond energy, the total interaction will be about 12% of the total solvation energy. For $z = 3$, the total interaction energy will be about 1.7 times the hydrogen bond energy, i.e., about -335 kJ/mole, slightly more than 7% of a typical $-4,600$ kJ/mole total solvation energy. For $z = 2$, no water breakdown will occur, and the energy of the shell will be about 19% of a typical $-1,900$ kJ/mole solvation energy. The effect for $z = 1$ is similar.

The fourth shell was also examined assuming single dipoles with moments approaching the vacuum value, and no Onsager or Lorentz cavities to give an upper interaction limit. Fourth cell packing was irregular, with distances varying from about 6.0 6.7 Å, 6.1 6.8 Å, 6.2 6.9 Å, and 6.7 7.4 Å for the greatest interactions for $z = 4$ to $z = 1$ respectively. The calculated ε_A, and permanent dipole interaction $-\phi_{ul,4}$ were 12.63, 13.97, 14.81, and 15.28, and 7.65kT, 4.06kT, 1.70kT and 0.31kt respectively. For $z = 4$, the induced dipole interaction was 0.40kT, so the total interaction is almost equal to the hydrogen bond energy and a dipole in the fourth shell should still just be attached to a more distant nearest neighbor. However, the x_A value (about 2.1) would suggest no free rotation. Multiplication of the dipole moments by the Lorentz cavity terms $(n^2 + 2)/3$ for the critical $z = 4$ case gave a dielectric constant of 20.29, $x_A = 1.617$ [$L(x_A) = 0.461$], increasing the interaction energy value $-\phi_{ul}$ to 7.89kT by increasing the apparent dipole moment. The addition of an Onsager cavity term $q_A = 3\varepsilon_A/(2\varepsilon_A + 1) \approx 1.5$ in the Langevin function increased ε_A to 29.27 with $x_A = 1.671$ [$L(x_A) = 0.475$] and gave a slightly higher $-\phi_{ul,4}$ value of 8.14 kT. However, $-\phi_{e,4}$ is decreased by the higher dielectric constant to 0.17kT, partly offsetting this effect. Finally, the use of Booth's second set of assumptions for the superdipole moment ($A = 1$, $B = 7/3$, Eq. 37) gives a local dielectric constant of 64.67 with $L(x_A) = 0.494$, $-\phi_{ul,4} = 8.22$kT and $-\phi_4 = -\phi_{ul,4} - \phi_{e,4}$ of 8.30kT, just less than the hydrogen bond strength.

The above shows the lack of sensitivity of $-\phi_4$ to the value of ε_A or the underlying assumptions, which is clear from the form of the expression $-P_{ul,A}\cdot A_A/2$, which mainly depends on the value of $L(x_A)$, which is almost exactly proportional to A_A to $x_A \approx 3$ (Eq. 50). The water structure may be just broken in the fourth shell for $z = 4+$, but not for $z = 1+, 2+, 3+$. Calculation shows that the corresponding dielectric constants for the fourth shell using Booth's assumptions are close to the bulk values for $z = 1+, 2+, 3+$ (76.20, 74.19, 70.59 respectively), with x_A values of 0.300, 0.720, and 1.173, $L(x_A)$ values of 0.099, 0.232, and

0.359, and $-\phi_4$ values of $0.32kT$, $1.80kT$, and $4.32kT$ at 298 K. These all correspond to the Dielectric Continuum range. The fifth shell for $z = 4$ (at about 8.4 Å) gives a dielectric constant of 73.72, $x_A = 0.790$, $L(x_A) = 0.253$, and $-\phi_4 = 2.15kT$. Again, this is in the Continuum range.

6. "Bottom-Up" Modeling of the Third Shell

Since the effects of the third shell are not negligible for high-z ions, it was approximately modeled using the same simultaneous equations as those for the second shell, i.e., to take a "bottom-up" calculation of the dielectric constants rather than using Booth's "top-down" method.[73,124,125] As expected, adding the third shell slightly reduced the net interactions in the second shell. Again, the principal conclusion was that the third shell for $z = 2$ would not show sufficient interaction to break the structure of water. Thus for $z = 1$, the second shell may considered the start of the Continuum Born-like term, where Booth's Langevin function should still apply. For $z = 2$ and the above dimensional assumptions, it is the third shell, and for $z = 3$ and 4, the fourth shell.

The effect of the first sphere fields on the interactions in the third shell was small. A spherical shell of charge should be equivalent to that of an equivalent single charge at its center. However, shells consist of dipoles with discrete lengths and thicknesses with equal and opposite charges on each side. The central charges should therefore cancel, giving no net external or external field. However, the shells considered here are neither continuous, and they interpenetrate geometrically. For more remote molecules within the continuum electrolyte, the tendency to see no dipole field from either successive shells towards the ion, or from shells farther away from the ion, will become more apparent (see Appendix). The local dipole field will then largely result from nearest neighbor dipoles within the same shell. We note that the Kirkwood-Booth theory of the dielectric constant[72,73] only involves nearest neighbors, and ignores longer-range interactions.

7. Simplified Modeling

If the innermost sphere can be treated independently of the second, it would simplify calculation. This can be tested by considering induced dipoles in the first shell independently of the second using Eq. (73), or via those in the second shell using two simultaneous equations based on

Eq. (72). The simultaneous calculation using the previous dimensions shows μ_i for the inner shell to be 1.469, 1.018, 0.570, and $0.268\mu_p$ for $z = 4+$, $3+$, $2+$, and $1+$ respectively. Disregarding the fields of the second trikisoctahedron shell gives 1.622, 1.172, 0.721, and $0.271\mu_p$ respectively. Thus, the second shell may be disregarded for $z = 1+$, but not for $z = 2+$, $3+$, $4+$. Accurate modeling for $z = 4+$ requires all three shells.

The attractive interaction per dipole is $-ze(\mu_p + \mu_i/2)/a^2$ and the total equilibrium interaction, including the repulsive permanent dipole-dipole and Mie repulsion terms, was typically about 85% of this value, assuming that the Mie repulsion term varies as $1/a^{12}$. This may be demonstrated by determining the equilibrium point by setting the derivative equal to zero.[93,98] Moelwyn-Hughes[93,98] used a simple analysis with a $1/x^9$ Mie term, but he overstated the μ_i terms by not considering the effects of the inner shell repulsive fields on their creation. If these are not included, the corresponding μ_i values at $a_o = 2.5$ Å are 2.196, 1.648, 1.098, $0.549\mu_p$ for $z = 4+$, $3+$, $2+$, and $1+$ respectively. The error in the equilibrium energy is small (ca. 1%) for $z = 1$, where the minor interactions with the second shell have little effect. Ignoring the second shell overstates the interaction energy by about 6% for $z = 2$, falling to 4.5% for $z = 4$. In view of the uncertainties introduced by minor order terms (e.g., crystal field interactions and induced electronic dipole-dipole dispersion) these may be ignored, or the μ_i terms for $z = 2+$, $3+$, $4+$ may be reduced by a nominal multiplier, e.g., 0.95. Finally, if $\varepsilon_A = A_A/(A_A + E_{p,A})$ can be estimated from the vector sum of the dipole fields, it is clear from Eqs. (68)-(70) that the overall interaction of each dipole in a given shell, with all dipoles assumed to be oriented along the ion displacement, dropping vector notation, and neglecting $+W_{A,p}$, is:

$$\Phi_{inner} = -\mu_p A_A(1 + 1/\varepsilon_A)/2 - \alpha_e A_A^2/2\varepsilon_A + \sum \phi_i \qquad (77)$$

The first term in this useful and easily-handled expression combines the ion-permanent dipole energy $-\mu_p A_A$ and its repulsive interaction with all permanent dipoles $+\mu_p E_{p,A}/2 = +\mu_p.A_A(1 - 1/\varepsilon_A)/2$ (compare Ref. 16).

246

A. J. Appleby

8. The "Continuum" Energy

The Born-like Continuum energy term may be reasonably accurately replaced by an integration (Eq. 19), the lower limit being a point between the inner edge and the center of the first shell of superdipoles. The upper limit has usually been considered to be the Debye length determined from the linearized Poisson-Boltzmann distribution,[44] but this cannot apply in the electrolyte concentrations used in kinetic experiments. For these, the distance a_n at which the sum of the surrounding positive and negative charges around an ion, plus is own charge, becomes zero is relevant. Milner[164] attempted to determine the virial of such systems, taking into account the interactions between all charges present. If n_v is the molecular salt concentration whose complete dissociation produces z_n cations and z_p anions of valences $+z_p$ and $-z_n$, the anionic and cationic charge concentration at distance 'a' from a selected cation are $-n_v z_p z_n \exp + q/a$ and $+n_v z_n z_p \exp - z'q/a$, where $q = z_n z_p e^2/\varepsilon_o kT$ and $z' = z_p/z_n$. For electroneutrality, the charge on the reference cation must equal the total surrounding charge, i.e.,

$$4\pi n_v z_n \int_{a_c}^{a_n} a^2 \left[\left(\exp - \frac{z'q}{a} \right) - \left(\exp + \frac{q}{a} \right) \right] \cdot da = 1 \qquad (78)$$

where a_c is the closest approach distance for cations and anions. For a selected anion, we have $z' = z_n/z_p$.* At high dilution the integral is $\sqrt{2}/\kappa$, where κ is the Debye reciprocal length, so Eq. (78) should be divided by $\sqrt{2}$ to give the correct result. Since it uses a non-linearized Boltzmann distribution and does not involve the Poisson equation, it should be applicable at higher concentrations than the Debye-Hückel theory. Activity coefficient calculations using this method[165] show a cube-root dependence of activity coefficient on concentration to about 0.1 M,[166,167] after which the Robinson-Stokes solvent activity corrections[168] become controlling.

For a 1:1 electrolyte at a randomly-chosen closest approach distance of 7.16 Å, Eq. (78) divided by $\sqrt{2}$ predicts an electroneutrality length of 10.4 Å at 0.1 M, where κ is 9.6 Å, and $(\kappa + a_c)$ would be 17.0 Å. If salt-like effective lattice spacings[134,166,167]

---

* This should be a summation, but numerical integration is sufficiently accurate.

occur in higher concentrations, then by analogy with the mean energy of an ion in a lattice, the electroneutrality distance will be given by a/α, where 'a' is the mean cation-anion distance and α is a Madelung constant.[160] Assuming a rock-salt-like lattice, 'a' would be 28.6 Å and 13.3 Å at 0.1 M and 1.0 M, and α is 1.748. The effective electroneutrality distances may then become very short in electrolytes used in practical experiments or devices, e.g., 16.3, 7.6, and 4.2 Å in 0.1, 1.0 and 6.0 M 1.1 electrolytes.*

As indicated in Section III-2, the Continuum may be modeled as spherical shells of five-molecule water superdipoles with an effective radius of 2.9 Å. Their real radius is the distance between nearest neighbors (2.9 Å), plus the radius of a water molecule (1.38 Å). They therefore interpenetrate or interlock as in gears in free rotation, accounting for the effective number of nearest neighbors at 298 K of about 4.4, rather than 4.[95,103,138] Similar interpenetration with the solvation shell of ions is likely, with an average 2.9 Å nearest-neighbor distance between outer superdipole molecules and ion solvation shell molecules. We note that the moments of inertia, therefore equipartition energy rotation rates, for 1+ ions and water superdipoles are similar, permitting coupling of rotations. However higher valency trikisoctahedral ions have considerably higher moments of inertia, give rotation rates about a factor of two less than superdipoles. This will encourage energy exchange, and also push the superdipoles somewhat farther out on average. Molecular models suggest that the superdipole centers lie at 4.9-5.0 Å beyond the Inner Sphere dipole centers for 2+ and 3+ ions, and at 3.8-3.9 Å for 1+ ions. These distances must be corrected to give the correct a_o value in the integral in Eq. (19) to approximate the Continuum interaction energy. With the effective electroneutrality distances given above, the first two Continuum shells, containing about 75% of the total energy summed to infinity, are the only ones of importance in 0.1 M 1:1 electrolytes. In 1.0 M 1:1 electrolytes, only the first shell, with 55-60% of the energy to infinity, is important. In 6.0 M 1:1 electrolytes, the Continuum energy effectively disappears.

As noted in Section IV-5, 4+ cations will have a layer of single somewhat oriented water dipoles of moderate ε_A around the

---

* Energy calculations for ionic lattices show $\varepsilon_A = 1$[160]. Molten salts can form transient dipoles and multipoles as ion pairs and clusters, but it is unlikely that these contribute to a dielectric constant. For charge transfer in molten salts, the equivalent of an FC process is the change in electroneutrality length as valence changes.

trikisoctahedron. This should be separately modeled, but a good approximation may be had by adding 5.0 Å to the inner shell center distance, correcting to give a_o, and integrating, since $(1 - 1/\varepsilon_A)$ for ε_A ≈ 10 is close to unity.

9. Solvation Energy Estimates

Calculations for higher-valency ions were performed early in the study. In the Rowlinson molecular model,[141] the vacuum dipole center lies at 0.293 Å from the oxygen center, whereas in the DP model,[143] it lies at 0.1225 Å. The water molecule structure used initially placed partial positive charges on each proton as in the Rowlinson model, but placed the negative charges at the molecule center,[16,159] which gives reasonable values for the solvation energy difference between positive and negative ions at constant oxygen distance and for the ion-oxygen distance for multivalent ions.[95,170] Both point charge and simple dipole models were used. The force constants and frequencies[71] calculated from the energy wells were satisfactory for the symmetrical breathing three-dimensional oscillator for 2+ ions, but those for 3+ ions give higher frequencies (see discussion, Section IV-4). The experimental values for 3+ ions must refer to cooperative modes between the first and second trikisoctahedron shells. In later work, the exact DP multipole model was used instead of a simple dipole model. Recomputing with DP multipoles in both the Verwey and dipole axial positions is desirable (see Appendix), but it is unlikely to radically change the model, because calculations were made to fit experimental values, and changes in dipole orientation would only result in slightly different bond-lengths for each assumed dipole moment value.

Some early computations for cations are given below. In all cases, the loss of two hydrogen bonding positions per water molecule was assumed (i.e., 20.9 kJ/mole of water), and the dipole is in the axial position. In the Verwey positions, only one hydrogen bonding position may be assumed to be lost. Solvation energies are the "quasi-absolute" values from Ref. 95. p. 106. Distances are ion to dipole center.

Monovalent cations, tetrahedral results:

- Li^+, −542.7 kJ/mole, first shell at 2.500 Å (Continuum −181.3 kJ/mole; 33.4%).
- Na^+, −428.0 kJ/mole, 2.782 Å (Continuum −168.7 kJ/mole; 39.4%)

- K^+, –344.3 kJ/mole, 3.068 Å (Continuum –156.7 kJ/mole; 45.5%)

- Rb^+, –323.0 kJ/mole, 3.157 Å (Continuum –154.5 kJ/mole; 47.7%)

- Cs^+, –298.7 kJ/mole, 3.269 Å (Continuum –151.0 kJ/mole; 50.6%)

  Monovalent cations, octahedral results:

- Li^+, –542.7 kJ/mole, 2.680 Å (Continuum. –173.1 kJ/mole; 31.9%)

- Na^+, –428.0 kJ/mole, 2.978 Å (Continuum –161.0 kcal; 37.6%)

- K^+, –344.3 kJ/mole, 3.267 Å (Born –150.7 kcal; 43.8%)

- Rb^+, –323.0 kJ/mole, 3.356 Å (Born –147.9 kJ/mole; 45.8%)

- Cs^+, –298.7 kJ/mole, 3.466 Å (Born –144.4 kJ/mole; 48.3%)

  Typical early results for multivalent ions:

- $z = 2$: Total solvation energy taken as –1852.3 kJ/mole. Inner octahedral shell at 2.641 Å, outer trikisoctahedral shell (8 water molecules) at 4.185 Å. Inner shell to outer shell dipole-dipole center distance 3.243 Å. Inner shell, 56.34%, Outer shell, 14.51%, Continuum, 29.15% of net solvation energy. Inner shell induced dipole 1.036 D, outer shell induced dipole, 0.200 D. Inner shell dielectric constant, 1.862. Outer shell dielectric constant. 3.960.

- $z = 3$: Total solvation energy taken as –4495.5 kJ/mole. Inner octahedral shell at 2.499 Å, outer trikisoctahedral shell at 3.951 Å. Dipole-dipole center distance 3.050 Å. Inner shell, 50.24%, Outer shell, 15.04%, Continuum, 34.75% of net solvation energy. Inner shell induced dipole 1.986 D, outer shell induced dipole, 0.372 D. Inner shell dielectric constant, 1.677. Outer shell dielectric constant. 3.587.

- $z = 4$: Total solvation energy taken as –8082.9 kJ/mole. Inner octahedral shell at 2.409 Å, outer trikisoctahedral shell (8 water molecules) at 3.800 Å. Dipole-dipole center distance 2.927 Å. Inner shell, 56.65%, Outer shell, 14.47%, Third shell (24 dipoles) and Continuum, 28.88% of net solvation energy. Inner shell induced dipole 2.990 D, outer shell induced dipole, 0.566

D. Inner shell dielectric constant, 1.599. Outer shell dielectric constant. 3.396.

Comparison of the above ion-dipole center distances with experimental EXAFS 2+ and 3+ ion-oxygen distances (e.g., 2.1 Å, 2.0 Å respectively)[94] show them to be too long. Taking 50% of the total net solvation energy to be in the innermost shell and using the DP multipole dimensions with ion dipole center distances of 2.205 Å and 2.095 Å for 2+ and 3+ respectively to correspond to the experimental ion-oxygen distances, after a careful correction of the dipole center distance to allow for electron displacement in forming the induced dipoles gave a result later used in estimating redox activation energies. Fitting this time used the vacuum value of the permanent dipole moment (1.86 D), and was based on Eq. (77). The counterfield values were split into two parts after a series of iterative processes. One part (29.4% of the total for 3+ ions, 34.6% for 2+) was associated with inner shell permanent dipole repulsions, proportional to $1/x^3$, and the second was associated with dipoles and superdipoles external to the inner sphere. The latter counterfield may be put equal to $A_A(1 - 1/\varepsilon_A)$, i.e., for most purposes, it may be assumed to be proportional to A_A. For 3+ ions with an assumed net solvation energy of 4468.4 kJ/mole, the inner sphere dielectric constant was 1.988, and for 2+ ions with an assumed net solvation energy of 1879.4 kJ/mole it was 2.515. Since energies were fitted to enthalpies, the values obtained for each solvating water molecule may be considered as enthalpies. These energies were −158.95kT and −72.09kT (T = 298.16 K) before correction for the ground state of the three-dimensional oscillator. The calculated force constants were 475,460 and 166,880 dynes/cm* for 3+ and 2+, giving frequencies of 668.8 cm^{-1} and 396.2 cm^{-1}, the latter in excellent agreement with experiment.[71] The corresponding hv values were 3.23kT and 1.91kT, i.e., the zero-point energies for the three-dimensional six-member oscillator are 4.85kT and 2.87kT.

This work suggested that the distances obtained for monovalent cations were also too long. For K^+, the inner and second shell were examined on this basis, and a fit could be made at a first shell ion-dipole center distance of 2.253 Å, with an induced dipole of 0.461 D ($\varepsilon_A = 3.094$), with a second shell just apparent at 4.560 Å, where the induced dipole was 0.050 D ($\varepsilon_A = 6.650$). This distance is certainly too

---

* 1 dyne/cm = 1 mN-m⁻¹.

short, and about 2.7 Å appears more probable, when the second shell largely disappears.

A modified Bernal-Fowler model appears to be a reasonably good approximation for monovalent ions, in which the first layer of water is fully oriented, the second layer has some orientation, and following layers may be treated as Continuum, with good accuracy. The Continuum energy term derived as described varies from about 46% for large ions to 35% for small ones.[170] It may be approximately fitted by log-log plots, e.g., if the total water quadrupole or dipole energy in the first shell is ($\Sigma\varepsilon$), the total solvation energy in the same units is $12.99(\Sigma\varepsilon)^{0.64}$ with ($\Sigma\varepsilon$) in kJ/mole [$7.758(\Sigma\varepsilon)^{0.64}$ in kcal/mole, $9.367(\Sigma\varepsilon)^{0.64}$ in kT] for univalent ions, before correction to account for lost hydrogen bonds. The same model is also a good approximation for trikisoctahedral 2+ – 4+ ions, provided second (and for 4+ ions third) oriented shells are considered. About 50% of the energy is then in the first shell, 14% in the second, and the rest in the Continuum. Whether the Verwey positions should be used requires further consideration.

V. THE STRUCTURE AND ENERGIES OF LIQUID WATER AND SOLVATED H_3O^+

1. Preliminary Approaches

The calculations described above for multivalent ions, or univalent ions with a single positive charge center (e.g. K^+) were relatively straightforword. However, in any solvated ion system without a definite single charge center, computational problems are more complex. For solvated protons, calculation must be matched to experimental information. The O   O in $H_3O^+-H_2O$ has been placed in the range 2.45 Å (based on the radii of O, H, and OH under strong-bond conditions, i.e., for H_2O, H_2 and O_2) to 2.5 Å (for the short hydrogen bonds in acid salts such as KH_2PO_4). These are discussed on p. 98 of Ref. 170, but 2.55 Å is also given on p. 118, with a possible preferred value of 2.49 Å in $HNO_3.3H_2O$. The length is determined by the requirement for a sufficiently high proton tunneling rate to permit conduction via "hopping".[171] Preliminary modeling used a flat configuration for the three dipoles solvating H_3O^+ with a single + charge close to the O, with three dipoles in the plane of the three H_3O^+

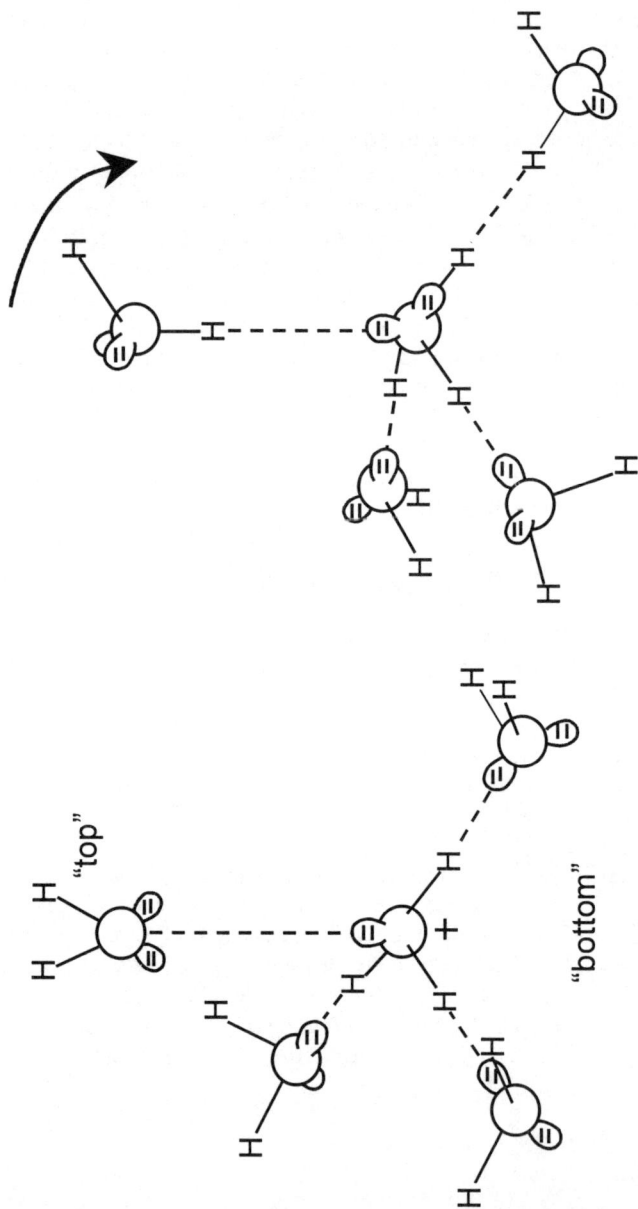

Figure 5. (Left): $H_9O_4^+ \cdot H_2O$ structure as finally determined (H_3O^+ bond angles 115°, bond length 1.02 Å). "Bottom" H_2O O...O distance 2.599 Å, "Top" O...O distance 3.213 Å. (Right): Corresponding $5H_2O$ structure showing rotation of "Top" H_2O.

protons. This results in the lowest repulsions, including those with the "top" H_2O (Figure 5). The H_3O^+ structure was taken as a flat pyramid with bond angles of 115°, bond lengths of 1.02 Å, and an estimated dipole moment of 1.49 D (p. 55-56 of Ref. 170). The latter was used to estimate the bonding electron positions[143b], with a lone-pair–oxygen distance as in water.

To estimate the H_3O^+–H_2O interactions, the Halliwell and Nyburg[170,172] values for the overall enthalpy of H_3O^+ solvation (1090.8 ± 10.5 kJ/mole, 262.7 ± 2.5 kcal/mole), an acceptable estimate for the proton affinity of water (711 ± 8 kJ/mole, 170 ± 2 kcal/mole, Ref. 168, p. 58), and the Continuum energy determined by the fitting procedure (Section IV-9) were used as a starting point. This resulted in a distance between O in H_3O^+ and the water dipole center of ca. 2.45 Å, i.e., an O   O distance of only 2.18 Å. If the proton charge is equally shared by the three H atoms, each of the three identical solvating H_2O molecules (called here "bottom" molecules) may be considered to lie next to 1/3 of a proton charge. First attempts to iterate using 2.49 to 2.45 Å O   O distances for the "bottom" H_2O molecules gave values ranging from –27.5kT to –38.0kT respectively. For the "top" H_2O, the values varied between –13.3 and –16.2kT, assuming O   O distances between H_3O^+ and the "bottom" H_2Os of 2.24 Å and 2.46 Å. The latter distances were chosen to represent a minimum of two O radii (2.24 Å) and the O   O distance corresponding to two water radii with the oxygens facing each other [2 x (1.38 – 0.15) Å]. At 2.46 Å, the induced dipole for each of the "bottom" H_2Os was about –4.1kT, assuming an effective dielectric constant determined separately from the dipole counterfields for K^+ ion of about 2.96. The "bottom" dipole-dipole repulsive energies (one half of two pairs each) were about +0.7kT, while the "top"–"bottom" repulsions (one half of one pair) were +1.4kT. The "top" water molecule lay at an equilibrium O   O distance of 2.67 Å from the O in H_3O^+. Its induced dipole energy was estimated as –1.2kT, and its repulsive energy (one half of three pairs of dipoles) was about +4.2kT. After correction for the equilibrium values of the inverse 12-power repulsions, obtained by differentiation, it was estimated that the "bottom" equilibrium energies should be adjusted by –0.5kT, and the "top" by +2.6kT.

These results may be of the correct order with a reasonable estimate of the magnitude of the continuum term using the scaling expression for monovalent ions. They are therefore in the range 430-520 kJ/mole (102-125 kcal/mole) before conventional correction of about 84 kJ/mole (20 kcal/mole) to account for broken hydrogen bonds, and before zero-point energy corrections. These are somewhat larger values than those derived by Halliwell and Nyburg[172] by extrapolation, including Outer Sphere contributions. The small induced electronic dipoles were ignored.

This *Version A* geometry placed the "bottom" dipoles in the plane of the three protons of the solvated H_3O^+ in an apparently preferred position minimizing dipole-dipole repulsions. Further calculation showed that the minimum energy position lay with the dipole axes below and at 46.8° to the perpendicular to the plane of the three protons in H_3O^+. The motions of the bottom H_2Os were rather loose, with a kT energy difference for 55.9° – 46.8° and 46.8° – 35.0°. The corresponding O   O distances were 2.62 Å and 2.48 Å, with 2.54 Å at the minimum, within the requirements of Ref. 170. The interaction energies for the "bottom" and "top" dipoles were –22.6kT and –33.9kT, corresponding to a total (including Continuum) interaction of –510 kJ/mole (–121 kcal/mole) before hydrogen bond and zero-point energy correction. This time, the "top" water molecule lay at an O   O distance of 2.93 Å due to the higher repulsions of the three "bottom" dipoles to the "top" dipole in the 46.8° rather than the planar 180° positions. In this *Version B* case, the effective dipole moment was taken as 2.138 D (see Appendix), with a DP dipole center at 0.1225 Å from the oxygen center.[143]

A problem with the *Version A and B* geometries was the inconsistency of the x^{-12} repulsive terms, which varied in a purely arbitrary manner by more than a factor of 3 for the "bottom" water molecules, and by a factor of 23 for the "top" water. Using experimental data for water to allow a common repulsive term to be calculated based on the average water O–O bond distance provided a different approach. The time-averaged three-dimensional energies (in kT units at 298 K) of water superdipoles may be:

$$u = (1 + 0.0574)Ar^{-12} - 1889.92r^{-6} - (0.4495)1889.92r^{-6} - 156.00/r^{-3}$$
$$(80)$$

where r (Å) is the average O   O bond distance in water, A is the repulsive term for neighboring molecules in an inverse 12 power field, and 0.0574 is the correction term for 4-fold coordination summed over all molecules.[173] The corresponding term for inverse 6-power induced dipole-induced dipole and electron dispersion force attractions is 0.4495.[173] The average permanent dipole-dipole interaction is discussed below. Setting the derivative equal to zero gives the mean energy u_o, in terms of the mean distance r_o:

$$u_o = -[(12 - 6)/12](1.4495)1889.92r_o^{-6} - [(12 - 3)/12]156.00r_o^{-3} \quad (81)$$

u_o (including the zero-point energy of the three-dimensional oscillating dipole) is known to be $-9.92kT$ from the latent heat of evaporation, and the r^{-6} multiplier may be readily calculated.[173] The mean three-dimensional dipole-dipole (r^{-3}) and repulsion (r^{-12}) terms may be obtained if r_o is assumed to be 2.707 Å, the mean O   O distance in liquid water estimated from its density. From the gas-phase moment, the dipole-dipole term is $-\{2 \cos\theta_a \cos\theta_b - \sin\theta_a \sin\theta_o b[\cos(\psi_a - \psi_b)]\}84.04kTr^{-3}$, where θ_a, θ_b are the angles measured in the same sense between the axes of the dipoles and the line joining their centers in a plane containing their centers, and ψ_a, ψ_b are the angles subtended by the axes and perpendiculars passing through their centers. The maximum value of the angle term is 2, and a comparison of the experimental value at r_o = 2.707 Å suggests that the vacuum moment applies, giving A = +823,986. However, this approach assumes that water molecules are distributed in a diamond lattice, so improvement was required.

A simple model for associated water was needed in which oxygen-oxygen repulsions in water could be equated to those in H_2O-H_3O^+, which required examination of the energy and associated structural changes on the addition of an unsolvated proton to a liquid water superdipole. Pople's modification[103] of Kirkwood's[72] static water dielectric constant model assumes restricted rotation with bent hydrogen bonding between tetrahedra. The dipole to central dipole $\cos\theta$ value, including second-and later-shell bending is given by the Langevin function of a rather large number, so $L(x_A) \approx 1$. The dipole moment of each nearest neighbor along the axis of the central dipole is $\mu_m\cos\theta$, where μ_m is the mean liquid water moment. Using Eqs. (31-35), the total moment induced by the four first shell dipoles along the length of each central dipole is $4 \times 2\alpha_e\mu_m\cos\theta/a^3$, where a is the mean dipole-dipole center distance. The total dipole $A\mu_v$ (Eq. 35) along this axis is $\mu_m = \mu_v + (2\alpha_e\mu_m)\Sigma n_i\cos\theta_i/a_i^3$, where n is the number of nearest neighbors, and i refers to the ith shell. Hence,

$A = 1/(1 - 2\alpha_e\Sigma n_i\cos\theta_i/a_i^3) \approx 1/(1 - 8\alpha_e\cos\theta/\varepsilon_A a^3).$* The superdipole moment $B\mu_v = \mu_m(1 + 4\cos\theta)$, so $B = A(1 + 4\cos\theta)$. Putting the experimental 298 K ε_o, n^2 values in Eq. (36), $X = AB = 2.765$. Thus, with $\alpha_e = 1.444 \times 10^{-24}$ cm^{-3} and a = 2.90 Å, $\cos\theta = 0.2737$.

Pople[103] assumed that the lone pairs and the –OH bonds were both at 105°. The DP angle[143] for the lone pairs is 122.2°, which gives a lone pair O–H bond angle of $\cos^{-1}[-\cos(105/2)\cos(122.2/2)]$. However, the bond is not quite aligned in the lowest energy *trans* and *cis* states (see below), being at +63.05° and –49.8° to the line of centers respectively. Hence, cos (122.2/2) should be replaced by an averaged cosine, e.g., 0.55. Thus, using a modification of Pople's expression, $\cos\theta = 0.55\cos52.5(Lg/kT)^2$, where g is the bending force constant of the hydrogen bond. This gives $Lg/kT = 0.904$, and g = 10.4, in excellent agreement with Pople's value obtained from the radial distribution function of water.[103] The contribution of the second and higher shells will reduce the effective value of $\cos\theta$ in the expression for A, giving a smaller g value. The g energy term is $F_{(\theta=0)} + g\cos\theta$, where θ is the deviation from the minimum value of $F_{(\theta=0)} + g$. For small θ, this corresponds to $+g\theta^2/2$ above the minimum, so g has the form, if not the dimensions, of a force constant. For small θ, the curvature of the calculated energy well for hydrogen bonds (see below) corresponds to g = 13.7kT, which becomes less at larger values of θ due to anharmonicity. Energies of ±kT were calculated at an average of 26° on either side of the most probable *trans* energy well, i.e., $Lg/kT = \cos26 = 0.899$, i.e., $g/kT = 11.25$, in good agreement with Pople's estimate from the radial distribution function.

From $\mu_m = \mu_v/(1 - 8\alpha_e\cos\theta/a^3)$ with $\mu_v = 1.86$ D, we obtain $\mu_m = 2.138$ D for the effective dipole moment in liquid water from the 298 K bulk dielectric constant. When this is used to estimate the cohesive energy between water molecules in approximately tetrahedral superdipoles at dipole-dipole or O   O distances of 2.9 Å, the results are about a factor of three too small. The simple dipole-dipole model for water was therefore replaced by a DP multipole-multipole point charge model for hydrogen bonding (c.f., Lih[173]), and the interactions

---

* Since ε_A is determined by the counterfields from the second and further shells, increasing incoherence with rising number of shells will tend to compensate for the increasing value of the summation as the number increases, so its effective value will be close to unity. An experimental value of $\cos\theta$ derived from the bulk dielectric constant is used here, which will give an effective value.

between two DP multipoles at a 2.900 Å O   O distance were calculated using the same approach for the interaction between positive and negative charges and DP multipoles (Appendix 2). A non-coulombic inverse 12-power repulsion based on the O   O distance was assumed.

The most consistent values showed energy minima with the nearest neighbor H–O–H plane lying at angles of +63.05° and –49.8° to the line joining the oxygens of neighboring water molecules. The nearest neighbor has one lone pair almost facing one O–H bond in the first molecule. In the first case, this energy minimum corresponds to the nearest neighbor H–O–H plane in the *trans* position relative to the non-bonding O–H bond in the first molecule. The combination of attractive and repulsive coulombic terms, combined with the assumption that the inverse 12-power repulsive potential operates between the O   O atom centers results in the bonding lone pair of the neighbor not being exactly in line with the line passing through O–H...O. It lies 1.95° below this line of centers, pointing in the direction of the non-bonding O–H bond in the first molecule. With the H–O–H plane in the neighbor in the *cis* position relative to this O-H bond, the bonding lone pair in the neighboring molecule lies at 11.30° below the O–H...O line. These are indicative of bending or kinking of the hydrogen bond.[103,158b] Before correction for second-nearest-neighbor dipole-dipole interactions (see below), the minima lie at –10.48kT in the *trans* position, and –8.73kT in the *cis* position. Removal of the mechanical constraint of a linear O–H...O coordinate will require considerable computing, but it will result in very little difference to the calculations, since libration of the H–O–H plane between +93° and +41° and –23° and –75° in the *trans* and *cis* cases respectively results in energy increases above the minimum of +kT. The maximum between the *trans* and *cis* states occurs with the H–O–H plane at 10.7° below the O–H...O line, pointing towards the non-bonding O–H in the first molecule. This maximum lies at only +3.04kT. The maximum with the H–O–H plane in the other direction lies at –1.8° below the O–H...O line, and is equal to –7.19kT. Thus, transitions from the *trans* to *cis* positions via libration will be easy. They will also be easy by free rotation around the O–H...O bond. As before, calculations showed that induced dipole terms could be neglected.

The distances between the dipole centers of the four outer water molecules (4.6 – 4.85 Å) were probably sufficient for the dipole

approximation of their electrostatic interactions to be reasonable.*
After some somewhat tedious calculations, these energies proved to
vary from +1.90kT (outer H_2Os of *Type A*, bonding with their O–H
bond, both in the *trans-trans* configuration) to –1.09kT (one *Type A*
outer H_2O and one *Type B* bonding via a lone pair, both *cis-cis*). This
permitted the most probable configurations and values of the average
hydrogen bond to be determined, after differentiation and using the
usual equilibrium assumption. All *trans* gave –9.65kT, all *cis* –9.25kT,
and *cis-trans-cis-trans* –9.48kT, while the others were A *cis,trans,* B
trans,trans –8.81kT; A *trans,trans,* B *cis, trans,* –9.16kT; A *cis,cis,* B
trans,trans, –8.12kT; A *trans,trans,* B, *cis,cis,* –8.19kT. Thus, this
study showed the mean H-bond energy to be –9.65kT, i.e. 23.9 kJ/mole
(5.72 kcal/mole). However, a further term to be estimated is the
induced dipole-induced dipole term, although it is debatable whether
this should be counted in the hydrogen bond calculation (the same is
true of dispersion forces). The induced dipole-induced dipole term is
$2\alpha_e\mu^2/a^6$, where μ is an effective dipole moment along that axis.

For two water molecules linked in the hydrogen bond positions, the
total coulombic attraction in the trans position at x = 3.061 Å (O   O
distance 2.9 Å) before correction for the inverse twelve-power
repulsion is –14.29kT (5.883 x 10^{-13} erg), which may be put equal to
$Q\mu^7/a^3$, where Q is a geometrical factor. This gives an effective $Q^{1/2}\mu$
value of 4.108 D, which can now be used to calculate $Q\alpha_e\mu^2/a^6$. With
α_e (assumed isotropic) equal to 1.444 x 10^{-24} cm^{-3}, $Q\alpha_e\mu^2/a^6$ is –0.72kT.
The inverse twelve-power repulsions corresponding to this term are
$+(12 - 6)bQ\alpha_e\mu^2/12a^7$, where a is the twelve-power repulsion distance
(assumed to be 2.9 Å). Hence the net change in overall hydrogen bond
energy due to the induced dipole-induced dipole term is only –0.34kT.
For a hydrogen bond between two water molecules, the value of the
displacement vector **A** is about three times less than that for the
corresponding vector at a water molecule solvating K^+. From the
discussion following Eq. (48), this is precisely the range of **A** in which
a rapid transition from the bulk value of the static dielectric constant to
dielectric saturation takes place, so that effective local dielectric

---

* Calculations for two water molecules with their dipole axes and one lone pair each
parallel (i.e., in parallel Verwey positions for positive ions, as would be possible in a
Helmholtz double layer) in which an accurate DP point charge model was compared with
a simple dipole model showed the latter to be 11.6% high at 2.0 Å separation, and 4.6%,
1.5%, 0% high at 2.25, 2.5, and 2.75 Å, and 1.4%, 2.6%, and 3.3% low at 3.0. 3.5. and
4.0 Å.

constant will be quite large.   Hence induced dipole-induced dipole effects may be neglected.

As is shown in Appendix 2, the DP multipole model[143] for water accounts very well for the differences in solvation energies of positive and negative ions. It also gives an excellent value of the hydrogen bond energy. This suggests that Stilinger and Raman's ST2 potential[144,157] frequently used in molecular dynamic calculations, is too over-simplified to give a good account of detailed interactions between water molecules. The same conclusion is also likely to be true of the central force potential (CFP).[145,158] A major concern is that these analyses permit "stacking" of water molecules around an ion, (e.g., in 9 or 10 random coordination in one geometric shell), which is counter-intuitive to what would be expected geometrically if all of the appropriate coulombic forces are considered.   The same argument applies to improved quantum simulations using quantum path integral and Born-Oppenheimer (BO) potential energy surface in the combined local-density-functional method (BO-LDA-MD) for H_3O^+ and $H_2O–H_3O^+$.[175]

2. H_3O^+ Solvation

The inverse twelve-power O   O repulsion constant (in units of kT-$Å^{-12}$ at 298 K) derived from the last hydrogen-bond calculation was 1.7599×10^6. This permitted a calculation of H_3O^+ solvation, regarding each component (H_3O^+ with bond lengths and angles 1.02 Å and 115°; H_2O in the Verwey position) as DP multipoles. For simplicity, the H_3O^+ lone pair–O distance was assumed to be at the same as in H_2O, and the bonding electrons are at the same proportionate distance along the bond. The preliminary result showed an exact alignment along the line of the O–H bond in H_3O^+ and the H_2O lone pairs, with a "bottom" H_2O coulombic attraction (corrected for the inverse twelve-power repulsions) equal to –33.72kT (at 2.653 Å O   O distance, with the H–O–H plane pointing towards the "top" position), and –29.55kT (at 2.671 Å, with the plane pointing in the opposite direction).

Before refining these values, and adding dipole-dipole repulsions and induced dipoles, the "top" H_2O was examined to see if it would influence the most probable "bottom" position. Because DP multipole point-charge calculations are tedious, some simplifying assumptions were made. The H_3O^+ lone pair points directly at the O in the "top" H_2O, whose lone pairs point downwards in either a symmetrical or Verwey position. However, the repulsions between the 6 H_3O^+ bond electrons and the 4 "top" H_2O lone pair electrons appear to steer the latter to a dipole axial position. The dipole approximation gives a good account of the interaction energies between multipoles in in-line dipole axial positions at distances beyond 3.00 Å, so the H_3O^+ ion may be regarded as a dipole with a superimposed charge. Its dipole moment was calculated to be 1.383 D, rather than 1.49 D (p. 55-56 of Ref. 170), with a dipole center at 0.040 Å from the O. The positive charge was regarded as being shared by each proton. With the "bottom" H–O–H plane pointing upwards, the net attractive energy was −12.71kT at an O O distance of 3.213 Å, whereas pointing downwards, the energy was −10.65kT at 3.220Å. Respective energies were (multipole attractions) −17.50kT and −17.41kT, (H_2O–H_3O^+ dipole-dipole repulsions) +3.75kT and +3.73kT, (induced dipole attractions, effective local ε_A = 2.96), −0.70kT and −0.69kT, and "top"–"bottom" dipole-dipole repulsions, +0.29kT, and +0.32kT, and (inverse twelve-power repulsions) +1.45kT and +1.42kT. Thus, all energies indicate that the "bottom" H–O–H plane facing upwards is favored. Induced dipole-induced dipole effects are only about −0.05kT, which requires an increase in twelve-power repulsions by +0.02kT. The net "top" energy is therefore −12.74kT.

A detailed model of the "bottom" H_2Os gave a net energy of −38.62kT at 2.599 Å. The energies were (multipole attractions) −53.82kT, ("bottom" H_2O H_2O dipole-dipole repulsions) +1.47kT, ("top" to "bottom" H_2O H_2O dipole-dipole repulsions), +0.10kT, (induced dipole attractions, effective local ε_A = 2.96), −4.84kT, (H_3O^+–H_2O induced dipole-induced dipole attractions, effective local ε_A = 2.96), −0.12kT and (inverse twelve-power repulsions) +18.60kT.[*] No further iterations in the energy of the "top" H_2O to account for minor changes in the "top" to "bottom" repulsive energies due to the change in "bottom" O O distance from 2.653 Å to 2.601 Å were considered necessary given the limits of accuracy of the calculation. Thus, the net solvation energy of the Inner Sphere is −128.6kT (−318.8 kJ/mole, −76.2 kcal/mole) before zero-point energy correction. The bonding of each of the three "bottom" H_2Os lies in a rather broad potential energy well. Their calculated stretching frequency force constant was 91,500 dynes/cm. The hv values for the non-degenerate joint three-dimensional symmetrical vibration of the "bottom" H_2Os

---

[*] The "bottom" H_2Os librate in a very shallow energy well with an amplitude of over 100°. This is because the dipole-dipole interactions go from positive to slightly negative on rotation through 90°. At the same time, the attractive interactions become less negative.

estimated from the properties of an almost flat regular pyramidal molecule were 1.42kT, with a ground state lying at 2.12kT. The well energy for the "bottom" H_2Os lay at the ground state of +2.12kT between 2.481 Å and 2.766 Å. The additional thermal energy required to attain a minimum distance of 2.45 Å for effective proton tunneling for conduction [Ref. 170, p. 98] was +1.58kT (3.90 kJ/mole, 0.93 kcal/mole). The stretching force constant for the "top" H_2O was 7,990 dyne/cm, with a ground state of 0.23kT (hv = 0.46kT). Corrected for the ground state energies, the Inner Sphere energy was –126.5kT (–313.5 kJ/mole, –74.9 kcal/mole). The correlation function gives a total energy including the Continuum of –207.4kT (514.0 kJ/mole, 122.9 kcal/mole). A Continuum integration with a lower limit of 3.9 Å gives –175.7 kJ/mole (–42.0 kcal/mole) for the Continuum energy, giving –489.2 kJ/mole (–116.9 kcal/mole) total. Both are before correction for lost hydrogen bonds. The latter is probably the most reliable value.

On "switching on" a proton in a five-molecule tetrahedral water cluster, two hydrogen bonds of *Type B* are converted into "bottom" H_3O^+–H_2O hydrogen bonds, their only major motion being a shortening of the bond, accompanied by minor sideways translations and rotations to account for changes in bond angle. One *Type A* water molecule detaches, rotates through about 107° and translates to allow for the change in angle as one lone pair on the central water molecule becomes a pair of bonding electrons. The "top" *Type A* water molecule detaches, rotates through about 120° and translates sideways (Figure 5). Taking bulk water as the ground state, the solvated H_3O^+ ion must have minus four hydrogen bonds (–95.7 kJ/mole, –22.9 kcal/mole) added to its Inner Sphere solvation energy. The net energy for the "top" H_2O is only –3.09kT, so it will be highly labile. Taking the outer Continuum energy into account using the best result given above gives a solvation energy of –393.5 kJ/mole (–94.0 kcal/mol). Assuming –711 kJ/mole (–170 kcal/mole) for the proton affinity of water, the result is in good agreement with Halliwell and Nyburg's value of –262.7 ± 2.5 kcal/mole.[170,172]

VI. SOLVATION AND CHARGE-TRANSFER

1. FC Redox Processes

Although hetero- and homogeneous electron transfer has normally been held to occur under FC conditions, especially for nonadiabatic or weakly coupled processes, there are suggestions in the literature[90] that "for adiabatic reactions with strong coupling between the redox couple and the metal surface, electron transfer occurs gradually as the system moves along the reaction coordinate." As we have seen in Section II-5, this appears to have been the intent of the original Hush theory,[50] which he recently reviewed in the context of later work.[176] If the flow of charge q during the reaction is rather slow ($\approx$ h/kT, 1.6×10^{-13} s at 298 K, as in normal transition state bond-breaking/making), the Continuum dipoles should be able to keep up their motion with the change of field, so that there will be no Continuum outer-sphere activation energy term. The transient superdipoles[72] in water have a $q_A\mu'$ value (Eq. 34) of ≈ 8.9 D and a moment of inertia $\approx 6.7 \times 10^{-38}$ g-cm^2. Remote from an ion, their rotational period is 8×10^{-12} s (frequency 1.25×10^{11} s^{-1}) at the equipartition energy value. Their lifetimes will be a few periods, perhaps 5×10^{-11} s. For superdipoles centered at 8-9 Å from z = 3 and z = 2 ions (corresponding approximately to the first superdipole shell, containing about 50% of the total Continuum energy), the Langevin energy interactions with the central dipole are 0.94-0.75kT and 0.63-0.5kT respectively, corresponding to Langevin angles of 72.8-74.3° and 78.3-80.6°. The time to rotate from the equilibrium 3+ to 2+ position and vice versa will be $(1.2-1.4) \times 10^{-13}$ s, approximately h/kT. FC conditions apply to complete electron transfers proceeding much more rapidly than this.

The 2+/3+ transition in the Marcus model or its successors involves only small Inner Sphere changes in solvation molecule coordinates in the radial direction to and from the ion. Electron transfer under FC or Born-Oppenheimer conditions demands that an activation energy in the outer Continuum should resist one-electron transfer via a continuum inertial term $\lambda = (e^2/2r')(1/n^2 - 1/\varepsilon_o)^*$ where r' is an effective intermediate reactant-product radius. To avoid error, λ can be

---

* Marcus (Appendix to Ref. 44) suggested substituting a mean *differential* dielectric constant $\partial\varepsilon_A/\partial A_A$ for ε_o (or ε_A) as dielectric saturation is approached.

estimated from the Continuum solvation energy $(z^2e^2/2r)(1 - 1/\varepsilon_o)$, which is about 35% of the total interaction energy for typical 2+/3+ ions (Section IV-9). This gives an average of about 170 kJ/mole for $(e^2/2r')(1 - 1/\varepsilon_o)$, and 96 kJ/mole for $\lambda = (e^2/2r')(1/n^2 - 1/\varepsilon_o)$. Hence, neglecting work-of assembly terms, the Marcus equilibrium inertial energy $\lambda/4$ is about 24 kJ/mole (0.25 eV, 9.7kT). In a 1.0M solution usually used for kinetic experiments, $\lambda/4$ will be about 50% of this (Section IV-8), far less than experimental equilibrium activation energies.[90,177]

We are therefore forced to search for an Inner Sphere activation energy term. A model of the reaction coordinate for the Inner Shells of 2+ and 3+ ions based on the latest Section IV-9 model is given in Figure 6. It shows a stretch of 0.11 Å in equilibrium ion-dipole center distance on going from the 3+ ion at 2.095 Å to the 2+ ion at 2.205 Å. These are equivalent to ion-oxygen distances of about 2.0 and 2.1 Å. The intersection of the terms is at 1.043kT per oscillator, at 2.140 Å. Before correction for the ground states of the three-dimensional oscillators, the equilibrium energies of each inner sphere water molecule were −159.70kT and −72.21kT at 298 K. The inner sphere dielectric constants for 3+ and 2+ were 1.988 and 2.515, the inner sphere repulsive counterfields being respectively 29.4% and 34.6% of the total counterfields. The corresponding force constants were 475,000 and 167,000 dyne/cm, giving frequencies of 2.07×10^{13} (669 cm^{-1}, $h\nu = 3.23$kT) and 1.19×10^{13} (396 cm^{-1}, $h\nu = 1.91$kT). Corrected for the $3h\nu/2$ ground states, the inner sphere energies are 953.3kT (2363 kJ/mole, 564.9 kcal/mole) and 430.4kT (1068 kJ/mole, 255.2 kcal/mole).

The 6 x 1.043kT barrier height must be reduced by about $(\sum 3h\nu/2)/2$ to give an overall height of only about +2.41kT. The barrier height in the second shell is small enough to be neglected. Including the maximum value (9.7kT) of the Continuum term, and before any reductions for coupling of reactant-product energies, the total Inner plus Outer λ value is +12.1kT per ion (0.31 eV, 30.0 kJ/mole, 7.2 kcal/mole). It should also be noted that even after correction for the oscillator ground states, the reactant and product energy wells are at equal enthalpy, not equal free energy. The $T\Delta S$ correction to place the curves at electrochemical equilibrium, will further reduce the barrier height.

Because of the small activation energy, the Tafel plots will show considerable curvature. An obvious problem is that the anharmonicity

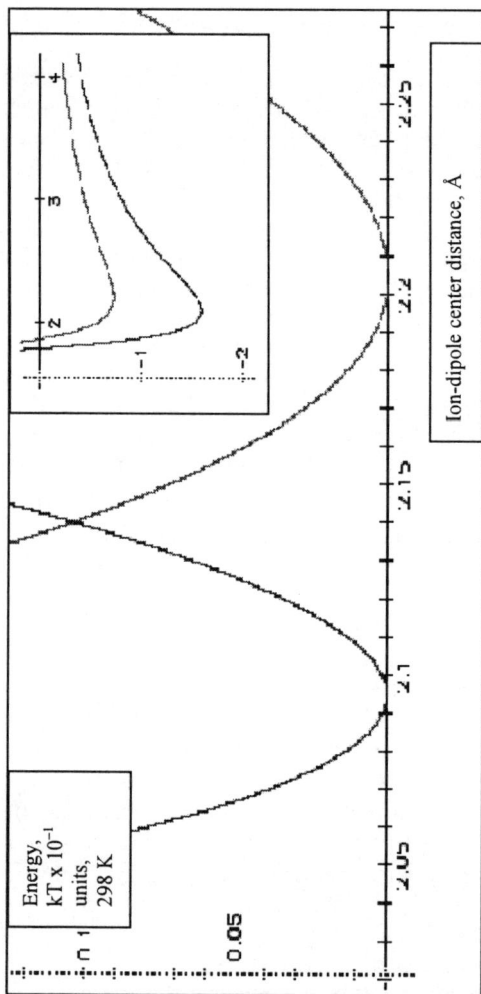

Figure 6. Calculated two-dimensional energy surface for heterogeneous Franck-Condon transition of each member of the three-dimensional 6-oscillator trikisoctahedral Inner Shells of 3+ and 2+ ions at equal potential energies (before ground state energy correction). Crossing point at +6.26kT at 298 K at 2.140 Å for the three-dimensional oscillator (+15.52 kJ/mole). Corrected for ground state energies (+4.84kT and +2.87kT), the crossing point is at +2.41kT (equilibrium potential energy of activation 5.97 kJ/mole). The Inset (x-axis kT x 10^{-2} units) shows absolute energies of 3+ (minimum −159.697kT at 2.095 Å) and 2+ (minimum −72.21kT at 2.205 Å), corresponding to ion-oxygen distances of about 2.0 and 2.1 Å.

argument for the inner sphere energy terms[71] cannot be used to straighten out the Tafel slope, since the form of the wells near the energy minimum is nearly parabolic, even though they are distorted upwards in the direction of the ion, and downwards away from it. We must seek an explanation elsewhere for the straight experimental Tafel lines in the careful work of Curtiss et al.[177]

2. FC Proton Transfer

Proton transfer is a more demanding case than simple electrochemical electron transfer (ECET)[90] redox processes. It involves coupled ion-electron transfer or electrochemical ion transfer (ECIT), recently described[90] as a process in which partial loss of the solvation sphere occurs, adsorption is involved, with "for univalent ions......the occupation number* changes gradually (adiabatically) as the ion approaches the surface," which is reminiscent of Hush.[50,176] Schmickler et al.[90,91,178] and others[179] have generally used the extended Anderson-Newns adsorption model[180,181] for such processes, using Kramers modifications[182] of transition state theory.

The period of a single water molecule rotating around the oxygen-protons axis is about 5×10^{-13} s. Hence the time for an incomplete rotation to assemble, e.g., the $H_9O_4^+$ ion or the solvation shell of a simple ion will be about 2×10^{-13} s, $\approx h/kT$. As Section VI-1 shows, the time for equilibrium Langevin angle change during or following a one-electron transfer is $1.2\text{-}1.4 \times 10^{-13}$ s, so the Continuum can keep up with this process under non-FC electron transfer conditions controlled by the motions of water disassembly and solvation shell assembly. The rate of the step $H_{ads} + H_2O \rightarrow H_3O^+$ is high under conditions where the Gibbs energy for the step is close to zero, i.e., the electrocatalyst (e.g., platinum) is at or near the top of the electrocatalytic volcano.[183] The corresponding activation energy must be low, which is in line with the requirements and activation energy for proton conduction[171] via rotation of solvated H_3O^+ and water supermolecules and proton "hopping" (Ref. 170, p. 96).

As in the redox case (Section VI-1), we will first discuss the reaction under FC conditions, corresponding in part to the Dogonadze-Levich model,[82] and somewhat analogous to Fawcett's EITC

---

* Sic, i.e., the electronic charge of the intermediate or transition state.

models.[184,185] It is helpful to consider the reaction in the anodic direction. From the model for proton solvation in Section V-2, the Continuum activation energy term $\lambda/4$ under FC conditions would be 98.8/4 = 24.7 kJ/mole (0.26 eV, 10.0kT). The crossing point between $H_2O–H_2O + H_{ads} \rightarrow H_3O^+–H_2O$ where the two water molecules on the left side are respectively of *Type A* and *B* (i.e., no rotation is required for assembly) was calculated in the same way as Figure 6. It occurs at a bond length of 2.710 Å, and is at 1.1kT above the minimum at $\Delta G = 0$. For water, the stretching force constant is about 16,800 dyne–cm^{-1} and hv for the three-dimensional oscillator is 0.61kT (126 cm^{-1}), so the ground state for each *Type A–B* molecule is at 0.31kT. The correction for the ground states of the stretching frequencies [(1/3)(3/2)(1.58+0.61)/2 per water molecule, i.e., –0.55kT] gives a total of +1.1kT activation energy for the two *Type A–B* molecules. The other two water molecules required to form $H_9O_4^+$ must either detach from their H-bonded state (requiring about +9.7kT each), or rotate to some compromise position between initial and final states. Calculation shows that 90° rotation from the minimum energy position to the approximate half-way point requires an energy equal to +8.23kT above the minimum. This option will be energetically preferred. Again, this must be corrected for the ground state energies. The bending frequency, as calculated here (c.f., Pople[103]) is $(1/2\pi)(g/I)^{1/2}$ or 282 cm^{-1} (hv = 1.36kT). The value for solvated H_3O^+ is probably about 50% higher, so the correction is –0.85kT per water molecule. Thus, the total energy to assemble the Inner Sphere in a compromise configuration between that of the reactants and the products is +15.9kT. With the Continuum term, we have a total activation energy under equilibrium conditions of +25.9kT (0.67 eV, 64.2 kJ/mole, 15.3 kcal/mole). If assembly of the solvation shell of the adsorbed proton under equilibrium conditions occurs from the solvation shell of the electrode, i.e., the Helmholtz double layer, the net energy of assembly will be the same, since the Helmholtz layer must be reconstituted after proton discharge. The value +25.9kT for proton discharge is much too large, as the results of Marković and Ross[186] (discussed below) indicate. Platinum is at the top of the $–H_{ads}$ Volcano, so the reaction takes place under equilibrium conditions close to zero overpotential.[183]

3. The Inertial Term

Libby[30] and Weiss[32] both stressed the inertial energy present in the Continuum as the energy barrier to FC electron transfer, which we have seen in Section VI-1 requires a relaxation time of about $(1.2-1.4) \times 10^{-13}$ s to accommodate itself to the change in charge. Weiss considered that the inertial energy was always the inertial energy difference between the solvent surrounding the reactants and products in their ground states, i.e., for a change of z to $z + 1$, $[(z + 1)^2 - z^2]e^2(1/n^2 - 1/\varepsilon_o)/2r'$. In contrast, Marcus[41-43] considered that the barrier could accommodate itself to a compromise position between reactants and products prior to electron transfer, so that his $\lambda = e^2(1/n^2 - 1/\varepsilon_o)/2r'$, corresponding to Weiss's barrier expression for a one-electron transfer, is reduced at equilibrium to $\lambda/4$ by intersection of similar parabolic energy terms for reactants and products. This arises from his use of $-(2m - 1)(\mathbf{E}^* - \mathbf{E})^2\alpha_u dV/2$ (Sections II-3 and III-2) integrated over the electrolyte volume outside the reactants, where the field term is the difference between the excited and final states in the same configuration, and m is the Lagrangian undetermined multiplier used in minimizing. How the system can rearrange itself to a compromise $\mathbf{E}^*$ value (not a thermal energy value) *prior to and in anticipation of* electron transfer requires discussion. If the solvent model is imagined to have coordinated harmonic thermal electron oscillations for reactants and products (ions of different valency for homogeneous redox processes, and a typical ion and one of infinite radius for electrode reactants) then such a model may be envisioned.

However, the Continuum model developed Section III-2 is molecular and has no place for such coupled oscillators. The remote rotating superdipoles can influence the free energy of a reacting ion if the average value of the cosine of the Langevin angle θ is permitted to change because of changes in the energy distribution in the solvent. Bockris and Sen examined this problem[4,187] using an equivalent to the expression $V = x_A \cos\theta$ for the potential energy V of remote dipole-ion charge interactions in the classical maximized entropy Maxwell-Boltzmann calculation for the mean energy [V] in the averaging expression:

$$\frac{d(V_i N_i)}{N} = \frac{V_i (\exp - V_i / kT)d \cos\theta}{\sum (\exp - V_i / kT)} \tag{82}$$

where $V_i N_i/N$ is the fraction of the number of molecules with a potential energy V_i in the angular interval $\cos\theta_i$ and $\cos\theta_i + d\cos\theta$. This was evaluated for classical small energy intervals by integrating the numerator and denominator as usual. The mean square energy $[V^2]$ was then evaluated by substituting V_i^2 for V_i in the left numerator. The standard deviation $\sigma^2 = [V^2] - [V]^2$ gave the fluctuation of the system from thermodynamic equilibrium about the mean value. The probability of a large number (N) dipoles having a cumulative random fluctuation from the mean is equal to the very small Gaussian distribution value $N\exp-\dfrac{\{V-[V]\}^2}{2\sigma^2}$, so the probability of providing sufficient energy to overcome any Born-Bjerrum inertial or Marcus activation energy barrier would be very small. This work has not received much attention.[188]   The probability may also be determined via the relationship between the Langevin angle and the total dipole energy from the Boltzmann distribution[103]:

$$<\cos\theta> = \int\cos\theta\,\sin\theta(\exp - V/kT)d\theta / \int\sin\theta(\exp - V/kT)d\theta \qquad (83)$$

We replace the energy in the numerator and denominator by $\exp c_i/kT\exp + u/kT$, where ε_i is the kinetic energy of the population of dipoles dN_i, which itself has a subset of energies depending on the value of $\cos\theta$ in the potential energy term u. The second and more distant shells contain a sufficient number of superdipoles to make the Stirling approximation and the Boltzmann distribution reasonably accurate. The $\exp-\varepsilon_i/kT$ terms cancel, so $<\cos\theta> = L(x_A)$ is independent of the kinetic energy. This again implies that activating the ion via electrostatic coupling in the electrolyte is improbable.

The Soviet school model considered that Outer Sphere or dielectric continuum dipoles perform small harmonic vibrations at a relatively low frequency.[58,59] The implication of vibrations is that the minimum of the potential energy of interaction $V = -B\cos\theta$, i.e., $B = -(A_A/\varepsilon_o)\mu'$ must be greater than $\varepsilon/2$, where ε is the relevent kinetic energy of the superdipole.[151] This may be taken as the equipartition energy for rotation in the two dimensions at right angles to the field vector. If we assume that the first layer of superdipoles is centered at 8.6 Å from a trivalent ion, with their structures intact, and the bulk dielectric constant still applicable, their maximum interaction with $z = 3$ will be $-0.6kT$, so they will be librating. In principle, such a model may be described by the electrostatic equivalent of the ferromagnetic effect, in which residual polarization is present because of order-disorder phenomena,[189] but such an explanation cannot be used for liquids. In any case, the next layer of superdipoles centered at a distance of 13.9 Å from the ion, will be rotating with a maximum interaction energy of $- 0.2kT$ for $z = 3$. These interactions are at an effective temperature greater than that of the electrostatic equivalent of the Curie point in ferromagnetism, so built-in polarization cannot occur. We should note that the period of rotation $\pi(2I/\varepsilon)^{1/2}[1 + B/2\varepsilon + (9/16)(B/\varepsilon)^2$], where ε is the total energy, will decrease from shell to shell as 'a' increases and B becomes less. However, at the same time, the maximum interaction between superdipole groups in adjacent shells remains constant at about $-0.4kT$, providing a coupling mechanism overriding the fall in B. It would be expected that nearest neighbor superdipole groups will rotate in opposite directions, so that they line up in the linear nose-to-tail position perpendicular to the field vector, and are in the parallel nose to tail position parallel to the field vector. Again, it is difficult to see how these motions influence the electronic energy of an ion.

In conclusion, the barrier height to FC change transfer, as envisioned by Weiss,[32] may be correct. In a 1.0M electrolyte, it would be about 50 kJ/mole (0.5 eV, 20kT, 12 kcal/mole). This appears to be of the correct order.[90,177] However, an alternative view is discussed below.

VII. NON-FC CHARGE TRANSFER

1. Water Molecule Rearrangement in Solvation Shell Assembly

The driving force for rearrangement of water molecules before instantaneous FC proton-electron charge transfer presents a problem. A five-molecule water superdipole assembly might acquire +25.9kT of thermal energy, but there will be a low probability that its subsequent rearrangement will direct it towards that for an intermediate state for FC electron transfer and proton discharge. This is not necessarily true for

the simple linear rearrangement required for redox valence change, but the "half-way" rearrangement of water molecules from their normal aqueous state to that for a solvation shell in a 0 to ±1 transition and vice versa must require some "steering" activity. The present discussion applies to such transitions.

For these, a different process is suggested, which may be related to Schmickler's concepts.[90,91,178] The activation energy for such a process may be less than that for FC electron transfer. If this so, the lower activation energy process with more probable rearrangement possibilities will be kinetically favored. In the suggested process, electron transfer occurs slowly over 10^{-12}-10^{-13} s so that the moving molecules during the assembly of the Inner Sphere solvation sheath are "steered" by the slow acquisition of charge, i.e., by the change in $\Psi\Psi^*$, where Ψ is the wave function of the transferring electron and Ψ^* is its complex conjugate. Such processes must be adiabatic since the transfer probability as a function of time must be high. The Outer Sphere or Continuum energy term will then disappear, since the inertial energy of its dipoles can keep up with the relatively slow change of field. A simple model for such a mechanism seems to be that of Hush.[50,176] He considered a "flow of charge" from state "ze" to "ze±e" during an apparently slow charge transfer process in which a sequence of ground states of intermediate charge occur. He only considered changes in ion-dipole interactions during the process (Section II-5), which resulted in activation energies much less than experimental values. After Marcus' early publications,[41] he therefore added an outer Continuum inertial term.[67] This effectively canceled his concept of slow flow of charge.

2. Non-FC Redox Electron Transfer

If an induced dipole term is added to Hush's[50] permanent dipole term, the resulting activation energies are illustrated by the following simple argument. The permanent- and induced-dipole terms along the reaction coordinate may be approximated by $-(Bz)^{m/2}r^{-m}$, where m lies between 2 and 4, with B constant. Combining with an inner sphere repulsion $+Ar^{-n}$, and minimizing using the derivative, the equilibrium energy well at $r_{o,z}$ is $-[(n - m)/n]Bz^{m/2}(r_{0,z}^{-m})$. A is approximately constant and equal to $(m/n)Bz^{m/2}(r_{o,z})^{n - m}$, so $r_{o,z - 1}/r_{o,z} = (z/z - 1)^{m/2(n-m)}$, where z, z - 1 are the initial and final valence states. Eliminating the r_0 terms gives $\phi_{z - 1} = \phi_z(z - 1/z)^{mn/2(n - m)}$, where the ϕ values are equilibrium potential energies. This energy expression is formally similar to Johnston's bond-energy-bond-order (BEBO) transition state model.[68] The energy in excess of ϕ_z along the reaction coordinate is $\phi' = -(\phi_z - \phi_{z-q} - y) + q(\phi_z - \phi_{z-1} - y)$, where z - q is an intermediate charge, and y contains the overpotential. ϕ_{z-q}, ϕ_{z-1} are $\phi_z(1 - q/z)^{mn/2(n-m)}$ and $\phi_z(1 - 1/z)^{mn/2(n-m)}$ respectively, with ϕ_z negative. Expansion as far as the quadratic binomial terms gives $\phi' = (\phi_z/z^2)p[q(1 - q)] + y(1 - q)$, where $p = mn(2m + mn - 2n)/8(n - m)^2$. The derivative is zero when $q = q^* = 1/2 - y/2(\phi_z/z^2)p$ The activation energy ϕ^* is therefore $(\phi_z/z^2)p/4 + y/2 + y^2/4(\phi_z/z^2)p$, which is identical in form to the Marcus expression,[41,42,44,50,67,176] and will therefore show the same type of Tafel curvature and change of symmetry factor with overpotential as that identified in Section II-8.

In early simulations of non-FC ECET, separate ion-permanent dipole and ion-induced dipole terms were used, along with Hush's assumption that A could be considered to be constant on going from reactants to products. This imposed restrictions introducing inaccuracy, so a model was used based on that in Section VI-1 for 2+ and 3+ ions. Reasonable assumptions made were (a) the ion-dipole center distance was proportional to the change in fractional charge on going from 2+ (2.205 Å) to 3+ (2.095 Å) which is reasonable since the z/x^2 and $(z/x^2)^2$ terms in the permanent and induced dipole energies can be expanded to linear expressions with good accuracy, and (b) any changes in the outer counter-field terms were also linear functions of the change in fractional charge.

The calculated non-FC barrier between the 2+ and 3+ states is shown in Figures 7A and 7B. The initial and final states are corrected this time for the ground states of the oscillators (Figure 7A). Motions along the reaction coordinate consist of the three-dimensional symmetrical stretching frequency and any independent stretching frequencies of the inner sphere oscillators, and those perpendicular to it. If ε_r, ε_i, and ε_p are the reactant, intermediate, and product ground state energies, it is reasonable to suppose $\varepsilon_i \approx (\varepsilon_r + \varepsilon_p)/2$ for motions perpendicular to the reaction coordinate. Along the reaction coordinate, it is assumed that motion is governed by the time constant of the Continuum superdipoles. The vibrational partition function in the transition state may therefore be replaced by $kT/h\nu_i$ in the usual way, where $1/\nu_i$ is the transition time for reactants to products, giving the preexponential frequency factor kT/h. If the barrier height with the two minima at equal energy is A, the minima may be adjusted so that the ground state energies are equal, then using the Brønsted rule, the

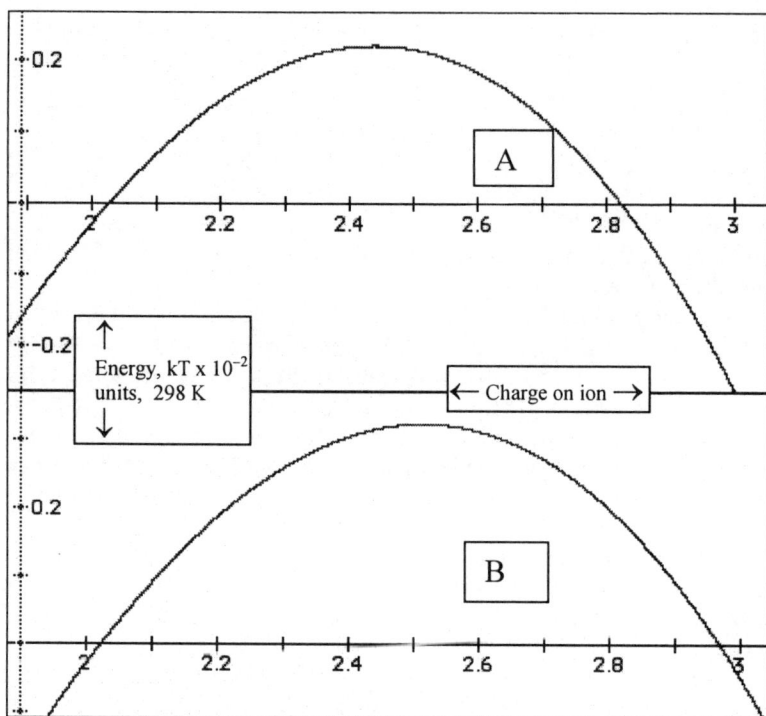

Figure 7. Calculated two-dimensional energy surfaces for heterogeneous non-Franck-Condon transition of the three-dimensional 6-oscillator trikisoctahedral Inner Shells of 3+ and 2+ ions at (A) equal potential energies after ground state energy correction; (B) at equal ground state free energies. Barrier height in B +22.9kT at 298 K (+56.9 kJ/mole, +0.59 eV, experimentally +0.59 eV[177]).

height of the barrier to the energy minimum for the intermediate becomes $(A - \varepsilon_r) + (\varepsilon_r - \varepsilon_p)/2 = A - (\varepsilon_r + \varepsilon_p)/2$. Alternatively, as in Figure 7, it may be done graphically. The barrier height is then 32.26kT. If kT/h appears in the preexponential, a further kT should be added to give the Arrhenius energy, E_{Arr}.

Since we are concerned with the Arrhenius activation energy under equilibrium conditions, a further correction should be made for the $T\Delta S$ change with the system in equilibrium under standard conditions, i.e., considering the relative standard entropies for for Fe^{2+}, Fe^{3+} equal to -137.7 and -315.9 J/mole-K,[190] a difference of -178.2 J/mole-K. The

Continuum entropies[98] relative to vacuum values are about -40.2 and -90.4 J/mole-K (Section III-1), a difference of -50.2 J/mole-K. We should point out that the assimilation of the oscillator potential energy to a portion of the 298 K experimental ion enthalpy may be somewhat in error (c.f., discussion for protons[191]), and the difference between the 2+ and 3+ entropies in Ref. 190 is 21% higher than for Latimer's original values.[192] In consequence, there is some uncertainty that the calculations truly represent equilibrium. However, the barrier height in Figure 7B is 56.9 kJ/mole (0.589 eV, 22.9kT, 13.6 kcal/mole), in very good agreement with the experimental value of 0.59 eV.[177] Other calculations showed that the activation energy is closely proportional to the oscillator energy in the transition state or to the mean solvation energy, in agreement with other work.[185] It is also roughly proportional to the mean charge on the transition state and to the change in bond-length at constant mean charge.

We should bear in mind that chemical processes in liquid media[193] (which should include at least some electrochemical processes) may follow the classical Berthoud-Hinshelwood expression for the reaction rate, in which internal classical vibrational modes reduce the potential energy of activation U_0 by a term involving some effective number 's' of classical oscillators per molecule to give E_{Arr}, i.e., $E_{Arr} = U_0 - skT$.[194,195] This will show an temperature-dependent E_{Arr}, which requires further study.

Assuming trikisoctahedral ions with average effective radii of 4.0 Å, that of the inner shell (Section IV-9) in collisions with single active water molecules (radius 1.4 Å), a collision frequency (Section II-8) of 1.8×10^{13} s^{-1} is obtained. Assuming an effective reaction layer thickness of twice the mean second trikisoctahedron shell radius, i.e., 10.8 Å, with an activation energy of 22.9kT at 298 K would yield an apparent rate constant of 2.2×10^{-4} cm-s^{-1} at this temperature. The use of kT/h as the frequency factor (6.2×10^{12} s^{-1} at ambient temperature) or the frequency ($\propto T^{1/2}$) of superdipole transition between redox states (7.7×10^{12} s^{-1}, c.f., the period $\approx 1.3 \; 10^{-13}$ s in Section VI-1) gives 7.6×10^{-5} cm-s^{-1} and 9.4×10^{-5} cm-s^{-1} respectively. The experimental value[177] is 5×10^{-5} cm-s^{-1}. Considering the assumptions made, the agreement may be considered good. As has been pointed out[91] the interpretation of kinetic data is often distorted by double layer effects.[196]

3. Proton Transfer and Other ECIT Processes

Before correction for ground states, the *Version C* model for proton solvation, giving importance only to ion-dipole averages (which may introduce some error) gave a total barrier height of only 6.81kT for all four water molecules (Figure 8). The ground states of the initial states were 2.35kT (one three-dimensional and one linear oscillator), and the estimated value for the hydrogen-bonded products was 1.45kT, giving a correction of 1.9kT. The same procedure as that for the redox case was used to estimate the ground state energies of the transition state as an increase in barrier height of 1.24kT, so the barrier height falls from 6.81kT to 6.15kT. The next question is the activation energy associated with the discharge process $H_{ads} + H_2O \rightarrow H_3O^+$, where H_{ads} has $\Delta G^o = 0$

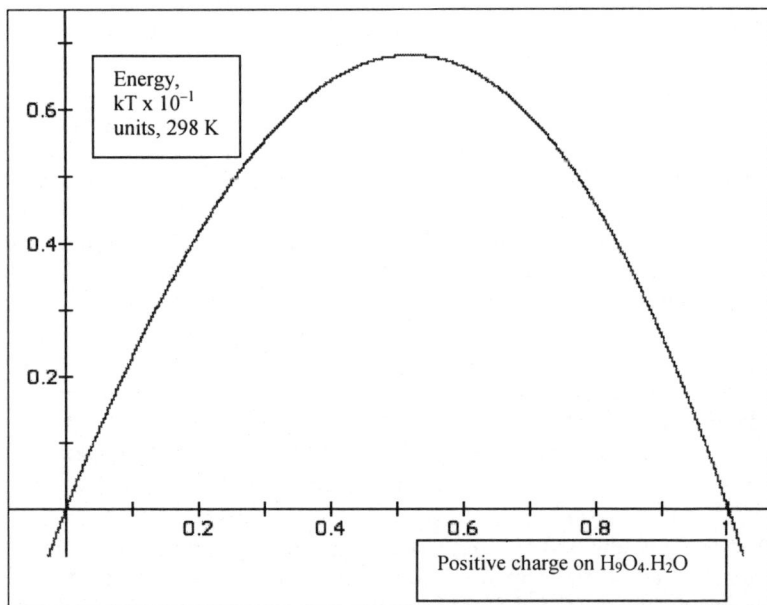

Figure 8. Calculated two-dimensional energy surfaces for heterogeneous non-Franck-Condon transition between $H_9O_4^+.H_2O + e^-$ and $5H_2O$ at equal potential energies, before correction for ground states. Barrier height +6.15kT (+15.2 kJ/mole, +0.16 eV) after correction for ground state energies.

at equilibrium. An attempt to examine this (along with the other hydrogen electrode steps) was made by Bockris and Srinivasan[197] using the semiempirical Eyring-Polanyi approximation[198] of the Heitler-London method to calculate the potential energy surface. This method is simple to use for normal chemical processes involving atom exchange, but it is not so clear how it should be used for electrochemical processes, i.e., how the terms for different species, particularly the electron, are to be accommodated at equilibrium. Attempts to apply it appear to show low barriers at equilibrium, although the assumption that the metal water interaction (the third interaction in a pseudo three-atom process M–H–H_2O process) can be ignored leads to incorrect results, as simulations carried out in the present work have shown. The reaction between the emerging H^+ and H_2O, a pseudo-two-atom process, is so strongly exothermic that a barrier would not be expected, as in, for example, $H + H \rightarrow H_2$.[199] Thus, the total activation energy of proton discharge under equilibrium conditions appears to be very low, even less than that for water dipole rotation (16.3 kJ/mole in liquid water, about four times less than in ice),[200,201] which partially contributes to it. Any contribution corresponding to the formation of $H_3O^+ + e^-$ from H and H_2O is similar mechanistically to the tunneling transfer of H^+ between water molecules.[171] The crossing of the FC potential energy surfaces for this process is shown on p. 99 of Ref. 170. The barrier is low enough and thin enough to be transparent to protons.

At overall H_2 and H^+_{aq} electrochemical equilibrium and under equilibrium conditions for the intermediate $-H_{ads}$ at the Volcano maximum, the barrier must be further corrected for the absolute entropy of H^+_{aq}, which from thermal cell studies[202] is about -20.9 J/mole-K.[203,204] Thus using the Brønsted rule, the barrier should be lowered by $-0.5T\Delta S$ or 1.26kT at 298 K, giving a final value of 12.1 kJ/mole (0.126 eV, 4.89kT, 2.90 kcal/mole), which may or may not represent the potential energy of activation (U_o) rather than the experimental Arrhenius energy, i.e., $U_o + kT$ for a transition-state process with a preexponential kT/h.

Marković and Ross[186] have studied the hydrogen oxidation reaction (HOR) on well-characterized Pt(110)(1x2), (100) and (111) surfaces in 0.05 M H_2SO_4 as a function of temperature. The basic Pt(110)(1x1) surface consists of parallel lines of atoms separated by the unit cell length 'l' (3.924 Å), with each atom in the line separated by nearest-neighbor distance $l/\sqrt{2}$. Each rectangle of four atoms has a further atom at its center, situated $l/2\sqrt{2}$ below the plane of the others, giving a ridge and furrow surface structure. The surface atoms have 6 missing nearest neighbors, whereas the furrow atoms have 2. This is an unstable configuration, and surface restructuring readily occurs, e.g., on annealing. The resulting (1x2) configuration has every other surface row missing, giving furrows twice as wide, with hexagonally close-packed (111) sides (4 missing nearest neighbors) at an angle of 54.74° to the perpendicular to the surface. To maintain numerical continuity, the surface has hexagonal steps to higher surface layers, giving terraces with missing row configurations. Unlike atoms on the (111) surface, which have 3 missing nearest neighbors, the surface atoms and furrow atoms are different in properties according to their number of closest neighbors. Pt(100)(1x1) is also known to reconstruct to a hexagonal form known as (5x20) from its LEED pattern under certain conditions, but the (1x1) surface was stable at the potentials examined.[186] Per l^2, the Pt(110)(1x2) surface ideally has 0.707 surface atoms, 1.414 atoms in the furrow sides, and 0.707 at the bottom, i.e., 2.818 in all. Ideally the surface packing density is 2.309 and 2.000 atoms per l^2 for the (111) and (100) planes.

The Tafel slopes for hydrogen oxidation obtained by Marković and Ross[186] at 274 K were A. (110)(1x2), RT/1.94F; B. (100) RT/0.49F (high current density), RT/1.47F (low current density); and C. RT/0.73F. The corresponding exchange currents and equilibrium activation energies were 0.98, 0.60, and 0.45 mA/cm² at 303 K and 9.5, 12, and 18 kJ/mole, respectively.[186] The authors considered that the corresponding mechanisms were A. dissociation (reverse combination) rds, rapid discharge (Tafel-Volmer, Process 1), B. electrochemical dissociation rds, rapid discharge (Heyrovsky-Volmer, Process 2), and C. either Process 1 or 2, respectively. A. is in good agreement with the expected value for the Tafel rds under high intermediate coverage conditions, with no other possible fit (low coverage results in a chemical limiting current). There are three possibilities for B: a switch from a second electron transfer step (Volmer) rate-determining under low coverage conditions to a first step (Heyrovsky) as overpotential increases, or the opposite at high coverage. Finally, it may represent a first electron transfer rds at high coverage, with a sudden switch to low coverage at an overpotential of +40 mV. C is considered later.

When filled with H_{upd}, the lines of atoms at the furrow tops of Pt(110)(1x2) are still unoccupied.[186] Even so, the weak structure shows a lattice expansion of 10% compared with 2% for the other low-index surfaces.[186] Unlike Pt(100), adjacent sites are available for Tafel dissociation, which is favored under high coverage conditions in the limited overpotential range (0.05 V) examined.

Following the reduction of a partial coverage (to about 0.33)[204] of adsorbed anions (bisulfate),[186,204] the well-prepared (100) surface is fully covered (to one H_{ads} per Pt atom) by under-potential-deposited hydrogen (H_{upd}) in the potential range of interest, which follows a Frumkin isotherm, i.e., its free energy of adsorption falls linearly with coverage as sites are occupied and sideways forces and lattice expansion progressively occur.[186] Alternatively, the isotherm may be regarded as a succession of overlapping fractional Langmuir isotherms, whose average free energy for each segment reduces with coverage. The last segment, completing the total coverage, represents a small fraction of the total area whose free energy at this differential coverage is zero or rather negative. This fractional part of H_{upd} should be the major contributor to the overall anodic reaction intermediate, i.e. only a small part of the overall surface area should be active. The marginal H_{upd} species, i.e., that with the weakest adsorption, may be the reaction intermediate at small anodic overpotentials at high local coverage, and the Heyrovsky rate-determining step switches to the Volmer step under high coverage conditions on the (100) surface as overpotential increases.

In the potential range of interest, the hexagonal (111) surface is filled by H_{upd} whose Frumkin peak is broad, indicating a large $+r\theta$ value, where r is the change in free energy of adsorption with the coverage θ. Over the 0.15 V examined, the electrode is in the medium coverage range. Examination of the complete kinetic equation for Heyrovsky rate-determining, Volmer in pseudoequilibrium under these conditions, including the preexponential θ, $(1 - \theta)$ terms and the exponential $\beta'r\theta$, $- (1+ \beta'r\theta)$ terms, where β' is the Brønsted slope,[183] shows that a $RT/0.73F$ Tafel slope may be readily fitted to this process over two decades. This provides an explanation for Case C, the (111) surface. Thus, on the (111) and (100) surfaces, the extrapolated Heyrovsky reaction appears to have activation energies of 18 and 12 kJ/mole respectively[186] under equilibrium conditions at high H_{upd} coverage, whereas the Tafel process has an activation energy of 9.5 kJ/mole under the same conditions on the Pt(110)(1x2) surface.

High coverages imply small experimental Arrhenius preexponential terms, which were (110) 0.043 A-cm^{-2}; (100) 0.070

A-cm^{-2}; (111) 0.570 A-cm^{-2},* assuming E_{Arr} to be temperature-independent. If one assumes a dihydrogen solubility $\approx 10^{-3}$ moles/liter, with kT/h at 303 K = 6.31 x 10^{12} and a reaction zone thickness of 2.8 Å corresponding to one water molecule diameter, we obtain a theoretical preexponential factor of 0.177 moles-cm^{-2}-s^{-1}, i.e., 3.41 x 10^4 A-cm^{-2}. Taking into account the surface packing densities, this indicates that on average only 1 surface atom pair in 7.9 x 10^5 surface atoms is active on the (110)(1x2) plane for Tafel dissociation; and 1 in 3.5 x 10^5 on the (100); and 1 in 4.5 x 10^4 on the (111) for the Heyrovsky reaction.

Under cathodic conditions, the low activation energies require correction for the corresponding overpotential deposited hydrogen (H_{opd}) cathodic reaction intermediate under high coverage conditions. These have been examined by Conway and coworkers[204,205] as part of a somewhat controversial study of hydrogen evolution in sulfuric acid on platinum single crystals, which apparently shows much higher cathodic rates compared with the Marković and Ross[186] micropolarization data,. The dη/dθ slope[204,205] at θ = 0.5 is equal to (RT/F)[(rθ/RT) + 4], giving r values of 5.7kT, 3.8kT, and 0.2kT for the (110), (100), and (111) faces. With β' $\approx$ 0.5, the cathodic process (Heyrovsky rds at high coverage, Tafel slope RT/2F) extrapolated from moderate to high overpotentials should have activation energies at equilibrium of about 18 and 17 kJ/mole on the (111) and (100) faces, that on the (110) being unknown. At cathodic overpotentials, assuming a temperature-independent β, a barrier of this height ($\approx$7.3kT) will only allow a change in rate of 3.2 decades from the exchange current value before the process becomes activationless.[206] This presents a difficulty, because data for platinum (which appear acceptable, because of the techniques used) show a linear Tafel behavior up to 10^2 A/cm^2.[207-209] The only apparent explanation is a basic difference between the exponential and preexponential terms on single crystal and polycrystalline platinum. Bowden's early work, dating before careful elimination of adsorbable impurities, reported an activation energy of 29-38 kJ/mole[15] on polycrystalline platinum, the higher value being on old electrode surfaces. At the lower activation energy value, this would allow 5.25 decades of rate change before the activationless condition occurs, making the results of Kabanow[207] and Bockris and Azzam[209]

---

* Bowden's early work, dating before careful elimination of adsorbable impurities, reported an activation energy of 29-38 kJ/mole[15] on polycrystalline platinum, the higher value being on old electrode surfaces. This suggests a preexponential value of about 0.5 A/cm^2.

plausible, though not mutually in agreement, since the latter showed hysteresis at very high current densities.

The hydrogen evolution reaction on mercury has an exchange current of about $10^{-12.3}$ A/cm^2,[210] an equilibrium activation energy of ca. 80 kJ/mole,[204] and therefore an Arrhenius prexponential factor of 50 A/cm^2, about 100 times greater than that on Pt(111). A convincingly straight composite Tafel plot[210] (potential independent $\beta = 0.60$) using results of Bowden and Grew[211] at low current density, of Bockris and Azzam,[209] and finally the apparently unpublished results of Nürnberg[212] extends from 10^{-9} A/cm^2 to $10^{+4.4}$ A/cm^2. The plot may be simulated without using the pseudoequilibrium assumption for H_{ads} coverage, since the back reactions are negligible at high cathodic overpotentials, where the forward rates for the electrochemical Volmer and Heyrovsky steps are high. Taking platinum to be at the top of the Volcano, the effective Heyrovsky rate constant on mercury will be increased by $\exp(1 - \beta')\Delta\Delta G_{ads}/RT$ compared with that on platinum, where β' is again the Bronsted slope, and $\Delta\Delta G_{ads}$ is the positive difference in the free energy of adsorption of H_{ads} between platinum and mercury. The rate constant of the Volmer reaction is correspondingly reduced by $-\beta'\Delta\Delta G_{ads}/RT$, and that of the Tafel reaction is increased by $2\beta'\Delta\Delta G_{ads}/RT$. A general expression for the rate of the overall Heyrovsky-Volmer process is $(k_1 k_2 - k_1'k_2')/(k_1 + k_2 + k_1' + k_2')$, where the back reactions have primes and the k values contain the $\Delta\Delta G_{ads}/RT$ and electrochemical rate terms, but not the θ and $(1 - \theta)$ terms, which are reflected in the denominator. A similar more complex (quadratic) expression may be written for the Tafel-Volmer mechanism.

Assuming that β' (for H_{ads}) is $(1 - \beta)$, where β is the symmetry factor for changes in $H^+ + e^-$ on the opposite side of the energy barrier, the best fit for $\Delta\Delta G_{ads}$ for mercury for the Tafel plot is $+68kT$ or about 168 kJ/mole more positive than the value for platinum, i.e., -83 kJ/mole, putting it very close to the low adsorption energy leg of Trassati's Volcano plot[213] shown by Conway.[204] The Volmer reaction should always rate-determining, with a potential-independent Tafel process to overpotentials of about -0.9 V, when it is overtaken by the Heyrovsky process as the fast step. The Brønsted slope of this leg, referring to the Volmer rds, is 0.32. On platinum, this rds has a rate of about 10^{-1} A/cm^2. Using this value results in a slightly steeper slope of 0.37, close to the 0.4 used in the simulation. It is also in good agreement with the presumed difference between the equilibrium activation energies on mercury and platinum (about 60 kJ/mole, or 0.36

x 168). An activationless process will occur a sufficiently high current density. Again assuming a potential-independent β, this would occur at $+10^{1.7}$ A/cm^2, or at higher values if some change of β at very high overpotentials take place. This may be occurring at the highest current densities shown in Ref. 210.

4. Tafel Plots for Redox ECET and ECIT Processes

The mechanisms discussed above for redox processes gives pseudo-parabolic V-i dependencies after binomial expansion of the energy terms, leading to Tafel curvature. The interesting result is the value of α, the transfer coefficient, which (in intervals of 4kT from equilibrium, or 0.1028 eV) is 0.432, 0.418, 0.403, 0.388, 0.373, and 0.359 (anodic); and 0.554, 0.539, 0.525, 0.511, 0.502, and 0.483 (cathodic). In these ranges, the effective charge on the transition state varies from +2.439 to +2.366, and from +2.432 to +2.524. The anodic and cathodic transfer coefficients at equilibrium correspond exactly to the fractional charge between the reactant and product at equilibrium, but deviate from it at higher overpotentials. This deviation is small in the anodic direction, but progressively much greater in the cathodic direction. Similarly, the sum of the anodic and cathodic transfer coefficients at constant rising current density deviates progressively from unity. An experimental anodic α for Fe^{2+} to Fe^{3+} at about 1 decade from equilibrium is 0.425.[90,177] The above results seem to be in good agreement with this, and Tafel behavior within experimental error occurs over about 5 orders of magnitude.

However, this model of kinetic behavior requires further discussion. The usual argument used to claim an experimental "straight" Tafel slope for the case of high overvoltage metals for hydrogen evolution at low coverage has assumed intersection of reactant and product energy curves, for example, those of Despic and Bockris[214] for metal deposition, extended later to the proton transfer case (Ref. 5a, p. 345, c.f., Bockris and Matthews[215]). The latter calculation is not very different from the intersection of Marcus continuum parabolic terms for reactants and products occurring high up one of the "legs" of the reactant parabola to give a rather constant Tafel slope of somewhat less than 2RT/F, c.f., Fischer.[216] However, such an intersection (after subtraction of the electron energy term on an absolute scale between reactants and products) implies that the reaction occurs under FC conditions, cf. Gurney, Ref. 8, and Figure 9A.

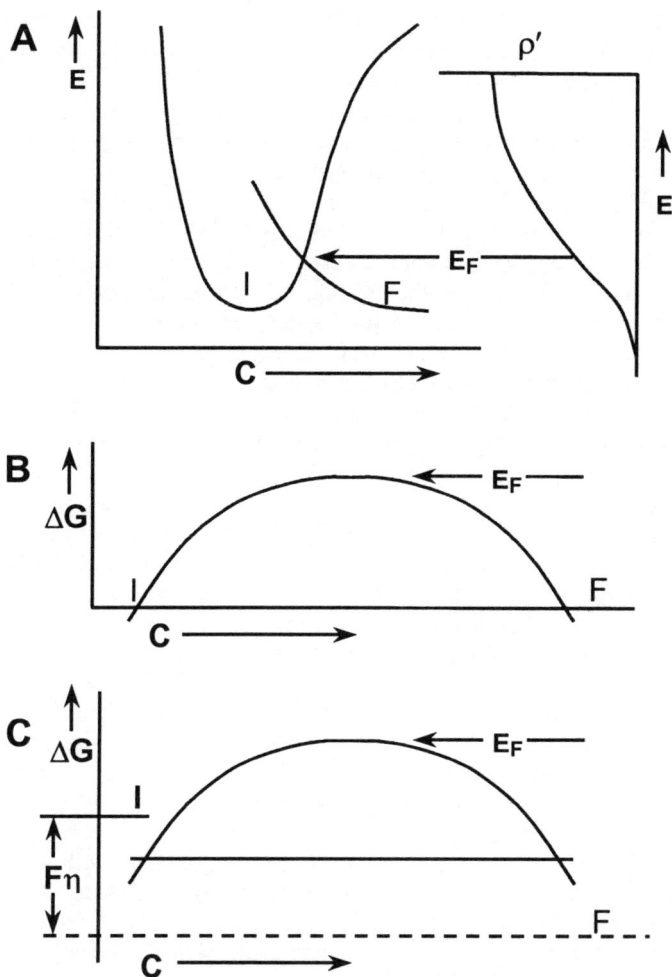

Figure 9. Illustration of how change in Fermi energy (E_F) acts on the energy barrier. E = potential energy; C = Nuclear configuration; ρ' = density of filled metallic states in metal electrode. (A). Gurney's model[8] with initial (I) and final (F) states at the same potential. (B). Non-Franck-Condon transfer in equilibrium. (C). Non-Franck-Condon transfer out of equilibrium.

Differing models for the transfer coefficient and symmetry factor, including a history of the terminology, were instructively reviewed by Bauer in 1968.[217] Parsons (1951),[218] c.f., Plonski (1969),[218] regarded it as equal to the fractional charge on the transition state, whose charge is mid-way between that of the initial and final states in a given step, and whose electrochemical potential is thereby governed by the solution potential. Vetter[219] regarded it (the *Durchtrittsfaktor*) as being the result of a *Durchtrittsreaktion,* in the sense of an unspecified stepping through from one state to another (c.f., Ref. 208, p. 151). Audubert[220] considered it to concern the transfer of kinetic energy to potential energy of an ion to overcome the barrier. However, it has also been associated with the way in which charge (electronic or ionic) is exchanged between the electrode and the solution, considered illustratively in Ref. 221 along with other interpretations, i.e., intersections of energy slopes.[214] Another view made it the work required in transferring charge through the double layer.[222] In some cases, it is the value of a parameter measured along the reaction coordinate, in others it is an energy at right angles to it. In equilibrium, the anodic and cathodic symmetry factors must equal unity, so that the Nernst equation emerges under reversible conditions.[217] However, away from equilibrium, this may not be so.

In a first approximation under non-FC conditions, the Tafel slope may be attributed to the action of the change in the free energy of electrons (i.e., the change in Fermi level of the electrode with respect to the electrostatic energy of the ions in the solution) on the fractional electronic change on the transition state. However, a persuasive case can be made that the experimental Volcano relations and Tafel slopes represent the same phenomenon.

The former must influence the barrier height following the Brønsted rule by acting on the ground state of adsorbates. Thus it would seem that the Tafel slope represents the same effect. The Brønsted rule should probably be examined using the Eyring-Polanyi-Heitler-London model of the potential energy surface (Section VII-3),[197,198] which yields cumbersome expressions in which the roles of the electron and overpotential are not clear. An approximation of the cross-section of the surface by using the BEBO model[68,210] with fitting of parameters to an assumed (experimental) activation energy under equilibrium conditions shows similar results to the model given above, i.e., translation of the barrier maximum towards the products as adsorption energy increases, with a corresponding change in $\beta' = (1 - n^*)^p$, where n^* is the bond order (for the reactants) at the maximum, and p is close to unity. Again, the sums of the transfer coefficients at the same cathodic and anodic current densities progressively recede from unity as current density increases. A better physical formulation of the system which attempts to explain the experimental linear Tafel slopes is required.

5. Linear Tafel and Brønsted Slopes

In equilibrium, the electrochemical potentials of the reactants and products are equal. When an overpotential is applied, the reaction becomes irreversible. The Gurney FC model of an electrochemical process involving neutralization of an ion under radiationless transfer conditions may be used to illustrate some general observations. Radiationless transfer under non-FC conditions only changes the length of the abcissa, since it allows nuclear movement during the completion time for electron transfer. Integration of the Gurney expression for the rate of, e.g., proton discharge[8] in the anodic and cathodic directions[223] results in two expressions which may be equated at equilibrium. In previous work, this was used to derive an exchange current[223] under these conditions, which contains (as expected from other less precise quasi-thermodynamic considerations[224]) no major electronic parameter of the electrode material (e.g., Fermi level or work function) apart from an (integrated) small preexponential electronic density-of-states term. However, it does show that the condition of radiationless electron transfer does involve the fact that the Fermi level of the electrode has an energy *identical with the barrier maximum.* Thus under equilibrium conditions, the free energy (very close, if not precisely equal to, the potential energy[225]) of the electrons is in equilibrium with the top of the thermal energy barrier, not with the ions in solution. When an overpotential is applied, the reaction becomes irreversible, and part of the free energy corresponding to the overpotential of the half-cell reactions becomes generated enthaply, i.e., $-T\Delta S_{irrev}$. The radiationless transfer condition still of course applies, and the Fermi level in the electrode must still be at the same energy level as the top of the thermal barrier (Figure 9). In effect, application of an overpotential or the driving of the overall reaction in either the anodic or cathodic directions must drive the thermal barrier up or down according to its average charge at the peak. Taking a simple case, neutral species on one side of the barrier will not be affected by the change in potential, whereas singly-charged species on the other side will have their electrochemical potential changed by $\pm eF$. The transition state will have its electrochemical potential changed by $\pm\beta eF$. This is illustrated in Figure 9, which gives some logic to the potential energy surface sections given without explanation by Gardiner and Lyons,[226] and Glasstone, Eyring, and Laidler.[227]

For redox processes, there is some evidence of Tafel slope curvature for certain processes under certain circumstances.[228,229] These may be a partial result of double layer effects.[196] In other cases experimental Tafel plots which are close to linear appear.[177,230] The controversial question of β possibly varying with temperature[231,232] will not be discussed here, although double layer effects[196] may often be responsible.

Because the observed Volcano relationships in combined ion-electron ECIT processes show similar linearity characteristics to the Tafel slopes, a similar explanation for this behavior appears probable. As discussed in the previous section, the BEBO approximation may be used to estimate the potential energy surface for normal bond-forming and bond-breaking processes. However, the Brønsted slope for these neutral molecule processes is only close to 0.5 when the enthalpy of reaction is zero. It deviates considerably from 0.5 at positive and negative enthalpy values, and at equal positive and negative enthalpy values the Brønsted slopes do not add up to unity. The argument is made above that under conditions of constant potential energy of adsorption of a neutral adsorbed intermediate (e.g., H_{ads}) under equilibrium conditions, the overpotential, i.e., the change in Fermi level of the electrode, only acts on the charge of the transition state, and may not influence it. Similarly, the change in potential energy of adsorption of a neutral reactant on the electrode surface at constant overpotential may also not change the charge on the transition state in a reaction of ECIT type. This is because this charge is governed only by the configuration of the forming or dissolving solvation shell, partly within, and partly without, the Helmholtz double layer. The physical model for this was sketched out by Conway and Bockris for metal deposition processes.[233] This configuration may be determined entirely by effective (in the Duncan-Pople sense) permanent dipole energies along the water molecule configurational vector, together with those resulting from the induced dipole. Setting the derivative of the simple electrostatic expressions for these energies equal to zero under equilibrium conditions determines the barrier maximum and the charge. This rather simple concept provides a consistent picture for ECIT processes involving adsorption.

VIII. CONCLUSIONS

It was at first apparent that one of the major problems in the understanding of the rates of different types of charge-transfer processes (those involving direct electron transfer, ECET, and ion plus electron transfer processes ECIT), was the failure to have a good model for solvation energy, both for the Inner and Outer Sphere. The former had a simple model,[16,92] but the latter had no molecular basis at all, only a conceptual electrostatic charging process[31] whose physics breaks down for electronic charges. Initial work looked at both of these concerns. The first was brought reasonably up to date, and the second was given a molecular basis. In addition, the whole problem of dielectric saturation was examined. After examining a number of alternative charge transfer processes, it appeared that activation energy effects concerning the Outer Sphere may in many cases be discounted, since many adiabatic reactions appear to proceed via non-FC processes in which the Outer Sphere dipoles or superdipoles have enough time to accommodate the changing charge on an ion as a function of time. The problems of the Outer Sphere and the approach to dielectric saturation may therefore be neglected in many cases. While an Outer Sphere explanation to certain charge transfer rate phenomena (particularly those which may be anadiabatic) is not discounted, non-Born-Oppenheimer or non-Franck-Condon phenomena may better fit the facts, as was suggested for proton transfer in earlier reviews.[234] Under adiabatic conditions, the process with the lowest free energy of activation will be the preferred one. If the activation energy contains an Outer Sphere term, it would be expected to change rapidly with ionic strength, reaching about 50% of the infinite dilution value at 1.0 M, and disappearing in very concentrated solutions, e.g., 6.0 M. This may be used as an experimental test.

A relatively constant Tafel slope for reactions not involving adsorption, and those involving adsorption with complete charge transfer across the double layer, distorted by second order effects, may also be explained in terms of a non-Franck-Condon process. Since adsorbed intermediates in charge transfer processes also show adsorption energies depending on potential in the same way as the potential energy barrier maxima, these should also follow the same phenomena.

APPENDIX

1. Applicability of $\varepsilon_A = 1 + 4n_o\alpha_{T,A}$

The inverse square law determines that the electric field inside closed shells of uniform charge distribution is zero. Hence, only the induced charge adjacent to a spherical condenser of radius 'a' with total charge q (equivalent to the charge on the central ion) need be considered as the opposing charge which reduces the field. If this charge is q', then the field $\mathbf{E}_A$ at the surface is $(q - q')/a^2 = \mathbf{A}_A$-$q'/a^2$. While the total induced polarization vector in any spherical shell with uniform dielectric properties surrounding a fixed molecule or charge necessarily vanishes,[72] the total scalar polarization in a shell of thickness 2s is $2sq' = P_A dV = P_A 4\pi[(a + 2s)^3 - a^3])/3 \approx 8\pi sa^2 P_A = 8\pi sa^2 n_o \alpha_{T,A} E_A$. Hence, $q'/a^2 = 4\pi n_o\alpha_{T,A}E_A$; $E_A = A_A/(1 + 4\pi n_o\alpha_{T,A})$; and $\varepsilon_A = 1 + 4\pi n_o\alpha_{T,A}$. When a approaches 2s, the exact expression for $dV = 4\pi[(a + 2s)^3 - a^3]/3$ should be used, so that ε_A becomes $1 + 4\pi(1 + 2s/a + 4s^2/3a^2)n_o\alpha_{T,A}$. This may be only approximate when the shells contain only small numbers of discrete charges, e.g., in molecular dipoles around an ion.

If we consider a uniform spherical shell of thickness 2s (the width of the volume of space containing a single molecular dipole), in which the moments are oriented by a central charge giving an electric displacement $\mathbf{A}_A$, it is clear from the inverse square law that the effective layers of equal positive and negative charge each produce equal and opposite fields at all points outside the shell. Thus, the only field outside the shell is that of the polarizing change at its center. If we consider an internal or external point located at a distance b (point b) from the center of such a shell of polarization of radius a, then the total moment in a ring of the shell located symmetrically around a line joining the center and point b is equal to $2\pi asin\theta P_A(2s)ad\theta$, where θ is the angle subtended at the center of the sphere by the center and circumference of the ring. The electric field due to the polarization in the ring at point b from the center of the sphere is given by

$$\phi = 2\pi asin\theta(\mathbf{P}_A)(2sad\theta)f(\theta)/d^3 \tag{84}$$

where d is the distance between point b and a point on the circumference of the selected ring, and $f(\theta)$ is a function of the angles between the polarization in the ring and that of an imaginary dipole lying at point b along the line joining point b and the center of the sphere. $f(\theta)$ is equal to $-(2\cos\theta - 3\sin\theta\sin\beta\cos\beta - 3\cos\theta\sin^2\beta)^{137}$ from the system geometry, where β is the angle subtended at the ring by point b and the center of the sphere. Since $\sin\beta = (b/d)\sin\theta$ and $\cos\beta = (r - b\cos\theta)/d$, $f(\theta)$ is $-(2\cos\theta - 3ab\sin^2\theta/d^2)$, where $d^2 = a^2 + b^2 - 2ab\cos\theta$. Hence:

$$\phi = -2p\mathbf{P_A}(2s)a^2 \int_0^{2\pi}\left(\frac{2\cos\theta\sin\theta}{d^3} - \frac{3ab\sin^3\theta}{d^5}\right)d\theta \qquad (85)$$

The field is zero for the special case where $b = 0$ and $a = r$. The general result is:

$$\phi = \left(\frac{\pi\mathbf{P_A}s}{b^2}\right)\left[d - \frac{2(a^2 + b^2)}{d} + \frac{(a^2 - b^2)}{d^3}\right]_{d=a-b}^{d=a+b} \qquad (86)$$

which is zero for all values of b except when $b = a$, when the lower limit is $d = 0$ and the expression is infinite, i.e., closed inner and outer shells of polarization do not contribute to the polarization field of the selected shell for which $b = a$. To determine the polarization field at the dipole center of a selected dipole in this shell, we integrate the value of d from a lower limit equal to the distance between the center of the selected dipole and the start of the next shell, i.e., the distance s. Thus the field due to the shell at $b = r$ without the field of the selected dipole is:

$$\phi = E_{p,A} = -(p\mathbf{P_A}s/a^2)[(4a^2/s) - s] \approx -4\pi\mathbf{P_A} \qquad (87)$$

Thus, the polarization field at the selected dipole is $E_{p,A} = -4\pi\mathbf{P_A}$, and the external field E_A at the dipole is $A_A-4\pi\mathbf{P_A}$. However, $\mathbf{P_A} = \alpha_{T,A}n_0E_A$, so $E_A = A_A/(1 + 4\pi n_0\alpha_{T,A})$, and ε_A equals $(1 + 4\pi n_0\alpha_{T,A})$ in radial geometry, provided that the number of dipoles in the shell considered is enough to make an integration (instead of a summation) sufficiently accurate. The same should be true for the polarization field at a selected solvated ion of e.g., diameter 2d, which may be several times 2s, but for which the shell thickness is 2d rather than 2s.

Equation (86) shows that $4a^2/s$ must be large compared with s, but this will be true even for the first shell of continuum dipoles. However, the treatment no longer applies at short distances, when A_A overwhelms

$4\pi\mathbf{P}_A$. The dipole field equation used in the derivation is reasonably accurate at distances of a few Å. The use of $2\pi a^2(2s)\sin\theta d\theta$ for the volume of each ring instead of $(2/3)\pi[(r+s)^3 - (r-s)^3](\cos\theta_1 - \cos\theta_2)$ will understate the volume, but the largest error occurs because of the non-continuous nature of the polarization distribution in the shell. The polarization in a shell of dipoles of radius a (measured to the dipole centers) was accurately estimated numerically for $s = a\sin(\pi/10)$, i.e., with a hydrogen-bond distance of 2.9 Å, a = 4.62 Å. This distance represents an approximate lower limit for the first continuum shell for univalent ions. The result obtained for the value of $E_{p,A}$ was $-3.16\pi\mathbf{P}_A$, 79% of the value for a = ∞. As expected, small residual fields were found outside and inside the shell. The summation improves as 'a' increases to higher values. The effect of this summation of polarization fields will be significant at short distances, and it means that Eq. (87) will therefore overstate the value of ε_A, and understate the values of $\mathbf{E}_A$ and the interaction energy between the central ion and the surrounding continuum dipoles.

2. An Appropriate Model for Water Molecule Orientation

An introduction to the problem of inner sphere orientation is given in Section IV-4. Solvated monovalent negative ions apparently have more negative solvation energies than positive ones when both have the same ion-water distance.[95,98,170,172] This has been explained by dipole[98] and simple quadrupole models.[95] Consider an ST2 potential quadrupole[142] often used in molecular dynamic calculations[142,144,145,157] with charges $+2e'$, $-2e'$ with polar coordinates y, θ, and z, γ respectively from the molecule center and a line along the dipole axis joining it and an ion (charge e) at distance 'a'. Assuming a "hard wall" interaction, the potential interactions between the quadrupole and the ions in the dipole axial position are:

$$\phi = -2ee'\{[(a-y\cos\theta)^2 + y^2\sin^2\theta]^{-1/2} - [(a-z\cos\gamma)^2 + z^2\sin^2\gamma]^{-1/2}$$
$$(-\text{ions}) \quad (88)$$

$$\phi = -2ee'\{[(a+z\cos\gamma)^2 + z^2\sin^2\gamma]^{-1/2} - [(a+y\cos\theta)^2 + y^2\sin^2\theta]^{-1/2}$$
$$(+\text{ions}) \quad (89)$$

Expanding to the third binomial term and ignoring all terms higher than a^2 gives:

$$\phi = -\left(\frac{2ee'}{a^2}\right)\left[y\cos\theta - z\cos\gamma) \pm \frac{3(y^2\cos^2\theta - z^2\cos^2\gamma)}{2a}\mu\frac{(y^2 - z^2)}{2a}\right]$$

$$(90)$$

for negative (+,–) and positive (–,+) ions. At constant distance, differences must be in these terms, since induced dipoles, Outer Sphere interactions, and repulsive terms should cancel.[170] Thus:

$$\phi^- - \phi^+ = -(2ee'/a^3)[y^2(3\cos^2\theta - 1) - z^2(3\cos^2\gamma - 1)] \quad (91)$$

A convenient monovalent ion-water center distance in Halliwell and Nyburg's correlations[170,172] is 2.924 Å i.e, $(a + 1.38)^{-3} = 0.04$ Å^{-3}, when the solvation energy difference between negative and positive ions is –41.8 kJ/mole, i.e., 16.86kT/c per water molecule at 298 K, where c is the coordination number. The modified[95] Rowlinson quadrupole[141] has a partial charge e' = +0.33e on each proton with –2 x 0.33e on the oxygen (vacuum moment 1.86 D, bond length and angle 0.964 Å and 105°). With the origin at the molecule center (0.15 Å from oxygen), y = 0.880 Å, cosθ = 0.4956, z = 0.15 Å and cosγ = –1, positive ions (–27.18kT) in the axial position[16,92,95] have enthalpies of solvation 3.67kT *more negative* than axial[95] negative ions (–23.51kT) at a = 2.924 Å. A more accurate point charge analysis gives –4.20kT difference (–26.03kT and –21.83kT for positive and negative ions). The only way to obtain a more negative solvation energy for negative ions compared with positive ions aligned along the dipole axis is by changing the off-axis charge. If θ is less than 53.57° at the same y value, the result is reversed, but this is unrealistic.

If we now consider the Verwey positions,[161] if the negative charges are considered to be on the oxygen,[95] a point charge calculation shows that for negative ions, the maximum interaction at 2.924 Å is –23.38kT with the ion-oxygen axis ±21.8° above and below the O–H bond direction, compared with –20.77kT when it is exactly aligned in the Verwey position (and –21.83kT on the dipole axis, above). For negative ions, the Verwey positions are stable, i.e, do indeed lie at marginal minima, but for positive ions, the only stable position is on the dipole axis at –26.03kT, and the Verwey positions (at 60.1° to the dipole axis) are unstable, and are significantly less negative. Considering the similar situation with the negative charges at the molecule center, for negative ions, the minima lie at –25.12kT and at

±14.2° above and below the O–H bond direction, with –21.03kT along the dipole axis. Again, for positive ions, the only stable position is the axial one, at –25.79kT.

Thus, using these simple quadrupole approaches, we may be led to believe that the electrostatic stabilization energy for positive ions will be more negative than that for negative ions due to the effects of the off-axis charges. A pre-quadrupole approach, ignoring off-axis charge,[98] gives –23.06kT and –28.05kT for positive and negative ions at a = 2.924 Å, i.e., the difference of –4.99ckT is approximately correct if c = 4. The real charges are indeed off-axis, and the DP multipole[143] should be a much more realistic molecular model. In it, the protons (+e, +e) are at +0.586 Å from the oxygen atom center (+6e) along the dipole axis, offset at ±0.764 Å (bond angle 105°). The lone pairs (–2e each) on the oxygen are located at right angles to and at ±0.275 Å above and below the proton-oxygen plane, at –0.158 Å along the dipole axis. The bond electrons (–2e each) may be located along the proton-oxygen bonds at +0.334 Å from the oxygen atom center along the dipole axis, offset at ±0.443 Å gives a dipole moment of 2.138 D (See Section V-1) to account for induction. The 8+, 8 charge separation means that the DP dipole length is very short (0.0557 Å for 2.138 D) compared with the 0.586 Å Rowlinson length, and its dipole center is at +0.1187 Å (at 2.138 D) from the oxygen compared with the Rowlinson value of +0.243. Thus, the dipole center for each lies on different sides of the water molecule center at +0.15 Å.

Appropriate coulombic potentials were used to obtain ion-DP multipole interactions at a = 2.924 Å. Instead of the "hard wall" model used above, a repulsive $+A/(a + q)^{12}$ interaction was assumed, where q is the ion-oxygen distance. The complete expression was differentiated to obtain the minimum potential at a = 2.924 Å. First the ions were assumed to lie along the dipole axis, giving a net interaction of –23.11kT for positive ions, and –19.99kT for negative ions, where the repulsive potentials were 16% and 17% of the total coulombic interaction respectively. Thus, the gross coulombic interactions for the DP model were –0.8kT (3.1%) and –1.6kT (7.3%) greater in this configuration than those for the corresponding point-charge Rowlinson model, which assumed a smaller dipole moment (1.86 D). Thus, in both point-charge calculations, positive ions situated at the same distance from the water molecules show greater interaction than negative ions. However, if the Rowlinson and DP models are regarded as simple dipoles, the first yields a greater interaction for negative ions, whereas the opposite is true for the second, due to the differing position of the dipole centers with regard to the molecule center. The Rowlinson dipole model gives an approximately correct answer, but it is too artificial to be useful.

Point-charge interactions for the ion-DP multipole were then examined as a function of a, and as a function of the angle of the molecule. At typical distances, the orientation giving the most negative gross electrostatic potential for negative ions was in line with each of the H–O bonds, i.e., the Verwey[161] position. However, this was not true for positive ions in the O-lone pair direction, unless 'a' is very short. In the Verwey positions at a = 2.924 Å, the net interaction energies were –18.16kT for positive ions, and –23.54kT for negative ions, with inverse 12-power repulsions calculated from the minima equal to 18.6% and 22.8% of the gross coulombic potentials respectively.

Positive ions in the dipole axis position have more negative potentials than those in the Verwey position. The most negative position should be favored, which would result in the Verwey position for negative ions, and the dipole axis position for positive ions. This would result in *similar* potentials for both negative and positive ions at a = 2.924 Å. One of the reasons why the Verwey position is preferred is the greater opportunity for retained hydrogen bonding, whether complete or "bent".[103] The hydrogen bond is about –10kT (–5.95 kcal/mole, 24.9 kJ/mole[103]). It is usually assumed[95] that two half hydrogen bonds are broken per water molecule of solvation, so that on the dipole axis positive and negative ions under the above conditions will have effective partial solvation potentials of only about –13.11kT and –9.99kT, before induction effects are taken into account. In the Verwey positions, one half-bond is broken, giving –13.16kT and –18.54kT. With a = 2.924 Å, the DP dipole centers in neighboring Verwey positions of least repulsion at c = 4 (one *cis* and one *trans* pair) are about 4.8 Å apart, which makes the dipole-dipole interaction a reasonable approximation. These orientations give dipole-dipole interactions of +0.88kT per dipole for positive ions compared with +1.52kT when aligned along the dipole axis. The effect of this is likely to be marginal.

REFERENCES

1. J. Franck, *Trans. Faraday Soc.* **21** (1926) 536; *Z. Phys. Chem.* **120** (1926) 144; E. U. Condon, *Phys. Rev.* **28** (1926) 1182; **32** (1928) 858.
2. M. Born and R. Oppenheimer, *Annal. Physik* **84** (1927) 457.
3. A. J. Appleby, J. O'M. Bockris, R. K. Sen, and B. E. Conway, in *MTP Int. Rev. Sci., Phys. Chem. Ser. 1*, Ed. by A. D. Buckingham and J. O'M. Bockris, Butterworth, London, 1972, Vol. 6, p. 1.
4. (a) J. O'M. Bockris and R. K. Sen, *J. Res. Inst. Catalysis, Hokkaido Univ.* **21** (1973) 55; (b). *Molec. Phys.* **29** (1975) 357.
5. (a) J. O'M. Bockris and S. U. M. Khan, *Quantum Electrochemistry*, Plenum, New York, 1979; (b). S. U. M. Khan and J. O'M. Bockris, in *Comprehensive Treatise of Electrochemistry*, Ed. by B. E. Conway, J. O'M. Bockris , E. Yeager, S. U. M. Khan, and R. E. White, Plenum, New York, 1983, Vol 7, p. 41; (c). J. O'M. Bockris and S. U. M. Khan, *Surface Electrochemistry, A Molecular Level Approach*, Plenum, New York, 1993.
6. R. A. Marcus, *Naval Res. Rev.* **45-46** (1993-94) 9.
7. J. Franck and F. Haber, *S. B. Preuss. Akad. Wiss.* **56** (1931-32) 869.
8. R. W. Gurney, *Proc. Roy. Soc. London* **A134** (1931) 137.
9. L. Landau, *Phys. Z. Sowietunion* **1** (1932) 88; **2** (1932) 46.
10. F. Hund, *Z. Physik.* **32** (1925) 1.
11. W. A. Caspari, *Z. Phys. Chem.* **30** (1899) 89.
12. J. Tafel, *Z. Phys. Chem.* **50** (1905) 641.
13. J. A. V. Butler, *Trans. Faraday Soc.,* **19** (1924) 729.
14. T. Erdy-Grúsz and M. Volmer, *Z. Phys. Chem.* **150** (1930) 203.
15. F. P. Bowden, *Proc. Roy. Soc. London* **A125** (1929) 446; **A126** (1929) 107.
16. J. D. Bernal and R. H. Fowler, *J. Chem. Phys.* **1** (1933) 515.
17. (a). J. Sherman, *Chem. Rev.* **11** (1932) 98; (b). E. A. Moelwyn-Hughes, *Physical Chemistry*, 2nd Revised Edition, Perganon Press, Oxford, 1965, p. 890.
18. G. Gamow, *Z. Physik.* **51** (1928) 204.
19. R. W. Gurney and E. U. Condon, *Phys. Rev.* **33** (1929) 127.
20. R. W. Gurney, *Trans. Faraday Soc.* **28** (1932) 447.
21. R. W. Gurney, *Proc. Roy. Soc. London* **A136** (1932) 378.
22. H. Eyring and M. Polanyi, *Z. Phys. Chem.* **B12** (1931) 279.

23. H. Eyring, *J. Chem. Phys.* **3** (1935) 107.
24. K. F. Herzfeld, *Ann. Physik.* **59** (1919) 635.
25. J. Horiuti and M. Polanyi, *Acta Physicochem. URSS* **2** (1935) 505.
26. J. A. V. Butler, *Proc. Roy. Soc. London,* **A157** (1936) 423.
27. D. R. Bates and H. S. W. Massey, *Phil. Trans. Roy. Soc. London* **A239** (1943) 269.
28. R. Platzmann and J. Franck, *L. Farkas Memorial Volume,* Research Council of Israel, Jerusalem, 1952, p. 21; *Z. Physik* **138** (1954) 411.
29. J. E. B. Randles, *Trans. Farad. Soc.* **48** (1952) 828.
30. W. F. Libby, *J. Phys. Chem.* **56** (1952) 863.
31. M. Born, *Z. Physik* **1** (1920) 45.
32. J. Weiss, *Proc. Roy. Soc. London* **A222** (1954) 128.
33. P. Debye, *Polar Molecules,* Chemical Catalog Company, New York, 1929.
34. L. Landau, Phys. *Z. Sowietunion* **3** (1933) 664.
35. N. F. Mott and R. W. Gurney, *Electronic Processes in Ionic Crystals,* Clarendon Press, Oxford, 1940.
36. R. J. Marcus, B. J. Zwolinski, and H. Eyring, *J. Phys. Chem.* **58** (1954) 432; B. J. Zwolinski, R. J. Marcus, and H. Eyring, *Chem. Rev.* **55** (1955) 157.
37. K. J. Laidler, *Can. J. Chem.* **37** (1959) 138.
38. E. Sacher and K. J. Laidler, *Trans. Farad. Soc.* **59** (1963) 396.
39. K. J. Laidler and E. Sacher, in *Modern Aspects of Electrochemistry,* Ed. by J. O'M. Bockris, Butterworth, London, 1964, Vol. 3, p. 1.
40. H. Taube, *Adv. Inorg. Chem. Radiochem.* **1** (1959) 1.
41. R. A. Marcus, *J. Chem. Phys.* **24** (1956) 966.
42. R. A. Marcus, *J. Chem. Phys.* **24** (1956) 979.
43. R. A. Marcus, (a) *J. Chem. Phys.* **38** (1963) 1335; (b) **38** (1963) 1734.
44. R. A. Marcus, *J. Chem. Phys.* **43** (1965) 679.
45. R. A. Marcus, *Rept. 12, Project NR 051-331,* Office of Naval Research, Washington, DC, 1957; *Can. J. Chem.* **37** (1959) 155.
46. R. A. Marcus, *Disc. Farad. Soc.* **29** (1960) 21.
47. R. A. Marcus, *Ann. Rev. Phys. Chem.* **15** (1964) 155.
48. R. A. Marcus, *J. Phys. Chem.* **67** (1963) 853.
49. R. A. Marcus, *J. Chem. Phys.* **41** (1964) 2624.
50. N. S. Hush, *Z. Elektrochem.* **61** (1957) 734; *J. Chem. Phys.* **28** (1958) 962.

51. Ref. 17b, p. 394.
52. C. Zener, *Proc. Roy. Soc. London* **A137** (1932) 696.
53. C. A. Coulson and K. Zalewski, *Proc. Roy. Soc. London* **A268** (1962) 437.
54. H. Gerischer, *Z. Phys. Chem.* **26** (1960) 223.
55. V. G. Levich and R. R. Dogonadze, *Dokl. Akad. Nauk. SSSR* **124** (1959) 123; **133** (1960) 158.
56. R. R. Dogonadze, *Dokl. Akad. Nauk SSSR,* **133** (1960) 1368; **142** (1962) 1108.
57. V. G. Levich, in *Advances in Electrochemistry and Electrochemical Engineering,* Ed. by P. Delahay and C. W. Tobias, Interscience, New York, 1966, Vol. 4, p. 249.
58. V. G. Levich, in *Advanced Treatise of Physical Chemistry,* Ed. by H. Eyring, D. Henderson, and Y. Jost, Academic Press, New York, 1971, Vol. IXB, p. 985.
59. R. R. Dogonadze, in *Reactions of Molecules at Electrodes,* Ed. by N. S. Hush, Wiley, New York, 1972, p. 135.
60. S. I. Pekar, *Untersuchung über die Electronentheorie der Kristalle,* Akademie Verlag, Berlin, 1954.
61. M. Lax, *J. Chem. Phys.* **20** (1952) 1752.
62. H. Fröhlich, *Adv. Phys.* **3** (1954) 325.
63. T. Holstein, *Ann. Phys.* **8** (1959) 325.
64. R. R. Dogonadze and Y. A. Chizmadznev, *Fiz. Tverd. Tela.* **3** (1961) 3712; R. R. Dogonadze, Y. A. Chizmadznev and A. A. Chernenko, *Ibid.,* **3** (1961) 3720.
65. R. R. Dogonadze, A. M. Kuznetsov, and A. A. Chernenko, *Usp. Khimii,* **34** (1965) 1779.
66. N. Sutin, *Ann. Rev. Nuclear Sci.* **12** (1962) 285.
67. N. S. Hush, *Trans. Farad. Soc.* **57** (1961) 557.
68. H. S. Johnston, *Adv. Chem. Phys.* **3** (1961) 131.
69. N. S. Hush, *Electrochim. Acta* **13** (1968) 1005.
70. R. Kubo, *Phys. Rev.* **86** (1952) 929; R. Kubo and Y. Toyozawa, *Prog. Theor. Phys.* **13** (1955) 160.
71. S. U. M. Khan and J. O'M. Bockris, *Chem. Phys. Lett.* **99** (1983) 83.
72. J. G. Kirkwood, *J. Chem. Phys.* **7** (1939) 911.
73. F. Booth, *J. Chem. Phys.* **19** (1951) 391.
74. P. George and J. S. Griffith, in *The Enzymes 1,* Ed. by P. D. Boyer, H. Lardy, and K. Myrback, Academic Press, New York, 1959, p. 347.
75. L. E. Orgel, *Proc. Solvay Conf., Brussels,* 1956, 289.

76. Refs. 5a, p. 161-3; 5c, p. 442-445.
77. Ref. 17b, p. 102.
78. M. Falk and P. A. Giguère, *Can. J. Chem.* **35** (1957) 1195; R. A. More O'Ferrall, G. W. Koeppl, and A. J. Kresge, *J. Am. Chem. Soc.* **93** (1971) 1.
79. J. O'M. Bockris, K. L. Mittal, and R. K. Sen, *Nature Phys. Sci.* **234** (1971) 118.
80. A. M. Kuznetsov, *J. Electroanal. Chem.* **159** (1983) 241.
81. R. A. Marcus, *J. Phys. Chem.* **72** (1968) 892; A. O. Cohen and R. A. Marcus, *Ibid.* **72** (1968) 4249.
82. R. R. Dogonadze, A. M. Kuznetzov, and V. G. Levich, *Electrochim. Acta* **13** (1968) 1025.
83. L. I. Krishtalik, *Elecktrokhimiya* **6** (1970) 507.
84. A. M. Kuznetzov, *J. Electroanal. Chem.* **65** (1975) 545.
85. R. R. Dogonadze, J. Ulstrup, and Y. I. Kharkats, *J. Chem. Soc. Faraday Trans. 2,* **68** (1972) 744.
86. Y. I. Kharkats and J. Ulstrup, *J. Electroanal. Chem.* **65** (1975) 555.
87. R. R. Dogonadze and A. M. Kuznetsov, in *Comprehensive Treatise of Electrochemistry*, Ed. by B. E. Conway, J. O'M. Bockris , E. Yeager, S. U. M. Khan, and R. E. White, Plenum, New York, 1983, Vol. 7, p. 1.
88. A. M. Kuznetsov, in *Modern Aspects of Electrochemistry,* Ed. by J. O'M. Bockris, R. E. White, and B. E. Conway, Plenum, New York, Vol. 20, p. 95.
89. E. D. German and A. M. Kuznetsov, in *Modern Aspects of Electrochemistry,* Ed. by R. E. White, B. E. Conway and J. O'M. Bockris, Plenum, New York, Vol. 24, p. 140.
90. M. T. M. Koper and W. Schmickler, in *Electrocatalysis,* Ed. by J. Lipkowski and P. N. Ross, Wiley-VCH, New York, 1998. p. 291.
91. M. T. M. Koper and W. Schmickler, *Chem. Phys.* **211** (1996) 123.
92. D. D. Eley and M. G. Evans, *Trans. Faraday Soc.* **34** (1938) 1093.
93. E. A. Moelwyn-Hughes, *Proc. Camb. Phil. Soc.* **45** (1948) 477.
94. B. S. Brunschwig, C. Creutz, D. H. Macartney, T-K Sham, and N. Sutin, *Farad. Disc. Chem. Soc.* **74** (1982) 113.
95. J. O'M. Bockris and A. K. N. Reddy, *Modern Electrochemistry,* Plenum, New York, 1970, Vol. 1, pp. 88-108, 171-174,
96. A. D. Buckingham, *Disc. Farad. Soc.* **34** (1957) 151.

97. B. E. Conway, in *Modern Aspects of Electrochemistry*, Ed. by B. E. Conway and R. E. White, Kluwer Academic/Plenum Publishers, New York, 2002, Vol. 35, p. 295.
98. Ref. 17b, p. 881-891.
99. B. E. Conway, *Ionic Hydration in Chemistry and Biophysics*, Elsevier, New York, 1981, p. 313.
100. Ref. 99, p. 214-221.
101. L. Onsager, *J. Am. Chem. Soc.* **58** (1936) 1486.
102. Ref. 17b, p. 308.
103. J. A. Pople, *Proc. Roy. Soc. London* **A205** (1951) 134.
104. R. R. Dogonadze and A. A. Kornyshev, *J. Chem. Soc. Faraday Trans. II* **70** (1974) 1121.
105. Ref. 17b, p. 310.
106. (a) Ref. 17b, p. 363-366; (b) Ref. 99, 221-222.
107. P. Debye, *Phys. Z.* **13** (1912) 97.
108. H. Frölich, (a) *Trans. Faraday Soc.* **44** (1948) 238; (b) *Theory of Dielectrics*, Oxford University Press, 1949.
109. G. Oster and J. G. Kirkwood, *J. Chem. Phys.* **11** (1943) 175.
110. F. E. Harris and B. J. Alder, *J. Chem. Phys.* **21** (1953) 1031.
111. G. H. Haggis, J. B. Hasted, and T. J. Buchanan, *J. Chem. Phys.* **20** (1952) 1452.
112. J. Malsch, *Physik. Z.* **29** (1928) 770; **30** (1929) 837.
113. Ref. 99, p. 290.
114. J. Van Vleck, *J. Chem. Phys.* **5** (1937) 556.
115. C. Böttcher, *Theory of Electric Polarization*, Elsevier, New York, 1952.
116. A. Schellman, *J. Chem. Phys.* **26** (1957) 1225.
117. D. C. Grahame, *J. Chem. Phys.* **18** (1950) 903.
118. H. Sack, *Phys. Z.* **27**, 206 (1926); **28** (1927) 299.
119. T. J. Webb, *J. Am. Chem. Soc.* **48** (1926) 2589.
120. B. E. Conway, J. O'M. Bockris and I. A. Ammar, *Trans. Faraday Soc.* **47** (1951) 746.
121. A. D. Buckingham, *J. Chem. Phys.* **25** (1956) 428.
122. D. M. Ritson and J. B. Hasted, *J. Chem. Phys.* **16** (1948) 11.
123. C. A. Coulson and D. Eisenberg, *Proc. Roy. Soc. London* **A291** (1966) 445, 454.
124. F. Booth, *J. Chem. Phys.* **19** (1951) 1327, 1615.
125. F. Booth, *J. Chem. Phys.* **23** (1955) 453.
126. Ref. 17b, p. 696.
127. Ref. 17b, p. 577.

128. A. D. Buckingham and J. A. Pople, *Proc. Phys. Soc.* **A68** (1955) 905; A. D. Buckingham, *Ibid.* **A68** (1955) 910.
129. P. Debye, *Marx Handbuch der Radiologie (Leipzig)* **6** (1925) 777.
130. D. C. Grahame, *J. Chem. Phys.* **21** (1953) 1054.
131. K. J. Laidler and C. Pegis, *Proc. Roy. Soc.* **A241** (1957) 80.
132. J. A. Schellman, *J. Chem. Phys.* **24** (1956) 912.
133. J. A. Stratton, *Electromagnetic Theory,* § 2.8, McGraw-Hill, New York, 1941.
134. H. S. Frank and P. T. Thompson, in *The Structure of Electrolytic Solutions,* Ed. by W. Hamer, Wiley, New York, 1960.
135. M. Mandel and P. Mazur, *Physica* **24** (1958) 116.
136. W. F. Brown, *Encyclopedia of Physics,* Springer-Verlag, Berlin, 1956, Vol. 17, p. 1; *Physica* **24** (1958) 695.
137. Ref. 17b, pp. 316, 338.
138. J. Morgan and B. E. Warren, *J. Chem. Phys.* **6** (1938) 666.
139. Ref. 17b, p. 306.
140. C. W. Kern and M. Karplus, in *Water, a Comprehensive Treatise,* Ed. by F. Franks, Plenum, NY, 1972, Vol. 1, Chap. 2.
141. J. S. Rowlinson, *Trans. Faraday Soc.* **47** (1951) 120.
142. H. L. Lemberg and F. H. Stillinger, *J. Chem. Phys.* **62** (1975) 1667.
143. (a). A. B. F. Duncan and J. A. Pople, *Trans. Faraday Soc.* **49** (1953) 217; (b). Ref. 17b, p. 479.
144. F. H. Stillinger and A. Rahman, *J. Chem. Phys.* **60** (1974) 1545.
145. A. Raman, F. H. Stilinger, and H. L. Lemberg, *J. Chem. Phys.* **63** (1975) 5223.
146. B. E. Conway and J. O'M. Bockris, *Electrochim. Acta* **3** (1961) 340.
147. Ref. 99, p. 584.
148. J. E. Desnoyers, R. E. Verrall, and B. E. Conway, *J. Chem. Phys.* **43** (1965) 243.
149. L. Pauling, *Phys. Rev.* **36** (1930) 430.
150. H. H. Nielsen, *Phys. Rev.* **40** (1932) 445.
151. Ref. 17b, p. 85-86.
152. J. O'M. Bockris and P. P. S. Saluja, *J. Phys. Chem.* **76** (1972) 2298.
153. L. E. Orgel, *J. Chem. Soc.* (1952) 4756.
154. J. O'M. Bockris and P. P. S. Saluja, *J. Phys. Chem.* **76** (1972) 2140.

155. J. O'M. Bockris and P. P. S. Saluja, *J. Electrochem. Soc.* **119** (1972) 1060.
156. Ref. 99, pp. 88-116.
157. F. H. Stilinger and A. Raman, *J. Chem. Phys.* **61** (1974) 4973; R. W. Impey, M. L. Klein, and I. R. McDonald, *J. Chem. Phys.* **74** (1981) 647.
158. (a) P. Bopp, G. Jancsó, and K. Heinzinger, *Chem. Phys. Lett.* **98** (1983) 129; (b) G. Pálinkás, P. Bopp, G. Janscó, and K. Heinzinger, *Z. Naturforsch.* **39a** (1984) 179; (c) G. Heinje, W. A. P. Luck, and K. Heinzinger, *J. Phys. Chem.* **91** (1987) 331; (d) E. Spohr, G. Pálinkás, K. Heinzinger, P. Bopp, and M. M. Probst, *J. Phys. Chem.* **92** (1988) 6754.
159. Ref. 99, p. 5.
160. Ref. 17b, p. 555.
161. E. J. W. Verwey, (a) *Rec. Trav. Chim. Pays Bas* **61** (1942) 127; (b) *Rec. Trav. Chim. Pays Bas* **60** (1941) 887.
162. P. Delahay, *Chem. Phys. Letters* **87** (1982) 607; **99** (1983) 87.
163. L. Pauling, *J. Am. Chem. Soc.* **49** (1927) 765.
164. S. R. Milner, *Phil. Mag.* **23** (1912) 551; **25** (1913) 742.
165. A. J. Appleby, to be published.
166. J. C. Ghosh, *J. Chem.Soc.* **113** (1918) 449, 707.
167. Ref. 95, pp. 267-272.
168. R. A. Robinson and R. H. Stokes, *Electrolyte Solutions,* 2nd Revised Edition, Butterworth, London, 1970.
169. N. Sutin, *Farad. Disc. Chem. Soc.* **74** (1982) 113.
170. B. E. Conway, in *Modern Aspects of Electrochemistry,* Ed. by J. O'M. Bockris and B. E. Conway, Butterworth, London, 1964, Vol. 3, p. 63-64.
171. B. E. Conway, J. O'M. Bockris, and H. Linton, *J. Chem. Phys.* **24** (1956) 834.
172. H. F. Halliwell and S. C. Nyburg, *Trans. Farad. Soc.* **59** (1963) 1126.
173. Ref. 17b, p. 337.
174. S. H. Lih, in *Physical Chemistry, An Advanced Treatise,* Ed. by H. Eyring, D. Henderson, and Y. Jost, Academic Press, New York, 1970, Vol. X, p. 439.
175. H.-P. Cheng, R. N. Barnett, and U. Landman, *Chem. Phys. Lett.* **237** (1995) 161.
176. N. S. Hush, *J. Electronal. Chem.* **460** (1999) 5.
177. L. A. Curtiss, J. W. Halley, N. C. Hung, Z. Nagy, Y. J. Rhee, and R. M. Yonco, *J. Electrochem. Soc.* **138** (1991) 2033.

178. W. Schmickler, *J. Electroanal. Chem.* **100** (1978) 277; **204** (1986) 31; *Chem. Phys. Lett.* **237** (1995) 152; *Electrochim. Acta* **41** (1996) 2329.
179. S. U. M. Khan, in *Modern Aspects of Electrochemistry*, Ed. by R. E. White, B. E. Conway and J. O'M. Bockris, Plenum, New York, 1993, Vol. 31, p. 71.
180. P. W. Anderson, *Phys. Rev.* **124** (1961) 41.
181. D. M. Newns, *Phys. Rev.* **178** (1969) 1123; J. P. Muscat and D. M. Newns, *Prog. Surf. Sci.* **9** (1978) 1.
182. H. A. Kramers, *Physica* **40** (1940) 284.
183. A. J. Appleby, in B. E. Conway, *Comprehensive Treatise of Electrochemistry*, Ed. by J. O'M. Bockris, E. Yeager, S. U. M. Khan, and R. E. White, Plenum, New York, 1983. Vol. 7, p. 173.
184. W. R. Fawcett and C. A. Foss, *J. Electroanal. Chem.* **250** (1988) 225.
185. W. R. Fawcett, *Langmuir* **5** (1989) 661.
186. N. M. Marković and P. N. Ross, in *Interfacial Electrochemistry, Theory, Experiment and Applications*, Ed. by A. Wieckowski, Marcel Dekker, New York, 1999, p. 821; *Surf. Sci. Repts.* **45** (2002) 117.
187. J. O'M. Bockris, R. K. Sen, and B. E. Conway, *Nature Phys. Sci.* **240** (1972) 143.
188. B. E. Conway, *Faraday Disc. Chem. Soc.* **74** (1982) 267.
189. Ref. 17b, p. 667-685.
190. *Handbook of Chemistry and Physics*, 70th Edition., Ed. by R. C. Weast, CRC Press Inc., Boca Raton, FL, 1989, p. D41.
191. Ref. 170, p. 59 (footnote).
192. W. M. Latimer, K. S. Pitzer, and W. V. Smith, *J. A. C. S.* **60** (1938) 1829.
193. Ref. 17b, p. 1240 et sequ.
194. C. N. Hinshelwood, *Proc. Roy. Soc. (London)*, **A113** (1927) 320; G. N. Lewis and D. F. Smith, *J. A. C. S.* **47** (1925) 1508; R. H. Fowler and E. K. Rideal, *Proc. Roy. Soc. (London)*, **A113** (1927) 570.
195. K. J. Laidler, *Theories of Chemical Reaction Rates*, McGraw-Hill, New York, 1969, pp. 106-129.
196. R. Parsons, in *Advances in Electrochemistry and Electrochemical Engineering*, Ed. by P. Delahay and C. W. Tobias, Interscience, New York, 1961, Vol. 1, p. 1; W. R. Fawcett, in *Electrocatalysis*, Ed. by J. Lipkowski and P. N. Ross,

Wiley-VCH, New York, 1998 p. 323; W. R. Fawcett, *J. Phys. Chem.* **93** (1989) 2675.

197. J. O'M. Bockris and S. Srinivasan, *J. Electrochem. Soc.* **111** (1964) 844, 853, 858.

198. S. Glasstone, K. J. Laidler, and H. Eyring, *The Theory of Rate Processes,* McGraw-Hill, New York, 1941.

199. Ref. 17b, p. 1177.

200. Ref. 170, p. 119.

201. R. P. Auty and R. H. Cole, *J. Chem. Phys.* **20** (1952) 1309; B. E. Conway, *Can. J. Chem.* **37** (1959) 613; M. Eigen, L. de Mayer, and H. C. Spatz, *Berichte Bunsengesell.* **68** (1964) 19.

202. B. E. Conway and J. O'M. Bockris, in *Modern Aspects of Electrochemistry,* J. O'M. Bockris and B. E. Conway, Butterworth, London, 1954, Vol. 1, p. 47.

203. Ref. 170, p. 66 (footnote).

204. B. E. Conway, in *Interfacial Electrochemistry, Theory, Experiment and Applications,* Ed. by A. Wieckowski, Marcel Dekker, New York, 1999, p. 131.

205. J. Barber, S. Morin, and B. E. Conway, *J. Electroanal. Chem.* **446** (1998) 125.

206. L. I. Krishtalik, *Electrochim. Acta* **13** (1968) 1045.

207. B. Kabanow, *Acta Physicochim. U. R. S. S.,* **5** (1936) 193.

208. H. A. Leibhavsky and E. J. Cairns, *Fuel Cells and Fuel Batteries,* Wiley, New York, 1968, p. 138,

209. J. O'M. Bockris and A. M. Azzam, *Trans. Faraday Soc.* **48** (1952) 145.

210. Ref. 3, p. 21; Ref. 5a, p. 228.

211. F. P. Bowden and K. W. Grew, *Discuss. Faraday Soc.* **2** (1947) 81, 91.

212. H. Nürnberg, *Studien mit Modernen Technik zur Kinetik Schneller Chemischer und Elektrochemischer Schritte von Protonen Transferprozessen,* Bonn, 1969, quoted by Levich, Ref. 58.

213. S. Trassatti, *J. Electroanal. Chem.* **39** (1977) 183.

214. A. R. Despic and J. O'M. Bockris, *J. Chem. Phys.* **32** (1960) 389.

215. J. O'M. Bockris and D. B. Matthews, *J. Chem. Phys.* **44** (1966) 298.

216. S. Fischer, quoted in P. P. Schmidt and J. Ulstrup, *Nature Phys. Sci.* **245** (1972) 126.

217. H. Bauer, *J. Electroanal. Chem.* **16** (1968) 419.

218. R. Parsons, *Trans. Faraday Soc.* **47** (1951) 1332; I. H. Plonski, *Rev. Roumaine de Chimie* **14** (1969) 569.
219. K. J. Vetter, *Z. Naturforsch.* **7a** (1952) 328; *Z. Elektrochem.* **59** (1955) 596; *Elekrochemischer Kinetik,* Springer-Verlag, Berlin, 1961, p. 101.
220. R. Audubert and S. Cornevin, *J. Chim. Phys.* **38** (1941) 46.
221. J. O'M. Bockris and Z. Nagy, *J. Chem. Education* **50**(1973)839.
222. V. Harff, *C. R. Acad. Sci. Paris Ser.* C **268** (1969) 1657; **269** (1969) 1352; **270** (1970) 1695.
223. A. J. Appleby, in *Modern Aspects of Electrochemistry,* Ed. by J. O'M. Bockris and B. E. Conway, Plenum, New York, 1974. Vol. 9, p. 369.
224. R. Parsons, *Surface Science* **2** (1964) 418; A. N. Frumkin, *Elektrokhimya* **1** (1965) 394.
225. Ref. 17b, p. 652.
226. H. J. Gardiner and L. E. Lyons, *Rev. Pure Appl. Chem.* **3** (1953) 134.
227. H. Eyring, S. Glasstone, and K. J. Laidler, *J. Chem. Phys.* **7** (1939) 1053.
228. R. Parsons and E. Passeron, *J. Electroanal. Chem.* **12** (1966) 524.
229. V. Marecek, Z. Samec, and J. Weber, *J. Electroanal. Chem.* **94** (1978) 169.
230. J. E. B. Randles, *Can. J. Chem.* **37** (1959) 238.
231. B. E. Conway, B. MacKinnon, and B. V. Tilak, *Trans. Faraday Soc.* **66** (1970) 1203; B. E. Conway, in *Modern Aspect of Electrochemistrey,* Ed. by B. E. Conway, J. O'M. Bockris, and R. E. White, Plenum, New York, 1986, Vol. 16, p. 103.
232. J. O'M. Bockris and A. Gochev, *J. Electroanal. Chem.* **214** (1986) 655.
233. B. E. Conway and J. O'M. Bockris, *Electrochim. Acta* **3** (1961) 140.
234. A. J. Appleby, *J. Electroanal. Chem.* **51** (1974) 1; **357** (1993) 117.

3

Electrosorption Valency and Partial Charge Transfer

Rolando Guidelli[*] and Wolfgang Schmickler[**]

[*]Dept. of Chemistry, Florence University, Via della Lastruccia 3
50019 Sesto Fiorentino, Firenze, Italy; [**]Abteilung Elektrochemie, Ulm University,
89069 Ulm, Germany

I. INTRODUCTION

By definition, electrochemical reactions involve charge transfer through the electrochemical interface. Therefore, the charge flowing in the elementary act is an important characteristic of a process. For a simple electron- or ion-transfer reaction, this charge is always a multiple of the unit charge: in the former case it is simply the number of electrons exchanged between the reactant and the electrode, in the latter case it is the valency of the ion that is incorporated into the electrode. But in an adsorption reaction the adsorbate need not exchange an integral number of electrons with the electrode. Since it stays on the surface, it can share electrons with the electrode, forming a polar or a covalent bond. However, the partial charge transferred by the adsorbate to the electrode is not clearly defined, since the division of the bonding electrons into parts pertaining to the adsorbate and to the electrode is arbitrary to some extent. In spite of this, the amount of charge flowing at constant potential can clearly be measured, and this

Modern Aspects of Electrochemistry, Number 38, edited by B. E. Conway *et al.* Kluwer Academic/Plenum Publishers, New York, 2005.

quantity is used to define the electrosorption valency, in analogy to the valency of a metal ion deposited on an electrode.

The definition of a quantity is one thing, its interpretation is another, and in the case of the electrosorption valency the interpretation has proved to be particularly difficult. Obviously, it must be related to the charge exchanged between electrode and adsorbate, or to the partial charge transfer. However, the latter cannot be measured and is only defined within a particular model. Nevertheless, it can be a useful quantity for understanding the meaning of the electrosorption valency, and for the interpretation of experimental data. A quantity that can be measured, though often with difficulties, is the dipole moment associated with an adsorbate. Its definition is based on other concepts than the electrosorption valency, but the two quantities are related, and one can be expressed through the other.

In this review, we will consider the adsorption of a single species; coadsorption phenomena will not be considered, since it is generally impossible to divide the flow of charge among several species. We will present the thermodynamics on which the concept of the electrosorption valency is based, discuss methods by which it can be measured, and explain its relation to the dipole moment and to partial charge transfer. The latter can be explained within an extension of the Anderson-Newns model for adsorption, which is useful for a semi-quantitative treatment of electrochemical adsorption. Our review of concepts and methods will be concluded by a survey of experimental data on thiol monolayers, which nowadays are adsorbates of particular interest.

II. THE ELECTROSORPTION VALENCY

When a species S^z is adsorbed at an electrode surface, it may form a purely physical bond (physisorption) or a much stronger chemical bond (chemisorption). In the latter case the species may transfer a positive or negative charge to the electrode, according to the formal adsorption reaction:

$$l(E) = l_Z - \frac{1}{F}\left[\left(\frac{\partial \sigma_M}{\partial \Gamma_S}\right)_E - \left(\frac{\partial \sigma_M}{\partial \Gamma_S}\right)_{E_z}\right]$$

$$= l_Z - \frac{1}{F}\int_{E_z}^{E}\left(\frac{\partial}{\partial E}\right)_{\Gamma_S}\left(\frac{\partial \sigma_M}{\partial \Gamma_S}\right)_E dE$$

$$= l_Z - \frac{1}{F}\int_{E_z}^{E}\left(\frac{\partial C_\infty}{\partial \Gamma_S}\right)_E dE \qquad (20)$$

Here λ is generally a positive or negative fractional number and e^- denotes the *negative* electronic charge. This process is called a "partial charge transfer" (pct) reaction. We will exclude the case in which a pct step is followed by one or more further pct steps, ultimately resulting in the overall transfer of a unitary charge $|e|$ with formation of a redox partner of S^z capable of existing in the bulk solution. The factor λ, called "partial charge transfer coefficient", was first introduced by Lorenz and Salié[1] in connection with an experimental research on the Tl-amalgam/Tl$^+$ ion electrode reaction.

Subsequently, Lorenz[2,3] correctly recognized that λ cannot be estimated without having recourse to some modelistic assumption. In this connection, he made a distinction between a "microscopic" pct coefficient λ and a "macroscopic" pct coefficient l', which he regarded as an experimentally accessible parameter. In what follows, the term "experimentally accessible parameter" will be exclusively referred to those parameters that can be determined without any modelistic assumption and are, therefore, thermodynamically significant. Later we will show that, strictly speaking, Lorenz's macroscopic pct coefficient l' is not thermodynamically significant since it relies on some mild modelistic assumptions. Subsequently, Vetter and Schultze[4,5] defined a different quantity:

$$l \equiv -\left(\partial \sigma_M / \partial \Gamma_S\right)_E / F \qquad (2)$$

where σ_M is the thermodynamic charge density on the metal, Γ_S is the surface excess of the given species S and E is the applied potential. As shall be shown in Sec. II.3, this quantity, called "electrosorption valency", is formally equal and opposite to Lorenz's macroscopic pct coefficient, l', but has the advantage of being thermodynamically significant by definition. Vetter and Schultze have also the further merit of having pointed out various alternative contributions to λ in addition to the possible contribution from pct.[4,6]

We can envisage the chemisorption of a species, with or without pct, as follows. Imagine adding, say, a specifically adsorbing anion to a solution containing a strong excess of a nonspecifically adsorbed electrolyte and imagine following its adsorption at the metal/water interface, at constant applied potential. If the anion were itself nonspecifically adsorbed, and hence had not to be deprived of its solvation sheath on the metal side in its position of closest approach to the electrode surface, then it would be accompanied in this approach by a nonspecifically adsorbed cation, and hence no flow of electrons along the external circuit would be observed. This situation is depicted schematically in Fig.1a, where x is the distance from the electrode surface, and $x = d$ is the distance of closest approach of the nonspecifically adsorbed ions to the electrode. The potential-distance profile is represented schematically by curve 1. Since, however, the anion is specifically adsorbed, it will bring its center of charge at a distance, $x=\beta$, from the electrode surface that is shorter than $x = d$, as shown in Fig.1b, thus allowing its lone-pair orbitals to come in direct contact with the electrode-metal surface atoms. This may require either the complete removal of one coordinating water molecule or a simple deformation of the solvation shell.[8] As a rule, an increase of ionic radius is accompanied by an increase of polarizability and electron-pair donicity and by a decrease of solvation energy; this favors a simple deformation of the solvation shell.[8] The approach of the center of charge of the anion to the electrode surface gives rise to a negative potential drop between the $x = \beta$ and $x = d$ planes, represented by curve 2 in Fig.1b. For a constant potential difference, $\phi^M - \phi_d$, across the inner layer, this negative contribution to $\phi_\beta - \phi_d$ must be compensated for by a positive contribution, which is produced by a positive shift, $\Delta\sigma_M$, in the charge density on the metal from σ_M^0 to σ_M. The contribution from the new charge density σ_M to the potential drop between $x = 0$ and $x = d$ is represented schematically by curve 3 in Fig.1b, while the overall potential-distance profile is represented by curve 1. If the anion also transfers a fraction λ of its negative charge to the electrode upon chemisorption, then the positive shift $\Delta\sigma_M$ will be greater, but the two contributions cannot be distinguished on the basis of purely thermodynamic arguments.

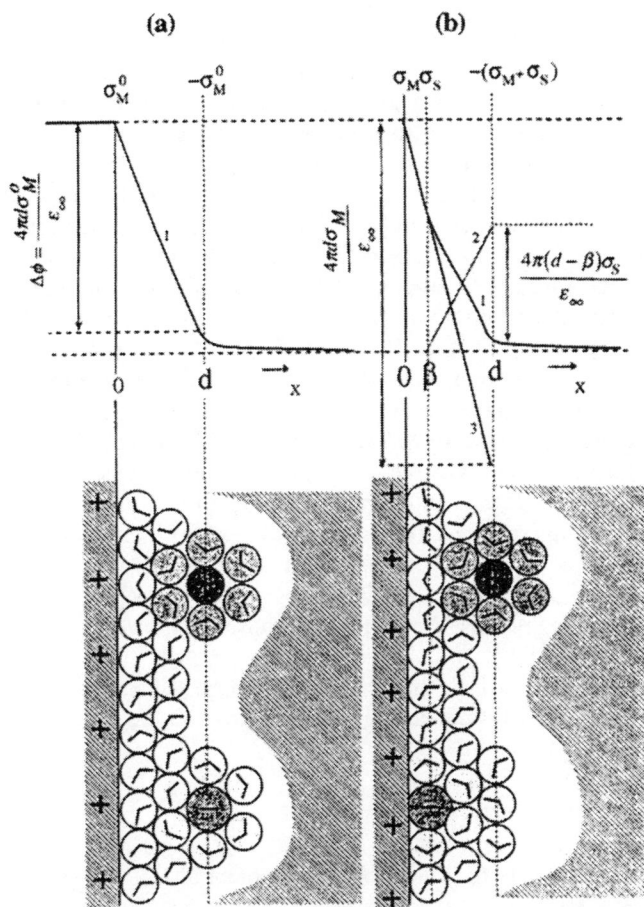

Figure 1. Schematic picture of the metal/solution interphase in the case of nonspecific (a) and specific (b) anionic adsorption. $x = 0$, $x = \beta$ and $x = d$ are the electrode surface plane, the plane of closest approach for the specifically adsorbed anions, and that for the nonspecifically adsorbed ions. Curve 1 represents the potential-distance profile. In (b), curve 1 results from the combination of curve 2, expressing the contribution from the charge density σ_S of the specifically adsorbed anions, and curve 3, expressing the contribution from the charge density σ_M on the metal. The potential difference, $\phi^M - \phi_d$, across the inner layer is the same in (a) and (b). (Reprinted from Ref.[7] with permission from the Am. Chem. Soc.)

1. Some Thermodynamic Considerations

The thermodynamics of electrified interfaces deals with "components" of the overall system, namely species whose concentration in their respective bulk phases can be varied independent of that of the other species; these are necessarily neutral components, that is, metals in the metal phase and salts and neutral molecular species in the solution phase. Extrathermodynamic charged "constituents" of the system, such as electrons and ions, can be temporarily introduced in the derivation of the thermodynamic Gibbs adsorption equation, but the imposition of the condition of electroneutrality of the interphase must ultimately lead to an expression that only deals with the bulk chemical potentials of neutral components. This is also the case for a bulk constituent S^z that is assumed to undergo a pct at the interface, leading to an adsorbed species $S^{z+\lambda}$ whose existence is only confined to the interfacial region. In fact, the thermodynamic treatment of electrified interfaces compares the real interfacial system with the ideal Gibbs model, in which the two homogeneous and neutral bulk phases are separated by a purely geometric "dividing surface". It is evident that in the Gibbs model a species assumed to exist only in the interfacial region has no right of citizenship. The situation is different for a redox couple Ox/Red undergoing a reversible electron transfer, because both Ox and Red are present in the bulk solution phase.

That pct cannot affect the Gibbs adsorption equation can be verified by applying a formal equilibrium condition to the pct reaction of Eq. (1), yielding:

$$\tilde{\mu}_S = \tilde{\mu}_{S,ads} + \lambda\tilde{\mu}_e \tag{3}$$

Here, $\tilde{\mu}_S$ and $\tilde{\mu}_{S,ads}$ are the electrochemical potentials of S in the bulk solution and in the adsorbed state. Let us apply the Gibbs adsorption equation to the interphase between a pure metal M and an aqueous solution containing molecular and ionic species denoted by the subscript j, in addition to water w and the species S. Choosing the neutral metal atoms M and the electrons e in excess with respect to metal atoms as the constituents of the metal phase, we may formally write:

$$-d\gamma = \Gamma_M \, d\mu_M + \Gamma_e \, d\tilde{\mu}_e + \Gamma_w \, d\mu_w + \Gamma_{S,non} \, d\tilde{\mu}_S + \Gamma_{S,ads} \, d\tilde{\mu}_{S,ads} + \sum_j \Gamma_j \, d\tilde{\mu}_j$$

$$(4)$$

Here, $\Gamma_{S,non}$ and $\Gamma_{S,ads}$ are the surface excesses of nonspecifically adsorbed S molecules and of those chemisorbed with pct, while γ is the interfacial tension. Let us assume for simplicity that water is in such a large excess with respect to the other components of the solution phase as to allow us to set $d\mu_w = 0$ at constant temperature T and pressure P. Noting that in the pure metal phase $d\mu_M$ is also equal to zero at constant T and P, and replacing $\tilde{\mu}_{S,ads}$ from Eq. (3) into Eq. (4), after rearrangement we get:

$$-d\gamma = \left(\Gamma_e - \lambda\Gamma_{S,ads}\right)d\tilde{\mu}_e + (\Gamma_{S,non} + \Gamma_{S,ads})d\tilde{\mu}_S + \sum_j \Gamma_j \, d\tilde{\mu}_j \qquad (5)$$

Upon formally separating the electrochemical potentials into their chemical and electrical contributions:

$$d\tilde{\mu}_e = -F d\phi^M \; ; \; d\tilde{\mu}_S = d\mu_S + zF d\phi^S \; ; \; d\tilde{\mu}_j = d\mu_j + z_j F d\phi^S \qquad (6)$$

where ϕ^M and ϕ^S are the electric potentials in the metal and in the solution phase, and taking into account the electroneutrality condition for the whole interphase:

$$-\Gamma_e + z\Gamma_{S,non} + (z+\lambda)\Gamma_{S,ads} + \sum_j z_j \Gamma_j = 0 \qquad (7)$$

Equation (5) becomes:

$$-d\gamma = \sigma_M \, d\!\left(\phi^M - \phi^S\right) + \Gamma_S \, d\mu_S + \sum_j \Gamma_j \, d\mu_j \qquad (8)$$

Here we have set:

$$\Gamma_S \equiv \Gamma_{S,non} + \Gamma_{S,ads} \; ; \; \sigma_M \equiv q_M + F\lambda\Gamma_{S,ads} \; ; \; q_M \equiv -F\Gamma_e \qquad (9)$$

To complete the thermodynamic treatment of the interphase, one should now eliminate the chemical potentials μ_j of any ionic species in favor of the chemical potentials of the corresponding salts. One should also combine the ideal polarized electrode consisting of the interphase under study with an indicator electrode reversible to one ionic species of the solution phase, in order to eliminate the extrathermodynamic potential difference $(\phi^M-\phi^S)$ in favor of the thermodynamic potential difference E between the leads of the two electrodes. However, for the present purposes, the still incomplete Gibbs adsorption relationship of Eq. (8) is sufficient to show that any trace of the proposed pct has disappeared. Thus, Γ_S is the total surface excess of the species S, independent of whether pct takes place or not. It should also be noted that the $\sigma_M d(\phi^M-\phi^S)$ term is thermodynamically significant and, therefore, experimentally accessible without having recourse to modelistic considerations. In fact, while the potential difference $(\phi^M - \phi^S)$ between two phases of different composition is experimentally inaccessible, its change $d(\phi^M - \phi^S)$ can be measured experimentally provided any change dE in the externally applied potential is entirely located in the interphase under study, that is, provided the indicator electrode is fully reversible. In this case we can set $(\phi^M - \phi^S) = E + $ constant, where the constant depends on the choice of the reference electrode. Therefore, the charge density, $\sigma_M = -(\partial\gamma/\partial E)_{\mu_i}$, is also thermodynamically significant. It can be regarded as the charge to be supplied to the electrode to keep the applied potential E constant when the electrode surface is increased by unity and the composition of the bulk phases is kept constant. This quantity is called "total charge density" by Frumkin[9] and "thermodynamic charge density" by Lorenz and Salié.[10]

The separation of σ_M into the two contributions $-F\Gamma_e$ and $F\lambda\Gamma_{S,ads}$ is modelistic and, therefore, extrathermodynamic.[11,12] However, it allows us to view σ_M as the sum of a charge density, $F\lambda\Gamma_{S,ads}$, stemming from pct to the electrode by the chemisorbed species S, and of the charge density $-F\Gamma_e$ on the metal side of the interphase, due to the surface excess of "free electrons". The latter is called "free charge density" by Frumkin[9] and "true charge density" by Lorenz and Salié[10]; in what follows it will be denoted by q_M. This separation of σ_M into two extrathermodynamic contributions may provide us with a satisfactory procedure to estimate the pct coefficient λ. Thus, if we may estimate the free charge density $q_M \equiv -F\Gamma_e$ extrathermodynamically, say, by using the modelistic Gouy-Chapman theory, then $F\lambda\Gamma_{S,ads}$ is obtained by subtracting q_M from the thermodynamic total charge density σ_M.

From Eq. (3) it is apparent that a pct reaction can be conveniently regarded as in equilibrium. It is therefore natural to provide some qualitative evidence for the presence of pct by measuring a thermodynamically significant, equilibrium quantity such as the electrosorption valency l of Eq. (2) introduced by Vetter and Schultze.[4] Strictly speaking, this quantity is thermodynamically significant only in the presence of such a large excess of supporting electrolyte as to justify the neglect of the contribution to Γ_S from the nonspecific adsorption localized in the diffuse layer. If this condition is not fulfilled, Vetter and Schultze[4,6] propose the definition:

$$l \equiv -\left(\partial \sigma_M / \partial \Gamma_{S,ads}\right)_{\Delta\phi} / F \qquad (10)$$

where $\Gamma_{S,ads}$ is the surface concentration of S in the inner layer, $\Delta\phi = E - E_z - \phi_d$, ϕ_d is the potential difference across the diffuse layer and E_z is the applied potential at which σ_M equals zero in the absence of specific adsorption. Even though $\Delta\phi$ is formally defined as the "absolute potential difference across the inner layer", it does not include the surface dipole potential due to the anisotropy of the interfacial forces, and is therefore an extrathermodynamic quantity, just as $\Gamma_{S,ads}$ is. In this respect, the definition in Eq. (10) is not strictly thermodynamic. Henceforth, we will confine ourselves to the thermodynamic definition of Eq. (2), tacitly assuming a strong excess of the supporting electrolyte, and consequently $\Gamma_S \cong \Gamma_{S,ads}$.

Equation (2) applies to both neutral and charged species. In the case of an ionic species of charge number z, it can be written in the form:

$$l = -z\left(\partial \sigma_M / \partial \sigma_S\right)_E \qquad (11)$$

where $\sigma_S \equiv zF\Gamma_S$ is the charge density due to the specifically adsorbed ions. The rationale behind this definition is the following: if the chemisorption of the ion at constant applied potential is accompanied by a flow of charge of equal magnitude but opposite sign along the external circuit, then the electrosorption valency coincides with the charge number z of the ion. This situation is encountered only under very particular conditions.

A further thermodynamic expression for l is possible.[4] Since the electrocapillary equation (Eq. 8) is a total differential equation, the second cross-partial-differential coefficients of γ are equal:

$$\left(\partial\Gamma_S / \partial E\right)_{\mu_S} = \left(\partial\sigma_M / \partial\mu_S\right)_E \quad (12)$$

Noting that σ_M depends upon Γ_S at constant E via μ_S, we may write:

$$\left(\partial\sigma_M / \partial\Gamma_S\right)_E = \left(\partial\sigma_M / \partial\mu_S\right)_E \left(\partial\mu_S / \partial\Gamma_S\right)_E \quad (13)$$

Combining Eqs. (12) and (13) we get:

$$\left(\partial\sigma_M / \partial\Gamma_S\right)_E = \left(\partial\Gamma_S / \partial E\right)_{\mu_S} \left(\partial\mu_S / \partial\Gamma_S\right)_E \quad (14)$$

Noting that $d\mu_S$ is a total differential, we can write:

$$d\mu_S = \left(\partial\mu_S / \partial\Gamma_S\right)_E d\Gamma_S + \left(\partial\mu_S / \partial E\right)_{\Gamma_S} dE \quad (15)$$

Rearranging terms at constant μ_S, i.e., for $d\mu_S = 0$, we obtain:

$$\left(\partial\mu_S / \partial\Gamma_S\right)_E\left(\partial\Gamma_S / \partial E\right)_{\mu_S} = -\left(\partial\mu_S / \partial E\right)_{\Gamma_S} \quad (16)$$

Comparing Eqs. (14) and (16) yields:

$$\left(\partial\mu_S / \partial E\right)_{\Gamma_S} = -\left(\partial\sigma_M / \partial\Gamma_S\right)_E = Fl \quad (17)$$

This equation provides an alternative definition of l, which can be obtained as a function of E from a thermodynamic analysis of adsorption data by using any of the two above definitions. Sometimes, both alternative procedures for the determination of l have been adopted in order to verify the self-consistency of data analysis.[13,14]

A further thermodynamic expression for l is obtained by writing the total differential of σ_M, regarded as a function of E and Γ_S:

$$d\sigma_M = \left(\partial\sigma_M / \partial E\right)_{\Gamma_S} dE + \left(\partial\sigma_M / \partial\Gamma_S\right)_E d\Gamma_S \quad (18)$$

Dividing by dE at constant σ_M (i.e. for $d\sigma_M = 0$), we obtain:

$$Fl \equiv -\left(\frac{\partial \sigma_M}{\partial \Gamma_S}\right)_E = \frac{C_\infty}{(\partial \Gamma_S/\partial E)_{\sigma_M}} \quad with: \; C_\infty \equiv \left(\frac{\partial \sigma_M}{\partial E}\right)_{\Gamma_S} \tag{19}$$

Another useful thermodynamic relationship that allows the potential dependence of l to be determined from the differential capacity C_∞ of the interphase at constant Γ_S is readily obtained from the very definition of l. Choosing E_z as the reference potential and denoting by l_Z the electrosorption valency at E_z, the l value at any other applied potential E is given by:

$$l(E) = l_Z - \frac{1}{F}\left[\left(\frac{\partial \sigma_M}{\partial \Gamma_S}\right)_E - \left(\frac{\partial \sigma_M}{\partial \Gamma_S}\right)_{E_z}\right]$$

$$= l_Z - \frac{1}{F}\int_{E_z}^{E}\left(\frac{\partial}{\partial E}\right)_{\Gamma_S}\left(\frac{\partial \sigma_M}{\partial \Gamma_S}\right)_E dE$$

$$= l_Z - \frac{1}{F}\int_{E_z}^{E}\left(\frac{\partial C_\infty}{\partial \Gamma_S}\right)_E dE \tag{20}$$

The quantity $C_\infty \equiv (\partial \sigma_M/\partial E)_{\Gamma_S}$ can be determined from a thermodynamic analysis of adsorption data. It can also be determined from direct differential capacity measurements at frequencies so high as to prevent the surface concentration Γ_S to follow the ac signal. Naturally, any partial charge λe associated with chemisorption does follow the ac signal, due to the extremely short relaxation time of electrons.

It should be noted that l may depend appreciably upon the Γ_S value at which the derivatives of Eq. (17) are calculated. Thus, adsorption isotherms can be expressed under the general form $a_S \exp(-\Delta G^0_{ads}/RT) = f(\Gamma_S)$, where a_S is the activity of the adsorbing species in the bulk solution, ΔG^0_{ads} is the standard Gibbs energy of adsorption, and $f(\Gamma_S)$ is some function of Γ_S that measures the activity of the adsorbed molecules. While the bulk activity a_S of the adsorbing species S can be identified with the corresponding bulk concentration c_S as a good approximation, $f(\Gamma_S)$ can be approximated to Γ_S only at low coverages. With an increase in coverage, $f(\Gamma_S)$ may deviate appreciably from Γ_S, due to the onset of strong interactions between the adsorbed S molecules. The activity coefficient, $f(\Gamma_S)/\Gamma_S$, may vary with a change in the electric field even at constant Γ_S; the activity coefficient of the residual adsorbed water molecules may also be affected. From the definition of l in Eq. (17) it follows that:

$$l = \frac{1}{F}\left(\frac{\partial \mu_S}{\partial E}\right)_{\Gamma_S} = \frac{RT}{F}\left(\frac{\partial \ln a_S}{\partial E}\right)_{\Gamma_S} = \frac{1}{F}\frac{d\Delta G_{ads}^0}{dE} + \frac{RT}{F}\left(\frac{\partial \ln f(\Gamma_S)}{\partial E}\right)_{\Gamma_S} \quad (21)$$

At low coverages, $f(\Gamma_S)$ practically coincides with Γ_S and consequently the second term in the last member of Eq. (21) vanishes. However, at higher Γ_S values, the activity coefficient of the adsorbed S molecules may vary with E at constant Γ_S to a different extent depending on the chosen Γ_S value. Therefore, a comparison between l values of different adsorbed species, or even that between l values for the same adsorbed species obtained by different procedures, should always be carried out at low coverages.

2. The Extrathermodynamic Contributions to the Electrosorption Valency

An insight into the extrathermodynamic, modelistic contributions to l can be gained by regarding chemisorption as a displacement of a number v of adsorbed solvent molecules from direct contact with the electrode surface by the given adsorbing species S. In an aqueous solution the adsorption equilibrium can be written:

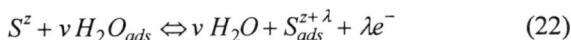

$$S^z + v\,H_2O_{ads} \Leftrightarrow v\,H_2O + S_{ads}^{z+\lambda} + \lambda e^- \quad (22)$$

where the subscript *ads* denotes species in the adsorbed state while its absence denotes species in the respective bulk phases. The equilibrium condition is:

$$\tilde{\mu}_S + v\mu_{w,ads} = v\mu_w + \tilde{\mu}_{S,ads} + \lambda\tilde{\mu}_e \quad (23)$$

where μ_w and $\mu_{w,ads}$ are the chemical potentials of water in the bulk solution and in the adsorbed state. The various chemical and electrochemical potentials of Eq. (23) can be formally separated into a purely chemical and an electrostatic contribution:

$$\tilde{\mu}_S = \mu_S + zF\phi^S \cong \mu_S^0 + RT \ln c_S + zF\phi^S \tag{24}$$

$$\mu_{w,ads} = \mu_{w,ads}^0 + RT \ln\left[f\left(\Gamma_{w,max} - v\Gamma_S\right)\right] + \langle m_w \rangle d\phi / dx$$

$$\tilde{\mu}_{S,ads} = \mu_{S,ads}^0 + RT \ln\left[f\left(\Gamma_S\right)\right] + \left(z + \lambda\right)F\phi_\beta + \langle m_S \rangle d\phi / dx$$

$$\tilde{\mu}_e = \mu_e - F\phi^M$$

Here, ϕ^M, ϕ^S and ϕ_β are the electric potentials in the metal, in the bulk solution and at the inner Helmholtz plane, $x = \beta$; $\Gamma_{w,max}$ is the surface concentration of the adsorbed water molecules in the absence of chemisorption and $f(\Gamma_{w,max} - v\Gamma_S)$ is their activity in the presence of chemisorption. In Eq. (24) the electrostatic contributions of the adsorbed molecules H_2O and S to their electrochemical potentials also include the electrostatic potential energy of their dipoles in the interfacial electric field, $-d\phi/dx$; $<m_w>$ and $<m_S>$ are average values of the normal components of these dipoles, regarded as positive when their positive end is directed towards the solution. By substituting Eq. (24) into Eq. (23), rearranging terms, and differentiating μ_S with respect to $(\phi^M - \phi^S)$ we obtain:

$$l = \frac{1}{F}\left[\frac{\partial\mu_S}{\partial\left(\phi^M - \phi^S\right)}\right]_{\Gamma_S}$$

$$= z\frac{\partial\left(\phi_\beta - \phi^S\right)}{\partial\left(\phi^M - \phi^S\right)} - \lambda\frac{\partial\left(\phi^M - \phi_\beta\right)}{\partial\left(\phi^M - \phi^S\right)} + \frac{1}{F}\frac{\partial\left(\langle m_S \rangle - v\langle m_w \rangle\right)}{\partial\left(\phi^M - \phi^S\right)}\frac{d\phi}{dx} \tag{25}$$

This equation holds strictly only if the activity coefficients $f(\Gamma_{w,max} - v\Gamma_S)/\Gamma_w$ and $f(\Gamma_S)/\Gamma_S$ do not change with varying the interfacial electric field at constant Γ_S. Assuming, for simplicity, that the electric field is constant across the whole compact layer $(0 < x < d)$, the partial derivatives in Eq. (25) can be replaced by ratios of finite increments, yielding:[4,6]

$$l = zg - \lambda(1 - g) + \kappa_S - \nu\kappa_w \qquad (26)$$

with:

$$g \equiv \frac{\phi_\beta - \phi^S}{\phi^M - \phi^S} \; ; \kappa_S \equiv \frac{1}{F} \frac{\langle m_S \rangle}{\phi^M - \phi^S} \frac{d\phi}{dx} \; ; \kappa_w \equiv \frac{1}{F} \frac{\langle m_w \rangle}{\phi^M - \phi^S} \frac{d\phi}{dx} \qquad (27)$$

3. The Partial Charge Transfer Coefficient of Lorenz and Salié

In their pioneering work on pct, Lorenz and Salié[1] adopted a kinetic approach, suggested by their initial experimental research on the Tl–amalgam/Tl$^+$ ion electrode reaction. In this system, pct is assumed to involve an adsorbed intermediate, while the reactant Ox and the final product Red are both present in the bulk of their respective phases. This led these authors to work out a general reaction scheme in which one or more adsorbed intermediates are interposed between Ox and Red.

Confining ourselves to the reaction scheme of Eq. (1), in which the intermediate S$^{z+\lambda}$ exists only in the adsorbed state, Lorenz's approach[10] starts from the expression for the adsorption current density j, which is given by $d\sigma_M/dt$. By expressing σ_M as the sum of the two extrathermodynamic contributions of Eq. (9), we get:

$$j = \frac{d\sigma_M}{dt} = \frac{dq_M}{dt} + F\lambda \frac{d\Gamma_S}{dt} \qquad (28)$$

Considering that the free charge density q_M is a function of both Γ_S and E, we can write:

$$\frac{dq_M}{dt} = \left(\frac{\partial q_M}{\partial E}\right)_{\Gamma_S} \frac{dE}{dt} + \left(\frac{\partial q_M}{\partial \Gamma_S}\right)_E \frac{d\Gamma_S}{dt} \qquad (29)$$

Substituting dq_M/dt from Eq. (29) into Eq. (28) yields:

$$j = \left(\frac{\partial q_M}{\partial E}\right)_{\Gamma_S} \frac{dE}{dt} + F\left[\lambda + \frac{1}{F}\left(\frac{\partial q_M}{\partial \Gamma_S}\right)_E\right]\frac{d\Gamma_S}{dt} = C'_\infty \frac{dE}{dt} + Fl'\frac{d\Gamma_S}{dt} \qquad (30)$$

with:

$$C'_{\infty} \equiv \left(\frac{\partial q_M}{\partial E}\right)_{\Gamma_S} = \left(\frac{\partial(\sigma_M - F\lambda\Gamma_S)}{\partial E}\right)_{\Gamma_S} \cong \left(\frac{\partial\sigma_M}{\partial E}\right)_{\Gamma_S} \equiv C_{\infty};$$

$$l' \equiv \lambda + \frac{1}{F}\left(\frac{\partial q_M}{\partial\Gamma_S}\right)_E$$

(31)

Lorenz refers to l' as the "macroscopic pct coefficient" and regards it as experimentally accessible. C'_{∞} can be measured by the extrapolated value of the differential capacity C at infinite frequency, when the surface concentration Γ_S is frozen, i.e., when it cannot keep up with the ac signal. Since, however, the partial charge associated with chemisorption does follow the ac signal, this is only true if λ can be regarded as potential independent. It can be readily seen that l' in Eq. (31) is the opposite of the electrosorption valency l. In fact, replacing σ_M from Eq. (9) into Eq. (2) yields:

$$l \equiv -\frac{1}{F}\left(\frac{\partial\sigma_M}{\partial\Gamma_S}\right)_E = -\frac{1}{F}\left(\frac{\partial q_M}{\partial\Gamma_S}\right)_E - \lambda = -l'$$

(32)

Equation (30) can be written in the form:

$$\frac{jdt}{dE} = \frac{d\sigma_M}{dE} \equiv C = C'_{\infty} + Fl'\frac{d\Gamma_S}{dE}$$

(33)

This equation can be used to determine l' if the differential capacity C is extrapolated to zero frequency, namely under the quasi-equilibrium conditions required to apply the electrocapillary equation for a thermodynamic estimate of $d\Gamma_S/dE$ (see Eq. 12):

$$(\partial\Gamma_S/\partial E)_{\mu_S} = (\partial\sigma_M/\partial\mu_S)_E \quad \text{with } \sigma_M = \int_{pzc}^{E} C(\omega \to 0)dE$$

(34)

Whence:

$$l' = -l = \frac{1}{F}\frac{C(\omega \to 0) - C'_{\infty}}{(\partial\Gamma_S/\partial E)_{\mu_S}}$$

(35)

This kinetic determination of the electrosorption valency l is based on the relatively mild assumption of a potential independent λ.

Lorenz and Salié[10] also propose an alternative kinetic approach to the determination of l'; this assumes a potential dependence of the adsorption and desorption rates of the pct process of Eq. (1) which is entirely analogous to that of the Butler-Volmer equation:

$$v_{ads} = \vec{k}c_S = \vec{k}_0 c_S \exp\left(-\beta l FE / RT\right)$$
$$v_{des} = \overleftarrow{k}\Gamma_S = \overleftarrow{k}_0 \Gamma_S \exp\left[(1-\beta)l FE / RT\right] \tag{36}$$

Here $\vec{k}$ and $\overleftarrow{k}$ are potential-dependent rate constants, while $\vec{k}_0$ and $\overleftarrow{k}_0$ are their values at $E = 0$. More precisely, Lorenz and Salié use l' in place of l, and obtain l' from the relationship:

$$l' = \frac{RT}{F}\left[\frac{\partial \ln\left(\vec{k}/\overleftarrow{k}\right)}{\partial E}\right] \tag{37}$$

Let us see under which conditions this expression can be regarded as acceptable. To this end we shall assume that the pct step proper is preceded by a chemisorption step without pct, regarded as in quasi–equilibrium:

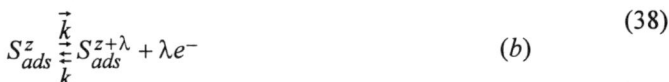

$$S^z + vH_2O_{ads} \xleftrightarrow{\ equilibrium\ } vH_2O + S^z_{ads} \qquad (a)$$
$$S^z_{ads} \underset{\overleftarrow{k}}{\overset{\vec{k}}{\rightleftharpoons}} S^{z+\lambda}_{ads} + \lambda e^- \qquad (b) \tag{38}$$

We will have:

$$\tilde{\mu}_S + v\mu_{w,ads} = v\mu_w + \tilde{\mu}_{S,ads} \tag{39}$$

with:

$$\tilde{\mu}_{S,ads} = \mu^0_{S,ads} + RT \ln \Gamma_S + zF\phi_\beta + \langle m_S \rangle d\phi / dx$$
$$\mu_{w,ads} = \mu^0_{w,ads} + RT \ln \Gamma_{w,max} + \langle m_w \rangle d\phi / dx \tag{40}$$

and the electrochemical potentials $\tilde{\mu}_S$ and $\tilde{\mu}_e$ still expressed by Eq. (24). Equation (40) refers to very low surface coverages by the adsorbing species S, and hence to $\Gamma_w \approx \Gamma_{w,max}$. For the rate-determining pct step (Eq. 38b) we will write:

$$\vec{v} = \vec{k}_0 \Gamma_S \exp\left[\frac{\beta\lambda F}{RT}\left(\phi^M - \phi_\beta\right)\right] \qquad (a)$$

$$\overleftarrow{v} = \overleftarrow{k}_0 \Gamma_S \exp\left[-\frac{(1-\beta)\lambda F}{RT}\left(\phi^M - \phi_\beta\right)\right] \qquad (b)$$

$$(41)$$

By expressing Γ_S in Eq. (41a) as a function of c_S via Eq. (39), with $\tilde{\mu}_{S,ads}$ and $\mu_{w,ads}$ given by Eq. (40) and $\tilde{\mu}_S$ and $\tilde{\mu}_e$ given by Eq. (24), the two rates of Eq. (41) take the form:

$$\vec{v} = \vec{k}_0' c_S \exp\left(-\frac{\Delta G^0}{RT}\right)\exp\left(\frac{\vec{\alpha} F E}{RT}\right) \qquad (a)$$

$$\overleftarrow{v} = \overleftarrow{k}_0' \Gamma_S \exp\left(-\frac{\overleftarrow{\alpha} F E}{RT}\right) \qquad (b)$$

$$(42)$$

with:

$$\Delta G^0 \equiv v\mu_w + \mu^0_{S,ads} - \mu^0_S - v\left(\mu^0_{w,ads} + RT \ln \Gamma_{w,max}\right) \qquad (43)$$

$$E = \phi^M - \phi^S + const.$$

and

$$\vec{\alpha} \equiv z\frac{\phi^S - \phi_\beta}{\phi^M - \phi^S} + \frac{\left(v\langle m_w\rangle - \langle m_S\rangle\right)d\phi/dx}{F\left(\phi^M - \phi^S\right)} + \beta\lambda\frac{\phi^M - \phi_\beta}{\phi^M - \phi^S}$$

$$\overleftarrow{\alpha} \equiv (1-\beta)\lambda\frac{\phi^M - \phi_\beta}{\phi^M - \phi^S}$$

$$(44)$$

By comparing this equation with Eqs. (26) and (27), it is readily seen that:

$$\overrightarrow{\alpha} + \overleftarrow{\alpha} = -l \tag{45}$$

Moreover, by equating $\overrightarrow{v}$ to $\overleftarrow{v}$ at equilibrium, we get:

$$E = \frac{RT}{Fl} \ln \frac{\overrightarrow{k_0}'}{\overleftarrow{k_0}'} - \frac{\Delta G^0}{Fl} + \frac{RT}{Fl} \ln \frac{c_S}{\Gamma_S} \tag{46}$$

This equation is consistent with the definition of l, in that, by differentiation, we obtain:

$$RT \left(\frac{\partial \ln c_S}{\partial E} \right)_{\Gamma_S} = \left(\frac{\partial \mu_S}{\partial E} \right)_{\Gamma_S} \equiv Fl \tag{47}$$

Conway and Angerstein-Kozlowska[15] replace the last term in the equilibrium relationship of Eq. (46) by $(RT/Fl) \ln[c_S(1-\theta_S)/(\Gamma_{S,max}\theta_S)]$, where $\theta_S = \Gamma_S/\Gamma_{S,max}$ is the surface coverage by the adsorbate molecules, thus introducing the $(1 - \theta_S)$ factor that accounts for the surface coverage by free adsorption sites according to a Langmuir isotherm. Differentiating the resulting expression for θ_S with respect to E they obtain an equation expressing the adsorption capacitance C as a function of E. This equation predicts an increase in the half-width of the peak-shaped curves of C vs. E with a decrease in l, as a natural consequence of the diminished sensitivity of C on E. If, in the absence of total charge transfer to the metal, the repulsion between the neighboring partially charged adsorbate molecules is accounted for, a further increase in the half-width of the C vs. E curves is predicted.

In spite of the above justification for the kinetic approach to the estimate of l, this has a number of drawbacks. First of all, there is no point in using a kinetic approach to determine a thermodynamic equilibrium quantity such as l. The justification of the validity of Eqs. (42) and (45) by the resulting equilibrium condition of Eq. (46) is far from rigorous, just as is the justification of the empirical Butler-Volmer equation by the thermodynamic Nernst equation. Moreover, the kinetic expressions of Eq. (41) involve a number of arbitrary assumptions. Thus, considering the adsorption step of Eq. (38a) in quasi-equilibrium under kinetic conditions cannot be taken for granted: a heterogeneous chemical step, such as a deformation of the solvation shell of the adsorbing species S or a complete displacement of one of its coordinating water molecule[8], may partially control the adsorption kinetics.

In addition, the condition of a low surface coverage required for the validity of Eq. (40) is often not satisfied in the kinetic measurements reported in the literature. Even the use of the equilibrium condition of Eq. (46) suffers from this limitation, especially with ordered and compact adlayers, which are difficult to investigate at low coverages. Thus, the use of the kinetic approach of Eqs. (42) and (45), and even that of the equilibrium relation of Eq. (46), at relatively high coverages may yield l values very close to unity, independent of the extent of pct. A slope of about $(-RT/F)$ for the plots of E versus ln $[X^-]$ at constant Γ_{X^-} on polycrystalline Ag, with $X = Cl$ and Br, led Schmidt and Stucki[16,17] to conclude that the electrosorption valency of these anions equals -1 in view of Eq. (46). Analogously, the use of Eq. (36) led Jovic et al.[18] to conclude that chloride adsorption on the low index faces of Ag occurs with an electrosorption valency of -1. These l values for Cl[-] and Br[-] adsorption on Ag are definitely higher than those obtained at low coverages (see later). The reason for these unitary values is as follows.[7] From the properties of partial derivatives we have:

$$\left(\frac{\partial E}{\partial \log c_S}\right)_{\sigma_S} = \left(\frac{\partial E}{\partial \log c_S}\right)_{\sigma_M} - \left(\frac{\partial E}{\partial \sigma_S}\right)_{c_S}\left(\frac{\partial \sigma_S}{\partial \log c_S}\right)_{\sigma_M} \qquad (48)$$

Since the partial derivative $\left(\partial E/\partial \sigma_S\right)_{c_S}$ for halide adsorption on the low index faces of Ag in the presence of an indifferent electrolyte is low[19], the potential shift with varying log c_S is expected to be practically the same at constant σ_S and at constant σ_M. Cross–differentiation of the electrocapillary equation yields the following expression for the Esin-Markov coefficient:

$$\left(\frac{\partial E}{\partial \ln c_S}\right)_{\sigma_M} = -RT\left(\frac{\partial \Gamma_S}{\partial \sigma_M}\right)_{c_S} = -\frac{RT}{zF}\left(\frac{\partial \sigma_S}{\partial \sigma_M}\right)_{c_S} \qquad (49)$$

Now, it is well known that σ_S vs. σ_M plots at constant c_S for adsorbing anions are roughly linear, with slopes slightly < -1. The fact that σ_S

tends to $-\sigma_M$ in dilute solutions of pure salts of a specifically adsorbed anion is well documented[20-22] and was also explained by Schmickler et al.[23] by simple model calculations. σ_M is relatively close to $-\sigma_S$, for sufficiently high values of $|\sigma_S|$, even in the presence of a constant excess of another electrolyte.[13,14,24] In fact, in this case, the relatively high ionic strength depresses the diffuse-layer charge, causing the charge σ_S to be neutralized mainly by the charge density σ_M on the metal. The slope of σ_S vs. σ_M plots at constant c_S is slightly <-1 even in mixtures of a salt of a specifically adsorbed anion with a nonspecifically adsorbed supporting electrolyte at constant ionic strength.[19,21,25-27] In view of Eq. (49), the slope of E vs. ln c_S plots at constant σ_M is therefore expected to be close to (RT/zF), independent of whether pct takes place, and similar conclusions hold for E vs. ln c_S plots at constant σ_S.

In conclusion, kinetic procedures based on Eqs. (42), (45) and (46) for the estimate of the electrosorption valency l are often expected to yield inaccurate results, especially if carried out at high surface coverages.

4. Hard Sphere Electrolyte Model for Specific Adsorption

In the concept of the electrosorption valency, double-layer properties and charge transfer are intricately mixed. This not only makes the interpretation of experimental values so difficult, but it has also prevented theorists from calculating the electrosorption valency of any particular system. Indeed, our present understanding of the double layer is only qualitative.[28,29] While pct can be estimated by quantum-chemical methods, these are really only suited to adsorption from the vacuum, and the effect of the solvent, which favors ionic adsorption, is hard to include. Therefore, simple models referring to limiting cases are quite useful for understanding the electrosorption valency. Here we consider the adsorption of charged hard spheres, which sheds some light on the double-layer aspects.

Carnie and Chan[30] treated an ensemble of hard-sphere ions and dipoles in contact with a hard wall. Specifically, they considered solvent molecules of radius r_w with a point dipole at their center, and two kinds of ions, positive and negative, both with the same radius r_i. One kind of ion can be adsorbed on the surface of the hard wall by a potential proportional to a Dirac delta function. Charge transfer between the metal (i.e. the hard wall) and the adsorbate was not considered. When both the charge density σ_M on the metal and the charge density σ_S due to the adsorbate are small, the following approximate expression for the total potential drop between the electrode and the bulk of the solution can be derived:

$$\Delta\phi_M^S = \frac{4\pi}{\epsilon\kappa}(\sigma_M + \sigma_S) + \left[\frac{4\pi r_i}{\epsilon} + \frac{4\pi r_w(\epsilon-1)}{\zeta}\frac{}{\epsilon}\right]\sigma_M$$
$$+ \frac{4\pi r_w(\epsilon-1)}{\zeta\epsilon(1+\zeta r_i/r_w)}\sigma_S$$

(50)

Here, κ is the Debye reciprocal length, ϵ is the bulk dielectric constant of the solvent, and ζ is determined by ϵ through the relation $\zeta^2(1+\zeta)^4 = \epsilon$, and characterizes the dielectric response of the solvent at the interface. For water at room temperature, $\epsilon \approx 78$, and $\zeta \cong 2.65$. The first term is the Gouy-Chapman term. As a rule, experiments are either conducted at such high electrolyte concentrations that this term is negligible, or else the data are corrected for double-layer effects; consequently, we shall disregard this term in the following. Also, to a good approximation, terms in $1/\epsilon$ can be neglected compared to terms of the order of unity. Note that $\Delta\phi_M^S$ in Eq. (50) equals zero when σ_M and σ_S are both equal to zero. In other words, the potential difference across the metal/solution interphase in the absence of specific adsorption and for $\sigma_M = 0$ is assumed to be zero, that is, no preferential orientation of the solvent dipoles is assumed at the potential of zero charge. In view of Eq. (11), differentiation of σ_M with respect to σ_S at constant $\Delta\phi_M^S$ in Eq. (50) yields the following simple expression for the electrosorption valency in the absence of pct:

$$l \equiv -z\left(\frac{\partial\sigma_M}{\partial\sigma_S}\right)_{\Delta\phi_M^S} = z\frac{r_w}{r_w + \zeta r_i}$$

(51)

This shows that, in the absence of charge transfer, l is determined by the relative sizes (cf. the notion of the thickness ratio introduced by Grahame[31]) of the solvent and the adsorbed ions, and by the dielectric response at the interface. In essence, this equation shows how much image charge flows onto the metal at constant potential when an ion is adsorbed. When the solvent molecules are smaller than the adsorbed

(a)   (b)

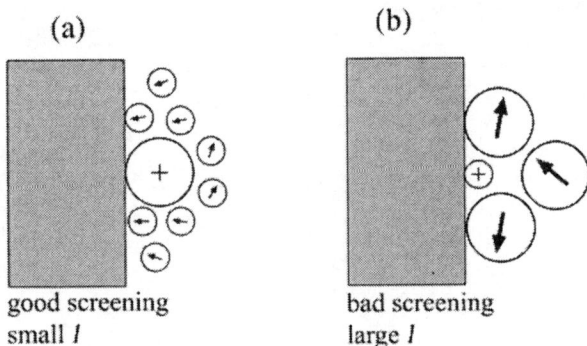

good screening   bad screening
small *l*   large *l*

Figure 2. Schematic picture of the screening of an adsorbed cation by
solvent molecules of different sizes with respect to that of the cation.

ion ($r_w \ll r_i$), they can screen the ionic charge well (see Fig. 2a), little
image charge flows, and *l* is small. Similarly, a large value for ζ
indicates good dielectric screening and reduces *l*. Conversely, small
ions cannot be screened by the solvent (Fig. 2b), and hence they have
large *l* values.

Even though this equation is based on an overly simple model, it
illustrates well the double-layer properties that govern the
electrosorption valency in the absence of pct. In particular, it shows that
a fractional value of *l* need not necessarily indicate pct. We shall return
to the hard-sphere electrolyte model when we discuss dipole moments
of adsorbates.

5. Experimental Procedures for the Determination of the Electrosorption Valency

Since *l* is a thermodynamic quantity, the most reliable procedures for
its determination are based on a thermodynamic analysis of adsorption
data, possibly at low coverages. Adsorption data to be analyzed by the
Gibbs adsorption equation can be obtained by measuring the interfacial
tension γ, the charge density σ_M or the differential capacity C. Direct γ
measurements are equilibrium measurements that can only be carried
out on mercury. Direct charge measurements are conveniently carried
out by the potential-step chronocoulometric technique, which can be

applied both on mercury[32,33] and on solid metals.[34,35] Direct differential capacity measurements are performed by phase-sensitive ac techniques; they are very accurate on mercury[21,25-27] but, on solid metals, they are hardly accurate enough to allow a thermodynamic analysis of the adsorption data. In the presence of frequency dispersion, often observed in connection with adsorption-desorption peaks of organic adsorbates, an extrapolation to zero frequency is required. Once the thermodynamic analysis of adsorption data is carried out, the same l value is to be expected by using any of the equivalent thermodynamic expressions for l in Eqs. (2), (17) and (21). However, this is true only if these expressions are all employed over the same low range of surface coverages.

When determining l as a function of potential, notable difficulties are encountered at potentials at which satisfactorily low surface coverages are only attained at very low bulk surfactant concentrations c_S, where the accuracy of adsorption measurements becomes dramatically low. As an example, Fig. 3 shows plots of σ_M versus Γ_S at constant potential for iodide adsorption on Ag(111) from KPF$_6$ solutions, obtained from potential-step chronocoulometric measurements of charge.[7] The gradual increase in adsorptivity of iodide ion with a positive shift in potential causes the accessible range of Γ_S values covered by the σ_M vs Γ_S plots to shift towards higher Γ_S values. In view of the downward concavity exhibited by these plots, the limiting slope of the experimentally accessible segment of the plots decreases as this segment shifts towards higher Γ_S values. This is more evident if the experimental point corresponding to the lowest accessible Γ_S value at each potential is joined to the point on the $\Gamma_S = 0$ axis at the same potential, i.e. the point obtained from the σ_M vs. E curve of the supporting electrolyte in the absence of iodide (see the dashed segments in Fig. 3). If we plot the l values measured by the slope of the plots in Fig. 3 against potential we obtain curve 1 in Fig. 4, which attains a maximum value and then decreases as E is shifted towards more positive values. The latter decrease is to be ascribed to the inaccuracy in the estimate of the limiting slope of the σ_M vs Γ_S plots at constant potential for $\Gamma_S \rightarrow 0$. A similar decrease was observed for bromide adsorption on Ag(111).[7] It was also observed, albeit to a lower extent, for chloride[13] and bromide[14] adsorption on Au(111), where the slope of the σ_M vs. Γ_S plots for the obtainment of l was estimated by including the points on the $\Gamma_S = 0$ axis.

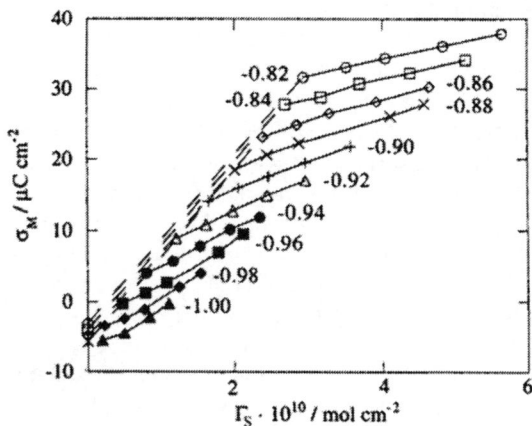

Figure 3. Plots of σ_M vs. Γ_S at constant potential for iodide adsorption on Ag(111) from KPF_6 solutions. Numbers on each curve denote potential values in V/SCE. (Reprinted from Ref.[7] with permission from the Am. Chem. Soc.)

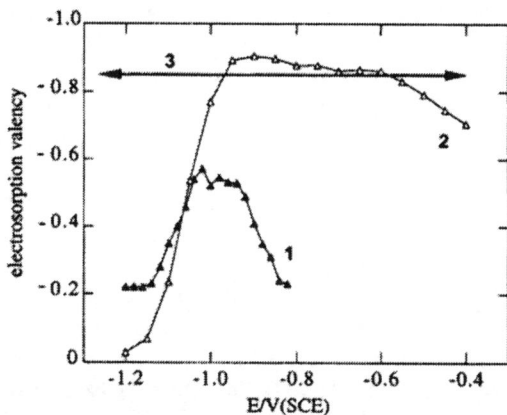

Figure 4. Plots of l (curve 1) and of $\bar{l}$ obtained from Eq. (53) (curve 2) against E for iodide adsorption on Ag(111). The horizontal arrow (curve 3) denotes the $\bar{l}$ value obtained from Eq. (52) over the potential range of stability of the iodide adlayer. (Reprinted from Ref.[7] with permission from the Am. Chem. Soc.)

In the case of strongly chemisorbed species, such as halide and sulfide ions on Ag(111), the electrosorption valency as a function of the applied potential can be conveniently estimated at low surface coverages by a chronocoulometric procedure that avoids the difficulties encountered with conventional adsorption measurements. It consists in adopting bulk anionic concentrations c_S in the range from 10^{-4} to 10^{-3} M in the presence of an excess of a nonspecifically adsorbed electrolyte, and in jumping from a fixed initial potential E_i, negative enough to exclude anionic adsorption, to a final potential E, where the anion is strongly adsorbed.[7] Under these conditions, the charge density $Q(t)$, following the potential jump, varies linearly with $t^{1/2}$ during the first 30-40 ms. The slopes of the $Q(t)$ vs. $t^{1/2}$ plots are proportional to the bulk concentration c_S, as shown in Fig. 5 for sulfide, iodide and bromide adsorption. The adsorption of the supporting anion takes place in a few milliseconds after the potential jump, because of its relatively high bulk concentration. Conversely, the adsorption of sulfide or halide ions takes place more slowly, because of their low bulk concentrations. The time dependence of the charge density $Q(t)$ is therefore determined by the slow adsorption of the latter anions.

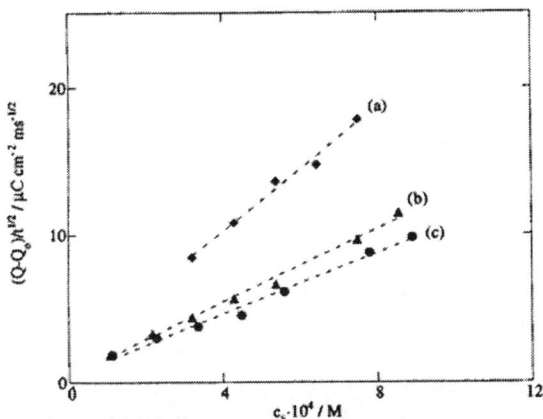

Figure 5. Plots of $(Q-Q_0)/t^{1/2}$ vs. c_S on Ag(111) from solutions of sulfide in 0.15M NaOH for a -1.40 V $\rightarrow$ -0.90 V jump (a), iodide in 0.15M NaOH for a -1.50 V $\rightarrow$ -0.50 V jump (b), and bromide in 0.1M KPF$_6$ for a -1.50 V$\rightarrow$-0.35 V jump (c). (Reprinted from Ref.[7] with permission from the Am. Chem. Soc.)

The fact that $Q(t)$ satisfies the integral form of the Cottrell equation indicates that, at least during the first 30-40 ms after the potential jump, the adsorption of halide and sulfide ions is diffusion controlled under limiting conditions, that is, with their volume concentration at the electrode surface much smaller than the corresponding bulk value. Under these conditions, the intercept of the $Q(t)$ vs. $t^{1/2}$ plots on the charge axis yields the charge density Q_0 following the potential jump $E_i \rightarrow E$ in the absence of the chemisorbed anion. Moreover, the surface charge density, $\sigma_S(t)$, of the chemisorbed anion at a given time t, as expressed by the Cottrell equation, equals $2zF(D_S t/\pi)^{1/2}c_S$, where D_S is its diffusion coefficient. In view of the definition of the electrosorption valency l for an ionic species in Eq. (11), the corresponding "integral value", $\bar{l}$, is given by the equation:

$$\bar{l} \equiv -z\left[\left(\sigma_M - \sigma_M^0\right)/\sigma_S\right]_E \qquad (52)$$

where σ_M^0 is the charge density at E in the absence of the adsorbate. The integral electrosorption valency $\bar{l}$ is clearly proportional to the slope of the $Q(t)$ vs. $t^{1/2}$ plot:

$$
\begin{aligned}
\bar{l} &= -z\left[\frac{\sigma_M(t) - \sigma_M^0}{\sigma_S(t)}\right]_E = -z\left[\frac{Q(t) - Q_0}{\sigma_S(t)}\right]_E \\
&= -z\frac{\pi^{1/2}}{2FD_S^{1/2}c_S}\left[\frac{Q(t) - Q_0}{t^{1/2}}\right]_E
\end{aligned}
\qquad (53)
$$

Curve 2 in Fig. 4 shows a plot of $\bar{l}$ for iodide adsorption on Ag(111) against E, as obtained by this equation, using the diffusion coefficient $D_S = 1.72\times10^{-5}$ cm^2 s^{-1} taken from the polarographic literature.[36] It is apparent that these $\bar{l}$ values differ appreciably from the l values obtained from the σ_M vs. Γ_S plots at constant E of Fig. 3. Both values increase rapidly over the potential range of the two-dimensional disorder-order phase transition yielding the $(\sqrt{3}\times\sqrt{3})R30^0$ iodide overlayer. However, when proceeding towards more positive potentials, over the range of stability of the overlayer, $\bar{l}$ attains a roughly constant value of about 0.85 while l decreases progressively.

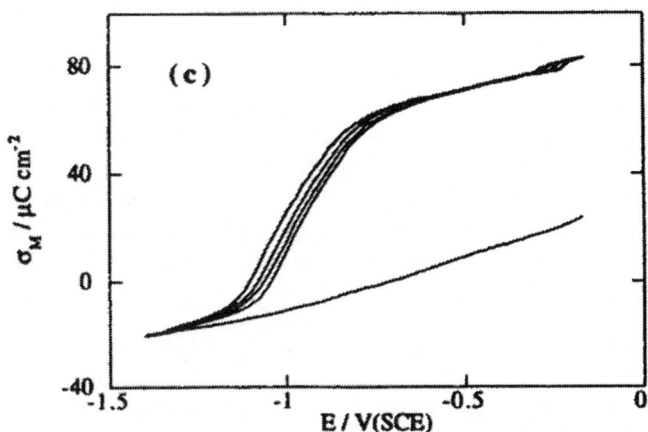

Figure 6. The upper curves are σ_M vs. E plots on Ag(111) obtained from solutions of 0.15 M NaOH containing KI of concentration 9.2×10^{-5}, 1.66×10^{-4}, 3×10^{-4} and 5×10^{-4} M, when proceeding from right to left. The lower curve is a plot of the charge density σ_M^0 for aqueous 0.15 KPF$_6$ against E. (Reprinted from Ref.[7] with permission from the Am. Chem. Soc.)

This is due to the fact that the electrosorption valency, $\bar{l}$, obtained from diffusion-controlled adsorption relies on measurements corresponding to the same small, initial amount of adsorbed material at all potentials. Consequently, this procedure does not suffer from the limitations involved in the use of Eq. (2) at high surface coverages.

In the case of chemisorbed ions yielding a well-defined ordered overlayer structure on a single-crystal electrode, the electrosorption valency, relative to the formation of the overlayer, can be obtained by recording the chronocoulometric charge density, σ_M, against E, both in the absence and in the presence of the chemisorbed ion. Figure 6 shows plots of σ_M versus E on Ag(111) from aqueous 0.15 M NaOH in the presence of different bulk iodide concentrations high enough to reach the limiting surface concentration, Γ_S, corresponding to a complete $(\sqrt{3} \times \sqrt{3})R30^0$ adlayer.[7] For comparison, the figure also shows a curve of the charge density σ_M^0 of the nonspecifically adsorbed KPF$_6$ electrolyte[37] against E; the KPF$_6$ concentration is equal to the overall salt concentration in the corresponding σ_M vs. E curves. Therefore, over the potential range of stability of the ordered overlayer, the vertical distance between the σ_M versus E curves and the corresponding σ_M^0 versus E curve provides the difference between the charge density in the presence of the overlayer and that in the absence of specific adsorption, at the same applied potential and ionic strength. In fact, in spite of the use of 0.15 M NaOH as supporting electrolyte in order to shift hydrogen discharge to more negative potentials, the formation of an ordered iodide adlayer prevents the chemisorption of hydroxyl ions. In view of Eq. (52), the ratio of the constant difference $\left(\sigma_M - \sigma_M^0\right)$ over the potential range of stability of the $(\sqrt{3}\times\sqrt{3})R30^0$ iodide adlayer to the charge density σ_S of the iodide ion, estimated from the adlayer structure, yields an $\bar{l}$ value of 0.86, in good agreement with that obtained from diffusion-controlled iodide adsorption over the same potential range (see the horizontal arrow in Fig. 4). The adsorption features of bromide and chloride ions on Ag(111) are similar to those of iodide.[7] The $\bar{l}$ values for these two ions, obtained by the same procedure, amount to 0.65 and 0.44, again in agreement with the values obtained from diffusion-controlled adsorption.

An increase in the bulk iodide concentration causes a gradual shift of the sigmoidal curves in Fig. 6 towards more negative potentials, while leaving the height of the plateau unaltered. In Fig. 7 this potential shift at constant σ_M on Ag(111) is plotted against log c_S, for sulfide, iodide, bromide and chloride. The slope of the resulting plots is just a measure of the Esin and Markov coefficient $\left(\partial E/\partial \mu_S\right)_{\sigma_M}$, where μ_S is the bulk chemical potential of the adsorbed anion. This slope equals -32 mV for sulfide, -74 mV for iodide, -64 mV for bromide and -85 mV for chloride.[7] We already pointed out in Section II.3 that, for these systems, the slope of μ_S vs. E plots at constant σ_M is very close to that at constant σ_S. The same reasonable assumption was made by Schultze[38] for the evaluation of l in cationic electrosorption systems. With this assumption, the reciprocal of the above Esin and Markov coefficients allows an estimate of l on the basis of its thermodynamic definition of Eq. (49). The resulting l values are -1.84 for sulfide, -0.81 for iodide, -0.94 for bromide and -0.70 for chloride.

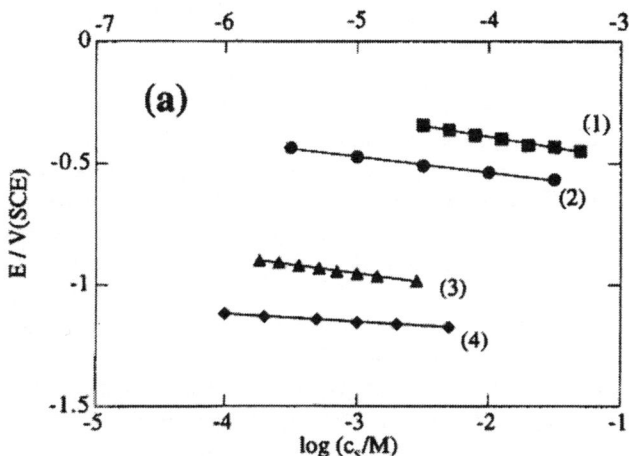

Figure 7. Plots of E vs. log c_S on Ag(111) from solutions of c_S M KCl + $(0.1 - c_S)$ M KPF$_6$ at $\sigma_M = +53$ μC cm^{-2} (1), bromide in 0.1 M KPF$_6$ at $\sigma_M = +57$ μC cm^{-2} (2), iodide in 0.15 M NaOH at $\sigma_M = +12$ μC cm^{-2} (3), and sulfide in 0.15 M NaOH at $\sigma_M = +59$ μC cm^{-2} (4); the lower log c_S axis refers to plots 1, 2 and 4, the upper one to plot 3. (Reprinted from Ref.[7] with permission from the Am. Chem. Soc.)

The l values for bromide and chloride are higher than the corresponding $\bar{l}$ values, and much higher than the l values obtained from the alternative thermodynamic definition of l in Eq. (11).

The reason for these discrepancies is quite probably to be ascribed to a strong dependence of the activity coefficient of Γ_S, $f(\Gamma_S)/\Gamma_S$, upon E and Γ_S at the high surface coverages employed for the estimate of l. In particular, l values obtained from μ_S versus E plots at constant Γ_S are affected by the potential dependence of the activity coefficient at constant Γ_S (see Eq. 21); conversely, the l values obtained from the dependence of σ_M upon Γ_S at constant E are affected by the Γ_S-dependence of the activity coefficient at constant E. These different dependences may have opposite effects on the l values obtained on the basis of the two alternative thermodynamic definitions. This may also explain the anomalously high l values for bromide and chloride adsorption on polycrystalline Ag obtained by Schmidt and Stucki[16] from μ_S versus E plots at constant Γ_S. For this reason, at high surface coverages, the procedures for the estimate of integral $\bar{l}$ values are to be preferred.

The determination of $\bar{l}$ has often been carried out during the underpotential deposition (upd) of a metal M from a solution of its ion M^z on a different metal of higher electronegativity. This is usually marked by a sharp voltammetric cathodic peak. In some cases (e.g, $Pb^{10,39,40}$ and Tl^{41} upd on Ag(hkl)) the quantity $-Q/(zF)$, where Q is the negative charge density under the peak corrected for the charging current in the pure supporting electrolyte, has been compared with the surface concentration Γ_S of the deposited metal, determined by some independent means.[42] An ingenious procedure for determining Q and Γ_S simultaneously makes use of a thin-layer cell, in which the electrolytic solution is interposed between an "indicator" electrode, where upd takes place, and a "generator" electrode, whose oxidation provides the ions to be underpotential-deposited on the indicator electrode. Indicator electrodes consisting of silver single-crystal faces [39,41,43] and $Pb^{39,43}$ or Tl^{41} generator electrodes have been employed. The oxidation current density, I_O, flowing at the generator electrode, and the reduction current density, I_R, flowing at the indicator electrode, are controlled independently by a bipotentiostat. The generator electrode is kept at a constant potential, thus creating in the thin solution layer a constant cation concentration expressed by the Nernst equation. On the other hand, the potential applied to the indicator electrode is scanned from an initial value, E_i, positive enough to avoid upd, to a final value, E_f, lying within the upd region. The integral of I_O over time, once divided by zF, yields the surface concentration Γ_S of the cations deposited on the indicator electrode. On the other hand, the integral, Q', of I_R over time yields the negative charge transferred from the indicator electrode to the metal ions, during their upd. To correct Q' for the capacitive charge flowing during the potential scan from E_i to E_f, the same scan is carried out while keeping the generator electrode at a potential negative enough to avoid its oxidation, and the resulting charge is subtracted from Q', yielding the corrected charge Q. Good agreement between Γ_S and $-Q/(zF)$ has been found. Note that Q is just equal to $(\sigma_M - \sigma_M^0)$ in Eq. (52). In view of this equation, such an agreement implies that $\bar{l}$ equals z, namely that charge transfer is complete.

The kinetic determination of l on the basis of Eqs. (37), (42) and (45) has frequently been reported in the literature,[38,44] often in connection with electrochemical impedance spectroscopy measurements.[45,46] As already pointed out, the l values obtained by the kinetic approach are poorly accurate, since they involve a number of arbitrary extrathermodynamic assumptions that are not required in the thermodynamic approach. The kinetic approach is unavoidable only if pct is assumed to involve an intermediate of an electrode reaction in which both the reactant and the final product are present in the corresponding bulk phases. In this case, however, fitting the kinetic data to the electrode reaction mechanism requires several adjustable parameters. At any rate, these systems are outside the scope of the present chapter.

III. THE PARTIAL CHARGE TRANSFER

The extent of pct from a chemisorbed molecule to an electrode is measured by the pct coefficient λ. Unfortunately, the estimate of λ is not straightforward, since it relies on more or less accurate modelistic assumptions. Thus, for instance, screening of the charge of the adsorbed ion by the inhomogeneous electron gas and by the solvent molecules produces an effect analogous to that of pct.[47-49]

The most common procedure for estimating λ consists in measuring the thermodynamically significant electrosorption valency l and in evaluating its extrathermodynamic contributions κ_S, $v\kappa_w$ and g of Eqs. (26) and (27) in order to extract the λ value. A more direct extrathermodynamic procedure that relies exclusively on the Gouy-Chapman theory can be applied to the important class of self-assembled thiol monolayers anchored to a metal surface.

1. Extrathermodynamic Estimate of the Partial Charge Transfer Coefficient from the Electrosorpion Valency

The $-v\kappa_w$ contribution to l in Eq. (26) from the desorbing water molecules can be roughly estimated by assuming that the potential ϕ varies linearly with the distance x in the compact layer enclosed between the electrode surface plane $x = 0$ and the outer Helmholtz plane $x = d$. In the presence of a strong excess of a nonspecifically adsorbed supporting electrolyte or upon correction for the potential difference across the diffuse layer, the electric potential ϕ^S in the bulk solution can practically be identified with that, ϕ_d, at the outer Helmholtz plane. The normal component $\langle m_w \rangle$ of the water dipole can be written as eL_w, where e is the absolute value of the electronic charge and L_w is the length of the normal component, taken as positive or negative depending on whether the dipole component is directed towards the solution or towards the electrode. The $-\nu\kappa_w$ contribution can therefore be written as:

$$
\begin{aligned}
-\nu\kappa_w &\equiv -\frac{\nu}{e}\frac{\langle m_w \rangle}{\phi^M - \phi^S}\frac{d\phi}{dx} \cong -\frac{\nu}{e}\frac{\langle m_w \rangle}{\phi^M - \phi_d}\frac{\phi_d - \phi^M}{d} \\
&= \nu\frac{\langle m_w \rangle}{ed} = \nu\frac{L_w}{d}
\end{aligned}
\tag{54}
$$

If the water dipoles are preferentially directed towards the solution, L_w is positive and the same is true for the $-\nu\kappa_w$ contribution to l. In fact, in this case the water dipoles make a negative contribution to the potential difference $(\phi^M - \phi_d)$ across the compact layer; consequently, upon their removal from the contact with the electrode by the adsorbing species S, a flow of electrons to the metal surface along the external circuit is required to keep the applied potential E constant. This implies a negative contribution to σ_M and a consequent positive contribution to $l = -(\partial\sigma_M/\partial\Gamma_S)_E/F$, in view of its definition. Naturally, if the water dipoles are preferentially directed towards the metal surface, they make a negative contribution to l. By analogous arguments, the contribution to l from the dipoles of the adsorbed species S is written as:

$$
\kappa_S \equiv \frac{1}{e}\frac{\langle m_S \rangle}{\phi^M - \phi^S}\frac{d\phi}{dx} \cong \frac{1}{e}\frac{\langle m_S \rangle}{\phi^M - \phi_d}\frac{\phi_d - \phi^M}{d} = -\frac{\langle m_S \rangle}{ed} = -\frac{L_S}{d}
\tag{55}
$$

If the adsorbate dipoles are preferentially directed towards the solution, the length L_S of their normal dipole component is to be taken as positive and their contribution to l is negative. In fact, in this case they make a negative contribution to the potential difference $(\phi^M - \phi_d)$ across the compact layer; consequently, upon their adsorption a flow of electrons away from the metal surface along the external circuit is required to keep the applied potential E constant. This implies a positive contribution to σ_M and a consequent negative contribution to l, in view of its definition. Naturally, an opposite effect on l is expected if the adsorbate dipoles are directed towards the electrode.

Under the assumption of a linear dependence of ϕ upon the distance x from the electrode surface, the g factor in Eq. (27) takes the form:

$$g = \frac{\phi_\beta - \phi^S}{\phi^M - \phi^S} \cong \frac{\phi_\beta - \phi_d}{\phi^M - \phi_d} = \frac{d - \beta}{d} \tag{56}$$

The $(d-\beta)/d$ ratio was called "thickness ratio" by Grahame[31], who regarded $z(d-\beta)/d$ as the only contribution to the thermodynamic quantity subsequently called electrosorption valency by Vetter and Schultze.[4,5] The more deeply the center of charge of an ionic adsorbate S of negative charge number z penetrates into the compact layer, the closer the g factor approaches unity, and the more the adsorbing anion shifts the potential difference ($\phi^M - \phi_d$) across the compact layer in the negative direction; consequently, upon anion adsorption, a flow of electrons away from the metal surface along the external circuit is required to keep the applied potential E constant. This implies a positive contribution to σ_M and a consequent negative contribution, zg, to l, in view of its definition. If the adsorbing anion S also transfers a part of its negative charge to the metal, an additional flow of free electrons away from the electrode is required to keep the applied potential constant. This implies a further positive contribution to σ_M and a consequent further negative contribution to l. This is accounted for by the $-\lambda(1 - g)$ term in the expression of Eq. (26) for l, since λ is positive for a negative charge transfer to the electrode in view of Eq. (1). The transfer of a partial negative charge to the electrode makes a negative contribution $-\lambda(1 - g)$ to l even if the adsorbate S is neutral and zg equals zero. Naturally, opposite contributions to l are created by an adsorbing cation or by the transfer of a partial positive charge from the adsorbate to the electrode.

The various contributions to l can be envisaged from a somewhat different viewpoint by expressing the potential difference ($\phi^M - \phi_d$) across the compact layer on the basis of a simple electrostatic model in which the double-layer region enclosed between the electrode surface plane, $x = 0$, and the inner Helmholtz plane, $x = \beta$, is ascribed a distortional dielectric constant, ε_β, while that between $x = \beta$ and the outer Helmholtz plane, $x = d$, is ascribed a generally higher distortional dielectric constant ε_γ, with $\gamma = d - \beta$:

$$\left(\phi^M - \phi_d\right) = \chi_e + 4\pi\left(\frac{\beta}{\varepsilon_\beta} + \frac{\gamma}{\varepsilon_\gamma}\right)\left(\sigma_M - \lambda F\Gamma_S\right) +$$

$$+ 4\pi\frac{\gamma}{\varepsilon_\gamma}\left[(z + \lambda)F\Gamma_S\right] - \frac{4\pi N_w\langle m_w\rangle}{\varepsilon_\beta} - \frac{4\pi N_S\langle m_S\rangle}{\varepsilon_\beta} \tag{57}$$

Here χ_e is the surface dipole potential due to electron spillover, N_w and $N_S = N_{Av}\Gamma_S$ are the numbers of water and adsorbate molecules per unit surface, and N_{Av} is the Avogadro number. The second term in Eq. (57) is the potential difference across the compact layer ($0 < x < d$) due to the free charge density, $(\sigma_M - \lambda F\Gamma_S)$, on the metal surface; the third term is the potential difference across the ($\beta < x < d$) layer due to the residual charge density, $(z + \lambda)F\Gamma_S$, of the adsorbate molecules at $x = \beta$; the last two terms are the surface dipole potentials due to the water and adsorbate molecules. Denoting by $N_{w,max}$ the number of adsorbed water molecules per unit surface in the absence of adsorbate, we have $N_{w,max} = N_w + \nu N_S$. Taking this relationship into account and assuming that χ_e, β, d, ε_β, ε_γ, λ, $N_{w,max}$, $\langle m_w\rangle$ and $\langle m_S\rangle$ are practically constant at constant potential, differentiation of σ_M with respect to Γ_S at constant $E \cong \text{const.} + (\phi^M - \phi_d)$ yields:

$$l = -\frac{1}{F}\left(\frac{\partial\sigma_M}{\partial\Gamma_S}\right)_E = zg - \lambda(1 - g) + \nu\frac{\langle m_w\rangle}{ed'} - \frac{\langle m_S\rangle}{ed'} \tag{58}$$

with:

$$g \equiv \frac{\gamma/\varepsilon_\gamma}{\beta/\varepsilon_\beta + \gamma/\varepsilon_\gamma} \quad ; \quad d' \equiv \beta + \frac{\varepsilon_\beta}{\varepsilon_\gamma}\gamma \tag{59}$$

This equation practically coincides with Eq. (26) in view of the expressions for κ_w, κ_S and g in Eqs. (54)-(56) and of Eq. (59) with $\varepsilon_\gamma = \varepsilon_\beta$.

In the chemisorption of inorganic ions, the two dipole contributions to l are generally regarded as negligible with respect to the first two terms of Eq. (26), yielding the approximate expression:[50]

$$l = gz - \lambda(1-g) \quad \rightarrow \quad \frac{l}{z} \cong g - \frac{\lambda}{z}(1-g) \tag{60}$$

In the case of a nonzero pct, chemisorbed anions generally transfer a negative charge to the electrode $(\lambda > 0)$, while chemisorbed cations transfer a positive charge $(\lambda < 0)$. Moreover, the transfer of a negative charge is favored by a positive shift in the applied potential E, while that of a positive charge is favored by a negative shift. In both cases $d\lambda/dE$ is positive, while dl/dE is negative in view of Eq. (60). In this respect, an appreciable negative value of dl/dE may be indicative of pct.

According to Pauling[51] the polarity of a chemical bond increases with an increase in the difference in the electronegativity of the two atoms involved in the bond formation. Pct is expected to increase with the covalent nature of the metal-adsorbate bond, and hence with a decrease in its polarity. Schultze and Koppitz[50] plotted the l/z ratio for different inorganic ions against the difference $|\chi^M - \chi^S|$ in the electronegativities of metal and adsorbate, as shown in Fig. 8. These l values refer to the potential of zero charge in the absence of chemisorption, E_z, and to low surface coverages θ, where the mutual interaction between the chemisorbed ions is expected not to affect l. For $|\chi^M - \chi^S| > 1$ the scattered points of the plot tend to a minimum limiting value of about 0.16. For these high differences in electronegativities it is reasonable to exclude pct $(\lambda = 0)$. This leads to the conclusion that g is $\cong 0.16$. If g is identified with the thickness ratio, $(d - \beta)/d$, this g value implies an improbable position of the chemisorbed ion very close to the outer boundary, $x = d$, of the compact layer. More probably, the approximation of a constant electric field $-d\phi/dx$ within the whole compact layer does not hold. Thus, for instance, if we regard ε_β as $<\varepsilon_\gamma$ in the more general expression of Eq. (59) for g, the electric field in the $(0 < x < \beta)$ layer is higher than that in the $(\beta < x < d)$ layer, and the 0.16 value for g may well be consistent with a position of the center of charge of the chemisorbed ion in the middle of the compact layer. For $|\chi^M - \chi^S| < 0.5$ the l/z values in Fig. 8 are very close to unity. These values suggest an almost

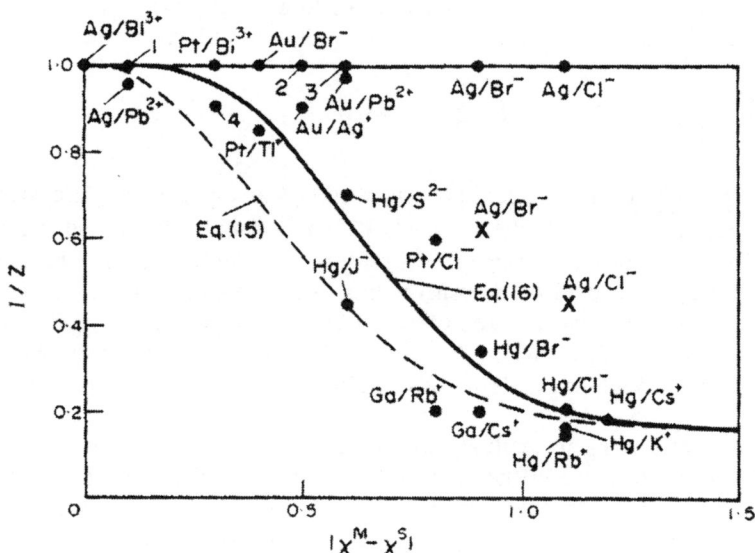

Figure 8. Plot of the l/z ratio against the absolute difference of electronegativities, $|\chi^M-\chi^S|$. 1 = Pt/H+, Cu/Pb^{2+}, Cu/Tl$^+$, Ag/Tl$^+$; 2 = Au/Cu$^+$, Au/Sb^{3+}, Au/Bi^{2+}; 3 = Au/Tl$^+$, Au/Cl$^-$; 4 = Pt/Cu^{2+}. The × markers are values taken from Ref.[7]. (Reprinted from Ref.[50] with permission from Pergamon Press)

total charge transfer, i.e. $\lambda \cong -z$. In view of Eq. (60) a l/z ratio equal to unity might also be explained by $g = 1$ in the absence of pct. However, a g value equal to unity implies that the center of charge of the adsorbing is on the metal side with respect to the compact layer. This can only be justified in the case of $\lambda = -z$, when the chemisorbed ions can be regarded as immersed in the inhomogeneous electron gas of the metal. For $0.5 <|\chi^M - \chi^S|< 1$, intermediate values of λ are expected. Particularly significant is the trend in the l/z values of ions of similar structure such as halide ions on mercury; this is suggestive of a corresponding trend in their λ values. A similar trend of halide ions is also observed on silver, once we exclude the unit values of l/z for bromide and chloride reported by Schmidt and Stucki.[16,17] The dashed curve in Fig. 8 was obtained by using the expression for l/z of Eq. (60), with $g = 0.16$ and $-\lambda/z = \exp[-a(\chi^M - \chi^S)^2]$.[50] The latter empirical relation, with $a = 0.25$, was proposed by Pauling[51] for charge transfer in a diatomic molecule in the gas phase; to account for the screening effect of water molecules favoring ionization, the a value used for the calculation of the dashed curve in Fig. 8 was set equal to 3. Better agreement with experimental data was obtained by setting $g = 0.16 - 0.84(\lambda/z)$ in place of $g = 0.16$. This yields the solid curve in Fig. 8. The above empirical relation for g postulates a progressive immersion of the adsorbed ions into the inhomogeneous electron gas with an increase in λ until ultimately, for $\lambda = -z$, the compact layer is entirely moved to the solution side of the adsorbed ions. Plots similar to that in Fig. 8 for aqueous solutions were obtained for methanol and DMF as solvents.[52,53] The main effect of the solvent is that of influencing the compact-layer thickness and, therefore, the g factor.

The electrosorption valency of several neutral aliphatic molecules on mercury was examined by Koppitz at al.[54] As usual, the experimental l values were referred to the potential of zero charge in the absence of specific adsorption, E_z, and to low surface coverages. With the remarkable exception of thiourea, all these molecules do not undergo pct ($\lambda = 0$). Since they are also neutral ($z = 0$), their electrosorption valency is exclusively determined by the two dipole terms:

$$l = \kappa_S - \nu\kappa_w \cong -\frac{\langle m_S \rangle}{ed} + \nu\frac{\langle m_w \rangle}{ed} \qquad (61)$$

The thickness d of the compact layer was estimated from the differential capacity C at full coverage, which was set equal to $\varepsilon/(4\pi d)$, with dielectric constant $\varepsilon = 5$. The orientation of the adsorbed molecule relative to the electrode surface was estimated from the d value and from the maximum surface concentration $\Gamma_{S,max}$. This was about equal to 6×10^{-10} mol cm^{-2} for a vertical orientation; lower values were considered to denote a flat orientation. The number, ν, of desorbed water molecules was estimated from the ratio of the calculated maximum surface concentration of water molecules ($\Gamma_{w,max} = 11.8\times10^{-10}$ mol cm^{-2}) to $\Gamma_{S,max}$; ν was found to be equal to 2 for molecules in the vertical orientation and somewhat higher for those in the flat orientation. Even though a reorientation of the adsorbed molecules with an increase in Γ_S or a tilted orientation at full coverage cannot be ruled out, for simplicity only a flat orientation and two opposite vertical orientations were considered. For the flat orientation, $<m_S>$ was set equal to zero and d in Eq. (55) was set equal to the diameter of the flat molecule. For the vertical orientation, d was set equal to the molecule length and $<m_S>$ was set equal to the total dipole moment of the molecule, with the positive or negative sign depending on whether the dipole was considered to point towards the solution or towards the electrode.

Molecules adsorbed in a flat orientation, such as ethanenitrile, propanenitrile, 2-butanol, 3-pentanol, dimethyl-formamide and several bifunctional aliphatic compounds, have small values of the electrosorption valency, l, in the range from 0.03 and 0.05, in spite of their large total dipole moments. It is therefore reasonable to ascribe these l values to the sole water dipole term, $-v\kappa_w$. Monofunctional aliphatic compounds with a vertical orientation and $l > 0.04$, such as n-propanol and allyl alcohol, were ascribed a positive κ_S term, i.e. a dipole pointing towards the metal. Molecules with a vertical orientation and a negative l value, such as chloroform, dibutylsulfide, urea and thiourea, were ascribed a negative κ_S term, i.e. a dipole pointing towards the solution; all these molecules have a strong dipole capable of interacting with the metal surface. The l values of the vertically oriented molecules are plotted in Fig. 9 against the corresponding κ_S terms, calculated as previously described. With the exclusion of dibutylsulfide and thiourea, the experimental points lie roughly on a straight line that crosses the vertical axis $\kappa_S = 0$ at $l = 0.03$. In view of Eq. (61) this value is equal to $-v\kappa_w$. Since for these compounds v equals 2, $<m_w>/(ed)$ amounts to about 0.015. In other words, the water molecules at the potential of zero charge in the absence of specific adsorption are slightly oriented with the oxygen turned towards the mercury surface, as expected.[55]

With the only exception of thiourea, all aliphatic compounds are characterized by positive dl/dE values.[54] Within the limits in which we can exclude an appreciable reorientation of the aliphatic molecules with a change in the applied potential E, the increase of l with an increase in E is to be ascribed to a gradual reorientation of the adsorbed water molecules. Water molecules have a particularly high dipole moment per unit volume, and are known to be readily orientable under the influence of an electric field. A progressive shift of E towards more positive values causes $<m_w>$ to pass gradually from negative to positive values. Consequently, the contribution to σ_M required to keep E constant upon removing v adsorbed water molecules by one

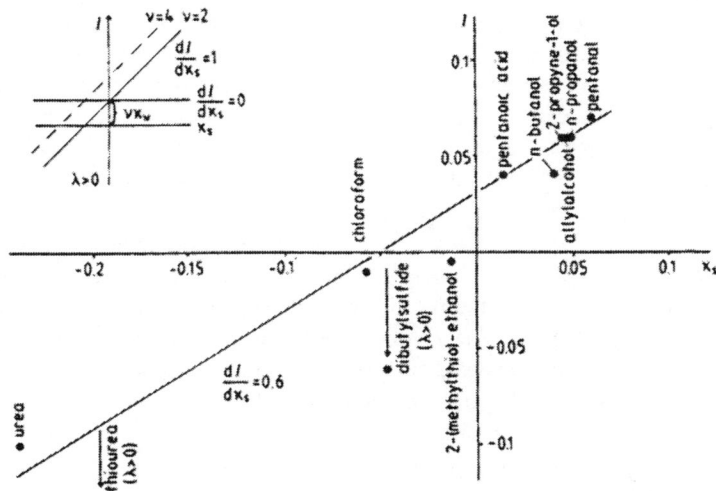

Figure 9. Dependence of the electrosorption valency l on the dipole terms κ_S. The inset shows the expectations from Eq. (61). Systems with expected charge transfer are marked by vertical arrows. (Reprinted from Ref.[54] with permission from Elsevier Science)

adsorbing S molecule passes from positive to negative values with a positive shift of E; this results in an increase of l, in view of Eq. (2).

The particular behavior of thiourea is to be ascribed to the transfer of a partial negative charge from the sulfur atom to the mercury. Hence, the more general expression of Eq. (58) for l must be applied, with $z = 0$. Thiourea has practically the same size and dipole moment as urea, which does not give rise to pct. Thus, these two molecules are characterized by the same dipole terms κ_S and $-v\kappa_w$, while their l values differ by the quantity -0.17, which must be exclusively ascribed to the $-\lambda(1 - g)$ term. Setting $g \cong 0.2$ (a value close to that estimated from adsorbed inorganic ions), a λ value of 0.2 is obtained for thiourea. As opposed to all other aliphatic compounds, thiourea has a negative dl/dE value, albeit small. This is probably due to an increase in the negative charge $-\lambda e$ transferred from the sulfur atom to the mercury with a positive shift of E. The resulting increasingly positive contribution to

σ_M required to keep the applied potential constant overcompensates the opposite contribution due to the removal of adsorbed water molecules.

Many aromatic and heterocyclic compounds undergo a passage from a flat to a vertical orientation with an increase in surface coverage.[56-61] A useful procedure for monitoring such a reorientation consists in measuring the shift in the applied potential E with an increase in the surface concentration, Γ_S, of the adsorbate, at constant charge density, σ_M, on the metal. Pyrazine, due to its rigid structure with a zero dipole moment, is adsorbed in a flat orientation at all charge densities. Consequently, the shift of E with Γ_S at constant σ_M depends exclusively on the orientation of the desorbing water molecules at the given charge density. The E vs Γ_S plots for pyrazine are linear, and exhibit a slope that passes gradually from positive to negative values, as σ_M passes from negative to positive values.[57] This is due to the desorbing water molecules passing from an average orientation with the hydrogens preferentially turned towards the metal to an opposite orientation, as σ_M is progressively increased. At low surface concentrations, the behavior of the E vs Γ_S plots for pyridine is similar to that for pyrazine. However, at Γ_S values that are higher the more negative σ_M is, these plots exhibit an abrupt change in slope, which becomes negative and practically constant at all charge densities.[57] This negative slope is due to a passage of the adsorbing pyridine molecules from a flat to a vertical orientation, with the nitrogen pointing toward the solution phase.

With aromatic and heterocyclic compounds undergoing a passage from a flat to a vertical orientation, the analysis of the adsorption data for the flat and for the vertical orientation provides two different l values at E_z.[61] The flat orientation is usually characterized by a minimum differential capacity C_{min} of 10-15 $\mu F \ cm^{-2}$ and a maximum surface concentration $\Gamma_{s,max}$ of $2\text{-}3 \times 10^{-10} \ mol \ cm^{-2}$, while the vertical orientation exhibits C_{min} and $\Gamma_{s,max}$ values in the range of 5-6 $\mu F \ cm^{-2}$ and $5\text{-}6 \times 10^{-10} \ mol \ cm^{-2}$. A neutral molecule in the flat orientation has a zero value of z and of the dipole moment normal component $<m_S>$. Hence, l is simply given by:

$$l = -\nu\kappa_w - \lambda(1-g) \qquad (62)$$

Setting $g \cong 0.2$, $\kappa_w = 0.015$ and $\nu = \Gamma_{w,max}/\Gamma_{S,max}$, this equation provides a rough estimate of λ. This is practically zero for pyridine compounds, while it is appreciable for benzene derivatives, such as toluene and phenol.

2. Partial Charge Transfer in Terms of the Anderson-Newns Model

As stated above, pct is only defined within a model since there is no unique way of dividing the electronic density into parts pertaining to the adsorbate and to the substrate. The Anderson-Newns model (for reviews, see Muscat and Newns[62] and Gadzuk[63]) offers a good way of understanding the basis of this phenomenon, and it can also be used for semi-quantitative estimates.[64,65] Let us consider what happens when a simple atom or ion approaches a metal electrode. When the particle is in the bulk of the solution, its valence orbital has a well-defined energy. As it moves towards the surface, several effects occur. (i) The particle loses a part of its solvation sphere. For an ion, this implies that the energy of the valence orbital is raised, since solvation favors the ionic state. (ii) In the case of an ion, image forces come into play; these lower the energy of the valence orbital, thus offsetting the effect of the desolvation. (iii) As the particle gets very close to the surface, chemical interactions come into play. These typically lower the energy of the orbital, and broaden it at the same time. This energy broadening is a result of the interaction with the conduction band of the metal. It can also be thought of as a lifetime broadening: when an electron is placed on the adsorbate orbital it can go over to the metal. Hence it has a finite lifetime τ, and according to the Heisenberg uncertainty principle, this entails an energy uncertainty, or orbital broadening, of $\Delta = h/(2\pi\tau)$. Thus, the adsorbate orbital is characterized by a density of states $\rho(\varepsilon)$ of width Δ (see Fig. 10). This is filled up to the Fermi level by an electronic density which the adsorbate shares with the metal. Therefore, in general, the valence orbital is only partially filled, and hence there is a pct as the particle is adsorbed. The situation in Fig.10, for example, corresponds to the adsorption of a halide ion. When the ion is in the bulk of the solution, the orbital lies below the Fermi level and is filled. When the particle is adsorbed, the center of the density of states lies still below the Fermi level, but its tail extends above. Therefore, the orbital is only partially filled, as some negative charge has been transferred to the metal. In passing, we note that the lower adsorbate levels likewise mix with the metal states, and thus contribute to the density of states; however, since they are filled, they do not contribute to the pct, and have therefore not been considered here.

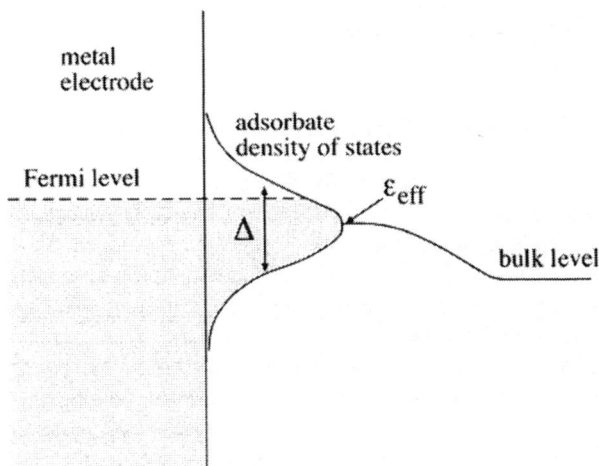

Figure 10. Shift and broadening of a valency level as it approaches
the electrode surface; this schematic curve is for the adsorption of
an anion.

These ideas can be cast into an Hamiltonian based on the
Anderson-Newns model and extended to the electrochemical situation
by one of us.[64] This Hamiltonian consists of several terms. We start
with the terms for the adsorbate orbital, which we label by a. This can
take up two electrons with opposite spins σ, which interact through a
Coulomb repulsion of magnitude U:

$$\varepsilon_a \sum_{\sigma} n_{a\sigma} + U n_{a,\sigma} n_{a,-\sigma} \tag{63}$$

where ε denotes an energy.

The electronic states on the metal are labeled by their
quasi-momentum k; they can exchange electrons with the adsorbate
orbital; the corresponding terms are:

$$\sum_{k,\sigma} \varepsilon_k n_{k,\sigma} + \sum_{k,\sigma} \left(V_k c_{k,\sigma}^+ c_{a,\sigma} + V_k^* c_{a,\sigma}^+ c_{k,\sigma} \right) \tag{64}$$

where V_k denotes the coupling between the metal and adsorbate states, c^+ denotes a creation and c an annihilation operator for the indicated states. The terms presented so far represent the Anderson-Newns model as it is used for adsorption *in vacuo*. In order to describe electrosorption from a solution, the interaction with the solvent must be introduced. The latter can be conveniently described as a bath of phonons, which couple linearly to the total charge:

$$\frac{1}{2}\sum_v \hbar\omega_v \left(p_v^2 + q_v^2\right) + \left(z - \sum_\sigma n_{a,\sigma}\right)\left(\sum_v \hbar\omega_v g_v q_v\right) \tag{65}$$

Here, p_v and q_v denote the dimensionless momenta and coordinates of the phonon mode with frequency ω_v, and z is the charge number of the ion when the valence orbital is empty. Thus, the first large bracket contains the total charge on the adsorbate. Finally, g_v denotes the coupling between adsorbate and solvent. Instead of using a phonon bath, the solvent can also be modeled as an ensemble of dipoles; within the linear response regime, this gives the same result.[66] From this Hamiltonian, two important quantities can be defined:

$$\Delta = \pi \sum_k |V_k|^2 \, \delta(\omega - \varepsilon_k) \qquad \eta = \frac{1}{2}\sum_v \hbar\omega_v g_v^2 \tag{66}$$

where Δ denotes the width that the adsorbate level acquires through its interaction with the metal, and η characterizes the interaction with the solvent, and has the same role as the concept of an energy of reorganization in the theory of electron-transfer reactions.

This Hamiltonian can be treated at various levels of sophistication. In the simplest approximation, the width Δ, which can be a function of the electronic energy ω, is taken as constant (wide–band approximation),[67] and the Coulomb interaction is treated at the restricted Hartree-Fock level, so that both spin states have the same occupation probability, $\langle n_{a,\sigma}\rangle = \langle n_{a,-\sigma}\rangle = \langle n\rangle$. In this case, the density of states of the adsorbate takes the form of a Lorenzian:

$$\rho(\omega) = \frac{1}{\pi}\frac{\Delta}{(\omega - \varepsilon_{eff})^2 + \Delta^2} \tag{67}$$

where the effective position of the adsorbate level is at:

$$\varepsilon_{eff} = \varepsilon_a + U\langle n\rangle + 2(z-2\langle n\rangle)\eta \qquad (68)$$

As expected, the Coulomb repulsion raises the energy of the adsorbate level. The interaction with the solvent is seen to favor the ionization of the adsorbate. Thus, when $z < 2\langle n\rangle$, the adsorbate is negatively charged, and the solvation further lowers the orbital and hence increases its negative charge. In contrast, when $z > 2\langle n\rangle$, the adsorbate is positively charged and the solvation effect raises the level further.

The occupation probability $\langle n_a\rangle$ has to be determined from the self-consistency relation:

$$\langle n\rangle = \int f(\omega)\,\rho(\omega)\,d\omega \qquad (69)$$

where $f(\omega)$ is the Fermi-Dirac distribution. Kornyshev and Schmickler[65] used this formalism to estimate the charge transfer coefficient in several simple systems. Naturally, a comparison with experimental values is not possible, but their work can be used to predict trends. In particular, for the adsorption of halides on mercury, it predicts that the absolute amount of charge transferred increases from fluoride to iodide, which is in line with experimental values of the electrosorption valency.[50]

The best modern methods for calculating charge transfer combine the Anderson-Newns model with quantum-chemical calculations, which yield the density of states in the absence of solvation, and with molecular dynamics simulations to obtain the interaction with the solvent. A good example is the work of Nazmutdinov and Spohr,[68] who have studied the adsorption of the iodide ion on a Pt(100) electrode from an aqueous solution. They have calculated the charge on the adsorbate both in the presence and in the absence of the solvent. In accord with the above considerations, they find that solvation favors charged adsorbates. Thus, the valence level of the adsorbed iodide ion is lowered by solvation, and the amount of electronic charge transferred from the ion to the electrode is reduced. In addition, these authors study the interaction of the ion with the metal and the solvent as a function of the distance from the electrode, and thus map out the potential-energy surface for the adsorption process. Unfortunately this elegant work, just like others in this field, is limited to the point of zero charge, since there is still no quantitative method to calculate the potential in the double-layer region other than at the potential of zero charge, where it is constant by definition.

3. Dipole Moment and Electrosorption Valency

Electrochemistry shares many concepts with surface science, and for the last two decades there has been an exchange of methods and ideas between these two neighboring disciplines. However, the electrosorption valency has no equivalent in surface science, since experiments at the solid/gas or solid/vacuum interface cannot be performed at constant potential. However, for low coverages, and near the potential of zero charge, the electrosorption valency can be related to the dipole moment of the adsorbate, which can be measured both in surface science and, though with greater difficulty, also in electrochemistry. In the following, we point out the relation between these two quantities.

(i) Adsorption from the Gas Phase

We begin by considering adsorption from the gas phase. In this case, the potential of the substrate is not a variable; the boundary condition is zero surface charge. When a species is adsorbed on the surface, any partial charge resident on the adsorbate is balanced by an image charge, so that the charge distribution results in a dipole moment, which can be measured by a change in the work function Φ. Let μ be the dipole moment consisting of a single adsorbate molecule with its image charge, and let N_S be its surface concentration in molecules per unit area. The change in the work function is then:

$$\Delta\Phi = -4\pi e N_S \mu \tag{70}$$

In the simplest interpretation, the dipole moment is caused by a partial charge q sitting at a distance d from the metal surface: $\mu = qd$. However, the image charge does not reside on the geometrical surface separating the substrate from the vacuum, but is spread out in front of it and may partially surround the adatom.[69] Therefore, it is difficult to infer the partial charge from the dipole moment – just as, in electrochemistry, the partial charge on an adsorbate cannot be measured.

At small coverages, the change $\Delta\Phi$ in the work function is proportional to the surface concentration. At higher coverages, the variation of Φ with N_S becomes less rapid; in some cases it may even pass through a maximum before reaching the value for a saturated monolayer. This apparent decrease of the dipole moment is caused by the Coulomb interaction between the adsorbed atoms, which makes large dipole moments unfavorable.[62]

(ii) Adsorption from an Electrolyte Solution

The dipole moment of an adsorbate can only be defined for an electrode at the potential of zero charge, where any excess charge residing on the adsorbate is balanced by an excess charge on the metal. A thought experiment for the definition and measurement of the dipole moment is the following:[49] before the adsorption, the electrode is kept at the potential of zero charge; in the elementary act, an ion is transferred from the bulk of the solution to the metal surface, while simultaneously the image charge flows onto the electrode. If the ion is partially or totally discharged, a fraction or all of the image charge is transferred to the ion. After the experiment, the total surface charge is zero; any excess charge residing on the adsorbate is balanced by the image charge. Hence there is no excess charge in the diffuse part of the double layer. The potential difference $\Delta\phi_M^S$ across the metal/solution interphase following this experiment, as measured relative to the potential of zero charge without adsorption, is equivalent to the change $\Delta\Phi$ in the work function during adsorption from the gas phase. Consequently, the surface dipole moment per adsorbed particle is given by:

$$\mu = -\Delta\phi_M^S \,/\!\left(4\,\pi N_S\right) \tag{71}$$

In the interpretation of the electrochemical dipole moment, it must be borne in mind that the adsorbate affects the surrounding solution. Thus, in the vicinity of an adsorbate with a dipole moment, the solvent dipoles will be oriented preferentially in the opposite direction, partially canceling the effect of the adsorbate dipole. The measurements give the total change in the surface dipole moment, containing the contributions from both the adsorbate and the solvent. Therefore, in electrochemistry it is even more difficult to deduce the partial charge on the adsorbate from the dipole moment than in surface science, and we may expect the dipole moments of electrochemical adsorbates to be substantially smaller. We shall return to this point below.

Real experiments do not have to be performed in the same way as the thought experiment described above. In the classical work of Grahame and Parsons[70] the electrode potential is plotted against the formal charge σ_S on the adsorbate layer (assuming that the ions have not been discharged) for various values of the total charge density σ_M on the metal. The electrode potential for $\sigma_S = -\sigma_M$, minus the potential of zero charge without adsorption, gives the desired electrode potential change $\Delta\phi_M^S$.

Unfortunately, this procedure sometimes requires an extrapolation of the experimental data that limits its accuracy. Schmickler and Guidelli[48] have gathered data for a number of systems. We shall report some of their data in the next section.

(iii) Relation to the Electrosorption Valency

The dipole moment is determined by the change of the potential at constant charge, while the electrosorption valency gives the change in the charge at constant potential. One might expect that the two quantities are related. This is indeed the case, but only at the potential of zero charge, since the dipole moment is not defined elsewhere. The relation was derived by Bange et al.[71] and by Schmickler.[49] For obvious reasons we follow the treatment of the latter author.

We consider a small amount of ions adsorbed at a potential close to the potential of zero charge. The deviation $\Delta\phi_M^S$ of the potential difference across the metal/solution interphase from its value at the potential of zero charge in the absence of adsorption depends on the total charge density σ_M on the metal and the adsorbed charge σ_S. In the linear response regime we may write:

$$\Delta\phi_M^S = A\sigma_M + B\sigma_S + (\sigma_M + \sigma_S)/C_{GC} \qquad (72)$$

Here C_{GC} denotes the Gouy-Chapman capacity, $A\sigma_M$ gives the first order deviation from the Gouy-Chapman theory in the absence of specific adsorption, and $B\sigma_S$ is the potential drop caused by the adsorbate layer. Note that this equation holds quite generally for small charge densities, and does not depend on any model.

Setting $\sigma_S = 0$ we obtain $A = 1/C_H$, where C_H is the Helmholtz or inner-layer capacity.[72] By definition, this is the first-order correction to the Gouy-Chapman theory. Setting $\sigma_S = -\sigma_M$ with $\sigma_S = zeN_S$, from Eqs. (71) and (72) we obtain for the dipole moment:

$$\mu = -ze(B - A)/(4\pi) \tag{73}$$

As mentioned before, the electrosorption valency is usually corrected for double-layer effects. By disregarding the Gouy-Chapman term in Eq. (72), the following expression for the electrosorption valency is obtained:

$$l \equiv -z\left(\frac{\partial \sigma_M}{\partial \sigma_S}\right)_{\Delta\phi_M^S} = z\frac{B}{A} \tag{74}$$

Substituting B from Eq. (74) into Eq. (73), the following relationship between the electrosorption valency and the dipole moment is obtained:

$$\mu = \frac{e}{4\pi C_H}(z - l) \tag{75}$$

which is valid for small amounts of adsorbate near the potential of zero charge. It follows that the dipole moment is zero if, and only if, $z = l$.

Table 1 summarizes a few values of the dipole moment μ and of the electrosorption valency l of halide and alkali metal ions adsorbed on mercury. The experimental values of l are relative to low coverages near the potential of zero charge and are taken from Schultze and Koppitz,[50] while the corresponding μ values were calculated from Eq. (75). The theoretical values in the last column are from a hard sphere electrolyte model. Further data can be found in the article by Schmickler [49] Note that, in the electrochemical environment, the dipole moments are much smaller than *in vacuo*, where they can reach values of the order of 7 D for the alkali metal ions. No doubt, this difference is caused by the screening of the adsorbate dipole by the solvent molecules.

Table 1
Electrosorption Valency l and
Dipole Moment μ of a Few
Simple Ions Adsorbed on
Mercury

Ion	l	μ / D	μ_{theor} / D
K^+	+0.16	+1.21	+0.72
Rb^+	+0.15	+1.22	+0.78
Cs^+	+0.2	+1.18	+0.86
Cl^-	-0.2	-1.15	-0.91
Br^-	-0.35	-0.95	-0.95
I^-	-0.45	-0.79	-1.01

(iv) Dipole Moments in the Hard-Sphere Electrolyte Model

The hard-sphere electrolyte model, presented in section II.4, can be used to estimate the dipole moment in the absence of charge transfer. The model can be improved by noting that, as mentioned before, the effective image plane of a metal usually sits at a distance x_{im} in front of the geometrical surface. From Eq. (73), in which B and A are obtained by comparing Eq. (72) with Eq. (50), we get:

$$\mu = ze\left[\frac{r_i}{\varepsilon} + \frac{r_i r_w (\varepsilon - 1)}{\varepsilon(r_w + \zeta r_i)} - x_{im}\right] \approx ze\left[\frac{r_w r_i}{r_w + \zeta r_i} - x_{im}\right] \quad (76)$$

Here, in the last step, terms of the order of $1/\varepsilon$ have been neglected. Once again this formula shows that a small solvent radius and a large dielectric constant reduce the dipole moment considerably. Thus, this model immediately explains why, in the electrochemical environment, the dipole moments are so much smaller than *in vacuo*. For mercury as a metal, the effective position x_{im} of the image plane can be estimated from the experimental Helmholtz capacity, and is about 0.28 Å from the geometrical surface. The last column in Table 1 gives estimates for the dipole moments based on this model. Considering the fact that pct is disregarded in this model, these estimates are satisfactory in that they give the right order of magnitude.

IV. ELECTROSORPTION VALENCY AND PARTIAL CHARGE TRANSFER COEFFICIENT IN SELF-ASSEMBLED THIOL MONOLAYERS

Self-assembled monolayers (SAMs) of alkanethiols and their derivatives on metal electrodes, especially gold and silver, have been extensively investigated during the last decade, and have provided a new class of materials with a wide range of applications (see Finklea[73] and references therein). Alkanethiol monolayers on gold are the most extensively studied group. Alkanethiol SAMs have a well-ordered structure and long-term stability, and are easily prepared. An in-depth understanding of the factors influencing the structure and stability of these monolayers is central to their application in electrochemistry and other areas. To gain insight into the interfacial processes leading to monolayer formation, the reductive desorption and the oxidative readsorption of alkanethiol SAMs have been investigated on gold polycrystalline and single-crystal electrodes[74-81] using cyclic voltammetry[67-70] and open-circuit potential measurements.[78] The influence of gold surface morphology, alkyl chain length and pH on the reductive removal and oxidative redeposition of thiol monolayers have been investigated and discussed in terms of compactness of the alkanethiol monolayer, solubility of the thiol molecules and intermolecular interactions between neighboring adsorbate molecules.

1. Self-assembled Alkanethiol Monolayers on Mercury

One important parameter of monolayers of alkanethiols, and more generally of thiols, is the amount of charge transferred from the sulfur atom of the sulphydryl group to the metal during their self-assembly. On the basis of the suggestion of Kolthoff[82] that thiols react very rapidly with mercury surfaces to form mercurous and mercuric adducts, it has often been taken for granted that the formation of thiol SAMs on mercury involves the total transfer of an electron from sulfur to mercury.[83-85] Thus, Muskal et al.,[84] by integrating the current under the reductive desorption peak of a n-decanethiol monolayer on Hg in acetonitrile, calculated a density of 9.2×10^{-10} mol cm^{-2} upon assuming total charge transfer; the differential capacitance of their alkanethiol films was anomalously high (3 to 6 μF cm^{-2}). Slovinski et al.,[83] by measuring the charge that accompanies the expansion of a mercury drop at constant applied potential, obtained a charge density of 80 ± 10

μC cm^{-2} from a n-octadecanethiol solution in methanol, and slightly lower values for shorter-chain alkanethiols. The above values should be regarded as true thermodynamic values of σ_M, since the procedure adopted is based on the thermodynamic definition of the total charge density σ_M. This procedure can only be applied in the presence of surfactant concentrations high enough to ensure an almost instantaneous adsorption equilibrium during drop expansion. Stevenson et al.[85] obtained three anodic voltammetric peaks on Hg from 0.5M NaOH aqueous solutions of n-alkanethiolates with 2 to 8 carbon atoms; the first, more negative peak was ascribed to oxidative physisorption of a thiolate submonolayer, the second to a further oxidative chemisorption leading to a compact monolayer, and the third to a bulk film of either mercurous or mercuric thiolate complex. On the basis of the negative shift in the midpoint between the peak potential of the more negative anodic peak and that of its cathodic counterpart by 59 mV per each unitary increment in the logarithm of the bulk thiolate concentration, these authors[85] concluded that one electron is transferred per thiolate physisorbed molecule, i.e. that total charge transfer takes place. The charge under the anodic adsorption peaks of long-chain alkanethiolates from aqueous 0.5M NaOH was reported to amount to ~ +95 μC cm^{-2} [85] and that under the cathodic desorption peak of n-decanethiol from acetonitrile was found to be equal to ~ -98 μC cm^{-2}.[84] Both charge values were not corrected for the capacitive contribution. If this can be reasonably considered to be of the order of 15-20 μC cm^{-2}, in both cases σ_M is close to 80 μC cm^{-2}.

The cross-sectional area of a linear hydrocarbon chain in close–packed monolayers of aliphatic compounds is estimated at about 0.20 nm^2.[83] This value agrees with that estimated from grazing-incidence X-ray diffraction and X-ray reflectivity measurements on n-alkanethiol SAMs on Hg.[86] Upon assuming the transfer of one electron per n-alkanethiol molecule in the monolayer, this cross sectional area corresponds to a total charge density σ_S of ~ +80 μC cm^{-2}. Therefore, experimental σ_M values close to ~ +80 μC cm^{-2} point to a close-packed monolayer of perpendicularly oriented n-alkanethiol molecules resulting from total charge transfer from the sulphydryl group to mercury. These values are normally reported with monolayers that are directly self-assembled from thiol solutions in organic solvents such as methanol[83] and acetonitrile,[84] or with thiolates dissolved in strongly alkaline aqueous solutions.[85] In these solutions the solubility of mercury adducts is not entirely negligible. Thus, the dissolution of mercury in organic solutions of thiols is significant, and can only be reduced in acidic media.[84] Strongly alkaline aqueous solutions may also favor a modest solubility of mercury adducts, as indicated by the formation of a massive deposit of mercury alkanethiolates from aqueous 0.5M NaOH at potentials as negative as -0.6 V versus a Ag/AgCl reference electrode.[85] Since such dissolution involves total charge transfer, σ_M values close to +80 μC cm^{-2} are neither surprising nor particularly revealing for an estimate of the pct coefficient λ. This is more conveniently determined under conditions in which any equilibrium between the surface compound resulting from an electron transfer from the adsorbate to the metal and the corresponding oxidized compound in the bulk solution can be excluded.

We shall, therefore, consider σ_M measurements on a mercury-supported n-octadecanethiol monolayer immersed in aqueous solutions of pH less than the $pK_a = 12$ of this thiol, where it is highly insoluble, after self-assembling the monolayer from an organic solvent. In this connection it is important to bear in mind that the only thermodynamically significant charge in adsorption processes involving partial or total charge transfer between the adsorbate and the electrode is the total charge σ_M, that is the charge to be supplied to the electrode to keep the applied potential E constant when the electrode surface is increased by unity and the composition of the bulk phases is kept constant. The method of choice to measure σ_M both on solid and liquid electrodes consists in starting with an uncoated electrode immersed in a solution of the supporting electrolyte alone and in stepping the applied potential from the potential of zero charge, E_z, to a final potential E_f negative enough to exclude the adsorption of the adsorbate under investigation, provided such a potential is experimentally accessible: the charge Q accompanying the potential step $E_z \rightarrow E_f$ is then just the common value, $\sigma_M(E_f)$, of the total charge density both in the presence and in the absence of the adsorbate. The charge accompanying a further potential step from any given initial potential E to E_f at an adsorbate-coated electrode, in the same supporting electrolyte, will be equal to $\sigma_M(E_f) - \sigma_M(E)$, thus allowing the estimate of the total charge density $\sigma_M(E)$ at E. Application of this procedure to a n-octadecanethiol monolayer on a mercury electrode immersed in aqueous 0.1M tetramethylammonium chloride, where this monolayer is completely desorbed at $E_f = -1.85$ V/SCE, yields a σ_M value of $\sim$ +56 μC cm^{-2} over the whole broad potential range of stability of the monolayer.[87] In fact, due to the very low differential capacity of the n-octadecanethiol monolayer, the changes in σ_M over this potential range lie within the limits of experimental accuracy.

The charge under the cathodic desorption peak of n–alkanethiols is not a precise measure of σ_M, since it includes the difference between the capacitive charge at a potential E_2 just negative of the desorption peak and that at a potential E_1 just positive of it. While the latter capacitive charge is very small, the former is not negligible. As a first approximation it can be identified with the charge density σ_M at E_2 on a bare mercury electrode, even though at potentials just negative of the desorption peak thiol desorption is not complete. In the case of a n-octadecanethiol monolayer on a mercury electrode immersed in aqueous 0.1M tetramethylammonium chloride, the charge under the desorption peak amounts to -70 μC cm^{-2}, while the σ_M value on bare mercury at a potential just negative of the peak equals -16 μC cm^{-2}.[87] This yields a total charge density of +54 μC cm^{-2} over the range of stability of the n-octadecanethiol monolayer, in excellent agreement with the value obtained by the more accurate potential-step chronocoulometric procedure (*vide supra*).

2. The Integral Electrosorption Valency of Metal-Supported Thiol Monolayers

Thiols are usually self-assembled on electrodes from solutions in which they are present in the neutral form. Upon self-assembly with pct, it is tacitly assumed that the sulphydryl groups that transfer a negative charge to the electrode undergo deprotonation, although this cannot be taken for granted. It is evident that, if such a deprotonation takes place, the movement of protons across the thiol monolayer of low differential capacity removes an appreciable positive potential difference across the monolayer. This must be compensated for by a flow of electrons away from the electrode surface along the external circuit in order to keep the applied potential constant. In other words, the magnitude of the total charge density σ_M depends to a large extent upon whether or not such a deprotonation takes place.

To better appreciate this point, let us express the potential difference $\Delta\phi_M^S = \phi^M - \phi^S$ across the interphase by taking into account both the pct coefficient λ and the degree of dissociation α of the sulphydryl groups of the thiols. In analogy with Eq. (57), $\Delta\phi_M^S$ can be approximately expressed as follows:[87]

$$\Delta\phi_M^S = 4\pi\left(\frac{\beta}{\varepsilon_\beta} + \frac{\gamma}{\varepsilon_\gamma}\right)\left(\sigma_M + \lambda\sigma_S\right)$$
$$+ \frac{4\pi\gamma}{\varepsilon_\gamma}\left[(1-\lambda)\sigma_S - (1-\alpha)\sigma_S\right] + \chi_e + \chi_S \tag{77}$$

Here, $\sigma_S = -F\Gamma_S$ is the charge density on the $-S^-$ groups of a hypothetical close-packed monolayer of thiolate anions, β is the distance of the center of charge of the sulfur atoms from the metal surface, $d = \beta + \gamma$ is the length of the thiol molecule, and ε_β and ε_γ are the dielectric constants accounting for the distortional polarization in the $(0 < x < \beta)$ and $(\beta < x < d)$ portions of the monolayer. Moreover, χ_e is the surface dipole potential due to electron spillover and χ_S is that due to any polar groups of the thiol molecule. For simplicity, the small potential difference ϕ_d across the diffuse layer is disregarded.

Three α values are particularly significant. If α equals zero, the transfer of a negative charge density $\lambda\sigma_S$ to the metal leaves a positive charge density $-\lambda\sigma_S$ in the sulfur atoms, and the diffuse-layer ions experience the total charge density σ_M. In the opposite case of $\alpha = 1$, the charge density in the sulfur atoms is negative and equal to $(1 - \lambda)\sigma_S$, and the diffuse-layer ions experience the charge $(\sigma_M + \sigma_S)$, as in the case of anionic adsorption. The most interesting case is represented by $\alpha = \lambda$; in this case the diffuse-layer ions experience a charge density $(\sigma_M + \alpha\sigma_S) = (\sigma_M + \lambda\sigma_S)$ and, therefore, monitor the pct.

The potential difference $\Delta\phi_M^S$ across the interface of a thiol-free, bare electrode in the absence of ionic specific adsorption at the same applied potential, $E = \Delta\phi_M^S + \text{constant}$, is likewise expressed by:

$$\Delta\phi_M^S = 4\pi\left(\frac{\beta}{\varepsilon_\beta} + \frac{\gamma}{\varepsilon_\gamma}\right)\sigma_M^0 + \chi_e + \chi_w \tag{78}$$

where σ_M^0 is the charge density on the bare electrode at E and χ_w is the surface dipole potential of the adsorbed water molecules. Even if the thiol monolayer is investigated in an aqueous electrolyte that is specifically adsorbed to some extent, the σ_M^0 value must refer to a truly bare electrode, namely an electrode immersed in an aqueous solution of a nonspecifically adsorbed electrolyte, such as NaF. This is because the thiol monolayer is expected to displace any electrolyte from direct contact with the electrode, independent of whether the electrolyte is specifically adsorbed or not in the absence of the thiol. Taking into account the definition of Eq. (52) for the integral electrosorption valency $\bar{l}$ and substituting σ_M and σ_M^0 from Eqs. (77) and (78), we obtain:

$$\bar{l} = -\left(\frac{\sigma_M - \sigma_M^0}{F\Gamma_S}\right)_E = \frac{\beta/\varepsilon_\beta}{\beta/\varepsilon_\beta + \gamma/\varepsilon_\gamma}(\alpha - \lambda) - \alpha$$
$$-\frac{\left(\Delta\phi_M^S - \chi_e - \chi_w\right)C_w}{\sigma_S} + \frac{\left(\Delta\phi_M^S - \chi_e - \chi_S\right)C_S}{\sigma_S} \tag{79}$$

Here:

$$C_w \equiv \left[4\pi\left(\frac{\beta}{\varepsilon_{\beta'}} + \frac{\gamma}{\varepsilon_{\gamma'}}\right)\right]^{-1} \quad ; \quad C_S \equiv \left[4\pi\left(\frac{\beta}{\varepsilon_\beta} + \frac{\gamma}{\varepsilon_\gamma}\right)\right]^{-1} \tag{80}$$

are the differential capacities of the bare and thiol-coated electrode. Even though the thiol molecules before adsorption are neutral, $F\Gamma_S$ has been set equal to $-\sigma_S$ by definition of the latter quantity. The expression of Eq. (80) for $\bar{l}$ is analogous to that of Eq. (58) for l.

If pct is not accompanied by deprotonation, α equals zero and, by momentarily neglecting the dipole terms in Eq. (79), we get:

$$\bar{l} = -\left(\frac{\beta/\varepsilon_\beta}{\beta/\varepsilon_\beta + \gamma/\varepsilon_\gamma}\right)\lambda \tag{81}$$

On the other hand, if the pct is assumed to be accompanied by deprotonation and by the movement of protons across the monolayer up to the bulk solution, α equals λ and $\bar{l}$ becomes:

$$\bar{l} = -\lambda \tag{82}$$

This assumption is indeed generally accepted. However, it is interesting to note that, since the β/d ratio is usually much less than unity, the movement of protons makes a major contribution to the positive charge density σ_M following thiol self-assembly.

3. Absolute Potential Difference between Mercury and an Aqueous Phase

To verify under which conditions the dipole terms in Eq. (79) can be disregarded, the extra-thermodynamic "absolute" potential difference $\Delta\phi_M^S$ must be estimated. On mercury an approximate estimate of ($\Delta\phi_M^S - \chi_e$) can be made by considering a mercury electrode coated with a self-assembled lipid monolayer having a neutral polar head, such as dioleoylphosphatidylcholine.[87] In this case the extrathermodynamic absolute potential difference $\Delta\phi_M^S$ across the whole mercury/solution interphase can be approximately expressed by the equation:

$$\Delta\phi_M^S = \chi_e + \chi_m + 4\pi\left(\frac{\beta}{\varepsilon_\beta} + \frac{\gamma}{\varepsilon_\gamma}\right)\sigma_M + \phi_d \tag{83}$$

Here β and γ are the thickness of the hydrocarbon tail and of the polar head region of the lipid monolayer, ε_β and ε_γ are the corresponding distortional dielectric constants, χ_e and χ_m are the surface dipole potentials due to the electron spillover and to the oriented polar heads, and ϕ_d is the potential difference across the diffuse layer. At ion concentrations that are not exceedingly low, ϕ_d can be disregarded as a good approximation. Moreover, the orientation of the polar heads of the lipid film is hardly affected by changes in σ_M. The differential capacity C of the electrode can, therefore, be written:

$$\frac{d\Delta\phi_M^S}{d\sigma_M} \equiv \frac{1}{C} = -4\pi x_{im} + 4\pi\left(\frac{\beta}{\varepsilon_\beta} + \frac{\gamma}{\varepsilon_\gamma}\right)$$
$$with: \frac{d\chi_e}{d\sigma_M} = -4\pi x_{im} ; \frac{d\chi_m}{d\sigma_M} \cong 0 \tag{84}$$

Here, the contribution $d\chi_e/d\sigma_M$ to the reciprocal of the interfacial capacity from the inhomogeneous electron gas is expressed on the basis of the jellium model by $-4\pi x_{im}$.[88] x_{im} is the "position of the effective image plane", i.e. the distance from the surface atoms of the metal at which the idealized, perfectly conducting metal surface plane should be placed to account for electron spillover; x_{im} is always positive, ranging from 0.3 to 1 Å, and hence makes a positive contribution to C. In the presence of a lipid monolayer, whose hydrocarbon tail has a length β of about 20 Å and a dielectric constant ε_β of about 2, this contribution is relatively small, and this is even more true for its possible changes with a change in σ_M. At any rate, this contribution is incorporated in the experimental value of the differential capacity C. Within the limits in which both the differential capacity C and the dipole potential χ_m due to the orientation of the polar heads at a given applied potential E can be regarded as independent of potential over the potential range from E to the potential of zero charge, from Eqs. (83) and (84) it follows that the absolute potential difference $\Delta\phi_M^S$ at E is given by:[89]

$$\Delta\phi_M^S - \chi_e = \sigma_M / C + \chi_m \tag{85}$$

In deriving this equation, ϕ_d and $4\pi x_{im}$ were disregarded on the basis of the previous considerations. As long as the metal is coated with a self-assembled film, the dipole potential χ_e due to the electron spillover can be regarded as constant *for the given metal* in view of its small rate of change, $-4\pi x_{im}$, with a change in the charge density σ_M on the metal. However, in passing from the bare metal in contact with an aqueous solution to the same metal coated with a self-assembled film, a certain change in χ_e can not be excluded. For the present purposes, the dipole potential χ_e for a given metal will be regarded as approximately constant with varying the solution side of the interphase. Within the limits of this approximation, the expression of $\Delta\phi_M^S - \chi_e$ in Eq. (85) can be regarded as an extrathermodynamic absolute potential difference across the whole interphase between the given metal and the bulk aqueous phase; it will be denoted by ψ. Both σ_M and C in Eq. (85) are experimentally accessible and thermodynamically significant. If χ_m can be determined independently on the basis of extrathermodynamic arguments, then the extrathermodynamic absolute potential value of $\psi(E)$ at the given applied potential E is obtained. Once $\psi(E)$ is known, it holds independent of the particular nature of the interphase interposed between the metal and the bulk aqueous phase. Moreover, the absolute value of the potential $\psi(E')$ at a different applied potential, E', differs from that, $\psi(E)$, at E by the difference, $(E' - E)$, between the two applied potentials relative to the same reference electrode.

Let us consider a bare mercury electrode in direct contact with an aqueous solution of 0.1 M tetramethylammonium chloride (TMACl) at the potential of zero charge, which equals –0.450 V/SCE.[90] If this electrode is coated with a self-assembled dioleoylphosphatidylcholine (DOPC) monolayer, whose polar heads are directed toward the aqueous phase, the charge density, σ_M, passes from the zero value to –0.75 μC cm^{-2}.[87] The latter σ_M value was determined by chronocoulometric potential-step measurements carried out as already described. It was also determined by an alternative procedure consisting in measuring the charge following a small contraction of the DOPC-coated mercury drop at –0.450 V/SCE, while keeping its neck in contact with the lipid reservoir on the surface of the aqueous electrolyte:[91] dividing this charge by the corresponding decrease in area yields σ_M. When applying Eq. (85) to bare mercury at $\sigma_M = 0$, the surface dipole potential is that, χ_w, due to the preferential orientation of the water molecules in direct contact with the bare mercury surface at the potential of zero charge. On the other hand, when applying this equation to DOPC-coated mercury at –0.450 V, we must consider that the differential capacity C over a potential range of about 600 mV, straddling –0.450 V, is practically constant and equal to 1.8 μF cm^{-2}. Moreover, the dipole potential χ_m due to the oriented polar heads of a DOPC monolayer, as estimated by different procedures both on black lipid membranes (BLMs)[92-95] and on a mercury electrode,[96] has been reported to assume values ranging from +150 and +250 mV, positive toward the hydrocarbon tails. Upon assigning to χ_m an average value of +200 mV, application of Eq. (85) to both bare and DOPC-coated mercury yields:

$$\psi(-0.450V \, / \, SCE) = \chi_w = \frac{\sigma_M}{C} + \chi_m \qquad (86)$$

$$= \frac{-0.75\,\mu C\,cm^{-2}}{1.8\,\mu F\,cm^{-2}} + (0.200 \pm 0.050)V \cong -(200 \pm 50)mV$$

It follows that the absolute potential difference ψ between mercury and the aqueous phase can be obtained by increasing the applied potential

E, measured vs. the SCE, by about 250 mV. From Eq. (86) it also follows that the surface dipole potential χ_w due to the water molecules in direct contact with bare mercury at the potential of zero charge is equal to -200 ± 50 mV. This value is somewhat more negative than that, -70 mV, estimated by Trasatti[55] from the shift, $\Delta E_{\sigma_M = 0}$, in the potential of zero charge caused by the adsorption of a number of aliphatic alcohols on Hg from aqueous solutions. However, the latter value relies on a model-dependent extrapolation of $\Delta E_{\sigma_M = 0}$ to full electrode coverage by the alcohol and on a scarcely accurate value of the surface dipole potential of water at the water/air interphase. In this respect, the χ_w value estimated herein seems more reliable. In fact, the electrode coverage by the self-assembled DOPC monolayer is certainly complete, as indicated by the differential capacity of this lipid film being practically twice that of a solvent-free BLM; moreover, its surface dipole potential is known with a fairly good accuracy.

This ψ scale was confirmed by adsorbing purple membrane (PM) fragments isolated from *Halobacterium Salinarium* on a mercury electrode coated with a DOPC monolayer.[97] Purple membrane fragments contain the light-driven proton pump bacteriorhodopsin, which pumps protons from the intracellular to the extracellular side of the membrane. Illumination with green light causes a capacitive quasi-stationary photocurrent that increases linearly from +0.030 to -0.200 V/SCE with a negative shift in the applied potential, attaining a zero value at ~ +10 mV/SCE. In view of the previous estimate of the absolute potential difference ψ across the mercury/water interphase, the potential of +10 mV/SCE at which the stationary photocurrent vanishes corresponds to a ψ value of (+250+10) Mv = 260 mV. This value compares favorably with the transmembrane potential to be applied to an oocyte plasma membrane incorporating bacteriorhodopsin in order to annihilate its stationary photocurrent, which amounts to +250 mV.[98]

4. Electrosorption Valency of Three Mercury-Supported Thiol Monolayers

After making an approximate estimate of the absolute potential difference ψ across the mercury/aqueous solution interphase, let us determine the electrosorption valency of three mercury-supported thiol monolayers for which σ_M was measured by potential-step chronocoulometry. As already reported in Sec. IV.1, *n*-octadecanthiol self-assembled monolayer on mercury has a σ_M value of +56 μC cm^{-2} at –0.450 V/SCE, where the supporting electrolyte has a zero charge density σ_M^0.[87] Assigning a calculated value of –80 μC cm^{-2} to $\sigma_S = -F\Gamma_S$, $\bar{l}$ equals (56/80) = 0.7 according to Eq. (79). In this case, the contribution of the dipole terms is entirely negligible. In fact, the value of the largest of these two terms, $-(\psi - \chi_w)C_w/\sigma_S$, as obtained by setting $\psi \cong$ (–0.450 + 0.250) V, $C_w \cong 21$ μF cm^{-2} and $\chi_w \cong -0.200$ V, is almost zero. If we assume that $\alpha = \lambda$, the $\bar{l}$ value of 0.70 equals $-\lambda$ according to Eq. (79). Unfortunately, we cannot confirm this reasonable assumption because the charge experienced by the diffuse-layer ions cannot be determined to an acceptable degree of accuracy by evaluating the change in the diffuse-layer potential ϕ_d as the supporting electrolyte concentration is varied. In fact, the lowest values attainable by the diffuse-layer capacity, C_d, using very low electrolyte concentrations, are still much higher than the capacity, $C_S = 0.70$ μF cm^{-2}, of the n-octadecanethiol monolayer, which is in series with C_d.

The situation is different with hydrophilic thiols forming self-assembled monolayers tethered to mercury. These thiols are often used to separate the metal surface from a lipid bilayer, self-assembled on top of them. In this way, the lipid bilayer is in contact with a hydrophilic region on the metal side and with an aqueous solution on the opposite side, and may conveniently incorporate ion channels and integral proteins in a functionally active state.[99-113] The monolayers of these hydrophilic "spacers" have a differential capacity C_S close to 10 μF cm^{-2}, which allows any changes in the diffuse-layer capacity C_d in series with it to be detected by impedance spectroscopy. As an example, Fig. 11 shows plots of the logarithm of the magnitude $|Z|$ of the electrode impedance versus the logarithm of the frequency f (Bode plot) and of the phase angle vs. log f at –1.000 V/SCE for a mercury-supported monolayer of a thiol-hexapeptide in contact with different concentrations of aqueous KCl.[114] The solid curves are least-squares fits to the simple equivalent circuit illustrated in inset (1) of the same figure: it consists of the electrolyte resistance R_S, with in series a $R_S C_S$ mesh representing the self-assembled thiol-hexapeptide monolayer and a further $R_d C_d$ mesh representing the diffuse layer. The fitting was carried out by requiring the R_s and C_S circuit elements to remain constant with varying the electrolyte concentration. The resulting C_S value equals 11 μF cm^{-2}, while R_s amounts to 0.14 MΩ cm^2. Inset (2) of Fig. 11 shows the reciprocal, $1/C_d$, of the experimental

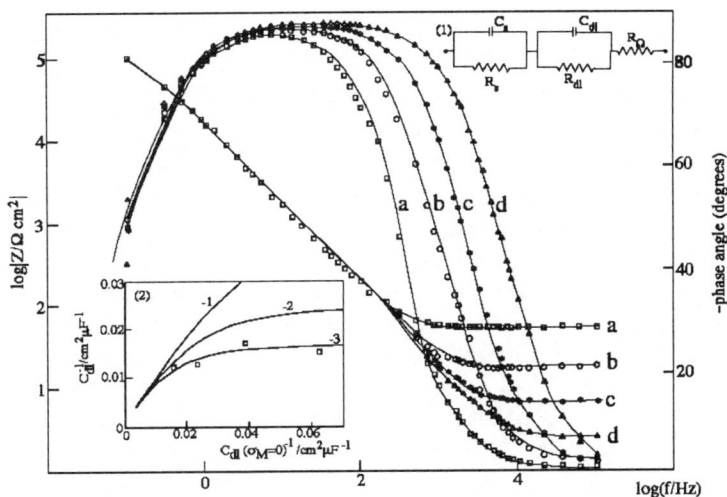

Figure 11. Plots of $\log|Z|$ and f vs. $\log f$ for a thiol-hexapeptide-coated mercury drop immersed in 5×10^{-3}M (**a**), 1.3×10^{-2}M (**b**), 3.6×10^{-2}M (**c**), and 0.1M (**d**) KCl, as obtained at -1.000 V over the frequency range from 0.1 to 10^5 Hz. At frequencies $<10^2$ Hz all Bode plots coincide; hence, only the experimental points for the lower KCl concentration were reported. The solid curves are least-squares fits to the simple equivalent circuit of inset (**1**), which consists of the electrolyte resistance R_Ω, with in series a R_sC_s mesh representing the self-assembled monolayer and a further $R_{dl}C_{dl}$ mesh representing the diffuse layer. $R_s = 0.14$ MΩ cm^2; $C_s = 11$ μF cm^{-2}; $R_{dl} = 4.53$ (**a**), 4.17 (**b**), 1.27 (**c**) and 0.87 KΩ cm^2 (**d**). $C_{dl} = 68$ (**a**), 61 (**b**), 80 (**c**) and 84 μF cm^{-2} (**d**). Inset (**2**) shows the reciprocal, $1/C_{dl}$, of the experimental diffuse-layer capacitance vs. the $1/C_{dl}(\sigma_M = 0)$ value corresponding to the same KCl concentration, as calculated on the basis of the Gouy-Chapman (GC) theory. The solid curves are $1/C_{dl}(\sigma_M)$ vs. $1/C_{dl}(\sigma_M = 0)$ plots calculated from the GC theory for different charge densities σ_M on the metal, whose values are reported on each curve. (Reprinted from Ref.[114] with permission from the Am. Chem. Soc.)

diffuse-layer capacitance against the $1/C_d(\sigma_M = 0)$ value corresponding to the same KCl concentration, as calculated on the basis of the Gouy-Chapman (GC) theory. The inset also shows $1/C_d(\sigma_M)$ vs. $1/C_d(\sigma_M = 0)$ plots calculated from the GC theory for different values of the charge density σ_M on the metal. The experimental points are in fairly good agreement with the $1/C_{dl}(\sigma_M)$ vs. $1/C_{dl}(\sigma_M=0)$ plot corresponding to a charge density of -3 μC cm^{-2} experienced by the diffuse layer ions. Note that this charge density is not the total charge density σ_M; in particular, if $\lambda = \alpha$, it equals the free charge density $q_M = \sigma_M + \lambda \sigma_S$, as pointed out in Section IV.2.

The total charge density σ_M on thiol-hexapeptide-coated mercury in aqueous 0.1 M KCl amounts to +17 μC cm^{-2}, while that, σ_M^0, on bare mercury at -1.000 V/SCE in a 0.1 M solution of the nonspecifically adsorbed NaF is about equal to -10 μC cm^{-2}. The σ_S value calculated from a space-filling model is about equal to -20 μC cm^{-2}. Thus, the $\bar{l}$ value amounts to $(-17-10)/20 = -1.35$. The $|\bar{l}|$ value being greater than unity is not surprising when we consider that in this case the dipole terms cannot be neglected. At -1.000 V/SCE the absolute potential difference ψ across the interphase amounts to $\sim(-1.000 + 0.250)$ V$= -0.750$ V. At this applied potential χ_w, which is about equal to -200 mV at the potential of zero charge and decreases in absolute value with a negative shift in potential, is negligible with respect to ψ, while C_w equals 16.3 μF cm^{-2}. Consequently, the water dipole contribution to $\bar{l}$, $-(\psi - \chi_w)C_w/\sigma_S$, equals ~ -0.61 (see Eq. 79). To estimate the contribution to $\bar{l}$ from the thiol dipole, the dipole potential χ_S of the thiol must first be determined. This can be obtained by differentiating Eq. (77) with respect to σ_M under the assumption that $\alpha \approx \lambda$ and that the dipole potential χ_S and the pct coefficient λ are not appreciably affected by a change in σ_M:

$$\frac{d\Delta\phi_M^S}{d\sigma_M} = \frac{1}{C_S} = 4\pi\left(\frac{\beta}{\varepsilon_\beta} + \frac{\gamma}{\varepsilon_\gamma}\right) - 4\pi x_{im} \approx 4\pi\left(\frac{\beta}{\varepsilon_\beta} + \frac{\gamma}{\varepsilon_\gamma}\right) \quad (87)$$

Combining Eqs. (77) and (87) yields:

$$\Delta\phi_M^S - \chi_e \equiv \psi = \frac{\sigma_M + \lambda\sigma_S}{C_S} + \chi_S \quad (88)$$

Noting that $\psi = -0.750$ V, $\sigma_M + \lambda\sigma_S = -3.0$ μC cm^{-2}, and $C_S = 11$ μF cm^{-2}, from Eq. (88) it follows that $\chi_S = -0.450$ V. The contribution to $\bar{l}$ from the thiol dipole, $(\psi - \chi_S)C_S/\sigma_S$, is therefore equal to +0.165. If we subtract the sum of the above two dipole contributions from the $\bar{l}$ value and assume that $\alpha = \lambda$, we obtain a pct coefficient λ close to unity in view of Eq. (79). In this case, the assumption of $\alpha = \lambda$ is confirmed by the independent measurement of the charge density,

–3 μC cm^{-2}, experienced by the diffuse-layer ions. With this assumption, this charge density is equal to $\sigma_M + \lambda\sigma_S = (17 - 20\ \lambda)$ μC cm^{-2}, from which it follows that λ equals +1. Note that the latter is a direct measurement of λ.

Another hydrophilic thiol used as a spacer anchored to mercury is triethyleneoxythiol.[115] The total charge density σ_M of triethyleneoxythiol-coated mercury, measured from 0.1 M KCl at –0.780 V by potential-step chronocoulometry, equals ~ +13 μC cm^{-2}. From the cross-sectional area of the triethyleneoxythiol molecule in its most extended configuration a σ_S value of –70 μC cm^{-2} is obtained. Noting that the charge density σ_M^0 on bare mercury from 0.1 M NaF at –0.780 V/SCE equals –7 μC cm^{-2}, Eq. (79) yields a value of –0.28 for the mean electrosorption valency $\bar{l}$. The charge density experienced by the diffuse-layer ions, measured by the same procedure as in Fig. 11, amounts to –3.5 μC cm^{-2}. Upon assuming that $\alpha = \lambda$, this charge equals $\sigma_M + \lambda\sigma_S = (13 - 70\ \lambda)$ μC cm^{-2}, from which it follows that λ equals +0.24. The slight difference between λ and $-\bar{l}$ is to be ascribed to the dipole terms in Eq. (79), which cannot be neglected. Noting that ψ equals ~(–0.785 + 0.250) V=-0.535 V, that χ_w is negligible with respect to ψ at this negative applied potential, and that C_w is about equal to 17 μF cm^{-2}, the water-dipole term, $-(\psi - \chi_w)C_w/\sigma_S$, equals –0.13. Noting that $\psi = -0.535$ V, $\sigma_M + \lambda\sigma_S = -3.5$ μC cm^{-2}, and $C_S = 11$ μF cm^{-2}, from Eq. (88) it follows that $\chi_S = -0.200$ V. The contribution to $\bar{l}$ from the thiol dipole, $(\psi - \chi_S)C_S/\sigma_S$, is therefore equal to +0.053. Subtracting the two dipole terms to $\bar{l}$ yields a λ value of 0.21, to be compared with the 0.24 value obtained from the charge density experienced by diffuse-layer ions. The fairly good agreement between the λ value obtained from the electrosorption valency and that obtained from diffuse-layer effects under the assumption that $\alpha = \lambda$ excludes the possibility that the low $\bar{l}$ value may be ascribed to $\alpha < \lambda$, namely to a lack of coupling between deprotonation and pct.

V. CONCLUSIONS

It has often been stated that it is a unique advantage of electrochemical reactions that they are accompanied by the flow of a current. In the cases of electron-transfer reactions and of metal deposition and dissolution, the interpretation of this current is straightforward, and it can be used to determine the reaction rate. For an adsorption process, however, the situation is more complicated: although the charge that flows in the elementary act can be measured, its *distribution* at the interface remains unknown. Thus, while the electrosorption valency is a useful and well-defined characteristic of an adsorption process, its interpretation is difficult since pct and double-layer screening are intricately mixed. It was the purpose of this review to present the foundations of this concept, which rivals "underpotential deposition" in the competition for the ugliest name of all electrochemical quantities, and to offer some ideas for its interpretation.

REFERENCES

[1] W. Lorenz and G. Salié, *Z. Phys. Chem.* **218** (1961) 259.

[2] W. Lorenz, *Z. Phys. Chem.* **218** (1961) 272.

[3] W. Lorenz and G. Krüger, *Z. Phys. Chem.* **221** (1962) 231.

[4] K. J. Vetter and J. W. Schultze, *Ber. Bunsenges. Physik. Chem.* **76** (1972) 920.

[5] K. J. Vetter and J. W. Schultze, *Ber. Bunsenges. Physik. Chem.* **76** (1972) 927.

[6] J. W. Schultze and K. J. Vetter, *J. Electroanal. Chem.* **44** (1973) 63.

[7] M. L. Foresti, M. Innocenti, F. Forni and R. Guidelli, *Langmuir* **14** (1998) 7008.

[8] B.E. Conway, *Electrochim. Acta* **40** (1995) 1501.

[9] A.N. Frumkin, O. Petry and B. Damaskin, *J. Electroanal. Chem.* **27** (1970) 81.

[10] W. Lorenz and G. Salié, *J. Electroanal. Chem.* **80** (1977) 1.

[11] P. Delahay, *J. Phys. Chem.* **70** (1966) 2067; 2373.

[12] K. Holub, G. Tessari and P. Delahay, *J. Phys. Chem.* **71** (1967) 2612.

[13] Z. Shi and J. Lipkowski, *J. Electroanal. Chem.* **403** (1996) 225.

[14] Z. Shi, J. Lipkowski, S. Mirwald and B. Pettinger, *J. Chem. Soc., Faraday Trans.* **92** (1996) 3737.

[15] B.E. Conway and H. Angerstein-Kozlowska, *J. Electroanal. Chem.* **113** (1980) 63.

[16] V. E. Schmidt and S. Stucki, *Ber. Bunsenges. Phys. Chem.* **77** (1973) 913.

[17] V. E. Schmidt and S. Stucki, *J. Electroanal. Chem.* **43** (1973) 425.

[18]B. M. Jovic, V. D. Jovic and D. M. Drazic, *J. Electroanal. Chem.* **399** (1995) 197.

[19]G. Valette, A. Hamelin and R. Parsons, *Z. Phys. Chem. N.F.* **113** (1978) 71.

[20]E. Dutkiewicz and R. Parsons, *J. Electroanal. Chem.* **11** (1966) 100.

[21]R. Payne, *Trans. Faraday Soc.* **64** (1968) 1638.

[22]A. R. Sears and F. Anson, *J. Electroanal. Chem.* **47** (1973) 521.

[23]W. Schmickler, D. Henderson and H. D. Hurwitz, *Z. Phys. Chem. N.F.* **160** (1988) 191.

[24]Z. Shi, J. Lipkowski, S. Mirwald and B. Pettinger, *J. Electroanal. Chem.* **396** (1995) 115.

[25]J. Lawrence, R. Parsons and R. Payne, *J. Electroanal. Chem.* **16** (1968) 193.

[26]C. V. D'Alkaine, E. R. Gonzalez and R. Parsons, *J. Electroanal. Chem.* **32** (1971) 57.

[27]G. J. Hills and R. M. Reeves, *J. Electroanal. Chem.* **42** (1973) 355.

[28]R. Guidelli and W. Schmickler, *Electrochim. Acta* **45** (2000) 2317.

[29]W. Schmickler, *Electrical Double-Layers: Theory and Simulation*, in *Encyclopedia of Electrochemistry*, Edited by A.J. Bard and M. Stratmann, Wiley-VCH, New York, 2002, Vol. I.

[30]S. L. Carnie and D. Y. C. Chan, *J. Chem. Phys.* **73** (1980) 2949.

[31]D. C. Grahame, *J. Amer. Chem. Soc.* **80** (1958) 4201.

[32]M. R. Moncelli, M. L. Foresti and R. Guidelli, *J. Electroanal. Chem.* **295** (1990) 225.

[33]M. R. Moncelli and R. Guidelli, *J. Electroanal. Chem.* **295** (1990) 239.

[34]J. Lipkowski and L. Stolberg in: *Adsorption of Molecules at Metal Electrodes*, J. Lipkowski and P.N. Ross, eds., p. 171, VCH, New York, 1992.

[35]M. L. Foresti, M. Innocenti and R. Guidelli, *J. Electroanal. Chem.* **376** (1994) 85.

[36]P. Beran and S. Bruckenstein, *Anal. Chem.* **40** (1968) 1044.

[37]G. Valette, *J. Electroanal. Chem.* **122** (1981) 285.

[38]J. W. Schultze, *Ber. Bunsenges. Phys. Chem.* **74** (1970) 705.

[39]H. Bort, K. Jüttner, W. J. Lorenz and E. Schmidt, *J. Electroanal. Chem.* **90** (1978) 413.

[40]K. Jüttner and W. J. Lorenz, *Z. Phys. Chem. N.F.* **122** (1980) 163.

[41]H. Siegenthaler, K. Jüttner, E. Schmidt and W. Lorenz, *J. Electrochim. Acta* **23** (1978) 1009.

368   R. Guidelli and W. Schmickler

[42]E. Schmidt, P. Beutler and W. J. Lorenz, *Ber. Bunsenges. Phys.Chem.* **75** (1971) 71.
[43] E. Schmidt and N. Wüthrich, *J. Electroanal Chem.* **28** (1970) 349.
[44]G. Salié and W. Lorenz, *Ber. Bunsenges. Phys. Chem.* **68** (1964) 197
[45]G. Salié, *J. Electroanal. Chem.* **447** (1998) 211.
[46]M. J. Walters, J. E. Garland, C. M. Pettit, D. S. Zimmerman, D. R. Marr and D. Roy, *J. Electroanal. Chem.* **499** (2001) 48.
[47]N. D. Lang, and A. R. Williams, *Phys. Rev. B* **18** (1978) 616.
[48]W. Schmickler and R. Guidelli, *J. Electroanal. Chem.* **235** (1987) 387.
[49]W. Schmickler, *J. Electroanal. Chem.* **249** (1988) 25.
[50]J. W. Schultze and F. D. Koppitz, *Electrochim. Acta* **21** (1976) 327.
[51]L. Pauling, *J. Am. Chem. Soc.* **54** (1932) 3570.
[52]F. D. Koppitz and J. W. Schultze, *Electrochim. Acta* **21** (1976) 337.
[53]F. D. Koppitz and J. W. Schultze, *Croatica Chem. Acta* **48** (1976) 643.
[54]F. D. Koppitz, J. W. Schultze and D. Rolle, *J. Electroanal. Chem.* **170** (1984) 5.
[55]S. Trasatti, in *Modern Aspects of Electrochemistry*, Edited by B. E. Conway and J. O'M. Bockris, Plenum Press, New York, 1979, Vol. 13, p.81.
[56]B.E. Conway, R.G. Barradas, P.G. Hamilton and J.M. Parry, *J. Electroanal. Chem.* **10** (1965) 485.
[57]B.E. Conway, J.G. Mathieson and H.P. Dhar, *J. Phys. Chem.* **78** (1974) 1226.
[58]Cl. Buess-Herman, L. Gierst and N. Vanlaethem-Meuree, *J. Electroanal. Chem.* **123** (1981) 1.
[59]Cl. Buess-Herman, N. Vanlaethem-Meuree, G. Quarin and L. Gierst, *J. Electroanal. Chem.* **123** (1981) 21.
[60]D. Rolle and J. W. Schultze, *Electrochim. Acta* **31** (1986) 991.
[61]D. Rolle and J. W. Schultze, *J. Electroanal. Chem.* **229** (1987) 141.
[62]J. P. Muscat and D. N. Newns, *J. Phys. C* **7** (1974) 2630.
[63]J. W. Gadzuk, in *Surface Physics of Crystalline Materials*, Edited by J. M. Blakely, Academic Press, New York, 1975.
[64]W. Schmickler, *J. Electroanal. Chem.* **100** (1979) 277.
[65]A. Kornyshev and W. Schmickler, *J. Electroanal. Chem.* **185** (1985) 253.
[66]W. Schmickler, *Ber. Bunsenges. Phys. Chem.* **80** (1976) 834.
[67]R. Brako and D. M. Newns, *Prog. Phys.* **2** (1989) 655.
[68]R. Nazmutdinov and E. Spohr, *J. Phys. Chem.* **98** (1994) 5956.

[69]N. D. Lang, *Surf. Sci.* **127** (1983) L118.

[70]D. C. Grahame and R. Parsons, *J. Am. Chem. Soc.* **83** (1961) 1291.

[71]K. Bange, B. Strachler, J. K. Sass, and R. Parsons, *J. Electroanal. Chem.* **229** (1987) 87.

[72]W. Schmickler, *Interfacial Electrochemistry*, Oxford University Press, New York, 1996.

[73]H. O. Finklea, in *Electroanalytical Chemistry*, Edited by A. J. Bard and I. Rubistein, Marcel Dekker, New York, 1996, Vol. 19, p. 109.

[74]C.-J. Zhong and M. D. Porter, *J. Electroanal. Chem.* **425** (1997) 147.

[75]C.-J. Zhong, N. T. Woods, G. B. Dawson and M. D. Porter, *Electrochemistry Communications* **1** (1999) 17.

[76]D.-F. Yang, C. P. Wilde and M. Morin, *Langmuir* **12** (1996) 6570.

[77]D.-F. Yang, C. P. Wilde and M. Morin, *Langmuir* **13** (1997) 243.

[78]P. Krysinski, R. V. Chamberlain II and M. Maida, *Langmuir* **10** (1994) 4286.

[79]F. P. Zamborini and R. M. Crooks, *Langmuir* **13** (1997) 122.

[80]L. Sun and R. M. Crooks, *Langmuir* **9** (1993) 1951.

[81]T. W. Schneider and D. A. Buttry, *J. Am. Chem. Soc.* **115** (1993) 12391.

[82]I. M. Kolthoff, W. Stricks and N. Tanaka, *J. Am. Chem. Soc.* **77** (1955) 5211.

[83]K. Slowinski, R. V. Chamberlain, C. J. Miller and M. Majda, *J. Am. Chem. Soc.* **119** (1997) 11910.

[84]N. Muskal, I. Turyan and D. Mandler, *J. Electroanal. Chem.* **409** (1996) 131.

[85]K. J. Stevenson, M. Mitchell and H. S. White, *J. Phys. Chem. B* **102** (1998) 1235.

[86]O. M. Magnussen, B. M. Ocko, M. Deutsch, M. J. Regan, P. S. Pershan, D. Abernathy, G. Grübel and J.-F. Legrand, *Nature* **384** (1996) 250.

[87]F. Tadini Buoninsegni, L. Becucci, M. R. Moncelli and R. Guidelli, *J. Electroanal.Chem.* **500** (2001) 395.

[88]W. Schmickler in: *Structure of Electrified Interfaces*, J. Lipkowski and P.N. Ross, eds., p. 201, VCH, New York, 1993.

[89]L. Becucci, M. R. Moncelli and R. Guidelli, *Langmuir* **19** (2003) 3386.

[90]F. M. Kimmerle and H. Ménard, *J. Electroanal. Chem.* **54** (1974) 101.

[91]L. Becucci, M. R. Moncelli and R. Guidelli, *J. Electroanal. Chem.* **413** (1996) 187.

[92]A. D. Pickar and R. Benz, *J. Membrane Biol.* **44** (1978) 353.

[93]R. F. Flewelling and W. L Hubbell, *Biophys. J.* **49** (1986) 541.

[94]K. Gawrisch, D. Ruston, J. Zimmerberg, V. A. Parsegian, R. P. Rand and N. Fuller, *Biophys. J.* **61** (1992) 1213.

[95]R. J. Clarke, *Biochim. Biophys. Acta* **1327** (1997) 269.

[96]L. Becucci, M. R. Moncelli, R. Herrero and R. Guidelli, *Langmuir* **16** (2000) 7694.

[97]A. Dolfi, G. Aloisi and R. Guidelli, *Bioelectrochemistry* **57** (2002) 155.

[98]G. Nagel, B. Kelety, B. Möckel, G. Büldt and E. Bamberg, *Biophys. J.* **74** (1998) 403.

[99]R. Guidelli, G. Aloisi, L. Becucci, A. Dolfi, M. R. Moncelli and F. Tadini Buoninsegni, *J. Electroanal. Chem.* **504** (2001) 1.

[100]C. Duschl, M. Liley, G. Corradin and H. Vogel, *Biophys. J.* **67** (1994) 1229.

[101]B. A. Cornell, V. L. B. Braach-Maksvytis, L. G. King, P. D. J. Osman, B. Raguse, L. Wieczorek and R. J. Pace, *Nature* **387** (1997) 580.

[102]L. M. Williams, S. D. Evans, T. M. Flynn, A. Marsh, P. F. Knowles, R. J. Bushby and N. Boden, *Langmuir* **13** (1997) 751.

[103]C. Steinem, A. Janshoff, W. P. Ulrich, M. Sieber and H.-J. Galla, *Biochim. Biophys. Acta* **1279** (1996) 169.

[104]C. Steinem, A. Janshoff, K. von dem Bruch, K. Reihs, J. Goossens and H.-J. Galla,*Bioelectrochem. Bioenerg.* **45** (1998) 17.

[105]B. Raguse, V. Braach-Maksvytis, B. A. Cornell, L. G. King, P. D. J. Osman, R. J. Pace and L. Wieczorek, *Langmuir* **14** (1998) 648.

[106]S. Heyse, O. P. Ernst, Z. Dienes, K. P. Hofmann and H. Vogel, *Biochemistry* **37** (1998) 507.

[107]S. Heyse, T. Stora, E. Schmid, J. H. Lakey and H. Vogel, *Biochim. Biophys. Acta* **85507** (1998) 319.

[108]A. Toby, A. Jenkins, R. J. Bushby, N. Boden, S. D. Evans, P. F. Knowles, Q. Liu, R. E. Miles and S. D. Ogier, *Langmuir* **14** (1998) 4675.

[109]R. Naumann, A. Jonczyk, R. Kopp, J. van Esch, H. Ringsdorf, W. Knoll and P. Gräber, *Angew. Chem. Int. Ed. Engl.* **34** (1995) 2056.

[110]N. Bunjes, E. K. Schmidt, A. Jonczyk, F. Rippmann, D. Beyer, H. Ringsdorf, P. Gräber, W. Knoll and R. Naumann, *Langmuir* **13** (1997) 6188.

[111]R. Naumann, A. Jonczyk, C. Hampel, H. Ringsdorf, W. Knoll, N. Bunjes and P. Gräber, *Bioelectrochem. Bioenerg.* **42** (1997) 241.

[112] E. K. Schmidt, T. Liebermann, M. Kreiter, A. Jonczyk, R. Naumann, A. Offenhausser, E. Neumann, A. Kukol, A. Maelicke and W. Knoll, *Biosens. Bioelectron.* **13** (1998) 858.

[113] R. Naumann, E. K. Schmidt, A. Jonczyk, K. Fendler, B. Kadenbach, T. Liebermann, A. Offenhäusser and W. Knoll, *Biosens. Bioelectron.* **14** (1999) 651.

[114] C. Peggion, F. Formaggio, C. Toniolo, L. Becucci,, M.R. Moncelli and R. Guidelli, *Langmuir* **17** (2001) 6585.

[115] L Becucci, R. Guidelli, Q. Liu, R. J. Bushby and S. D. Evans, *J. Phys. Chem.B.* **106** (2002) 10410.

4

Phosphoric Acid Fuel Cells (PAFCs) for Utilities: Electrocatalyst Crystallite Design, Carbon Support, and Matrix Materials Challenges

Paul Stonehart[*] and Douglas Wheeler[**]

[*]Stonehart Associates Inc. 17 Cottage Read, Madison, Connecticut 06443-1220 U.S.A.
and [**]UTC (United Technologies Corporation) Fuel Cells,
195 Governor's Highway, South Windsor, Connecticut 06074 U.S.A.

I. INTRODUCTION

Now that phosphoric acid fuel cells are commercial, this review covers the technology details for the electrochemical cells. It is concerned with the applicability of phosphoric acid electrolyte fuel-cell systems towards power generation for Utilities. Due to the constraints regarding the use of an hydrocarbon fuel source and the high throughput of the air electrode, aqueous alkaline systems are not viable. This leaves molten carbonate fuel cells (MCFCs), solid oxide fuel cells (SOFCs), and acid systems as candidates for Utility power generation. In the latter category, two candidates are under production, those being phosphoric acid fuel cells (PAFCs) and acidic solid-polymer electrolyte membrane (*e.g.*, Nafion®) fuel cells (PEMFCs) With each acid electrolyte system there is a high degree of commonality regarding the structure of the electrocatalysts, decay of catalyst activity, and poisoning problems. The various

Modern Aspects of Electrochemistry, Number 38, edited by B. E. Conway *et al.* Kluwer Academic/Plenum Publishers, New York, 2005.

electrolytes dictate different operating characteristics in order to maintain the most advantageous power regime.

The leading organizations in the United States for development of the technology base in fuel-cell applications, as a viable, economic method of generating electricity, has been aided by the research for large stationary power generation funded by the U.S. Government, the Electric Power Research Institute and the Gas Research Institute. Minimum lifetimes of 40,000 hours are required for stationary systems in order to compete with conventional power generating methods. Since the generation of electricity by fuel-cells occurs *via* electrochemical reactions, the mechanical stability and maintenance of activity of the electrodes are the most important factor in maintaining performance. The primary objective of this review is to examine the most relevant fuel-cell literature to determine those factors which contribute to the operation and, most importantly, degradation of fuel-cell electrode and cell components.

Fuel-cells are inherently simple engines, where the only reactions of significance are the oxidation of hydrogen molecules and the reduction of oxygen molecules. At any one time, there are three driving forces that must be met for commercialization of any energy product: they are high performance, long life and low cost. The fuel-cell system can be designed easily to achieve any two of these parameters, whilst sacrificing the third. To achieve all three at the same time presents a challenge, which when met, assures commercial success.

It is necessary to discuss four scientific topics for phosphoric acid fuel cells. Those interconnected topics are: the design of the precious metal electrocatalyst; properties of the phosphoric acid; properties of the matrix, and those of the carbon catalyst support.

II. THE ROLE OF ELECTROCATALYSIS IN PHOSPHORIC ACID FUEL CELLS (PAFCS)

In the gas-phase, *catalysis* is understood to be the maximizing of reaction rates (by using highly dispersed materials) in difficult environments for the production of preferred products. Anything else is the physics and chemistry of metal surfaces. This gas-phase analogy to fuel-cells is seen with the great advances for fuel-cells technology involving the preparation of *highly dispersed* noble metal catalysts on specifically textured carbon supports for both phosphoric acid and polymer electrolyte fuel-cells. These ultra-fine materials, having metal crystallite and alloy dimensions of 1-2 nm, do not have the same metallurgical properties as bulk materials. The principal reason for preparing ultra-fine catalyst materials is to increase the heterogeneous catalyst reaction surface. Of secondary importance, is the lowering of the apparent cost of the catalyst, particularly where precious metals are being used. The latter point is somewhat trivial since precious metal reclamation processes for spent catalysts are now extremely efficient.

For fuel-cell technology development, it has been important to understand the characteristics and operation of highly dispersed platinum and platinum alloy electrocatalysts. A series of papers on platinum crystallite size determinations in acid environments for oxygen reduction and hydrogen oxidation was published together by Bett, Stonehart, Kinoshita and co-workers.[5] The conclusion from these studies was that the specific activity for oxygen reduction on the platinum surface was independent of the size of the platinum crystallite and that there were no crystallite size effects.

A later paper by Bregoli[6] on supported Pt electrocatalysts for PAFC cathodes suggested that as the platinum became more highly dispersed on the carbon support (i.e. the platinum crystallites becaming smaller), so then the specific activity of those crystallites for oxygen reduction in phosphoric acid became less but the mass activity appeared to be constant, irrespective of the crystallite size. This argued against developing techniques for more efficiently dispersing the platinum electrocatalyst for the PAFC. It is important to examine this postulate.

III. SPECIFIC AND MASS ACTIVITIES FOR OXYGEN REDUCTION ON PLATINUM IN PHOSPHORIC ACID

The most important operational features are the relationships between the crystallite sizes of the platinum electrocatalysts to the specific (A.real m^{-2} Pt) and the mass (Ag^{-1} Pt) activities. These features are most directly applicable to the efficiency and utilization of the catalyst in operating fuel-cells.

As noted previously, the earliest paper to examine the specific activities of platinum black and platinum supported on carbon was that of Bett, and co-workers[1] where they showed that over a large range of crystallite sizes, the specific activity was constant at about 18 μA real cm^{-2} of platinum (0.18 Am^{-2} Pt) at 900 mV in sulphuric acid at 50 and 70 °C. There was no difference between the activities of unsupported platinum

black and platinum supported on carbon up to a platinum surface area of
nearly 100 $m^2 g^{-1}$ Pt.

The aforementioned paper by Bregoli[6] called these earlier results into
question, since it was implied from his paper that the mass activity
appeared to be constant and the specific activity decreased as the
crystallite size became smaller. Watanabe and co-workers;[7] (their Figure
6) further examined this question and showed apparent constant specific
activities for supported platinum on carbon in hot phosphoric acid out to
210 $m^2 g^{-1}$ Pt. The subsequent work of Buchanan, et al.[8] showed a wide
variety in the mass activities for dispersed platinum crystallites in the
range of 50-130 $m^2 g^{-1}$ Pt but they claimed they were not able to confirm
the earlier results of Watanabe, et al.[7] A later paper by Buchanan et al.[9]
identified mass activities for platinum in the range of 58-82 $m^2 g^{-1}$ Pt
where they indicated a slope of the line for the mass activity versus the
electrochemical area to be exactly the same as the slope of the line
previously published by Watanabe, et al.[7] i.e., the specific activity was
constant at 0.6 A m^{-2} Pt surface.

In all experimental situations, some judgment must be made
concerning the quality of the results.[10] In this, it is important to
understand the validity of the measurements, and the limits of error in the
analytical determinations. In the case of *electrocatalysis*, the reaction rate
on a catalyst surface is the most significant factor. Measurements of the
reaction rate can easily be *lower* than the true reaction rate value but *never
higher*. For this reason, the highest values must be given greater
consideration than the lowest values.

Reasons for low reaction rate values are manifold:

- Poisoning of the electrocatalyst surface
- Flooding of a porous electrode structure
- Low utilization of the electrocatalyst
  (not contacted by the electrolyte)

It is important then to re-examine the Buchanan et al. results[8] and
identify those that are not reliable. There is a wide scatter of activity
results, particularly at higher dispersions. For this reason, the results have
been winnowed to exclude those obtained on catalyst dispersions greater
than 90 $m^2 g^{-1}$, or where the results have been obtained on similar
crystallite sizes where low reduction activities were obtained. Those
values are shown in Figure 1a.

It is also important to include the Bregoli[6] results into this analysis,
since that work provided much of the impetus to resolve these issues.

Again, careful examination of his data suggests that all of the values below 80 $m^2 g^{-1}$ Pt should be considered but, with higher surface areas, there are large experimental errors. Accordingly, only those values lower than 80 $m^2 g^{-1}$Pt have been inserted into Fig. 1a.

The results of Watanabe *et al.* and both sets of those of Buchanan *et al.* are also shown in Figs. 1a and 1b. It can be seen that when the supposedly dubious experimental values are excluded, the results agree

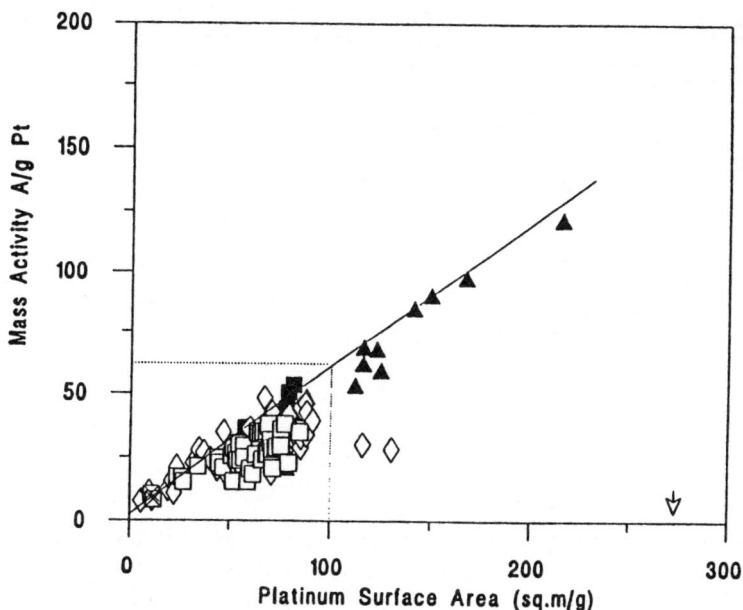

(a)

Figure 1. Compilation of platinum mass activities as a function of platinum B.E.T. surface area: [▲] Watanabe *et al.*[7]; [◊] Buchanan *et al.*[8]; [■] Buchanan *et al.*[9]; and [□] Bregoli[6]. The solid line is 0.6A.m^{-2} constant specific activity platinum. The broad arrow on the abscissa denotes the maximum surface area for a platinum crystallite when all of the atoms are located at the surface (275 $m^2 g^{-1}$ Pt). Phosphoric acid at 190 °C and 0.9 V vs. hydrogen in the same electrolyte. (a) Data up 210 $m^2 g^{-1}$ Pt. (b) Data below 100 $m^2 g^{-1}$ Pt; [☒] Bregoli[6] results on unsupported platinum black.

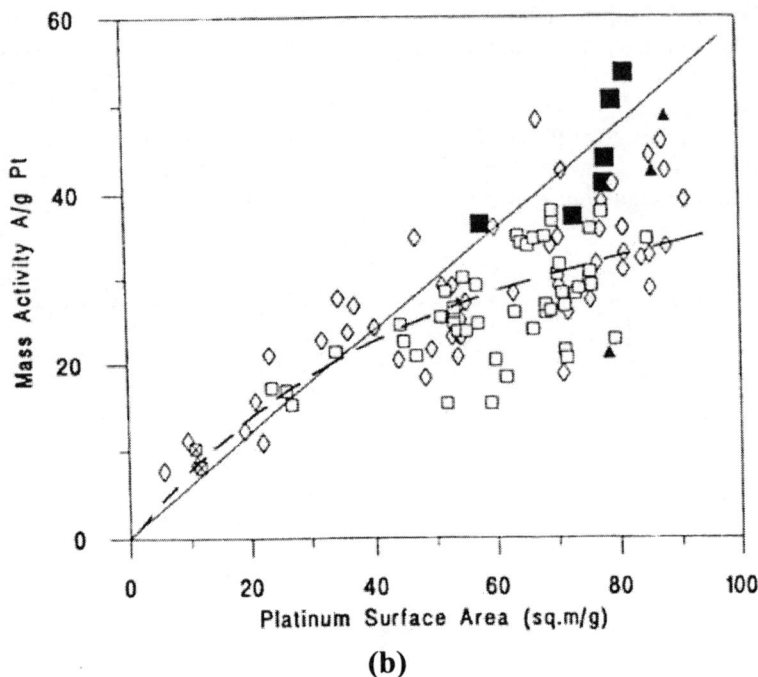

Figure 1. Continuation.

with the projections of Watanabe *et al.* and show that, over the total crystallite size range, the specific activity is constant at 0.6 A m^{-2} Pt at 190 °C for 1 atmosphere partial pressure of oxygen. All of the high data conform to a linear relationship between the mass activities and the platinum surface areas; which means that over the entire crystallite size range, the specific activity is a constant. It is concluded that the published results and conclusions of Watanabe, Sei and Stonehart[7] are supported by subsequent data from both of the Johnson Matthey group[8,9] and from the Bregoli results.[6] It is further concluded therefore, that there are no platinum crystallite size effects.[11] The measurements below 100 $m^2 g^{-1}$ Pt are shown in Fig. 1b since there is such a large quantity of conflicting information at these low surface area values. Included in this figure is the dashed line denoting the average values as shown by Bregoli.[6]

Further confirmation on the absence of crystallite size effects for oxygen reduction was given by the work of Mizuhata *et al.*[12] Here, they examined the specific activities of highly dispersed platinum particles on three carbon supports, *viz.* 560, 250, and 50 $m^2 g^{-1}$ Pt. It was found that for polymer electrolyte membrane electrodes at 0.9 V and 90 °C using oxygen at 5.84 atm. pressure; the specific activity for oxygen reduction, on a wide range Pt crystallites supported on 560 $m^2 g^{-1}$ carbon, was invariant at a value 0.95 A m^{-2} Pt. The Pt crystallite sizes ranged from 1.3 nm diameter (lowest value) to 4.3 nm diameter (highest value). For platinum crystallite diameters less than 3 nm, with increasingly closer packing for Pt crystallites on the 50 $m^2 g^{-1}$ carbon surface, the apparent specific activities progressively decreased. Correlating these results to mass activity functions, the authors showed that the mass activities increased linearly with crystallite size from 35 $m^2 g^{-1}$ Pt to 145 $m^2 g^{-1}$ Pt. at both gas pressures of 5.84 atm, and at 2.94 atm oxygen. The latter gas pressure gave a constant specific activity of 0.6 A m^{-2} Pt. At atmospheric pressure, the specific activities were 0.5 A m^{-2}Pt for Pt crystallites up to 60 $m^2 g^{-1}$ Pt, above which the mass activity was constant between 35 $m^2 g^{-1}$ Pt and 145 $m^2 g^{-1}$Pt *i.e.*, the specific activities decreased with increased crystallite packing on the carbon surface.

Notwithstanding the foregoing, there still is no universal agreement on whether or not there are crystallite size effects for oxygen reduction. Kinoshita[13,14] and Giordano and Kinoshita and co-workers[15,16] had suggested that as the crystallite sizes change, so the surface morphology changes in such a way that the specific activity changed as a function of the crystallite size. Both Kinoshita[14] and Mukerjee[17] have provided carefully worded reviews of the data and have concluded that *there are* crystallite size effects based on the changing morphology of platinum crystallites as they become larger. It is argued that the reduction of oxygen may be different on the (100) and (111) crystallite faces, and that the ratio of those faces in an individual crystallite changes with crystallite size.

The subject of crystallite morphology in the real world is not so simple. No matter how diligent is the skill for the catalyst preparation, there is inevitably some broad distribution of crystallite sizes. Within those sizes are crystallites having incomplete faces, initially producing an excess of edge sites; an extensive discussion on these matters was made by Bett *et al.*[1]

Under operation in a fuel cell, it is well known that, at short times, the crystallites change in size, with growth of the largest crystallites being at the expense of the smallest crystallites. At short times (less than 50 hours,[18] their Figure 10, admittedly under the more stressful conditions of 210 °C and 0.8 V) alloy components in the surfaces of the crystallites, such as cobalt, lose 30 % of the cobalt content in the first 10 hours, after which the residual cobalt content in the alloy appears to be constant. For the platinum, however, there is conservation of electrocatalyst mass; *i.e.* no platinum metal loss was detected from the electrocatalyst over this time period.

Ito *et al.*[19] were the first to recognize the problems of solubility of platinum in hot phosphoric acid and a more detailed study was carried out by Bindra *et al.*[20] They found that in hot phosphoric acid, an equilibrium of the type:

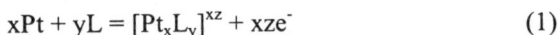

$$xPt + yL = [Pt_xL_y]^{xz} + xze^-$$ (1)

was obtained, where L is the complexing ligand and z, the valence state of Pt in the complex. This gives rise to a Nernstian equation:

$$E = E_0 + RT/xzF.\ln[\gamma c/(a_L)^y]$$ (2)

where a_L is the activity of the ligand and γ and c are the activity constant and the concentration of the platinum complex, respectively. At constant phosphoric acid concentration and temperature, a plot of log C_{Pt} versus E must be a straight line with a slope of 2.303RT/xzF. The function is shown in Fig. 2 with a slope of 49 mV/decade at 176 and 196 °C. From the Tafel slopes at these temperatures. it was deduced that x = 1 and z = 2, suggesting that the Pt is probably in a divalent state in the complex.

This study was on a smooth platinum foil using gravimetric analyses and it was found that the equilibrium concentrations of platinum in the phosphoric acid were achieved rapidly, *i.e.,* within one hour. Since it is to be expected that the kinetics of dissolution would be modified with high surface-area platinum crystallites, this equilibrium would be achieved more rapidly. Consequently, it is clear that, at the higher potentials, dissolution of the platinum is of great concern so that operating procedures must be established to prevent exposure under hot open-circuit conditions.

Error bars in Fig. 2 for the data at the lowest potentials are indicated, reflecting the limits for the gravimetric analyses but also it is realized that at the 0.65 V operating potentials of the PAFC, platinum dissolution will still occur with a projected equilibrium solubility of 1×10^{-10} moles

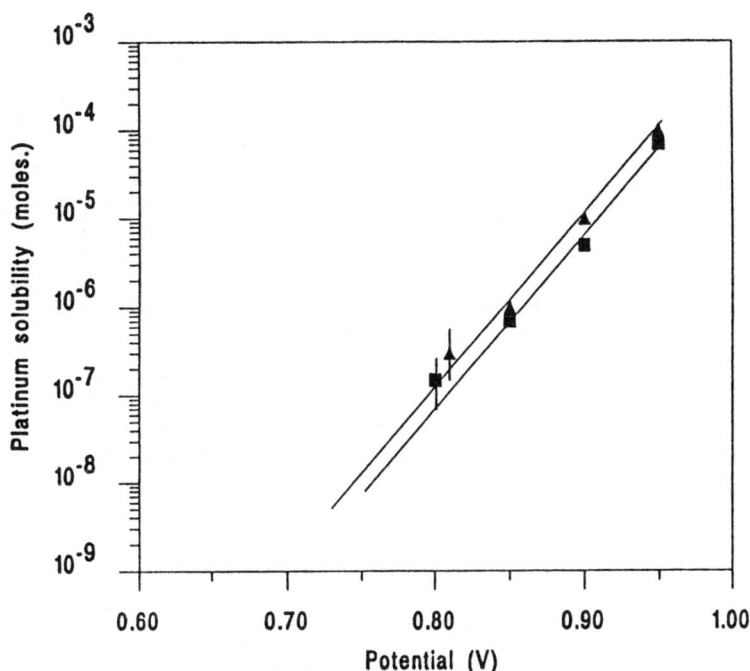

Figure 2. Platinum dissolution in 96 % phosphoric acid. 176 °C [■], 196 °C [▲].

Pt, which is not trivial over the projected lifetime for a PAFC. Migration of the platinum from the cathode towards the anode is due to the platinum being deposited, not on the anode catalyst, but rather on the silicon carbide matrix, adjacent to the anode, as a consequence of the small solubility of hydrogen in solution at the electrode/matrix interface.

Initially, crystallite growth occurs rapidly either by ion dissolution/reprecipitation (Ostwald ripening) or by surface atom diffusion, due to the requirement for lowering the surface energy of any individual crystallite. This thermodynamic driving force will tend to eliminate the incomplete faces but with the drive to lower the surface energy, the crystallites also will strive towards sphericity. This means, that to all intents and purposes, the *ratios* of the (111) and (100) faces should be approximately the same. Bett *et al.*[1] noted that as the platinum crystallite sizes grew, the size distribution increased. If this is so, then notwithstanding any difference in the specific activities for the individual faces, the *mean* of any crystallite activity for oxygen reduction should stay the same.

Clearly, there is great difficulty in preparing high surface area, mono-dispersed electrocatalysts on carbon supports, when any electrochemical process to diagnose the crystallite size or activity in these difficult environments perturbs the very physical properties of the electrocatalysts.

IV. THE STONEHART THEORY OF CRYSTALLITE SEPARATION

The early Bregoli work [6] was carried out on catalysts that preserved the mass content of the platinum on the same Vulcan XC-72R carbon support in the catalyst formulation. This meant that as smaller platinum catalyst particles were produced, they were more closely situated in relation to each other on the carbon surface. As the electrocatalysts lost surface area ("sintering") and grew larger, so the crystallite separations increased, provided that no platinum was lost from the electrocatalyst (*i.e.*, the platinum mass was maintained). Stonehart and Baris[21, 22] had examined the problem of the apparently diminishing specific activities with increasing platinum crystallite surface areas for oxygen reduction, shown by Bregoli[6] and concluded that the one critical feature that had been overlooked was the changing distance between the platinum crystallites (crystallite separation) on the carbon support surface.

A description of the rationale for the crystallite separation effect was then given by Stonehart.[23] In that work he described comparisons for the oxygen reduction activity in phosphoric acid at 180 °C for equally loaded platinum particles supported on Shawinigan acetylene black, to those on Vulcan XC-72R. The only difference in the sets of experiments was the difference in surface area of the carbon support. The apparent specific activities for reduction of oxygen with the platinum on the acetylene black were significantly lower than similarly sized platinum crystallites supported on the Vulcan XC-72R shown by Bregoli.[6] Extrapolation of the medians for both sets of data to zero surface areas gave a similar specific activity value of 0.8 Am^{-2} Pt. Extrapolation of the high points gave 0.8 Am^{-2} Pt for the Vulcan XC-72R data and 0.65 Am^{-2} Pt for the acetylene black.

Subsequently, an equation was provided[7] to describe the uni-dimensional intercrystallite distance on the carbon support surface; *viz.*

$$X = \sqrt{([\pi\sigma d^3 S_c(100 - y)]/3\sqrt{3}y)} \qquad (3)$$

where : X = intercrystallite distance (nm); σ = density of supported platinum catalyst particles ($g.nm^{-3}$); d = mean diameter of platinum catalyst particles (nm); S_c = specific surface area of the carbon support ($nm^2 g^{-1}$); and y = platinum catalyst loading on the carbon black support (wt%). These results show that when the crystallite separation was greater than 17 nm, the apparent specific activity became a constant. The above equation was used by Buchanan et al.,[8] and by Giordano et al.,[15,24] who

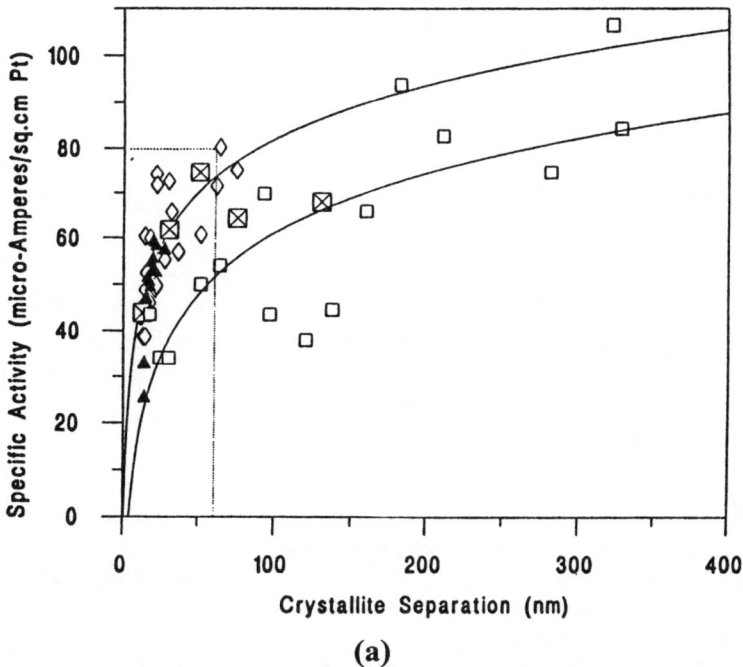

(a)

Figure 3a. Platinum on Carbon catalysts in 190 °C phosphoric acid. Compilation of specific activities for oxygen reduction versus platinum crystallite separation from Watanabe[7] et al. [▲]; Giordano et al. –maximum values only[24][⊠]; Buchanan et al.[8][◊]; and Giordano et al.[15,16][□]. Function of upper line is a best fit to the Buchanan values and the lower line is a best fit to the Giordano values. (a) Data for crystallite separations up to 325 mn. (b) Data below 60 nm crystallite separations.

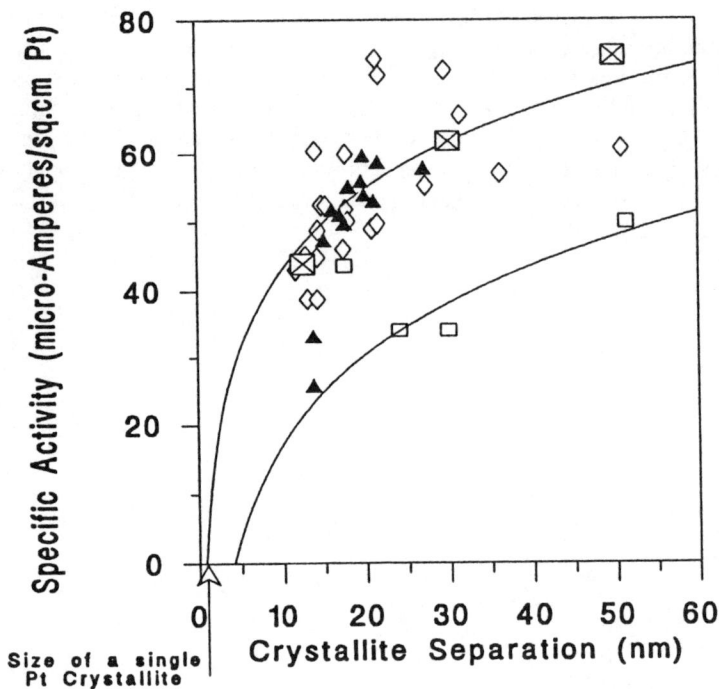

Figure 3b. Continuation.

were unable to support the veracity of the work by Watanabe, Sei and Stonehart.[7] In a separate development, Kinoshita[13] postulated that the changing crystallite facets for platinum with particle size might account for the specific activity variations but none of the experimental results agreed with that postulate.

Giordano et al.[15] recognized the thrust of the Stonehart theory[7,11,21,23] of crystallite separations, and increased the platinum distribution by courageously increasing the carbon support surface areas to very large values together with lowering the platinum loading on the carbon support, as previously demonstrated by Watanabe et al.,[7] with the concomitant results shown in Fig. 3. Later work by Giordano et al.[15] showed a comparison of their results to those of Buchanan et al.,[8] who also tested the crystallite separation theory. On plotting crystallite separation values to over 300 nm, Giordano et al.[15] show a doubling in the apparent specific activity compared to their crystallites with closest separation.

Figure 3 is shown as a compendium of these results, with the individual computations for the best fit of the Giordano and Buchanan results, compared to the Watanabe results. There will, of course, be some activity, symptomatic of an ideal smooth platinum surface but with a significantly reduced surface area. Based upon this analysis, and given the uncertainty of analyzing the results from other workers, it would seem that all of the data are bounded by the lower curve of Fig. 3 (best fit to the Giordano et al. results) and the upper curve (best fit to the Buchanan et al. results). There is reasonable agreement with the function by all of the different results produced by this multiplicity of workers, bearing in mind that it is a determination of the heterogeneous reaction rate on highly dispersed catalyst surfaces, and incorporating all of the errors in materials analysis, differences in electrode preparation, and testing.

V. CALCULATION FOR HEMI-SPHERICAL DIFFUSION TO A CRYSTALLITE

Previously, Stonehart[11] had described the mechanism for diffusion of oxygen to an individual platinum crystallite on the carbon support surface. In this instance, at low current-densities, it was anticipated that hemispherical diffusion conditions operate, since the base of the platinum crystallite is shielded by the carbon support and thereby is inactive.

The number of moles, dN_r diffusing across a boundary to a spherical surface of $4\pi r^2$ in the time dt, is given by Fick's First Law :

$$dN_r = 4\pi r^2 \, D(\delta C/\delta r)_r \, dt \qquad (4)$$

So the flux at r is:

$$f_r = dN_r \, /4\pi r^2 \, dt = D(\delta C/\delta r)_r \qquad (5)$$

The area at the exterior of the diffusion annulus surrounding the crystallite at $(r + dr)$ is $4\pi(r + dr)^2$ and in time, t, the number of moles diffusing across this surface in time dt is:

$$dN_{r+dr} = 4\pi(r+dr)^2 D(\delta C/\delta r)_{r+dr}\, dt \tag{6}$$

The concentration gradient at the crystallite surface at any time t is:

$$(\delta C/\delta r)r = r_0 t = C(1/r_0 + 1/\sqrt{\pi Dt}) \tag{7}$$

so the steady value of the current to the crystallite is given by

$$i_t = nFADC(1/r_0 + 1/\sqrt{\pi Dt}) \tag{8}$$

where A is the area of the spherical crystallite, $4\pi r_0^2$.

Owing to the interference of the neighboring crystallites which limits the availability of oxygen, then $C \neq C_\infty$ but is modified by the inter-crystallite distance, as well as the electrode potential. Thence the maximum diffusion radius; $(r+dr)_{max} = X/2$; where X is defined above. In this case

$$C = f.C_\infty(1 - 2/X) \tag{9}$$

The indeterminate function, f is not defined due to the open porous electrode structure, and the gas-fed backing of the electrode, leading to asymmetry, but it can be seen that as the inter-crystallite distance increases, then the apparent specific activity for the platinum catalyst becomes the true specific activity, and $C = C_\infty$.

Then we can expect a power function for the apparent specific activity of the platinum catalyst, having the form:

$$i_t = nF4\pi r_0^2 D(f.C_\infty(1-2/X))^m(1/r_0 + 1/\sqrt{\pi Dt}) \tag{10}$$

where m is a representation of spatial three-dimensionality.

Although the above treatment has been developed for particles as spheres, in reality, especially where diffusion is concerned, hemi-spheres ($4\pi r_0^2/2$) are the operands, with the characteristic X/2 between them. Then at long times, or with attainment of steady-state,

$$i = nF2\pi r_0 D. f.C_\infty(1-2/X)^m \tag{11}$$

When the dimensions of the crystallite are small compared to those of the diffusion field, then the crystallite appears as a point source. As the diffusion fields expand, they coalesce (as the current-density increases)

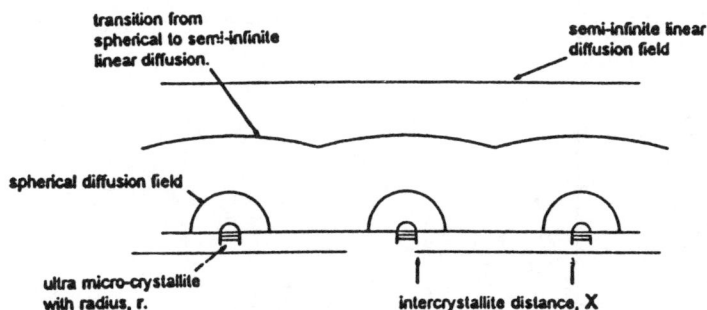

Figure 4. Schematic of the development of the diffusion layer as a function of current density and intercrystallite distance for the supported electrocatalyst particles.

until the composite electrocatalyst surface exhibits a quasi-planar semi-infinite diffusion field, shown schematically in Fig. 4.

Figure 5 shows the influence for taking the logarithm of the specific activity versus the logarithm of $1/X$, so that the slope defines the value of the exponent, m, which is of course, -3. All of the data fall in a linear band, limited at the upper level by the results of Buchanan et al.,[8] and at the lower by the results of Giordano et al.[15,16] The results of Watanabe-Stonehart fit between the boundaries with a directive that indicates a transition from spherical to semi-infinite diffusion due to closeness of approach of the crystallites. The size of an individual platinum crystallite also is indicated in Fig. 5. Extrapolation of the high points to zero crystallite separation gives 20 μA cm^{-2} Pt, which is close to that value obtained by Bett et al.[1] in sulphuric acid.

Buchanan et al. and Giordano et al. demonstrate that as the crystallite separation is increased to very large values, the apparent specific activity for the Pt catalyst increases. These authors use the equations developed by Watanabe, Sei, and Stonehart[7] but an examination of the experimental results of Buchanan et al. and of Giordano et al. show very great variations. There was no theoretical support for the linear relationship of crystallite separation with specific activity, drawn by Giordano et al., or for the intercept at zero separation. From the analyses presented here, measurements of oxygen reduction activity for platinum on porous electrodes in PAFCs at 0.9 V are not always diffusion-free and cross comparisons amongst all of the published experimental results shows

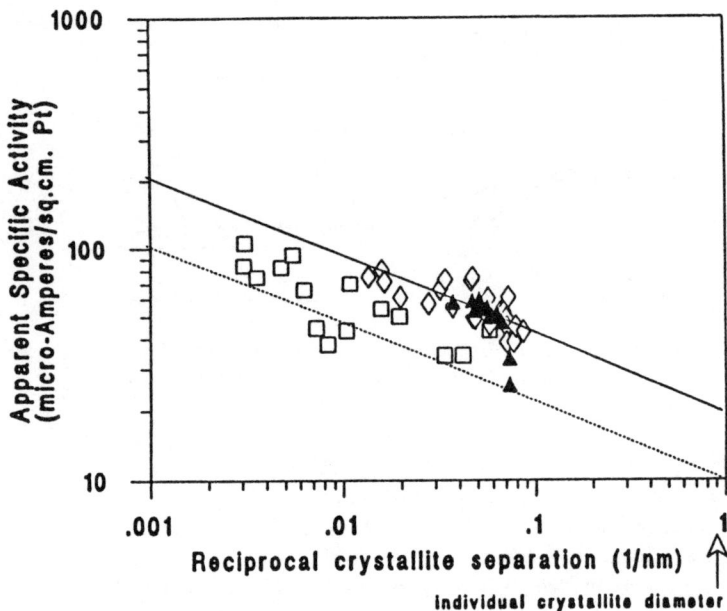

Figure 5. Power function of log specific activities versus the reciprocal of the platinum crystallite separation on the carbon surface. Watanabe et al.[7] [▲]; Buchanan et al.[8] [◊]; and Giordano et al.[15,16] [□]. The broad arrow on the abscissa denotes the average diameter of a single platinum crystallite.

good agreement with the data of Watanabe, Sei, and Stonehart. A best fit is obtained for a power function of the crystallite separation, in accordance with that expected for hemi-spherical diffusion.

The theory of crystallite separation for oxygen reduction on platinum was tested by Mizuhata et al.[12] in PEMFC electrodes. They found that there was universal agreement for carbon supports with surface areas of 560, 250, and 50 m²g⁻¹, and oxygen pressures of 5.84 atmospheres at 90°C, irrespective of the platinum crystallite size. When the platinum crystallite separation was below 15 nm, the apparent specific activities progressively declined (their Figure 11) and above 15 nm the specific activities became effectively constant. This is in remarkable agreement with the previous findings of Watanabe, Sei, and Stonehart[7] (their Figure

9) for oxygen reduction on platinum who found the critical platinum crystallite separation value to be 17 nm in phosphoric acid at 190 °C, below which the apparent specific activities progressively declined and above which the specific activities became effectively constant. For practical purposes, there is a trade-off between increasing the inter-crystallite distance on the carbon surface (to maximize the platinum crystallite specific activity by lowering the platinum loading in the electrode) and the electrode thickness, since as the loading becomes smaller, then the electrode thickness must increase to obtain a sufficient mass of platinum electrocatalyst in order to achieve the required current-density. Unfortunately, thicker electrodes then lead to progressive ohmic losses within the electrode structures.

Again, notwithstanding these results, whether or not there are crystallite separation effects is still a thorny problem. What is needed is a further careful examination of high surface-area mono-disperse platinum electrocatalysts on carbon surfaces where the separation of the crystallites is controlled. This means making a series of electrocatalysts at different loadings, establishing the crystallite sizes and evaluating the specific activities for oxygen reduction. After cross-plotting the results to remove

Table 1
Procedures and Patent Numbers for Catalyzation of Carbon Supports by Platinum

Reference	Catalyzation	Carbon Support
Bregoli[6]	Colloidal Pt 10%, see USP 4,044,193	Vulcan XC-72 250 $m^2 g^{-1}$
Stonehart et al.[21]	Sodium thiosulfate, see USP 4,956,331	Shawningan Acetylene black 65 $m^2 g^{-1}$ Vulcan XC-72 250 $m^2 g^{-1}$
Watanabe et al.[7]	Sodium thiosulfate, see USP 4,956,331	60-1300 $m^2 g^{-1}$ carbons
Giordano et al.[15]	Sodium dithionite, (also with ethanol) 10% 20% Pt see USP 4,136,059	Ketjen Black 950 $m^2 g^{-1}$
Buchanan et al.[8]	Johnson Matthey, see USP 5,068,161	Vulcan XC-72, acetylene black; Graphitized furnace blacks. 60-225 $m^2 g^{-1}$
Mizuhata et al.[12]	Ion exchange; H_2 reduction with Pt-trinitrodiamine	Furnace back, 560 $m^2 g^{-1}$ Acetylene black 50 and 250 $m^2 g^{-1}$

any ambiguities regarding crystallite separations and crystallite sizes, there may be a further resolution to this quandary.

It is worthwhile to identify those different catalyzation techniques employed by these workers, together with the carbon supports, as given in Table 1.

VI. DEVELOPMENT OF PLATINUM ALLOY OXYGEN-REDUCTION ELECTROCATALYSTS

As part of the early work to find alloys of platinum with higher reactivity for oxygen reduction than platinum alone, International Fuel Cells (now UTC Fuel Cells, LLC.) developed some platinum-refractory-metal binary-alloy electrocatalysts. The preferred alloy was a platinum-vanadium combination that had higher specific activity than platinum alone.[25] The mechanism for this catalytic enhancement was not understood, and post-test analyses[26] at Los Alamos National Laboratory showed that for this binary-alloy, the vanadium component was rapidly leached out, leaving behind only the platinum. The fuel- cell also manifested this catalyst degradation as a loss of performance with time. In this instance, as the vanadium was lost from the alloy, so the performance of the catalyst reverted to that of the platinum catalyst in the absence of vanadium. This process occurs fairly rapidly in terms of the fuel-cell lifetime, i.e., within 1-2000 hours. Such a performance loss means that this Pt-V alloy combination may not be important commercially but it does pose the question, why does the electrocatalytic enhancement for oxygen reduction occur?

The next advance in development of PAFC binary-alloy cathode electrocatalysts was the use of Pt-Cr alloys.[27] In this patent, it was disclosed that with the platinum-vanadium alloy in 99 % phosphoric acid at 194 °C and at an electrode potential 0.9 volts, over 67 % by weight of the vanadium had dissolved in the first 36 hours. In the case of Pt-Cr, only 37 % had dissolved under the same conditions. It is not clear from the descriptions in these patents whether or not there is any unreacted vanadium or chromium present in the catalyst because it is not identified that all of the vanadium or all of the chromium was initially alloyed with the platinum. It is conceivable that significant amounts of the non-noble metal components are not fully reacted.

It was proposed by Jalan and Taylor[28] that the improvement in oxygen reduction activity for platinum by producing binary-alloys of platinum with non-noble metals, such as vanadium, carbon, silver, etc., lay with the changes in the inter-atomic distances as measured by X-ray diffraction analyses. Luczak and co-workers expanded upon this concept to include cobalt and ternaries of Pt-Co-Cr and Pt-Co-V. The results are shown in Fig. 6. Since the alloying process requires that the electrocatalyst be exposed to 900 °C, then inevitably the alloy crystallites become larger due to crystallite growth during processing. It is particularly curious that a long linear relationship for the specific activity over a large change in the platinum nearest-neighbour dimension is observed.

Although interesting, the answer for the electrocatalytic enhancement on alloys may lay elsewhere, since X-ray analyses are bulk analytical

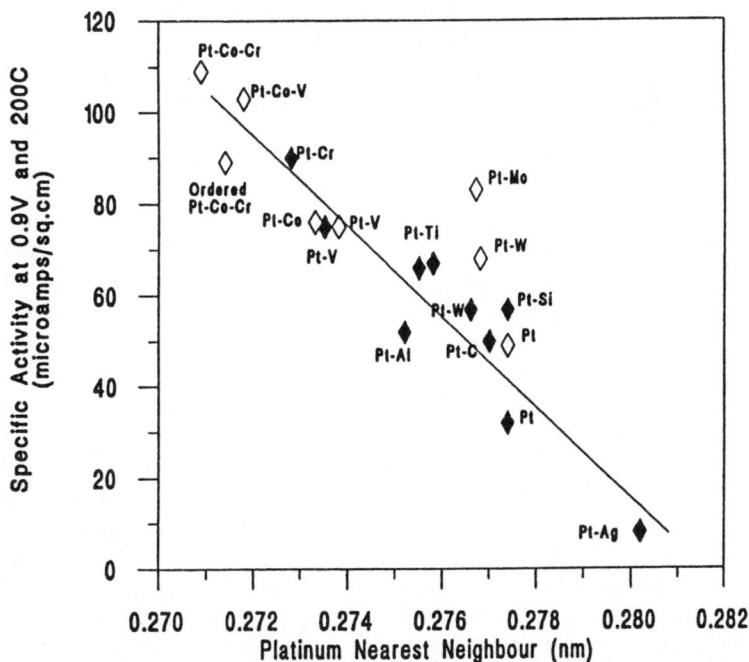

Figure 6. X-ray diffraction platinum atom nearest neighbour distances versus specific activity for oxygen reduction at 0.9 V in phosphoric acid at 200 °C: (♦) Jalan and Taylor[28] and (◊) Luczak.

techniques and give no information on the surface atom arrangements on an individual crystallite. It is recognized that both surface segregation and demetallization of the alloy by the least noble metal component in the phosphoric acid environment occurs rapidly. It is, however, beyond the scope of this review to cover platinum and platinum alloy crystallite sizes growth ("sintering") in fuel- cell environments at this time, or to go into the surface structures of these nano-dimensioned particles.

A series of papers to understand the Pt-Cr alloy system were published by Paffett, Gottesfeld and co-workers at Los Alamos National Laboratory.[29-32] These workers applied a series of sophisticated analytical techniques to the study of Pt-Cr alloy electrodes that were well characterized as bulk materials, as opposed to high surface-area particles. The general conclusion from this work was that the surface of the Pt-Cr alloy was rapidly leached with loss of the chromium, leaving behind a roughened platinum surface. This roughened platinum surface then gave a higher oxygen reduction rate than smooth platinum. It was concluded that the "catalytic enhancement" for the Pt-Cr alloys was due to this surface roughening of the platinum, rather like a "Raney" metal, so that no other intrinsic catalytic improvement was involved.

Ross,[33] and Beard and Ross[34] had also been interested in electrocatalytic properties of Pt-3d transition metal binary-alloys, with a view that stable intermetallics could be formed. It was also their view that the catalytic enhancement shown by Pt-V, Pt-Cr, and latterly Pt-Co was due to the surface roughening of the platinum crystallites caused by leaching of the non-platinum elements from the surface. In the case of the Pt-Co alloy, they believed that a more stable alloy is formed that protects against further alloy degradation.

In the patent area, the Pt-Cr binary alloy, referenced previously,[27] was followed by the first potential ternary alloy by the addition of cobalt to the Pt-Cr.[35] The addition of cobalt provided higher performance than that exhibited by the Pt-Cr alone. It was recognized that high performance for a short time was not sufficient for commercial utilization of alloys (witness the higher performance but short life-time for Pt-V). A secondary patent[36] was issued to International Fuel Cells (IFC) (now UTC Fuel Cells, LLC.) where the lifetime of the Pt-Co-Cr was extended by a long heat treatment procedure during the catalyst preparation. It is dissolved in the patent that due to this heat treatment, the ternary alloy becomes an ordered structure and that the lifetime of the catalyst is extended beyond 9000 hours in hot phosphoric acid fuel-cell environments without the catalyst degradation.

In a related development, workers at Englehard had proposed Pt-Ga as a binary alloy[37] to enhance the catalytic activity of the platinum. Although this patent application was in 1982, it took a long while to appear, and the issue date is actually subsequent to many other patents that were filed later and issued in the interim. Englehard also applied for a European patent on a Pt-Fe combination.[38] At this time there does not appear to be a U.S. patent that is issued on a Pt-Fe binary formulation. Nippon Englehard applied for new alloy patents, one of which was Pt-Cu[39] and the other was Pt-Fe-Co.[40] It would appear that the development of Pt-Fe-Co was spurred by the Pt-Fe patent applied for previously by Englehard [38] and in view of the Pt-Cr-Co patent that had been subsequently issued to IFC.[35]

In the case of the Pt-Cu[39] patent, although the performances quoted for oxygen reduction are high, this catalyst gives some pause for thought, since it would be expected that copper could leach from the alloy and dissolve in the phosphoric acid. In that event, it would be electrodeposited at the fuel cell anode causing site selective poisoning for the hydrogen molecule oxidation reaction. Copper is a well known poison for platinum catalysts in hydrogen oxidation. It remains to be seen whether this cautionary advice is borne out in actual fuel cell operation. Gallium additions to the Pt-Cr-Co[41] alloy combinations have been claimed by IFC which are extensions of their Pt-Cr-Co[36] issued previously. Further to this is the Pt-Co-Ga ternary alloy[42] that has been applied and assigned to IFC.

It would appear from this patent literature that the development trends for PAFC cathode alloy electrocatalysts are from platinum, to Pt-refractory metal (vanadium); to Pt-Cr, to Pt-Cr-Co through to Pt-Fe-Co and then the various gallium additions to these combinations. It would seem, particularly in the case of gallium additions, that the gallium should induce porosity into the platinum alloys, since it would be expected to leach out easily.

A summary of the PAFC cathode alloy patents is shown in Table 2.

It is difficult, however, to accept that the performance increase is caused by a surface roughening of the metal alloy crystallite as proposed by Los Alamos and Berkeley, since it would be expected that the surface rearrangements at these temperatures would be very fast, particularly for the nano-dimensions involved. It is a worthwhile exercise to translate the dimensions for solid state diffusion of a metal atom in a metal matrix from the usual $m^2 hour^{-1}$ to $nm^2 s^{-1}$ (*i.e.,* 1 $m^2 hour^{-1}$ is $1.6 \times 10^{16} nm^2 s^{-1}$). It is clear that atomic rearrangements on a nanometer dimensioned metal surface occur very rapidly, especially with the added inducements of liquid-phase enhanced surface diffusion. Moreover, considering the previous discussion on crystallite sizes and growth of individual crystallites in the phosphoric acid fuel-cell environment, it is clear that the particles are rearranging rapidly,[18] which precludes any long term atomic dimensioned roughening on an individual crystallite surface. Nevertheless the stabilities and performance enhancements for these alloys for oxygen reduction are seen over the long term.

Gallium is a curious additive, since the element leaches out rapidly and, after 168 hours, most of it is gone excepting a residuum of 0.7 atom % but the catalyst stability remains for over 2000 hours.[41,42] In this regard, it is reasonable to speculate on the mechanism by which the

Table 2
Cathode Catalyst Performances For Those U.S. Patents Referenced

Reference	Alloy Combination	Atom Ratio	Performance 190-200 °C 0.5 mg Pt		
			Air @ 160mA	200mA (mV)	Mass Activity
[25] USP 4,202,934	(Pt-V)	Not defined		720	
[27] USP 4,316,944	(Pt-Cr)	3:1		735	
[37] USP 4,822,699	(Pt-Ga)	3:1	700-684[a]	742	
[35] USP 4,447,506	(Pt-Cr-Co)	8:1:1			47 mA.mg^{-1} Pt @ 0.9 V on oxygen
[38] EPA 84303984.3	(Pt-Rh-Fe)	5:1.5:3.5		788	
[39] USP 4,716,087	(Pt-Cu)	1:1		795	
[40] USP 4,794,054	(Pt-Fe-Co)	2:1:1		725[b]	
[41] USP 4,806,515	(Pt-Ga-Co/Ni/Cr)	5:2:3		714-725[b]	
[42] USP 4,880,711	(Pt-Ga-Cr/Co)	5:2:2/3		725	
[43] USP 5,013,618[b]	(Pt-Ir-Cr) (Pt-Ir-Fe)	5:3:2 5:2:3		725	

[a] Performance degraded to 684mV over 200 hours. Then stable for 1350 hours.
[b] Performance shows long-term stability of 670mV for 2000 hours.
[c] Low surface area loss and long-term stability.

gallium affects the catalyst structure. It is more likely that the gallium is incorporated into the carbon support in much the same way as boron is known to do. This will provide trap sites for the alloy crystallites and retard surface area loss.

In order to evaluate the performance advantages of alloying platinum with 3d transition metals, Buchanan, Keck, et al.[8] examined platinum with first-row transition elements to examine whether or not there were significant improvements in the performance of the alloys in hot phosphoric acid as oxygen reduction catalysts. They concluded that alloying platinum with the first-row transition elements significantly increases the performances above that for platinum alone.

In their work they did not distinguish between the alloying elements, since they expressed the view that, generally, the specific activities of all these platinum alloys were similar. The superiority of the alloys that they actually prepared, produced an improvement about 1.5 times the activity of platinum alone. Of greater importance was their finding that the alloys showed greater stability against "sintering" and degradation than did platinum alone. This translated into an increased lifetime with higher performance for the alloys. Although they tested their alloy combinations for the critical intercrystallite distance referred to by Watanabe et al.,[7] they were not able to determine that this effect was controllable under the conditions used to prepare their catalysts.

Most recently, the preferred ternaries Pt-Ir-Cr and Pt-Ir-Fe [43] have appeared. In the case of the alloy $Pt_{50}Ir_{30}Fe_{20}$ (the numerical subscripts refer to the atom percentages of the elements), a mass activity of 48 Ag^{-1} under oxygen at 0.9 V in 99 % H_3PO_4 at 177 °C was achieved, compared to 44.8 Ag^{-1} for the $Pt_{50}Ir_{30}Cr_{20}$ and 38 Ag^{-1} for the reference alloy $Pt_{50}Co_{30}Cr_{20}$ under the same conditions. More importantly, lifetimes for fuel-cell catalyst and the abilities of the electrocatalysts to resist surface area losses, are of overriding importance and in these cases, the $Pt_{50}Ir_{30}Cr_{20}$ alloy showed greater stability than $Pt_{50}Co_{30}Cr_{20}$ over 200 hours at 200 ASF and 218 °C; the former losing only 51 % of the initial surface area, whereas the latter loses 67 % under these stressful conditions. At 205 °C and 0.75 V in 99 % H_3PO_4 over 168 hours, the $Pt_{50}Ir_{25}Cr_{25}$ alloy lost only 13 % of the original surface area, the $Pt_{50}Ir_{25}Co_{25}$ lost 18.9 % whereas the reference $Pt_{50}Co_{30}Cr_{20}$ alloy lost 35.7 % of the original surface area under the same conditions. This becomes understandable, given that the platinum-iridium phase diagram shows mutual solubility at all concentrations and no evidence of intermetallic phases. The surface energy of iridium is 2,200 ergs cm^{-2} compared to 1,800 ergs.cm^{-2} for platinum and thus it will harden (that is experience increase of the metal interatomic bond energies) the alloys considerably over those other patented alloy combinations without iridium, leading to greater operational lifetime under fuel-cell conditions.

The mechanisms for electrocatalyst surface area loss are by a) crystallite migration or b) atom or ion dissolution and reprecipitation, either to the electrolyte or over the carbon surface. It is well known that surface diffusion of atoms on the individual crystallites can provide for mobility (much like the treads on a military tank). In either case, small crystallite become annihilated and fewer but larger crystallites are produced.[18] In either event, these processes lead to demetallization of the less noble components in the alloy.

Previously, the first reviews on alloy electrocatalysts for oxygen reduction in phosphoric acid fuel-cells[44-46] concentrated on those patents that had been issued in the United States, since that was where most of the early work had been done. Subsequently, similar alloy work has been done in Japan, and that work is reflected in the Japanese patent literature shown in Table 3, whence corresponding alloy-combination atom ratios and the air/oxygen performance values are given in Table 3a.

It is not surprising that the alloy combinations being claimed as improvements are similar to those that have been claimed in the United States previously. In most instances, quaternary alloy combinations have been claimed that utilize platinum with one or more of the 4th period–group 8 transition metals, together with a 4th element that modifies the metallurgical behavior of the alloy.

Alloy catalyst developments for oxygen reduction in phosphoric acid have further concentrated on forming discretely ordered structures of the form Pt-M or Pt_3-M where M is a non-platinoid element, commonly identified as Brewer-Engel materials. This is achieved by heat-treating the ternary and quaternary alloys at specific elevated temperatures related to the bulk phase diagrams for those alloys heating for an extensive period of time with subsequent quenching to freeze the ordered structures, is employed.

The purpose for forming ordered alloys has been not so much to increase the alloy performance but to increase the *lifetime* of the electrocatalyst. On-going research is geared towards identifying the surface features of these ordered alloys and the mechanisms of their operation.

Table 4 records the US patents that have been issued on various PAFC cathode-catalyst ordered alloys.

Table 3
Japanese PAFC Patents

Kokai Number	Application Date	Issue Date	Assignee
S61-8851 Vinod M. Jalam Fuel Cell and Electrocatalyst (Pt-Ni-Co)	6 June 1985	16 Jan 1986	Giner Inc.
S62-163746 Takashi Itoh, Sigemitsu Matsuzawa, Katsuaki Katoh Pt Alloy Electrocatalyst and Acid Fuel Cell Electrode (Pt-Fe-Co)	13 Jan 1986	20 July 1987	Nippon Englehard
S62-269751 Takashi Itoh, Sigemitsu Matsuzawa, Katsuaki Katoh Platinum-Copper Alloy Electrocatalyst and Acid Fuel Cell Electrode (Pt-Cu)	16 May 1986	24 Nov 1987	Nippon Englehard
H2-61961 Takashi Itoh, Sigemitsu Matsuzawa Supported Platinum Alloy Electrocatalyst (Pt-Fe-Cu)	26 Aug 1989	1 Mar 1990	N.E. Chemcat
H2-236960 Takashi Itoh, Katsuaki Katoh Platinum Alloy Electrocatalyst (Pt-Fe-Co-Cu)	9 Mar 1989	19 Sept 1990	N.E. Chemcat
H4-87260 Takashi Itoh, Katsuaki Katoh, Shinji Kamitomai Supported Platinum Quarternary Alloy Electrocatalyst (Pt-Co-Ni-Cu)	31 July 1990	19 Mar 1992	N.E. Chemcat

Table 3
Continuation

Kokai Number	Application Date	Issue Date	Assignee
H4-135642 Paul Stonehart, Kazunori Tsurumi, Toshihide Nakamura, Akira Sato Platinum Alloy Catalyst and Process for Preparing (Pt-Ni-Co-Mn)	26 Sept 1990	11 May 1992	Tanaka Kikinzoku Kogyo & Stonehart Associates
H4-141236 Paul Stonehart Platinum Alloy Catalyst and Process of Preparing (Pt-Co-Ni-Au & Pt-Co-Ni-Mn-Au)	29 Sept 1990	14 May 1992	Tanaka Kikinzoku Kogyo & Stonehart Associates
H4-141233 Paul Stonehart, Masahiro Watanabe, Kazunori Tsurumi, Noriaki Hara, Toshihide Nakamura, Nobuo Yamamoto Electrocatalyst for Anode (Pt-Ru-Pd)	29 Sept 1990	14 May 1992	Tanaka Kikinzoku Kogyo & Stonehart Associates
H4-141235 Paul Stonehart, Masahiro Watanabe, Kazunori Tsurumi, Noriaki Hara, Toshihide Nakamura, Nobuo Yamamoto Electrocatalyst for Anode (Pt-Co-Ni)	29 Sept 1990	14 May 1992	Tanaka Kikinzoku Kogyo & Stonehart Associates

Table 3a
Japanese PAFC Alloy Performances from Table 3

Kokai Number Alloy Combination	Atom ratio	Performance 190-200 °C 0.5mgPt.cm^{-2}		
		Air @ 160mA	or 200 mA	O$_2$ @ 200mA mV)
S61-8851 (Pt-Co-Ni)	2:1:1		740	805
S62-163746 (Pt-Fe-Co)	2:1:1		795	
S62-269751 (Pt-Cu) (Pt-Co-Ni)	1:1 2:1:1		788 774	
H2-61961 (Pt-Fe-Cu) (Pt-Fe-Co) (Pt-Cu) (Pt-Co-Ni)	2:1:1 2:1:1 1:1 2:1:1	750 750 740 733		
H2-236960 (Pt-Co-Ni-Cu)	3:1:1:1	750		
H4-87260 (Pt-Fe-Co-Cu) (Pt-Co-Ni)	3:1:1:1 2:1:1	750 734		
		Performance decay (mV.1000^{-1} hrs)	Mass activity (mA.mg^{-1}Pt)	
H4-135642 (Pt-Ni-Co-Mn) (Pt-Co-Ni)	2:1:1	5	58	
H4-141236 (Pt-Co-Ni-Au) (Pt-Co-Ni-Mn-Au)		2	58	

The voltage change on going from air to Oxygen at 200 mA is +65mV.
The voltage change on going from 160mA to 200mA is -10mV.
Tafel slope is 90mV at 190°C (2.303RT/F).

It became obvious that long-term stability of high surface area electrocatalysts was as important, or even more important than short-term activity. Luczak[36] and Landsman pioneered the heat treatment of ternary alloy electrocatalysts in order to provide an ordered crystallite structure. This work was followed in Japan by Itoh and Katoh, and subsequently by

Watanabe *et al.*, also shown in Table 4. As is shown in Fig. 7, the ordering of the Pt-Co-Cr alloy does indeed provide for an enhanced electrode catalytic reduction of the oxygen molecule in operating PAFCs.

The stability of these alloys over long terms are of greater importance and previously it was noted that the Pt-V alloy was particularly fugitive. Analyses of the ordered Pt-Co-Cr alloy were obtained after 9000 hours of testing in hot phosphoric acid at 200 °C. It is particularly significant that even though the catalyst undergoes demetallization and surface area loss ("sintering"), performance gains are maintained such that after 9000 hours, the ordered alloy electrocatalyst still outperforms the initial performance of a pure platinum catalyst, as shown in Table 5.

The beneficial effects on the cathode performance by alloying and on producing an ordered alloy structure are shown in Fig 7.

Here, the performances of cathodes operating on pure oxygen are recorded at a potential of 0.9 V. The catalyst loadings are 0.75 mg Pt.cm^{-2} electrode surface. It is symptomatic that the reaction rate increases with temperature but the rate of increase is greater for the Pt-V alloy compared to that for pure platinum alone, while that of the ordered ternary Pt-Co-Cr alloy shows an even greater enhancement, shown in Table 5.

VII. THE MATRIX FOR PAFC'S

Phosphoric acid at 190-200 °C is a particularly ugly medium with respect to materials stabilities. An inert matrix is thus needed to maintain a thin-layer separation of the electrodes and many materials have been tested. Fortuitously, it was found that silicon carbide was inert and sufficiently stable to last for the lifetime of the fuel-cell. Since silicon carbide in the fuel-cell is a micro-fine rubble, attempts to bond it into a porous matrix structure using PTFE showed initial success. Over time, however, the cells showed an increase in resistance, leading to a loss of performance. Diagnostic tests showed that small bubbles of nitrogen were entrained within the matrix, coming from the air feed and were causing the matrix structure to have an effect comparable to deep sea diver's "bends", and expressing the phosphoric acid from the matrix pores. This was due to the wet-proofing nature of the PTFE. Substituting[48] a binder of polyethersulphone overcame this problem and showed improved strength, wetability and bubble pressure (resistance to gas cross-over), compared

Table 4
US-PAFC Ordered Alloy Patents

U.S. Patent Number	Application Date	Issue Date	Assignee
4,677,092 Frank J. Luczak & Douglas A. Landsman[36] *Ordered Ternary Alloy Fuel Cell Catalysts Containing Platinum and Cobalt and Method for Making the Catalysts (Pt-Co-Cr)*	23 Dec 1985	30 June 1987	International Fuel Cells
4,711,829 Frank J. Luczak & Douglas A. Landsman[47] *Ordered Ternary Alloy Fuel Cell Catalysts Containing Platinum and Cobalt (Pt-Co-Cr)*	16 Mar 1987	8 Dec 1987	International Fuel Cells
5,024,905 Takashi Itoh & Katsuaki Katoh *Platinum Alloy Electrocatalyst (Pt-Fe-Co-Cu)*	9 Mar 1990	18 June 1991	N.E. Chemcat (Japan)
5,189,005 Masahiro Watanabe, Paul Stonehart, Kazunori Tsurumi, Nobuo Yamamoto, Noriaki Hara, & Toshihide Nakamura *Electrocatalyst and Process of Preparing Same (Pt-Co-Ni)*	3 Apr 1992	23 Feb 1993	Tanaka Kikinzoku Kogyo & Stonehart Associates Inc.

Table 5
Lattice Dimensions, Compositions, and Catalytic Activities of Pt-Cr-Co Cathode Alloys Before and After Operation for 9000 Hours at 200 °C

Time hours	Alloy Composition Atom % Pt	Lattice Parameter nm	Physical Structure	Catalytic Activity mA.mg^{-1} Pt
0	50	0.3807	Ordered	39
0	50	0.3807	Random	39
0	100	0.3923	-	13
9000	60	0.3831	Ordered	23
9000	85	0.3895	Random	19

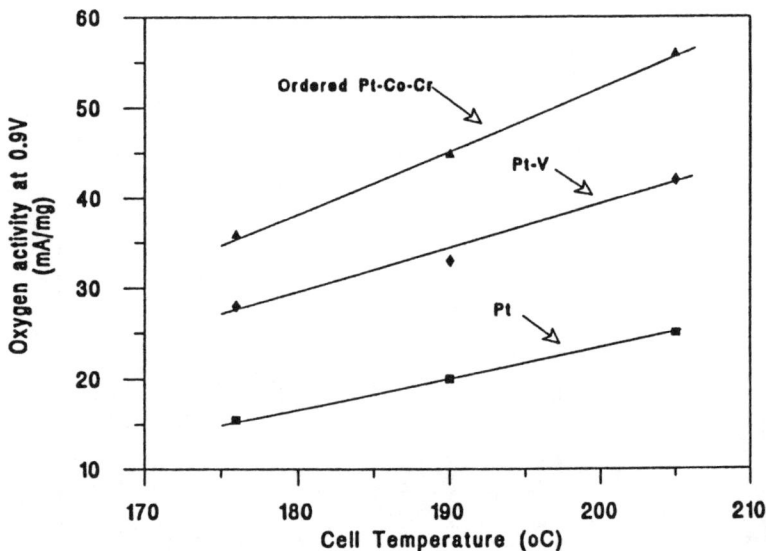

Figure 7. Platinum, platinum alloy, and ordered platinum alloy catalyst mass oxygen reduction activities at 0.9 V. 0.75 mg Platinum loading on the electrode. Influence of temperature.

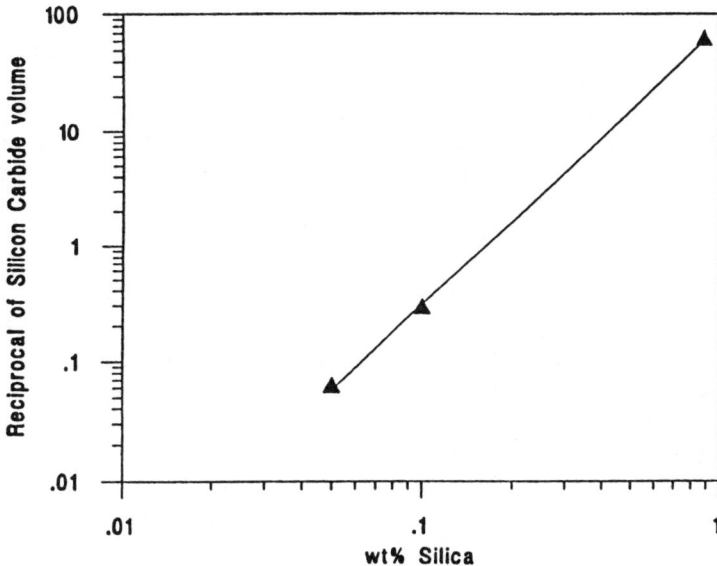

Figure 8. Weight % unreacted silica in silicon carbide particles as a function of silicon carbide volume.

to previous matrices. In the preferred formulation, the composite matrix has at least 38 % porosity and contains between 70-90 wt% SiC, with the balance being the hydrophilic polyethersulphone.

Traces of unreacted silica (SiO_2) in the SiC will produce a fluffy white precipitate, about which little is known. Certainly it is known that when hot concentrated phosphoric acid is in contact with glass hardware, the rapid formation of a fluffy gelatinous white precipitate is seen. From an X-ray diffraction analysis of the precipitate from operating fuel-cells, it was determined that the principal component was $Si_3(PO_4)_4$, although metallographic and SEM/EDS analyses support the presence of the other silico-phosphate complexes in varying amounts.

The amount of silica present in the silicon carbide depends upon the SiC particle sizes, these being 0.5 μm, 3 μm, and 5 μm. Washing the SiC with HF removed most of the silica, resulting in soluble-silica analyses of 0.89, 0.10, and 0.05 weight percent respectively, shown in Fig. 8.

Although there is a paucity of data, plotting a logarithmic function for the reciprocal of the SiC particle volume $(4/3\pi r^3)$ *versus* the logarithm of wt% silica found in the silicon carbide, provides a good linear relationship as shown in Fig. 8, allowing for a predictive solution. The smaller the SiC particle size the greater is the amount of free silica and the greater the phosphate formation rate in hot phosphoric acid. It appears that the selection and cleaning of the silicon carbide is critical to the long term operation of PAFCs and further preparation of SiC particles containing a minimum of silica is required.

VIII. CARBON ELECTROCATALYST SUPPORTS

There are many considerations that must be taken into account when choosing a particular carbon, or carbon structure, as an electrocatalyst support. In hot phosphoric acid at cathodic potentials, the carbon surface is capable of being oxidized to carbon dioxide. The degree of oxidation will depend on the pretreatment of the carbon (for instance, the degree of graphitization), on the carbon precursor, and the provenance. There are two important parameters that will govern the primary oxidation rate for any given carbon material in an electrochemical environment. These are electrode potential (the carbon corrosion is an electrochemical process and therefore will increase rapidly as the electrode potential is raised) and temperature.

An early detailed work in hot phosphoric acid is that of Kinoshita and Bett[49,50] who reported corrosion rates for four different carbons in phosphoric acid at temperatures up to 160 °C. All of the carbons tested showed similar behavior as a function of time. At constant potential the corrosion currents were relatively large but declined rapidly with time. These authors concluded that the rate of oxidation was dependent on the surface micro-structure for each specific carbon sample. Graphitized carbons showed lower specific oxidation rates than ungraphitized carbons.

Following from the previous work of Kinoshita and Bett,[49,50] a detailed analysis of carbon corrosion in hot phosphoric acid was carried out by Stonehart and MacDonald.[68]

The reactions contributing to this oxidation current are the formation of surface oxides on the carbon (quinones, lactones, carboxylic acids, etc.) and the evolution of carbon dioxide. Binder et al.[51] had previously examined the oxidation of numerous carbons in KOH, H_2SO_4, and H_3PO_4 but not at temperatures as high as those in the work of Kinoshita and Bett.

Binder *et al.*[51] had observed that the reaction products were carbon dioxide and carbon surface oxides and that the reaction rate was proportional to the BET surface area. These early results however, were not substantiated by Kinoshita and Bett.

Recent trends in the studies of carbon surface oxides have been to identify the specific properties of various oxygen complexes (*e.g.,* acidic CO_2 complex; non-acidic CO_2 complex; and CO complex) and how they modify the carbon surface properties.[52-54] It is considered that the electrochemical corrosion of carbon electrocatalyst supports in phosphoric acid proceeds through the formation of carbon surface oxides at preferred sites. These sites on the basal plane are generally at edges, dislocations and discontinuities in the layer planes of microcrystalline carbons. In particular, reactive sites (usually referred to as unsaturated sites) have been identified.[55] It is thought that once an unsaturated site has been formed, continuous propagation of the carbon corrosion (oxidation) will proceed until the crystallite orientation has been consumed. The review by Thomas[56] deals with the subject of the influence of dislocations in graphitic carbons on this oxidation. He points out that basal dislocations do not influence the oxidation but non-basal dislocations are inextricably associated with "active sites" in oxidation at the basal surfaces.

Heckman and Harling[57] examined the gas-phase oxidation of carbon black micro-structures and showed that oxidative attack of carbon crystallites was concentrated on the small crystallites, at the edges of layer planes and at lattice defects. Partial graphitization of a carbon black, so that only the outermost surface layers are well-ordered, causes oxidative corrosion within the core of the carbon particle, leaving an outer "shell". Consequently, similar behavior can be expected for ungraphitized and partially graphitized carbons in electrochemical environments.

Donnet and Ehrbuger[58] examined the solution oxidation of furnace blacks by ozone in the absence of potential control. They concluded that there were two simultaneous oxidation mechanisms: the formation of carbon dioxide and the formation of partial oxidation products (essentially poly-carboxylic acids). Later, Panzer and Elving[59] reviewed the nature of surface compounds on graphite electrodes (mainly pyrolytic graphite). They pointed out that Levy[60] had suggested that pyrolytic graphite become oxidized preferentially on the **c** planes (basal) rather than at the **ab** planes (edges). Later, Levy and Wong[61] claimed that oxidation of the pyrolytic graphite occurred in the **a** direction. Traces of both light and heavy metals, causing dislocations, increase the rates of oxidation on pyrolytic graphites.

The foregoing discussion serves to show that disordered carbon structures are oxidized more readily than well-ordered graphite planes and that dislocations and active sites provide nucleation points for attack of the carbon crystallite. Another factor that must be considered is that dispersed electrocatalysts, such as platinum, on the carbon surface are not benign. The electrocatalysts interact with the carbon causing local oxidation or corrosion, $i.e.$, the platinum catalyzes the corrosion of the carbon itself. In the presence of oxygen, which is the condition under which the electrocatalyst will operate, reduction intermediates from the oxygen ($e.g.$, HO_2^-) can have an accelerated corrosion effect.

An alternative to using commercially available carbon for electrocatalyst carbon substrates is to build a specific carbon structure having controlled properties. Thus, carbons have been prepared by the controlled pyrolysis of polyacrylonitrile (PAN) and contain surface nitrogen groups that act as peroxide decomposing agents.[62]

Simple treatments, such as heating the carbon in steam or in an oxidizing environment, change the surface structure. The surface porosity increases [63] due to operation of the Boudouard reaction:

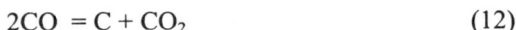

$$2CO = C + CO_2 \qquad\qquad (12)$$

which is favored towards the right-hand side at temperatures below 700°C. In this way, reactive end-groups would be removed and pores or dislocations in the carbon surface enlarged.[64] Other surface treatments that have been considered are the introduction of carbon-sulphur surface complexes. Puri and co-workers[65] have shown that ungraphitized carbons, after treatment with hydrogen sulphide or carbon disulphide at 600 °C, develop porosity and, moreover, the catalytic activity for hydrogen peroxide decomposition increases markedly on fixation of a small amount of sulphur (about 5 %) in the carbon surface.

Once a carbon support has been prepared. it is desirable to post-treat it to modify the surface structure in order to confer certain properties. Since the electrocatalyst is to be platinum on the carbon, even dispersion of the platinum crystallites over the carbon surface and minimal loss of surface area during fuel cell operation are important concerns.

One mechanism for surface area loss is crystallite migration, for which Kinoshita et al.[66] concluded that the mechanism of surface area loss was two-dimensional Ostwald ripening by means of ad-atom migration on the carbon surface. Nevertheless, trap sites for the migrating ad-atoms on the surface of the carbon can produce nucleation points for generation of new crystallites. In order to prevent crystallite migration, Bett and co-workers[67] had previously invoked the concept of trap sites on carbon surfaces to retard platinum crystallite migration in the gas phase. Physical properties of the carbon substrate, such as electronic conductivity, surface area, and surface morphology are important, since the former can contribute to resistive losses in the electrocatalyst structure and the latter may determine the sites on which the platinum electrocatalyst crystallites may be located. Both the initial deposition of the platinum crystallites on the carbon substrate and the subsequent surface area loss of the platinum crystallites under operating conditions by a surface migration mechanism can be influenced by the surface carbon structure.

There are several carbons that are candidates for PAFCs. Vulcan XC-72R (Cabot), a furnace black, has a low resistivity and a specific surface area of about 254 m^2 g^{-1}. Acetylene black (Shawinigan) is a very clean carbon with a structure that is very different from Vulcan XC-72R. The initial carbon content of the Shawinigan material is 99.5 % compared to 98 % for the Vulcan XC-72R. The initial specific surface area is lower (65 m^2 g^{-1}) but this type of carbon has been used frequently in fuel-cell electrode structures. Monarch 800 (Cabot), another furnace black, has a BET surface area that is comparable with that of Vulcan XC-72R but has a lower oxygen content. The greatest difference between the two is the oil absorption number, which is a measure of the carbon structure. The dibutylphthalate oil absorption for Vulcan XC-72R is 1.85 x 10^{-3} g^{-1} against 0.75 x 10^{-3} g^{-1} for the Monarch 800. Darco KBB is an active carbon from a cellulosic precursor, and has a very high BET surface area (1540 m^2 g^{-1}). A carbon material, BP800-1600C-Cl2 is made from "Black Pearls" (Cabot Corp.) and has been treated in chlorine at 1600 °C. This procedure removes all of the metals volatile chlorides that otherwise contribute to the ash. Consequently, this carbon is very clean and has undergone some slight degree of reorganization. The "Black Pearls" carbon is analogous to the Monarch 800 except that it has been previously pelletized. Vulcan XC-72R, after heat treatment to 2500 °C for two hours under argon, has a BET surface area of 65 $m^2 g^{-1}$, and is used as a graphitized carbon. These carbons provided a spectrum of specific surface areas which enable diagnoses to be made of the mode of corrosive attack of the carbons in the hot phosphoric acid; whether it is at disordered carbon groups, edge sites, or incomplete faces. Kinoshita and Bett [49,50], for Vulcan XC-72, found a corrosion current of 1.3 x 10^{-4} A m^{-2} carbon surface at 1.0V and 135 °C after 100 minutes. Using a value of 200 $m^2 g^{-1}$ for the carbon surface area, this becomes 2.6 x 10^{-2} A g^{-1} carbon.

In long time tests at 210 °C and 1.0V, for both the XC-72R and the acetylene black in the as-received condition, the rate of corrosion of the XC-72R is much greater than that of the acetylene black out to 500 minutes, then the Vulcan-XC72R corrosion current decreases below that of the acetylene black. This is considered to be due to significantly greater amounts of carbon being removed from the XC-72R, causing a decrease in the measured current, equivalent to the corrosion reaction rate. The acetylene black after 500 minutes, also exhibits the start of a steady corrosion current that does not decrease with time. Since it appears that the decrease in carbon corrosion with time is due principally to the progressive removal of the most highly disordered carbon at the surface, and attack at the micropores and dislocations, it should be surmised that the activation energy of the reaction will appear to change and so will the form of the corrosion equation. Thus, the steady corrosion currents at long times will represent more closely to the fundamental underlying characteristics of the carbon structure rather than will the ephemeral results that are seen due to surface disorder.

The magnitude of carbon loss that is observed is of some concern, since at 1000 minutes, which is only 16 hours, the Vulcan XC-72R shows a weight loss of 7 % at open-circuit values. It was found that at the highest potentials for studying the corrosion of the carbons, gas evolution within the porous electrode structure was distorting the data.

Tafel slopes were obtained for the carbons at differing times. It was found that, in general, the Tafel slopes became more acute as the carbon was removed from the surface. (The value of dE/dlog i became less). In addition, the Tafel slopes differed according to the type of carbon and the history of the carbons. Heat treating Vulcan XC-72 showed a linear weight loss with increase in heat-treatment temperature. At temperatures of 750, 1350, 1800, and 2500 °C the weight losses were 0.75, 3.5, 5, and 8 wt%, respectively. This is due probably to removal of low molecular weight fragments and reduction of surface oxide compounds (lactones, phenols, carboxylic acids, etc.).

The corrosion currents for Vulcan XC-72R after heat treating at 2700 °C and 3000 °C are shown in Fig. 9. There is a continual decrease in the corrosion currents over the first 4000 minutes to very low corrosion values.   Reduction in the corrosion rate is due both to the lower specificsurface area of the carbon and the change in the Tafel slope. The specific corrosion rate at long times suggests that the numbers of active sites on the surface are comparable on a per square cm basis. If a parallel is drawn with the oxidation of carbon in the gas phase, which is by attack at active sites, then there is no suggestion that atoms in the basal plane are active. Corrosion attack must be at edges and dislocations.

The literature teaches that heat treatments of high surface-area carbons produce significant changes in both the surface area and the surface oxide content in the temperature range between 400°C and 1000 °C. For an oxidized carbon, significant increases in the BET surface area are seen with heat treatments at 800 °C. It is thought that the removal of oxide opens up micro-pores giving the surface area enhancement. The removal of oxides improves the inter-particle contact and thus makes the carbon more conductive. Restructuring of the carbon surface is expected to occur at temperatures above 1400 °C with progressive graphitization at higher temperatures.

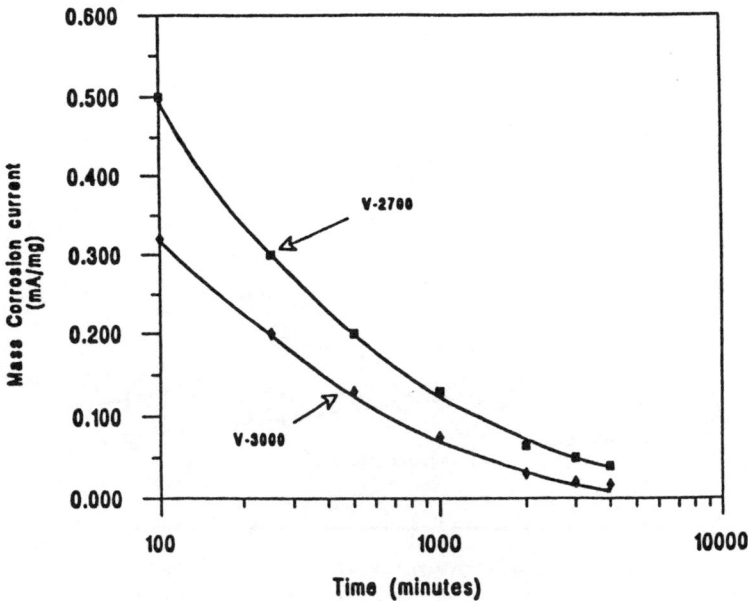

Figure 9. Corrosion rates 1.0 V in 100 % phosphoric acid at 0.8 V for graphitized Vulcan XC-72R carbon at two graphitization temperatures: 2700 °C [■] and 3000°C [◆].

The changes in the Tafel slopes observed during corrosion of the Vulcan XC-72R are probably associated with the degree of disorder of the carbon crystallite surfaces. Shawinigan acetylene black was also heat treated at temperatures up to 2500 °C. but for such materials there were no large changes in the Tafel slopes. By and large, the Tafel values were close to 100 mV. Neither the Tafel slopes nor the corrosion currents changed over long periods of time (out to 6000 minutes). It was concluded that with heat treatment, the Shawinigan crystallites become more ordered and probably had fewer dislocations. There is some evidence that slits and dislocations in the carbon surface are opened up during the corrosion process but further work is needed to confirm this.

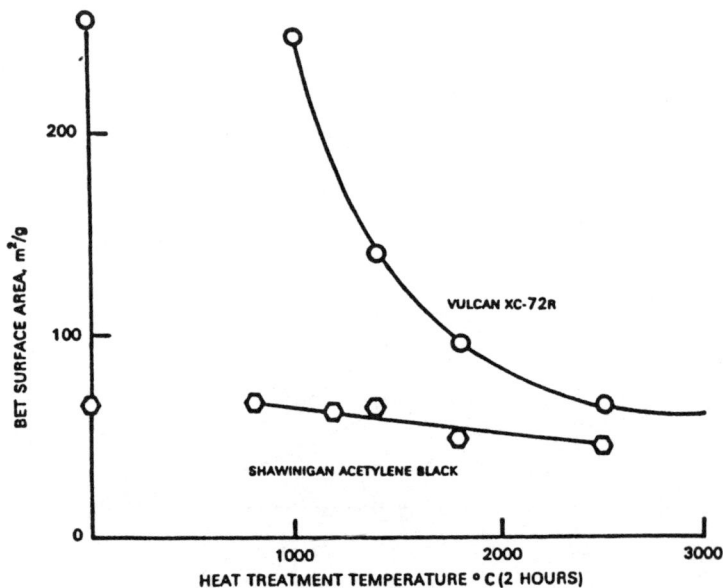

Figure 10. B.E.T. Surface areas of heat treated Vulcan XC-72R and Shawinigan acetylene black carbons.

For Vulcan XC-72R, the specific corrosion rates (A.real m^{-2}) at
1.0 V and 180 °C in H_3PO_4 were essentially independent of heat-treatment
temperature; however, since the Tafel slopes decrease with increasing
heat-treatment temperature, heat treatment affords increased life at the
fuel-cell cathode operating potential (0.7 V). Little difference in either the
Tafel slope or specific corrosion rate at 1.0 V was observed between as-
received and heat-treated samples of Shawinigan acetylene black.

From the plots of BET specific surface areas as a function of heat-
treatment temperature for both Vulcan XC-72R and Shawinigan acetylene
black, shown in Fig. 10, it is obvious that the latter black has experienced
a much higher temperature than that for Vulcan XC-72R. It is well known
that Shawinigan acetylene black is formed at a high temperature, which
leads to significant surface area diminution.

Differences in the electrochemical behavior of carbons are seen with
respect to their structural properties as reflected by lattice parameters.
Heat treatment was found to affect the structures of heat-treated samples
of Vulcan XC-72R: d_o (the spacing between graphitic layers, $c/2$, where
c is the unit cell) was found to vary with heat-treatment temperature. In
Fig. 11, the lattice parameter, d_o, is plotted for the various heat-treated

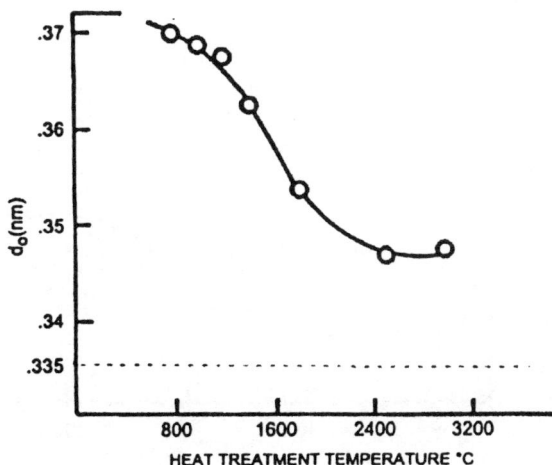

Figure 11. Changes of Vulcan XC-72R lattice parameter with heat
treatment.

Vulcan samples. Since d_o for pure graphite is 0.335 nm, it is apparent that heat-treated Vulcan never becomes fully graphitized but rather is constrained due to carbon ribbon formation, much like that seen for glassy carbons.

Figure 12 shows the change in the Tafel slope with lattice spacing. This indicates the degree of disorder on the carbon surface--the larger the Tafel slope, the greater the degree of disorder. The plot in Fig. 12 is linear with the exception of the as-received material (The as-received Vulcan XC- 72R is turbostratic, with little long range order).

If this plot is extrapolated to d_o = 0.335 nm for a fully graphitic structure, the Tafel slope is exactly RT/F, for this carbon, at this temperature. This supports the belief that the Tafel slope is sensitive to the d_o lattice dimension of the carbon. For any carbon that is not fully graphitized, the observed Tafel slopes result from mixtures of contributions of the two types of surfaces, disordered and ordered. As the amount of disorder decreases, the Tafel slope more closely approaches the expected values for the basal structure of the carbon.

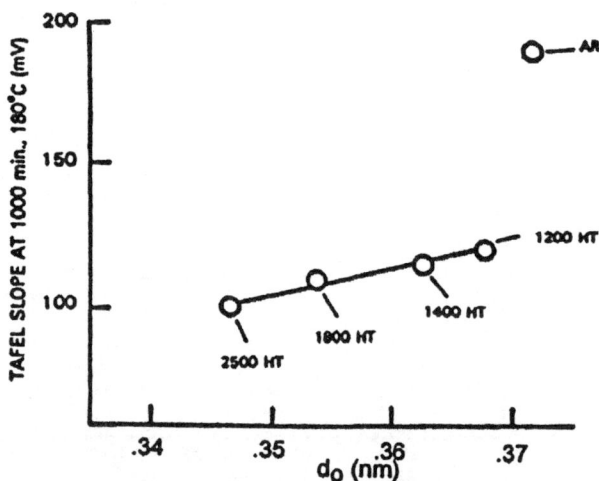

Figure 12. Carbon corrosion Tafel slopes versus d_o for Vulcan XC-72R heat treated (HT) at various temperatures. AR is "as received".

Since the process of corrosion of carbon is expected to have an equation of the form:

$$C + 2H_2O = CO_2 + 4H^+ + 4e^-$$ (13)

Then, in non-aqueous electrolytes, the corrosion of carbon is suppressed. Under conditions where the phosphoric acid concentration is very high, it might be supposed that the electrolyte could be considered as a non-aqueous solution, in which case the corrosion rate would be kinetically limited by the water activity. Arrhenius plots of corrosion currents at 1.0 V at various acid concentrations are shown in Fig. 13. It can be seen that there are two distinct slopes, one for acid concentrations greater than 97 wt% and one of acid concentrations lower than 93 wt%.

The preliminary results indicated that there is indeed a marked effect of water concentration on the corrosion rate. In further experiments at 97, 93 and 85 % H$_3$PO$_4$, it appeared that the activation energies were one

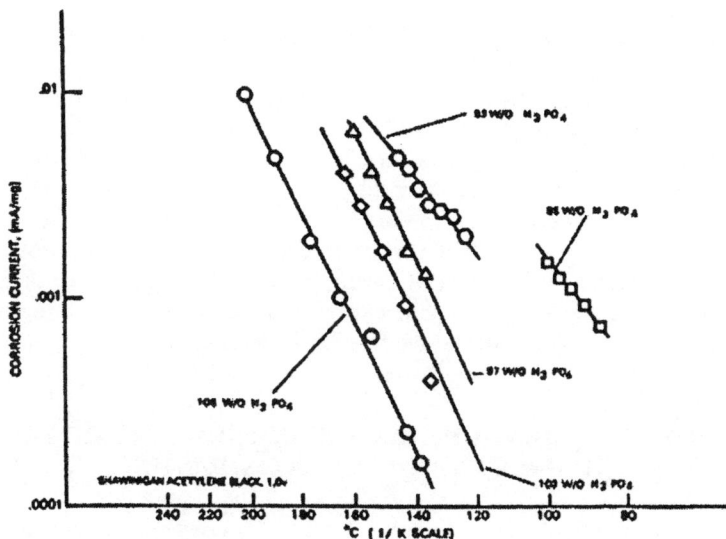

Figure 13. Arrhenius plots for the corrosion rates of Shawinigan acetylene black at 1.0 V in phosphoric acid at various concentrations.

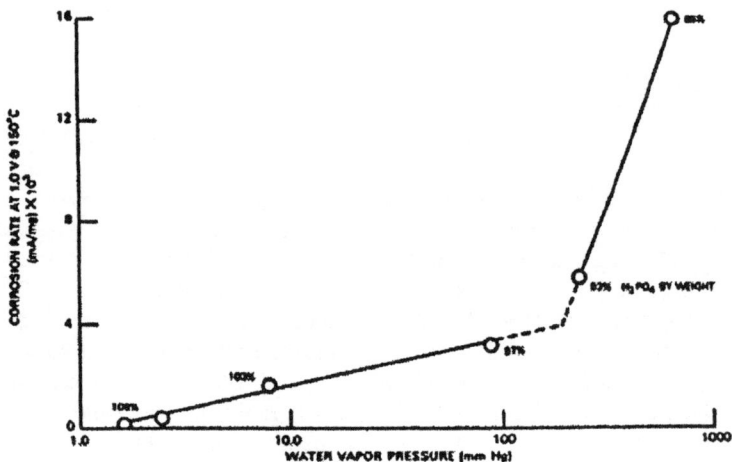

Figure 14. Effect of water vapour pressure on the corrosion rate of Shawinigan acetylene black as a function of phosphoric acid concentration.

value at acid concentrations of 97 % and above ($25.8 \pm$ kcal mole^{-1}), and another value ($14.2 \pm$ kcal mole^{-1}) at H_3PO_4 concentrations of 93 % and below. This is significant, since the change in phosphoric acid form from *ortho* to *pyro* occurs at about 96 %. Corrosion rates for the Shawinigan acetylene black at 150 °C as a function of the logarithm of the partial pressure of water are plotted in Fig. 14. The break in the slope is persuasive that the structure of the phosphoric acid electrolyte plays an important role in the carbon corrosion kinetics. These results provide further insight into the effect of electrolyte on the corrosion rate of the carbon substrate. It appears that two effects are operating: the water concentration in the acid, and the form of phosphoric acid.

IX. DEVELOPMENT OF ALLOY ELECTROCATALYSTS FOR HYDROGEN MOLECULE OXIDATION

Alloys for the oxidation of hydrogen in PAFCs are not as important as advanced alloys for oxygen reduction. At the start of this review, it was emphasized that cost of the catalyst is a large factor in translating materials from academic curiosities to commercial reality. Most of the emphasis in reducing platinum electrocatalyst costs has been on increasing the surface area. Substitution of the platinum for another, less expensive material, is equally important.

Under concentration control, the reversible hydrogen electrode exhibits Nernstian reversibility. This provides for a potential shift of 29.75 mV at room temperature, which translates to a shift of 46.8 mV at 200 °C for each decade of change in hydrogen concentration. Under fuel-cell operating conditions with highly dispersed electrocatalysts, it is possible to approach the kinetic rate; determined by the dual-site dissociation of the hydrogen molecule, *viz*.:

$$H_2 = 2H\bullet \text{ (Tafel reacion)} \tag{14}$$

which is followed by the more rapid ionization step:

$$2H\bullet = 2H^+ + 2e^-\text{(Volmer reaction)} \tag{15}$$

Whereas the rate-determining step for hydrogen molecule oxidation now is recognized[69,70] to be the dissociative chemisorption of the hydrogen molecule on dual sites at the platinum surface, the rate of this step is so high that in most electrochemical environments platinum electrocatalysts are almost always operating under diffusion control.

There are no "Tafel Slopes" for hydrogen oxidation, since the rate determining step shown above is a purely chemical reaction (dissociation) and is the same as that measured for hydrogen-deuterium exchange in the gas-phase. Under these conditions, therefore, blocked sites act as inhibitors to the hydrogen molecule oxidation reaction. Hence, those sites with adsorbed hydrogen atoms are denied for the hydrogen molecule dissociation step. In this regard, the hydrogen molecule oxidation reaction rate increases with potential due to the decrease in coverage by hydrogen atoms in the equilibrium adsorbed hydrogen atom isotherm.

Carbon monoxide is produced from fuel processing of hydrocarbons, *e.g.* reforming, and is strongly adsorbed as a surface metal carbonyl, inhibiting the dual-site hydrogen molecule dissociation step.

Adsorption of carbon monoxide on platinum is not irreversible, since on switching carbon monoxide gas concentrations or with switching to pure hydrogen, the electrode potential rapidly relaxes and reverts to those values previously obtained for the absence of carbon monoxide. This shows that the kinetics of adsorption/desorption are rapid and reversible but the equilibrium is strongly favored to approach full coverage by the carbon monoxide.

As the temperature of the phosphoric acid fuel cell increases, so the oxidation rate for hydrogen in the presence of carbon monoxide increases, by many orders of magnitude. This is not due to the rate of the hydrogen oxidation reaction increasing with temperature but the equilibrium adsorption isotherm for the carbon monoxide shifts to a situation where more bare platinum sites become available for the dual site hydrogen molecule dissociation step. The influences of temperature on the hydrogen molecule oxidation rates are shown as Arrhenius plots in Figure 15 for overvoltages of 10 mV and 25 mV from equilibrium in the same electrolyte and at the same temperature. A voltage shift from equilibrium is necessary to produce a concentration driving force for the oxidation reaction to be observed.

The poisoning influence of carbon monoxide diminishes as the temperature increases and at a 1 % and 2 % CO concentration, the hydrogen molecule oxidation rate approaches that of an unpoisoned platinum electrocatalyst at 200 °C.

The hydrogen fuel for the PAFC is produced from hydrocarbons by fuel processing. It is worthwhile describing the gas-phase catalytic processes used to generate the hydrogen for the PAFC since the product compositions will determine how the PAFC anode operates. Methane or methanol are the easiest fuels to process, since reforming does not require breaking of a carbon-carbon bond.

There are three general catalytic processes that can be used to strip hydrogen from single carbon molecules. These processes are known as reforming reactions and they convert the hydrocarbon into a mixture known as "synthesis gas", where hydrogen and carbon monoxide molecules contain the original fuel value. The principal reactions are:

a)   Steam Reforming (SR), which is an endothermic reaction:

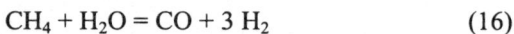

$$CH_4 + H_2O = CO + 3 H_2 \qquad\qquad (16)$$

Steam reforming runs at 650-1000°C, typically on a nickel catalyst supported on alumina and typically runs at about 95 % of completion.

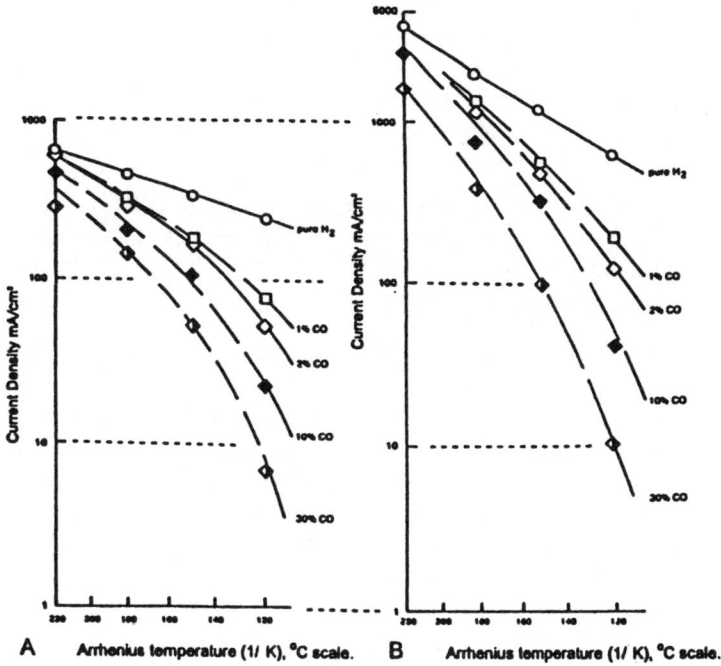

Figure 15. Arrhenius plots for hydrogen oxidation and hydrogen oxidation in the presence of carbon monoxide. A results at 10 mV; and B results at 25 mV polarization versus the reversible hydrogen electrode in the same electrolyte and at the same temperature. 0.5 mg Pt on carbon per square centimeter of electrode.

b) Partial Oxidation Reforming (POX) which is an exothermic reaction:

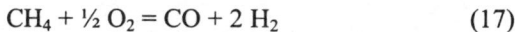

$$CH_4 + \tfrac{1}{2} O_2 = CO + 2 H_2 \qquad (17)$$

Partial oxidation runs at 700-1000 °C, typically on a platinum or rhodium catalyst supported on alumina or other oxides; and

c) Autothermal Reforming (ATR) which combines steam reforming and partial oxidation reactions to produce a roughly thermo-neutral reaction:

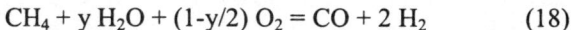

$$CH_4 + y\, H_2O + (1\text{-}y/2)\, O_2 = CO + 2 H_2 \qquad (18)$$

Auto thermal reforming has a theoretical thermally neutral value for $y = 1.115$. auto thermal reforming typically runs at 700-1000 °C on platinum or rhodium catalysts supported on alumina or other oxides. Since auto thermal reforming reactions are thermally neutral, then feeds must be pre-heated to within 200 °C of the reactor outlet temperature.

In practice, all three reforming reactions are run with an excess of water and/or oxygen, for a combination of kinetic and thermodynamic considerations. Excess water reacts via the Water Gas Shift (WGS) mechanism to produce more hydrogen:

$$CO + H_2O = CO_2 + H_2 \tag{19}$$

Excess oxygen reacts to form fully oxidized products, thereby reducing the ultimate hydrogen yield. In the initial stages of both partial oxidation and auto thermal reforming, there is a tendency to reduced selectivity, $i.e.$, generating more fully oxidized species. The best catalysts show higher selectivity to less oxidized products:

$$CO + \tfrac{1}{2} O_2 = CO_2 \tag{20}$$

$$H_2 + \tfrac{1}{2} O_2 = H_2O \tag{21}$$

Water gas shift reactions are typically in dynamic equilibrium at the exit of all three reactors. As a result of thermodynamics, lower exit temperatures favor lower CO and higher hydrogen in the product.

It can be seen from the above reactions, significant quantities of CO are produced, which poison the anode platinum catalysts. As a result, one or more dedicated water gas shift reactors always follow the reformer. The water gas shift reactor converts most of the CO, thereby generating more hydrogen. The CO conversion is limited by water gas shift equilibrium, and conversion increases with decreasing temperatures. When the reformer product has less than about 4% CO, a single stage water gas shift reactor is sufficient. This low temperature water gas shift typically operates at 180-230 °C on Cu/ZnO catalysts supported on alumina. When the reformer product has more than about 4% CO, the low temperature water gas shift is preceded by a high temperature reactor stage and typically is conducted on a Fe/Cr oxide catalyst operating at 300-450 °C.

Sulphur compounds in the fuel (0.05-0.01 %) must be removed because they would poison the reformer catalyst. Sulphur compounds are removed by using a hydrodesulphurization catalyst (HDS), typically noble metals on alumina, to convert the organic sulphur to hydrogen sulphide. Then a zinc oxide adsorbent bed is used to trap the hydrogen sulphide:

$$H_2S + ZnO = ZnS + H_2O \qquad (22)$$

These gas-phase fuel processing reactions have significantly different operating temperatures and thermal requirements (exothermic versus endothermic), so that thermal matching of the various gas streams through heat exchangers is an engineering requirement.

The methane feedstock after fuel processing entering the fuel cell stack contains 19.2 % water; 63.6 % hydrogen; 0.9 % unreacted methane; 0.6 % carbon monoxide; and 15.2 % carbon dioxide, so that once the PAFC operating temperature exceeds 165 °C, CO poisoning is a lesser problem. The gas compositions exiting the PAFC stack are 26.8 % water; 32 % hydrogen; 2.1 % unreacted methane; 1.3 % carbon monoxide; and 37.9 % carbon dioxide. The hydrogen and methane in the exit gas are recycled and burned to provide heat for the steam reformer.

The operating temperature of the PAFC stack has gradually evolved from 165 °C; to 180 °C; and then to the present 190 °C. Several factors favor operating the stack at higher temperatures:

a)   CO is concentrated at the end of the fuel cell stack, increasing CO poisoning,

b)   Water gas shift catalyst activity declines over time, increasing CO in the PAFC feed.

c)   Oxygen reduction activity is lower at lower temperatures.

The ability to operate the PAFC stack at 190 °C has in large part enabled a reduction in Pt loading to 0.25 mg Pt/cm^2 electrode area. In addition, the efficacy for increasing the temperature removes the need for producing more advanced alloy electrocatalysts to overcome the carbon monoxide poisoning problem. That is not the case for PEMFCs operating at lower temperatures below 100 °C.

Stonehart examined highly dispersed platinum and alloys of Pt-Pd[70-73] and found that with sophisticated catalyst preparation techniques, it was possible to maintain very small crystallites of these binary alloys on carbon supports, and that the alloys were more active than Pt alone for hydrogen oxidation in the presence of both carbon monoxide and hydrogen sulphide catalyst poisons. Not only was there a significant cost advantage but also the resultant alloys were significantly harder than either of the alloy components. This meant that the alloy particles were resistant to surface area loss by surface atom migration or crystallite migration. Further improvements are achieved by alloying the electrocatalyst crystallite structure as shown with Pt-Co-Ni and the Pt-Ru-Pd in Table 3.

X. CONCLUSIONS

This review identifies the progress that has occurred over many years for developing the electrocatalyst and matrix technology towards commercial realization of phosphoric acid fuel-cells.

For oxygen reduction, nano-dimensioned platinum electrocatalyst particles supported on carbon are employed and have specific activities in phosphoric acid at 190 °C of 0.6 A m^{-2} real Pt surface. Careful examination of the literature shows that there is no "crystallite size effect" in the catalyst activity. Loss of apparent activity is, however, seen when the Pt crystallite separation on the carbon support surface becomes too small. The critical separation between crystallites is 17 nm, below which the apparent specific activity decreases. A similar result is seen in PEMFCs. Dramatically increasing the separation leads to a slow increase in the apparent specific activity due to the expansion of the three-dimensional diffusional boundary layer to an individual crystallite. Initially, binary and then ternary alloys for oxygen reduction showed activity enhancements but the non-noble metal components become progressively leached out, although these activity enhancements are retained for many thousands of hours. The ternary alloys of Pt-Cr-Co show sufficient lifetime to be commercially viable, although the more advanced Pt-Ir-Cr and Pt-Ir-Fe ones do show increased stabilities beyond that of Pt-Cr-Co. Attempts at understanding the mode of operation of these alloys have come to the point of either treating it as a simplistic "roughening" of the alloy surfaces, due to leaching of the non-platinum elements, or in terms of the interatomic distances of the alloys, but these aspects bear further examination. It is not reasonably expected that surface roughening is a major operating principle for the explanation of the observed enhanced electrocatalysis due to the dissolution of the non-noble components and rapid rearrangements of the alloy surfaces ("sintering") during fuel-cell operation. Dissolution of the platinum itself at cathode potentials also limits the lifetime of the cells.

In the anode area, dilution of the platinum by any less expensive catalyst material is attractive, especially since the performances of these alloys in PAFCs have been demonstrated to be superior to platinum alone in the presence of carbon monoxide. In addition, consideration must be made for the fuel compositions and hydrogen utilizations arising from the fuel processing chemistries. With regard to the matrix, silicon carbide of a critical size can be bonded with polyethersulphone providing one answer to the demand for long term stability, although new materials are continually being sought.

Finally, the understanding of the chemistry of carbon and its stability in hot phosphoric acid in relation to its use as electrocatalyst supports, has led to the use of highly graphitic carbons, where the fundamental electrochemistry now is well defined.

ACKNOWLEDGMENTS

John Bett, friend and colleague, was one of the first to bring his expertise in gas-phase catalysis to high surface area electrochemistry, and who, together with Kimio Kinoshita and other co-workers at Pratt and Whitney Aircraft, performed pioneering work on supported electrocatalysts in these difficult reaction environments. Doug Landsman and Frank Luczak worked on advanced alloys, together with Paul Plasse and Aaron Gaskin, who carried out much of the work. Dick Breault has made many contributions to the technology, especially towards the silicon carbide matrix. Dick Bellows also is especially acknowledged for his sagacity on fuel processing. Colleagues and friends at Yamanashi University, under the direction of Masahiro Watanabe, together with co-workers at Tanaka Kikinzoku Kogyo K.K. learned and then contributed much to the field; especially Kazunori Tsurumi, Takayuki Nakamura and many others. Joy MacDonald should have a special note, since she was the one who performed much of the very elegant work towards solving the difficult electrochemistries on high surface-area carbons in these demanding systems. Over the years, our funding from the United States Department of Energy, the National Aeronautics and Space Administration, the Electric Power Research Institute, and the Gas Research Institute was critical in promoting these sciences for advanced fuel-cell technologies. Finally, it is with pleasure that we acknowledge permission from UTC

Fuel Cells, LLC., to publish many results contained in internal memoranda, and to the Board of Directors of Stonehart Associates Inc. for funding this study.

REFERENCES

[1]   J. Bett, J. Lundquist, E. Washington, and P. Stonehart *Electrochimica Acta*, **18**, (1973) 343
[2]   P. Stonehart and J. Lundquist, *Electrochimica Acta*, **18**, (1973) 349
[3]   P. Stonehart and J. Lundquist, *Electrochimica Acta*, **18**, (1973) 907
[4]   K. Kinoshita and P. Stonehart, *Electrochimica Acta*, **20**, (1975) 101
[5]   K. Kinoshita, D. Ferrier and P. Stonehart, *Electrochimica Acta*, **23**, (1977) 45
[6]   L.J. Bregoli, Electrochimica Acta, **23**, (1978) 498
[7]   M. Watanabe, H. Sei, and P. Stonehart , *J. Electroanal. Chem.* **261**, (1989)375.
[8]   J.S. Buchanan, L. Keck, J. Lee, G.A. Hards, and N. Scholey, *International Fuel Cell Workshop*. Tokyo 1989, pp. 29.
[9]   J.S. Buchanan, G.S. Hards, L. Keck, and R.J. Potter, *Fuel Cell Seminar* 1992 pp. 505.
[10] Edward R. Tufte, *Envisioning Information*, Graphics Press, Cheshire, Connecticut ISBN 0-961-3921-1-8, 1990.
[11] P. Stonehart, in *Electrochemistry and Clean Energy*, Ed. by J.A.G. Drake, ISBN 0-85186-472-4, The Royal Society of Chemistry, Special Publication #146, London England, 1994, pp 16-32.
[12] M. Mizuhata, K. Yasuda, K. Oguro, and H. Takenaka, *Denki Kagaku*, (1997) 692.
[13] K. Kinoshita, *J. Electrochem.* Soc. **137**, (1990) 845
[14] K. Kinoshita, *Electrochemical Oxygen Technology*, Wiley-Interscience,. ISBN 0-471-57043-5, 1992, pp.43.
[15] N. Giordano, E. Passalacqua, L. Pino, A.S. Arrico, V. Antonucci, M. Vivaldi, and K. Kinoshita, *Electrochim. Acta*, **36**, (1991) 1979.
[16] N. Giordano, E. Passalacqua, L. Pino, V. Alderucci, and P.L. Antonucci, *International Fuel Cell Conference*, Tokyo, II-A-3, 1992, p. 25.
[17] S. Mukerjee, 'Reviews of Applied Electrochemistry" No. 23, in *J. Applied Electrochem.* **20** (1990) 537.
[18] M. Watanabe, K. Tsurumi, T. Mizukami, T. Nakamura, and P. Stonehart, *J. Electrochem.* Soc. **141** (1994) 2659.
[19] K. Ito, H. Suzuki, T. Iida, and Y. Yamada, *Bull. Nagoya Inst. Tech.*, **25** (1973) 377.
[20] P. Bindra, S.J. Clouser, and E. Yeager, *J. Electrochem. Soc.* **126** (1979) 1631.
[21] P. Stonehart, J. Baris, *Preparation and Evaluation of Electrocatalysts for Phosphoric Acid Fuel Cells,* (DEN-AC03-78ET15365), U.S. Department of Energy, Washington, DC., 1980.
[22] P. Stonehart, J. Baris, J. Hochmuth, and P. Pagliaro, *Preparation and Evaluation of Advanced Catalysts for Phosphoric Acid Fuel Cells,* (NASA CR-168223: DEN3-176), U.S. Department of Energy, 1984.
[23] P. Stonehart, *Deutsche Bunsengesellschaft für Physikalische Chemie.* **94** (1990) 913.
[24] E. Passalacqua, L. Pino, M. Vivaldi, N. Giordano, M. Scagliotti, and N. Ricci, *Fuel Cell Seminar,* 1990, p. 355.
[25] V.M. Jalan USP 4,202,934 *Noble Metal-Vanadium Alloy Catalyst and Method for Making,* File date 03 July 1978, Issue date 13 Mar 1980.

[26] L. Borodovsky, J. G. Beery, and M. Paffett, *Nuclear Instrumentation and Methods in Physics Research* (1987).

[27] D. A. Landsman, F. J. Luczak (Pt-Cr) USP 4,316,944 *Noble Metal-Chromium Alloy Catalysts and Electrochemical Cell*, File date 18 June 1980, Issue date 23 Feb. 1982.

[28] V. Jalan and E. J. Taylor, *J. Electrochem. Soc.* **130** (1983) 2299.

[29] K.R. Daube, M. Paffett, S. Gottesfeld, and C. Campbell, *J. Vac. Sci. Technol.* A 4, (1986) 1617.

[30] S. Gottesfeld, M.T. Paffett, and A. Redondo, *J. Electroanal. Chem.* **205** (1986) 163.

[31] M.T. Paffett, K.R. Daube, S. Gottesfeld, and C. T. Campbell, *J. Electroanal. Chem.* **220**, (1987) 269.

[32] M. T. Paffett, J. G. Beery, and S. Gottesfeld, *J. Electrochem. Soc.* **135** (1988) 1431.

[33] P. N. Ross, *Extended Abstracts*, Electrochemical Soc. Meeting, Los Angeles, Vol. 89-1,1989, p. 659.

[34] B. C. Beard and P. N. Ross, Jr., *J. Electrochem. Soc.* **137** (1990) 3368.

[35] F. J. Luczak, D. A. Landsman, (Pt-Cr-Co) USP 4,447,506 *Ternary Fuel Cell Catalysts containing Platinum, Cobalt, and Chromium*, File date 17 Jan 1983, Issue date 8 May 1984.

[36] Frank J. Luczak and Douglas A.Landsman,USP4,677,092, *Ordered Ternary Alloy Fuel Cell Catalysts Containing Platinum and Cobalt and Method for making the Catalysts. (Pt-Co-Cr)*, File date 23 Dec 1985, Issue date 30 Jun 1987.

[37] Chung-Zong Wan (Pt-Ga) USP 4,822,699, *Electrocatalyst and Fuel Cell Electrode Using the Same; Platinum-Gallium Alloy on Conductive Carrier*, File date 20 Dec 1982, Issue date 18 Mar 1989.

[38] Chung-Zong Wan (Pt-Fe) EPA 84303984.3, *Platinum-Iron Electrocatalyst and Fuel Cell Electrode Using the Same*, File date 13 June 1984, Issue date 27 Dec 1984.

[39] T. Ito, S. Matsuzawa, K. Kato (Pt-Cu) USP 4,716,087, *Platinum-Copper Alloy Electrocatalyst and Acid-Electrolyte Fuel Cell Electrode Using the Same*, File date 10 Dec 1986, Issue date 29 Dec 1987.

[40] T. Ito, K. Kato, S. Matsuzawa (Pt-Fe-Co) USP 4,794,054 *Platinum Alloy Electrocatalyst and Acid-Electrolyte Fuel Cell Electrode Using The Same; Iron-Cobalt on Conductive Carrier*, File date 24 June 1987, Issue date 27 Dec1988.

[41] F. J. Luczak, D. A. Landsman (Pt-Ga-Co/Cr) USP 4,806,515, *Ternary Fuel Cell Catalyst Containing Platinum and Gallium*, File date 16 Nov 1987, Issue date 21 Feb 1989.

[42] D. A. Landsman, F. J. Luczak (Pt-Ga-Cr/Co) USP 4,880,711, *Ternary Fuel Cell Catalyst Containing Platinum and Gallium*, File date 19 Oct 1988, Issue date 14 Nov 1989.

[43] F. J. Luczak [33] (Pt-Ir-Cr) (Pt-Ir-Fe) USP 5,013,618, *Ternary Alloy Fuel Cell Catalysts and Phosphoric Acid Fuel Cell Containing the Catalysts*, File date 5 Sept 1989, Issue date 7 May 1991.

[44] P. Stonehart, *Deutsche Bunsengesellschaft für Physikalische Chemie.* **94**, (1990) 913.

[45] P. Stonehart, Extended Abstracts 31st Meeting ISE, Venice, Italy, 1980, pp. 96.

[46] P. Stonehart, "Reviews of Applied Electrochemistry 32" *J. Appl. Electrochem.* **22**,(1992) 995.

[47] Frank J. Luczak and Douglas A. Landsman, USP 4,711,829, *Ordered Ternary Fuel Cell Catalysts Containing Platinum and Cobalt*, (Pt-Co-Cr), File date 16 Mar 1987, Issue date 8 Dec 1987.

[48] J.C. Trocciola, J. Powers, and R.G. Martin, USP 4,695,518 (1987) *Silicon Carbide Matrix for Fuel Cells.*

[49] K. Kinoshita and J.A.S. Bett, *Carbon* **11** (1973) 237.

[50] K. Kinoshita and J.A.S. Bett in *Corrosion Problems in Energy Conversion and Generation*, Ed. by C. S. Tedmon Jr., The Electrochemical Soc., Princeton, N.J., 1974,

p. 43.
[51] H. Binder, A. Kohling, K. Richter and G. Sandstede, *Electrochim. Acta* **9** (1964) 255.
[52] B. R. Puri, *Carbon* **4**, (1966) 391.
[53] S. S. Barton, G. L. Boulton and B. H. Harrison, *Carbon* **10**, (1972) 395; **13**, (1975) 283.
[54] P. L. Walker Jr. and J. Janov, *J. Colloid Interface Sci.* **28** (1968) 449.
[55] B. R. Puri, *J. Indian Chem. Soc.* **51** (1974) 62.
[56] J.M. Thomas, *Enhanced Reactivity at Dislocations in Solids* in *Advances in Catalysis* **12** (1969) 293.
[57] F.A. Heckman and D.F. Harling, *Rubber Chemistry and Technology* **39** (1966) 1.
[58] J.B. Donnet and P. Ehrburger, *Carbon* **8** (1970) 697.
[59] R.E. Panzer and P.J. Elving, *Electrochim. Acta* **20** (1975) 635.
[60] M. Levy, *Ind. Eng. Chem. Prod. Res. Develop.* **1** (1962) 19.
[61] M. Levy and P. Wong, *J. Electrochem. Soc.* **111** (1964) 1088.
[62] G. Luft, K. Mund, G. Richter, R. Schultze and F. von Sturm, *Siemens Forsh. u Entwickl. Ber.* **3** (1974) 177.
[63] K. V. Kordesch and E. M. King, U. S. Patent 3,077,507, Feb. 12, 1963.
[64] J. M. Thomas, E. L. Evans and J. O. Williams, *Proc. Roy. Soc.* A **331** (1972) 417.
[65] B. R. Puri, B. C. Kaistha and O. P. Mahajan, *J. Indian Chem. Soc.* **50** (1973) 473.
[66] K. Kinoshita, J. A. S. Bett and P. Stonehart, in *Sintering and Catalysis,* Ed. by G. C. Kuczynski, Plenum Publishing Corp., N.Y., 1976, pp.117.
[67] J.A.S. Bett, K. Kinoshita and P. Stonehart, *J. Cat.* **35** (1974) 307.
[68] P. Stonehart and J. P. MacDonald, *Stability of Acid Fuel Cell Cathode Materials,* Electric Power Research Institute Report EM-1664, 1981.
[69] W. Vogel, J. Lundquist, P. N. Ross, and P. Stonehart, *Electrochim. Acta* **20** (1975) 79.
[70] P. Stonehart and P. N. Ross, *Catal. Rev.* **12**(1975) 1.
[71] P. Stonehart, *Proceedings of the 4th World Hydrogen Energy Conference*, California, 1982, pp.1149.
[72] P. Stonehart, *J. Hydrogen Energy* **9** (1984) 921.
[73] P. Stonehart, USP 4,407,906 (1983), *Fuel Cell with Pt-Pd Electrocatalyst Electrodes,* File date 3, Nov. 1981, Issue date 4 Oct. 1983.

5

Nanostructural Analysis of Bright Metal Surfaces in Relation to their Reflectivities

N. D. Nikolić,[1] Z. Rakočević[2] and K. I. Popov[3]

[1]ICTM–Institute of Electrochemistry, Belgrade, Serbia and Montenegro
[2]Vinča Institute of Nuclear Sciences, Belgrade, Serbia and Montenegro
[3]Faculty of Technology and Metallurgy, University of Belgrade, Serbia and Montenegro

I. INTRODUCTION

Electroplating processes are widely used in many branches of industry, the most important of which are: the automobile industries, building and construction industries and electrical and electronics industries. These processes are often used for decorative purposes.

In order for metal coatings to be used for decorative purposes, it is necessary that their surfaces be smooth and bright. In everyday experience the optical properties of the metals are associated with high reflectivity and low transmission. For this reason, the terms "mirror bright", "semi bright", and "high bright" are often used when considering electrodeposited-metal coatings.

However, it is known that there are no precise definitions of these terms. Brightness of a metal surface is usually defined as its reflecting power, measured by the amount of light specularly reflected off the surface, *i.e.,* at an angle equal and opposite to that of the incidence light

Modern Aspects of Electrochemistry, Number 38, edited by B. E. Conway *et al.* Kluwer Academic/Plenum Publishers, New York, 2005.

with respect to the normal to the geometrical surface. A more precise definition, not involving the actual reflectivity of the surface, would be the ratio between specularly and diffusely reflected light.[1,2] The ratio of specular to total reflectivity can also be used for an estimation of the brightness of a surface.[3]

Structural details which must be fulfilled in order for light from metal surfaces to be to a high degree specularly reflected have not been systematized yet. One of the reasons for the absence of systematized criteria of these characteristics is, among other things, the lack of an appropriate corresponding technique for the examination of the structure of bright metal surfaces. The technique of scanning electron microscopy (SEM), which is usually used for the examination of the structure of metal coatings,[2,4] gives only the information that bright metal coatings have a smaller grain size than mat coatings obtained from a plating bath in the absence of leveling and brightening agents. Over the last few years, the techniques of scanning tunneling microscopy (STM) and atomic force microscopy (AFM) have proven to be very valuable techniques for the examination of metal deposits.[5,6] The value of these techniques lies in their ability to provide *local* information concerning the metal deposit, which is unsurpassed by any other technique owing to the resolution. These techniques enable a precise analysis of the topography of a metal surface, *i.e.*, examination of the submicro structures of metal surfaces to be made. Also, the digital images obtained by these techniques can easily be analyzed by powerful software packages.

II. REAL SYSTEMS

The (total) reflected light from an ideally flat (geometrical) surface is exclusively mirror reflected light, *i.e.*, the light is reflected at an angle equal and opposite to that of the incident light with respect to the normal to the geometrical surface. On the other hand, any (every) real surface possesses surface irregularities, which lead to scattering of the light. Then, reflected light from a metal surface includes diffusely reflected, not only specularly reflected light.

According to Bockris and Razumney,[1] if the brightness of a surface is defined as the ability to reflect light only in a direction making an angle with the normal equal to that of the incident beam (*i.e.*, approach to an "ideal mirror"), then the surface will be increasingly bright as the scattering introduced in the reflected light by surface irregularities decreases, that is, as deviations of the actual surface from the ideal, geometrical boundaries are eliminated.

The system which is the nearest approximation to an ideally flat surface, from which reflection of light is exclusively mirror reflection, is the surface of a silver mirror or of mercury. For that reason, reflection and structural characteristics of a silver mirror surface were firstly examined in order to define the reference standard for the comparison of reflection and structural characteristics of different metal surfaces.[7,8] Silver mirror surface is obtained by chemical deposition of silver onto a glass by the silver mirror reaction.[9]

The opposite case of that of a silver mirror surface are metal coatings obtained from solution without brightening addition agents. These metal surfaces are usually known as mat surfaces.

1. Limiting Cases

(i) Silver Mirror Surface

(a) Reflection analysis

The ideal (theoretical) reflectance of silver[10] and the degrees of total, mirror and diffuse reflection as a function of visible light wavelength for a silver mirror surface are given in Fig. 1. Figure 1 shows clearly that the reflection of light from this surface is mostly mirror reflection and the degree of diffuse reflection is very small (up to 2 %). The degree of mirror reflection from a silver mirror is also very close to the ideal reflectance of silver.

(b) Structural analysis

The 3D (three-dimensional) STM image (700 x 700) nm of a silver mirror surface is shown in Fig. 2, from which it can be seen that the surface of such a mirror is very smooth.

The line section analysis of this surface is shown in Fig. 3, which indicates that this surface consisted of relatively flat and mutually parallel parts. The STM software measurements showed that distances between adjacent, relatively flat, parts were several atomic diameters of silver.[11]

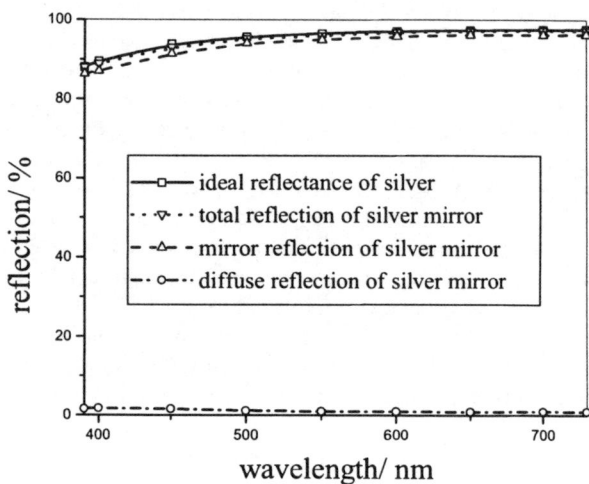

Figure 1. The dependence of the degrees of reflection on the visible light wavelength for the ideal reflectance of silver (□), the total (∇), mirror (Δ) and diffuse (o) reflections of a silver mirror. (Reprinted from Ref.[7] with permission from the Serbian Chemical Society.)

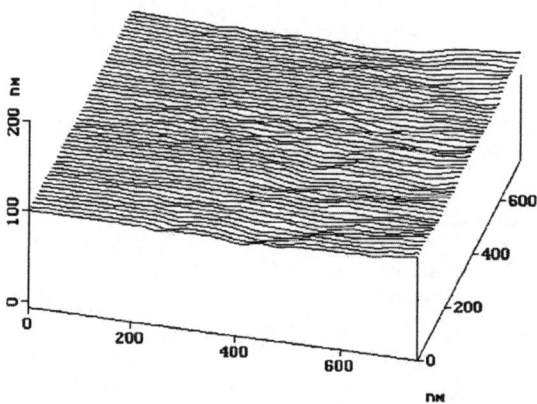

Figure 2. 3D STM image of a silver mirror surface. Scan size: (700 x 700) nm. (Reprinted from Ref.[7] with permission from the Serbian Chemical Society.)

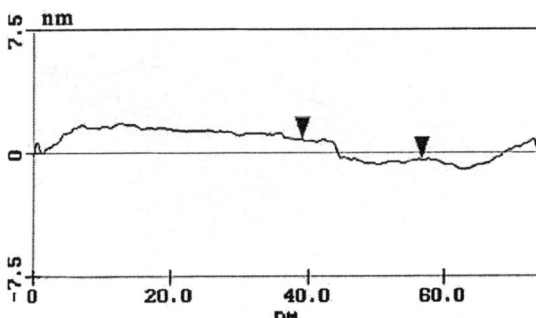

Figure 3. Line section analysis of a portion of the STM surface shown in Fig. 2. The vertical distance between the markers represents 1.246 nm. (Reprinted from Ref. [7] with permission from the Serbian Chemical Society.)

The line section analysis of a relatively flat part of the surface is shown in Fig. 4. It was shown by STM software data processing that the roughness of the flat part of the surface is less than the atomic diameter of silver, *i.e.*, the flat parts are smooth on the atomic scale. The atomic arrangement of a flat part of a surface is shown in Fig. 5.

Finally, the reflection of light from the silver mirror surface is mostly mirror reflection which is very close to the ideal reflectance of silver. The structural characteristics of this surface, which enable a high degree of

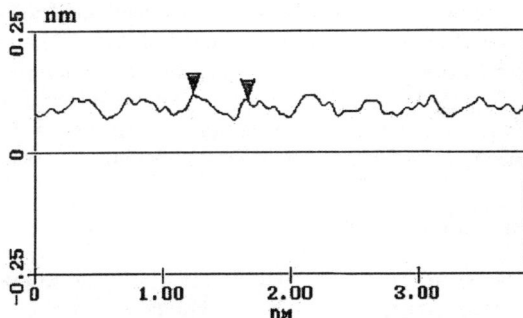

Figure 4. Line section analysis of the flat part of the surface shown in Fig. 3. (Reprinted from Ref. [7] with permission from the Serbian Chemical Society.)

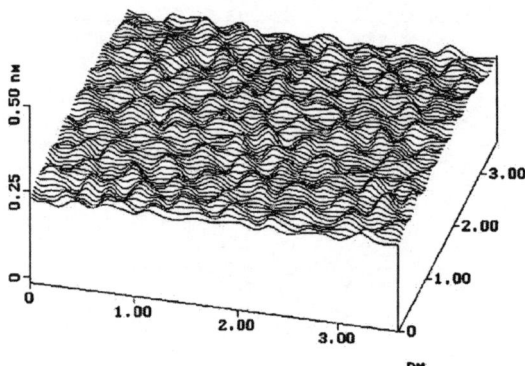

Figure 5. 3D STM image of a silver mirror surface. Scan size:
(3.50 x 3.50) nm. (Reprinted from Ref.[7] with permission from
the Serbian Chemical Society.)

mirror reflection, are that the surface should have flat and mutually
parallel parts which are smooth on the atomic scale, with adjacent flat
parts being separated by several atomic diameters of silver.

(*ii*) *Mat surfaces*

(*a*) *Reflection analysis*

Figure 6 shows the ideal reflectance of copper[10] and the degrees of
total, mirror and diffuse reflection as a function of visible light wavelength
for the copper coating obtained from a pure sulfate solution (composition:
240 g l^{-1} CuSO$_4$·5 H$_2$O + 60 g l^{-1} H$_2$SO$_4$). It can be seen from Fig. 6 that the
reflection of light from this coating is mostly of the diffuse kind. The extent
of mirror reflection from this coating reaches a maximum of 6 %.

(*b*) *Structural analysis*

It can be seen from Fig. 7 that surface of the copper coating
obtained from a pure sulfate solution is relativelly rough.

It can be observed from Figs. 8a and 8b that she surface areas of
lateral parts of the surface are larger than the surface areas of flat parts
of the same surface.

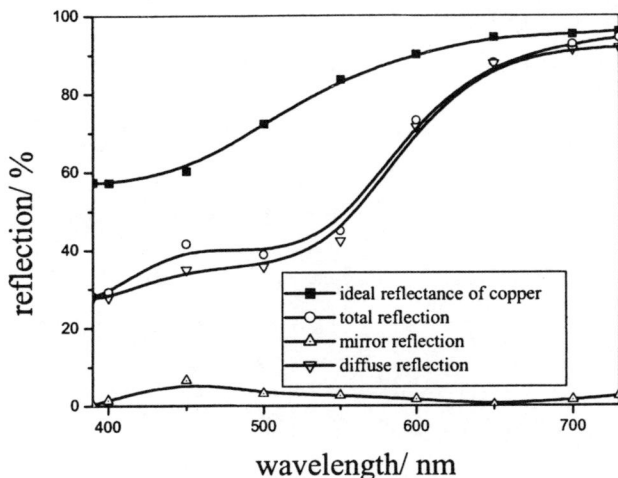

Figure 6. The dependence of the degrees of reflection on the wavelength of visible light for the ideal reflectance of copper (■), the total (o), mirror (Δ) and diffuse (∇) reflections of the copper coating electrodeposited from a pure acid sulfate solution. (Reprinted from Ref.[12] with permission from Elsevier.)

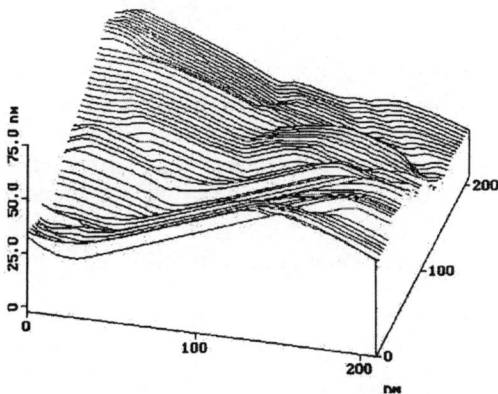

Figure 7. 3D STM image of the copper coating electrodeposited from a pure sulfate solution. (Reprinted from Ref. [13] with permission from Elsevier.)

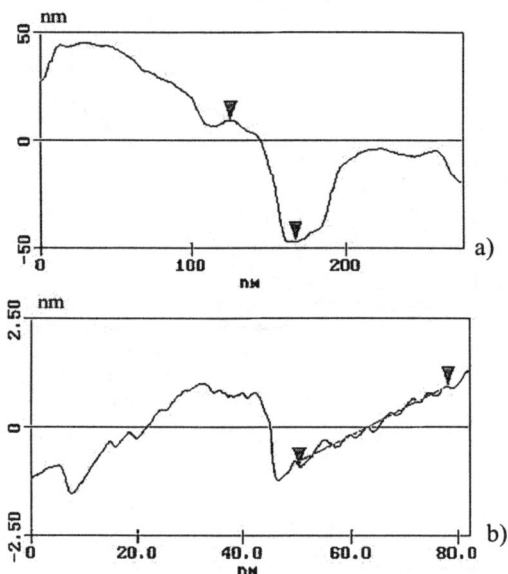

Figure 8. The line section analysis of the copper coating obtained from a pure acid sulfate solution. Scan sizes: a) (300 x 300) nm, b) (80 x 80) nm. (Reprinted from Ref. [12] with permission from Elsevier.)

Finally, the reflection of light from the copper coating obtained from a pure sulfate solution is mostly diffuse reflection. The structural characteristics of this copper surfaces which enabled a high degree of diffuse reflection with a negligible degree of mirror reflection are that the lateral parts of the surface were larger than the flat parts.

2. Systems in Metal Finishing

(i) Polished Copper Surfaces

(a) Reflection analysis

The reflection characteristics for the copper surface polished mechanically and the copper surface polished both mechanically and electrochemically[12] are shown in Fig. 9a.

The degrees of mirror and diffuse reflection of the same copper surfaces as a function of wavelength of visible light are shown in Fig. 9b. It can be seen from this figure that electropolishing of the mechanically polished copper surface led to an increase of the degree of mirror reflection and a decrease of the degree of diffuse reflection. The degree of mirror reflection of the copper surface, polished both mechanically and electrochemically, is approximately 15 - 20 % greater than the degree of mirror reflection of the copper surface polished only mechanically, while the degree of diffuse reflection is smaller by 2-10 % than the degree of diffuse reflection of the mechanically polished copper surface. Also, it can be seen from Figs. 9a and 9b that the degrees of mirror reflection of the copper surface, polished both mechanically and electrochemically, approach very nearly the ideal reflectance of copper.

(b) Structural analysis

Figure 10 shows the copper surface polished mechanically (Fig. 10a) and the copper surface polished both mechanically and electrochemically (Fig. 10b).

Figures 10a and 10b show that the topographies of these copper surfaces differ from one another. Electropolishing of the copper surface, first polished mechanically, led to a decrease of the roughness of this copper surface (Figs. 10a and 10b). Analysis of the topographies of the copper surfaces and the dependences of the degrees of mirror and diffuse reflection as a function of the wavelength of visible light showed that a decrease of the roughness of copper surfaces is accompanied by a decrease of the degree of diffuse reflection and by an increase of the degree of mirror reflection.[12]

It can also be seen from Fig. 10b that the topography of the copper surface, polished both mechanically and electrochemically, consists of small and mutually parallel parts of the surface.

Figure 11 shows the line sections analysis of regions of the copper surfaces shown in Fig. 10. It was shown by the STM data processing that the distance between two flat parts of the copper surface, polished both mechanically and electrochemically, is several atomic diameters of copper.[11] The same distance for the copper surface polished only mechanically is approximately 70 atomic diameters of copper. Also, it can be seen from Figs. 11a and 11b that these relatively flat parts of the surface are to a higher degree mutually parallel for the copper surface

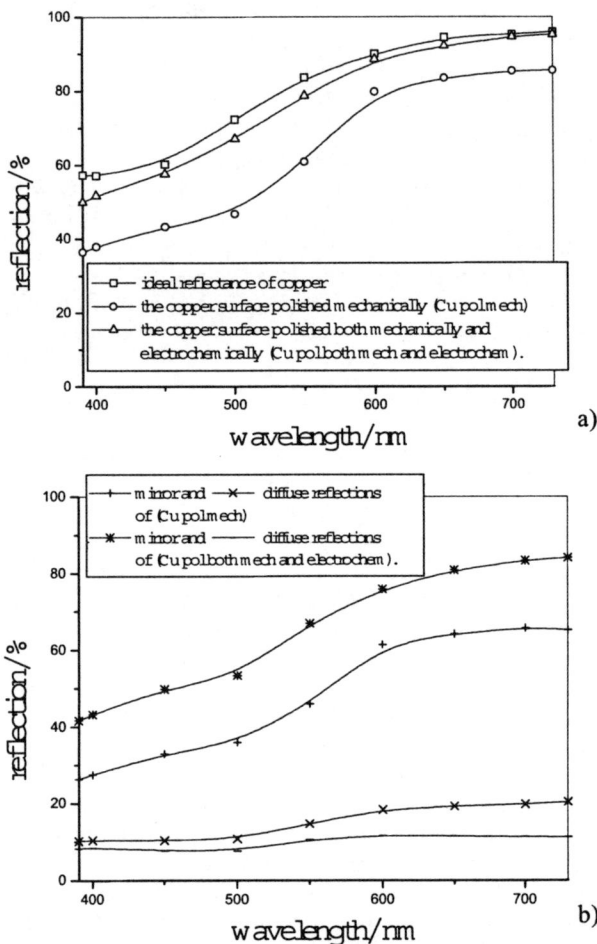

Figure 9. The dependence of the degrees of reflection on the wavelength of visible light for: a) the ideal reflectance of copper (□) and the total reflections of the copper surface polished mechanically (o), the copper surface polished both mechanically and electrochemically (Δ), b) mirror (+) and diffuse (x) reflections of the copper surface polished mechanically; mirror (*) and diffuse (-) reflections of the copper surface polished both mechanically and electrochemically. (Reprinted from Ref. [12] with permission from Elsevier.)

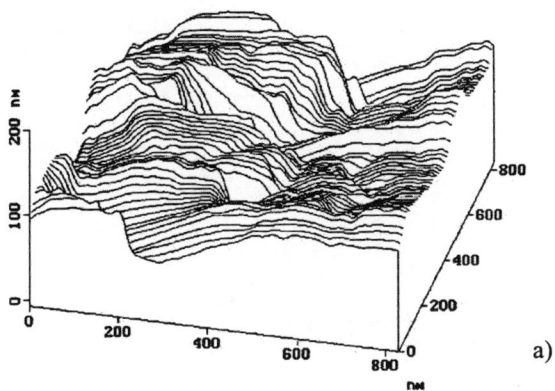

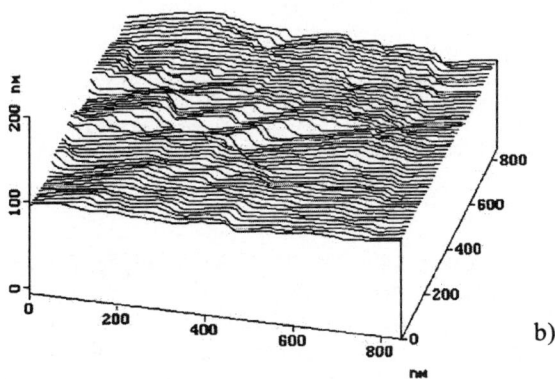

Figure 10. 3D STM images of: a) the copper surface polished mechanically, b) the copper surface polished both mechanically and electrochemically. Scan size: (880 x 880) nm. (Reprinted from Ref.[12] with permission from Elsevier.)

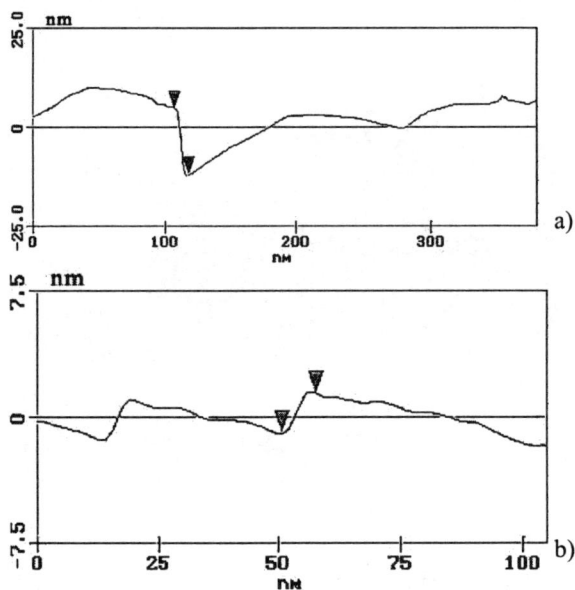

Figure 11. The line sections analysis from the regions of the STM surfaces shown in Fig. 10: a) the copper surface polished mechanically, b) the copper surface polished both mechanically and electrochemically. The vertical distances between markers represent: a) 16.780 nm, b) 2.394 nm. (Reprinted from Ref. [12] with permission from Elsevier.)

polished both mechanically and electrochemically than for the copper surface polished only mechanically.

The line sections analysis of the flat parts of the copper surfaces is shown in Fig. 12. The roughness of the flat parts of the copper surface, polished both mechanically and electrochemically, is very small and less than the atomic diameter of copper. For this reason, it can be said that these flat parts of the surface are smooth on the atomic level. The roughness of the flat parts of the copper surface, only polished mechanically, is less than 2 atomic diameters of copper.[12]

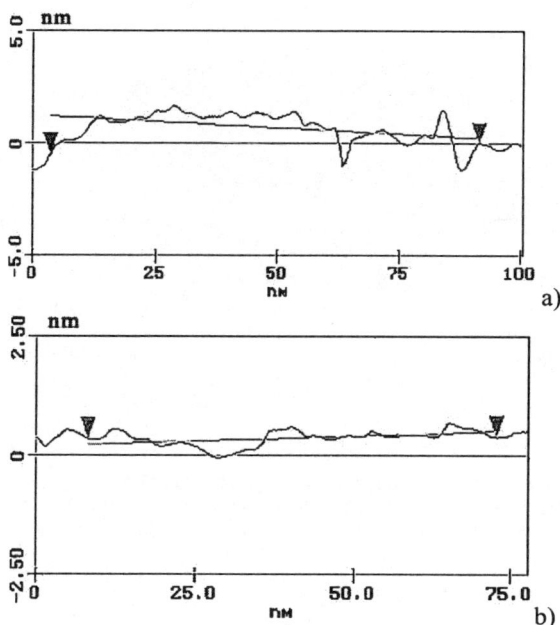

Figure 12. The line sections analysis of the flat parts of surfaces: a) the copper surface polished mechanically, b) the copper surface polished both mechanically and electrochemically. The roughnesses of the observed surfaces were: a) 0.444 nm, b) 0.111. (Reprinted from Ref.[12] with permission from Elsevier.)

The atomic arrangement of a flat part of the copper surface polished both mechanically and electrochemically is shown in Fig. 13.[14]

X-ray diffraction (XRD) analysis of the copper surface, polished both mechanically and electrochemically, is shown in Fig. 14. As can be seen from Fig. 14, a relatively disordered structure of this copper surface arises, with an increased ratio of copper crystallites oriented in (200), (220) and (311) planes.

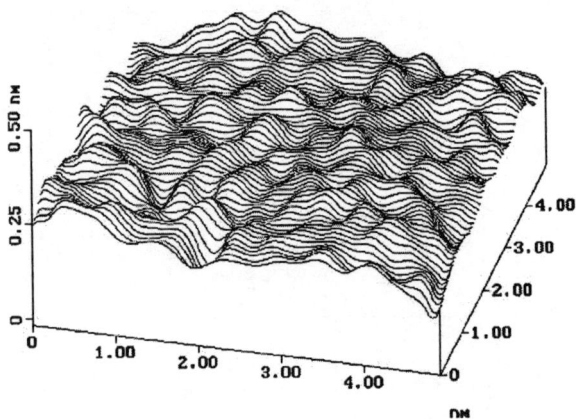

Figure 13. 3D STM image of the copper surface polished both mechanically and electrochemically. Scan size: (5.0 x 5.0) nm. (Reprinted from Ref.[14] with permission from the Serbian Chemical Society.)

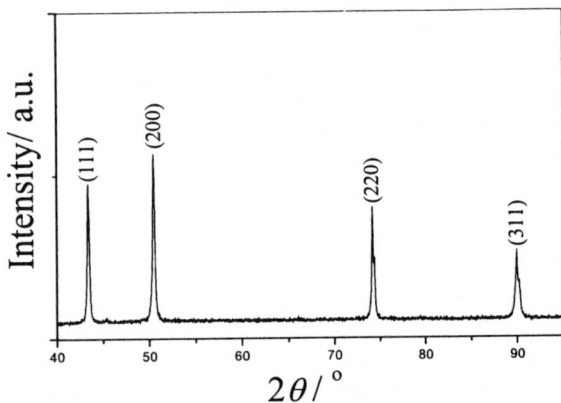

Figure 14. XRD pattern of the copper surface polished both mechanically and electrochemically. (Reprinted from Ref. [12] with permission from Elsevier.)

(ii) Copper Coatings

(a) Reflection analysis

The total reflection as a function of the wavelength of light in the visible range for the copper coating obtained in the presence of thiourea (*solution Cu I:* 240 g l^{-1} $CuSO_4 \cdot 5$ H_2O + 60 g l^{-1} H_2SO_4 + 0.050 g l^{-1} thiourea) of 20 μm thickness[15] and the copper coating obtained in the presence of modified polyglycol ether (Lutron HF 1), PEG 6000 and 3 - mercapto propane sulfonate (*solution Cu II:* 240 g l^{-1} $CuSO_4 \cdot 5$ H_2O + 60 g l^{-1} H_2SO_4 + 0.124 g l^{-1} NaCl + 1.0 g l^{-1} modified polyglycol ether (Lutron HF 1) + 1.0 g l^{-1} poly(ethylene glycol) M_n = 6000 (PEG 6000) + 1.5 mg l^{-1} 3 -mercapto propane sulfonate) of the same thickness[15] are shown in Fig. 15a. The degrees of mirror and diffuse reflections of the same copper coatings as a function of wavelength of visible light are shown in Fig. 15b.

Figure 16 shows reflection characteristics of the same copper coatings, but having a 25 μm thickness.[12, 13]

It can be observed from Figs. 15 and 16 that the reflection characteristics of these copper coatings were very similar to the same characteristics of the copper surface polished both mechanically and electrochemically. Both the copper coatings exhibited high degrees of mirror reflection which approach very nearly the ideal reflectance of copper for the wavelengths above 600 nm. The degrees of diffuse reflection of these copper coatings have approximately the same values.

(b) Structural analysis

The 3D STM image of the 25 μm thick copper coating obtained from *solution Cu II* is shown in Fig. 17. Figure 18 shows 3D STM images at higher magnification (*i.e.*, smaller scan size) of the copper coating obtained from *solution Cu I* and the copper coating obtained from *solution Cu II*, thicknesses of 20 μm (Figs. 18a and 18b, respectively) and 25 μm (Figs. 18c and 18d, respectively).

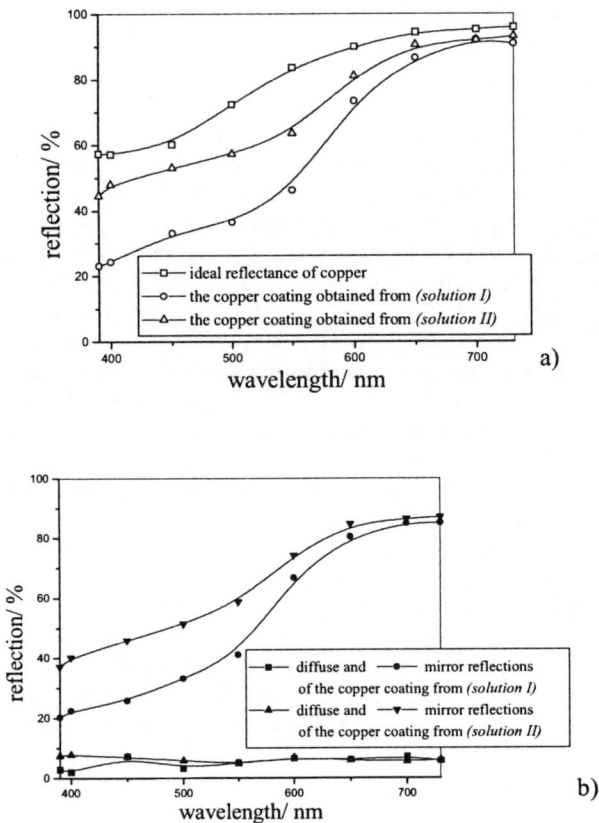

Figure 15. The dependence of the degrees of reflection on the wavelength of visible light for: a) the ideal reflectance of copper (□) and the total reflections of the copper coating electrodeposited from *solution Cu I* (O), the copper coating electrodeposited from *solution Cu II* (Δ), b) mirror (•) and diffuse (■) reflections of the copper coating electrodeposited from *solution Cu I*, mirror (▼) and diffuse (▲) reflections of the copper coating electrodeposited from *solution Cu II*. (Reprinted from Ref. [15] with permission from Springer–Verlag.)

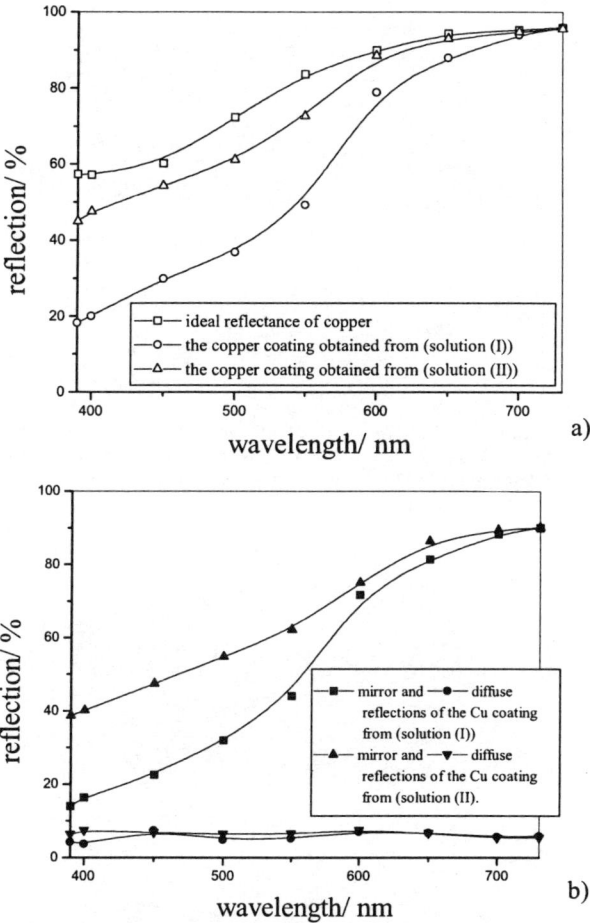

Figure 16. The dependence of the degrees of reflection on the wavelength of visible light for: a) the ideal reflectance of copper (□) and the total reflections of the copper coating electrodeposited from *solution Cu I* (O),[13] the copper coating electrodeposited from *solution Cu II* (Δ),[12] b) mirror (■) and diffuse (●) reflections of the copper coating electrodeposited from *solution Cu I,*[13] mirror (▲) and diffuse (▼) reflections of the copper coating electrodeposited from *solution Cu II.*[12] (Reprinted from Ref. [12] and Ref. [13] with permission from Elsevier.)

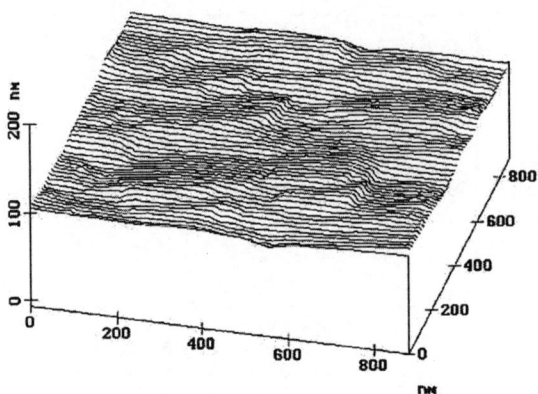

Figure 17. 3D STM image of 25 μm thick the copper coating electrodeposited from *solution Cu II*. Scan size: (880 x 880) nm. (Reprinted from Ref. [12] with permission from Elsevier.)

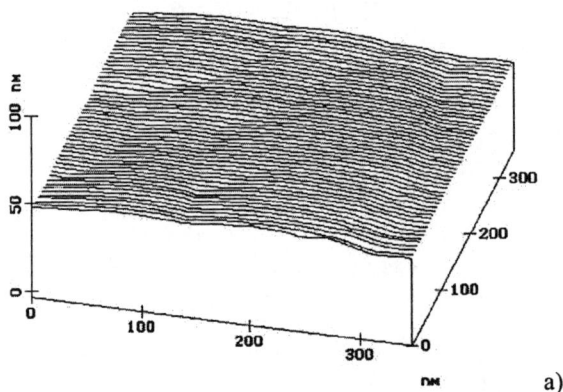

a)

Figure 18. 3 D STM images of copper coatings electrodeposited from: a) *solution Cu I*, thickness of the coating: $\delta = 20$ μm,[15] b) *solution Cu II*, $\delta = 20$ μm,[15] c) *solution Cu I*, $\delta = 25$ μm,[13] d) *solution II*, $\delta = 25$ μm.[16] (Reprinted from Refs.[13,15,16] with permission from Elsevier, Springer–Verlag and Union of Engineers and Technicians for Protecting of Materials of Serbia, respectively.)

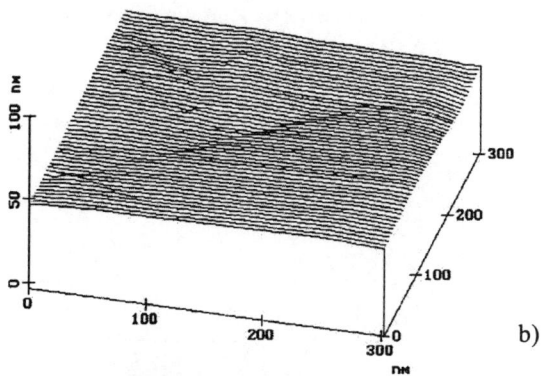

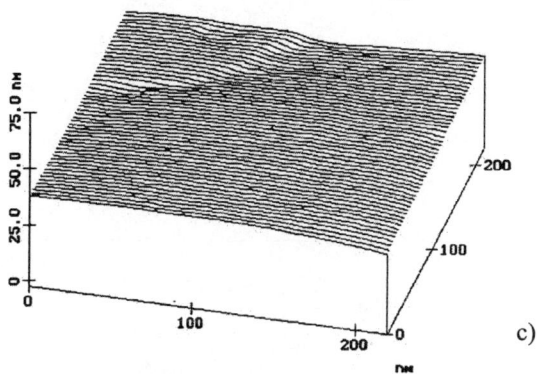

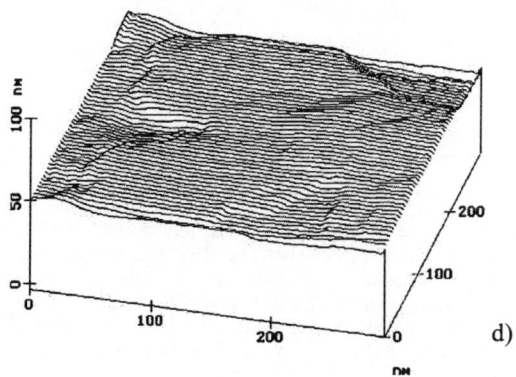

Figure 18. Continuation

Figure 19 shows the line sections analysis of portions of the copper surfaces shown in Fig. 18, while the line sections analysis of the flat parts of these copper surfaces are shown in Fig. 20.

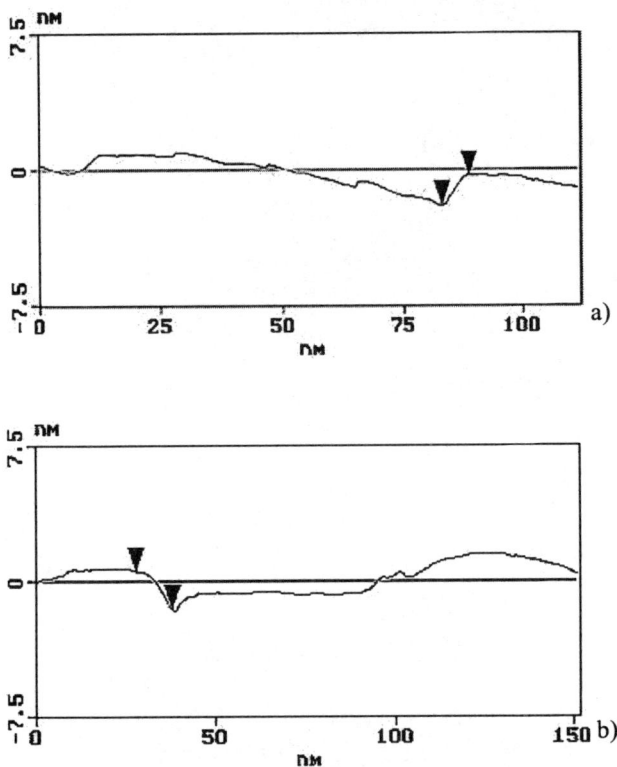

Figure 19. The line sections analysis from the portion of the STM surfaces of the copper coatings shown in Fig. 18 electrodeposited from: a) *solution Cu I*, $\delta = 20$ μm,[15] b) *solution Cu II*, $\delta = 20$ μm,[15] c) *solution Cu I*, $\delta = 25$ μm,[13] d) *solution Cu II*, $\delta = 25$ μm.[12] The distances between markers represent: a) 1.656 nm, b) 2.136 nm, c) 1.443 nm, d) 1.177 nm. (Reprinted from Refs.[12,13,15] with permission from Elsevier and Springer–Verlag.)

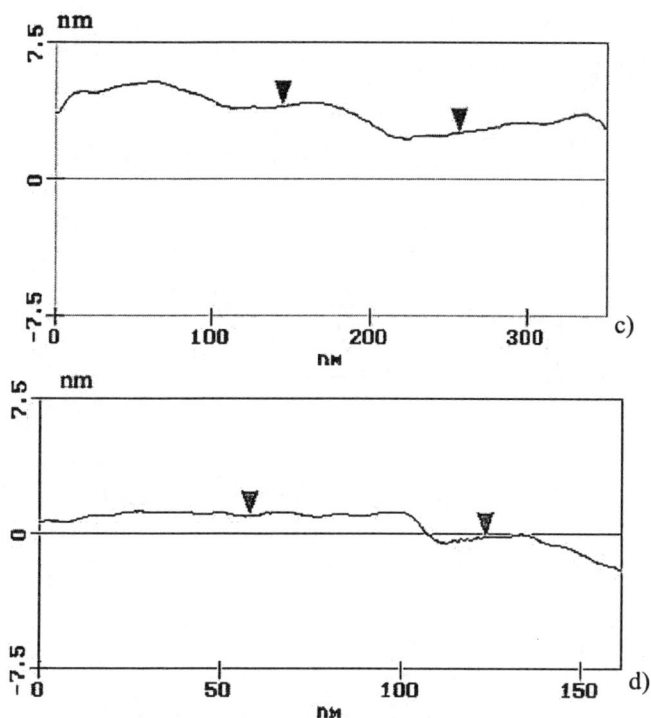

Figure 19. Continuation

Similar reflection and structural characteristics to them showed and 40 μm thick the copper coating obtained from *solution Cu I*. The line section analysis of this copper coating is shown in Fig. 21. The atomically flat parts of these copper coatings are shown in Fig. 22.

The mean size of these atomically flat parts was estimated by STM software measurements, using option of the determination of the autocovariance (ACVF) and power spectral density (PSD) functions. The estimated mean sizes of the flat parts are given in Table 1.

Structural characteristics of the formed copper coatings were very similar to the same characteristics of the copper surface polished both mechanically and electrochemically. The surfaces of these copper coatings consisted of flat and mutually parallel parts of the surface. The

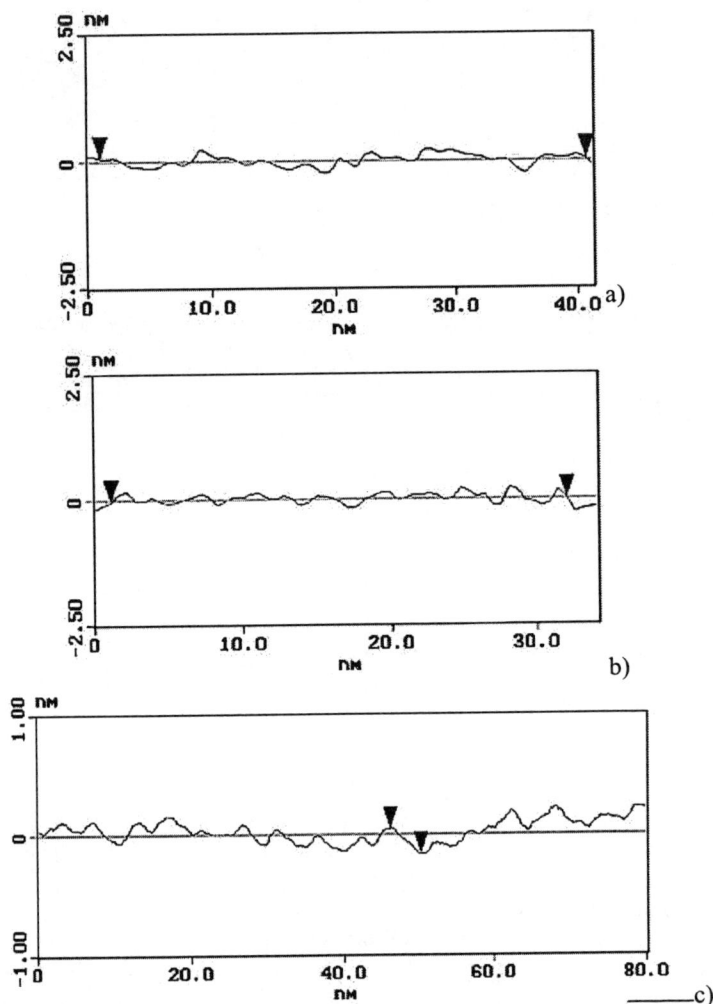

Figure 20. The line sections analysis of the flat parts of surfaces of the copper coatings electrodeposited from: a) *solution Cu I*, δ = 20 μm,[15] b) *solution Cu II*, δ = 20 μm,[15] c) *solution Cu I*, δ = 25 μm,[13] d) *solution Cu II*, δ = 25 μm.[12] (Reprinted from Refs.[12,13,15] with permission from Elsevier and Springer–Verlag.)

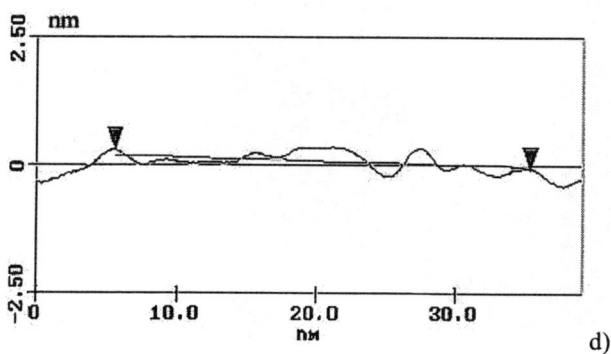

Figure 20. Continuation

d)

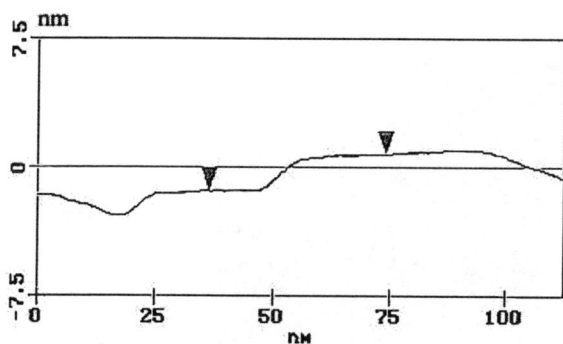

Figure 21. The line section analysis from a portion of 40 μm thick the copper coating electrodeposited from *solution Cu I*. (Reprinted from Ref. [16] with permission from Union of Engineers and Technicians for Protecting of Materials of Serbia.)

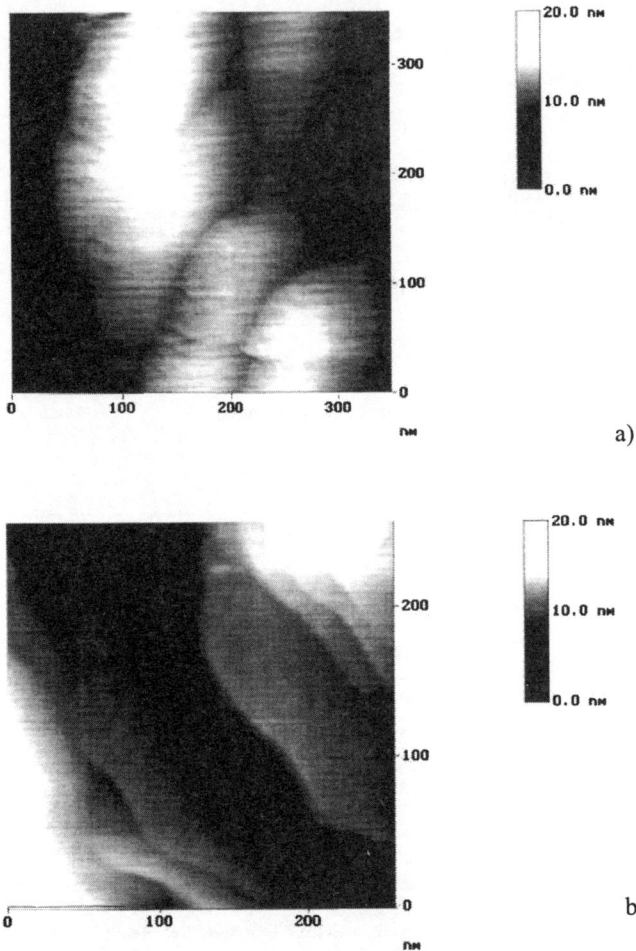

Figure 22. STM images of copper coatings electrodeposited from: a) *solution Cu I, δ = 20 μm,*[15] b) *solution Cu II, δ = 20 μm,*[15] c) *solution Cu I, δ = 25 μm,*[13] d) *solution Cu II, δ = 25 μm.*[17] (Reprinted from Refs.[13,15] with permission from Elsevier and Springer–Verlag, respectively.)

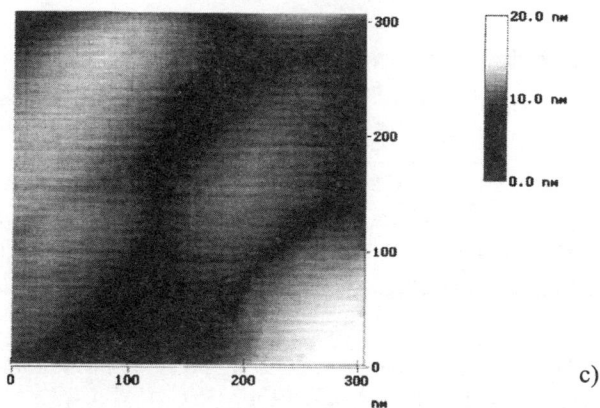

c)

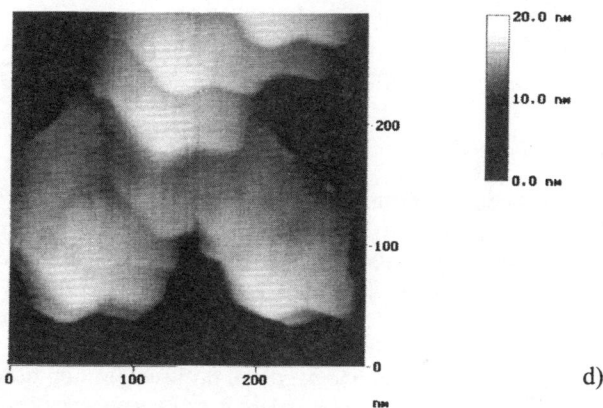

d)

Figure 22. *Continuation.*

distances between adjacent flat parts were several atomic diameters of copper.[11] The flat parts of the surfaces were smooth on the atomic scale.

X-Ray diffraction (XRD) patterns of 20 μm thicks copper coatings obtained from from *solution Cu I* and *solution Cu II* are shown in Figs. 23a and 23b, respectively. From Figs. 23a and 23b can be seen that the copper surfaces exhibited different a preferred orientation. The copper coating electrodeposited from *solution Cu I* showed (111) preferred ori

Table 1
The Estimated Mean Sizes of Atomically Flat Parts of Copper Coatings

STM	The estimated mean sizes of atomically flat parts of copper coatings electrodeposited from:			
	Cu I Solution		Cu II Solution	
Functions:	$\delta = 20\ \mu m$	$\delta = 25\ \mu m$	$\delta = 20\ \mu m$	$\delta = 25\ \mu m$
ACVF/ nm²	(160 x 200)	(140 x 240)	(140 x 160)	(160 x 170)
PSD/ nm	229	243	166	198

entation (Fig. 23a) while the copper coating obtained from *solution Cu II* showed (200) preferred orientation (Fig. 23b).

Finally, it can be concluded on the basis of structural analysis of these copper coatings that structural characteristics did not depend on thickness of the coatings or preferred orientations.

(*iii*) *Zinc Coatings*

(*a*) *Reflection analysis*

The reflection characteristics for the zinc coatings,[18] having thicknesses of 20, 25 and 60 μm, electrodeposited onto copper cathodes, obtained from the acid sulfate solution containing the dextrin/salicyl aldehyde mixture (*solution Zn I:* 300 g l⁻¹ $ZnSO_4 \cdot 7\ H_2O$ + 30 g l⁻¹ $Al_2(SO_4)_3 \cdot 18\ H_2O$ + 15 g l⁻¹ NaCl + 30 g l⁻¹ H_3BO_3 + 3.0 g l⁻¹ dextrin + 2.8 ml l⁻¹ salicyl aldehyde), are shown in Fig. 24. The degree of mirror reflection reached 85 %, while the degree of diffuse reflection from these coatings is quite small (up to 5%). Also, it can be observed from this figure that the degree of mirror reflection increases with increasing thickness of the zinc coatings.

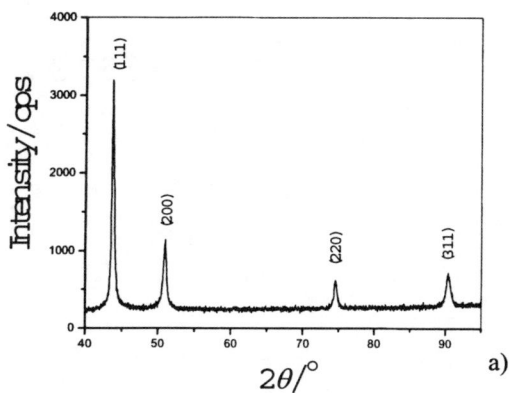

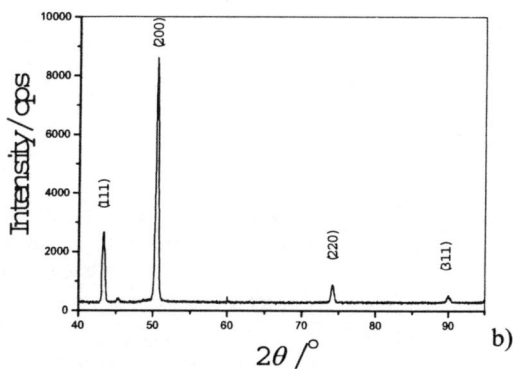

Figure 23. XRD patterns of 20 μm thicks copper coatings electrodeposited from: a) *solution Cu I*, b) *solution Cu II*. (Reprinted from Ref. [15] with permission from Springer–Verlag.)

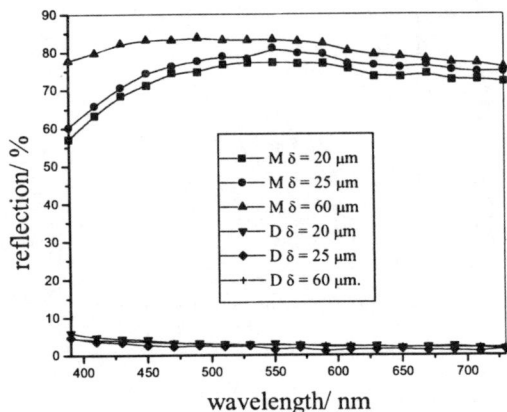

Figure 24. The dependence of the degree of reflection on the visible light wavelength for zinc coatings obtained from solution containing the dextrin/salicyl aldehyde mixture (*solution Zn I*). M: mirror reflection, D: diffuse reflection. (Reprinted from Ref.[18] with permission from the Serbian Chemical Society.)

(*b*) *Structural analysis*

It can be seen from Fig. 25 that the zinc coating obtained from *solution Zn I* is relativelly smooth, but without large flat and mutually parallel parts of the surface, which were characteristic of the previously observed copper surfaces (see Section II.2(*i*) and Section II.2(*ii*). The surfaces of zinc coatings of thicknesses 20 and 60 μm were very similar to that of the zinc coating of thickness 25 μm and, consequently, the STM images of these zinc coatings are not presented.

Analysis of the zinc coatings at even higher magnification led to the morphologies shown in Fig.26, where the light tones represent high areas in the STM images. The analyzed surface was (50 x 50) nm. As can be seen from Fig. 26, surfaces of these zinc coatings are covered with hexagonal zinc crystals.

The line sections analysis of these later morphologies are shown in Fig. 27, from which it can be observed that the top surfaces of the hexagonal zinc crystals are relatively flat and mutually parallel.

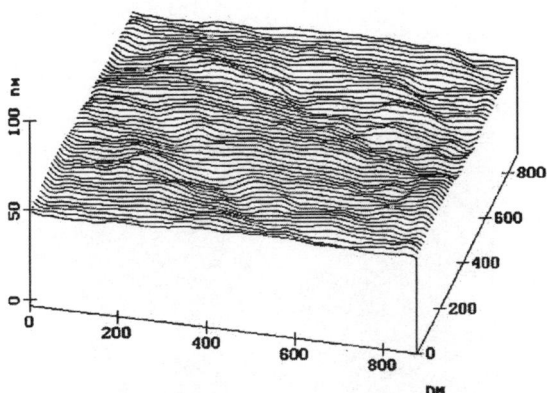

Figure 25. 3D STM image of the zinc coating obtained from *solution Zn I*. Scan size: (880 x 880) nm. (Reprinted from Ref.[18] with permission from the Serbian Chemical Society.)

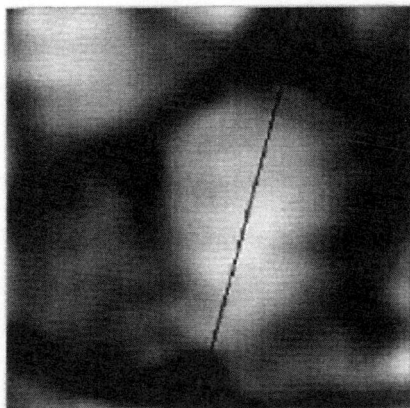

a)

Figure 26. STM images of zinc coatings and the average roughness of the observed hexagonal zinc crystals: a) 20 μm, R_a = 0.252 nm, b) 25 μm, R_a = 0.264 nm, c) 60 μm, R_a = 0.185 nm. Scan size: (50 x 50) nm. (Reprinted from Ref.[18] with permission from the Serbian Chemical Society.)

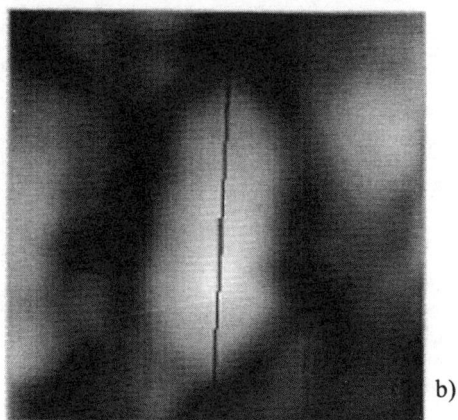

b)

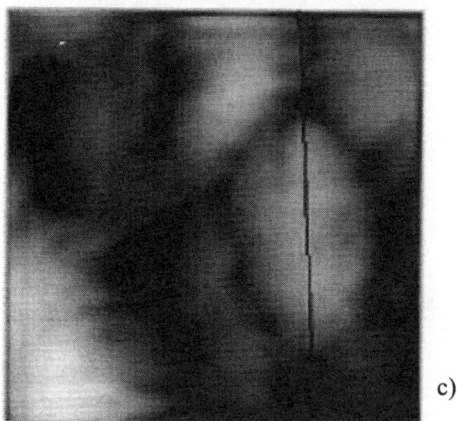

c)

Figure 26. Continuation

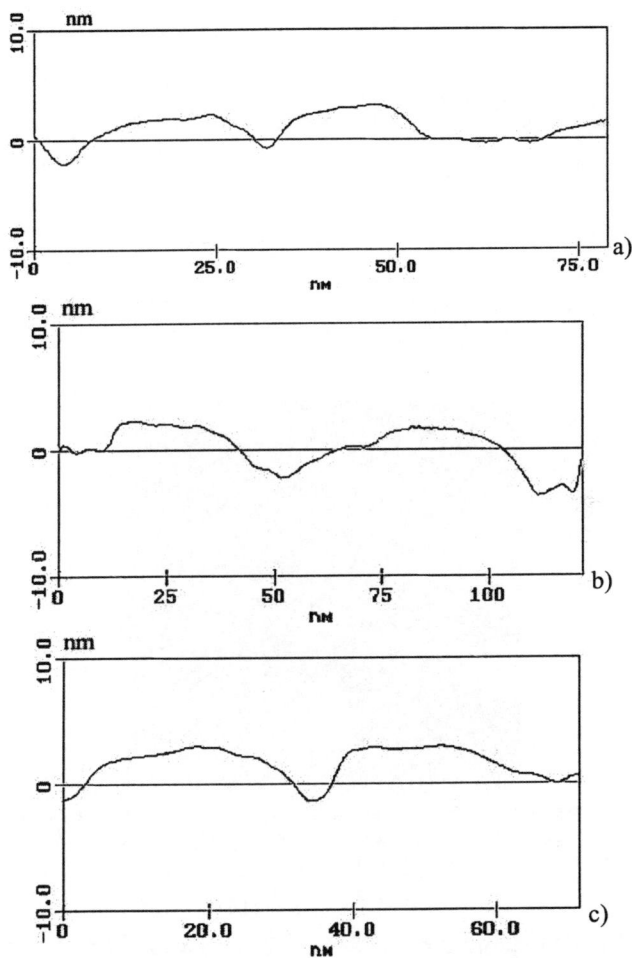

Figure 27. Typical surface profile of zinc coatings from (50 x 50) nm STM images: a) 20 μm, b) 25 μm, c) 60 μm. (Reprinted from Ref.[18] with permission from the Serbian Chemical Society.)

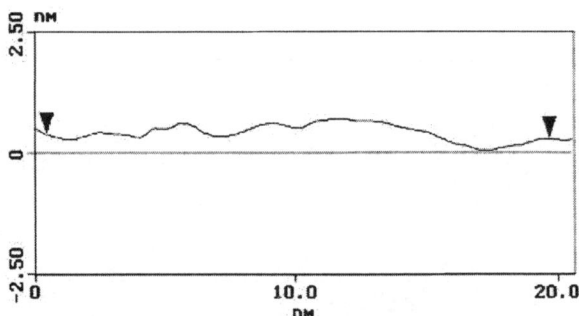

Figure 28. The line sections analysis a surface of hexagonal zinc crystal. (Reprinted from Ref.[15] with permission from Springer–Verlag.)

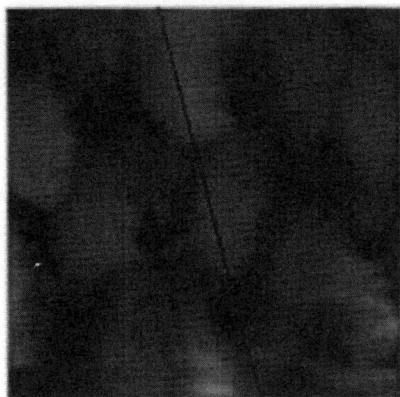

Figure 29. STM image of zinc coating obtained from *solution Zn I*; cathode: zinc. Scan size: (60 x 60) nm. (Reprinted from Ref.[19] with permission from Union of Engineers and Technicians for Protecting of Materials of Serbia.)

It is also shown by the STM software measurements that the roughness of the hexagonal zinc crystals from Fig. 28 is on the atomic level, *i.e.,* less than the atomic diameter of zinc.[11]

The mean size of this hexagonal zinc crystal (estimated by the STM software data processing) was by approximately (18 x 20) nm^2 by the ACVF and about 20 nm by the PSD functions.

Figures 29 and 30 show structural analysis of the 25 μm thick zinc coating electrodeposited in the presence of the same brightening addition agents, but onto a zinc cathode.[19] The degrees of mirror and diffuse reflection of this coating were very close to the same degrees for the zinc coating electrodeposited onto a copper cathode. It can be seen from Fig. 29 that the surface of this zinc coating is again covered with hexagonal zinc crystals. From Fig. 30 it can be seen that the top planes of these zinc crystals are flat and mutually parallel. These zinc crystals exhibit a smoothness on the atomic scale.

Structural STM analysis of the zinc coating obtained from solution containing the dextrin/furfural mixture are shown in Figs. 31 and 32.[13] Reflection characteristics of zinc coating obtained from the *solution Zn II (solution Zn II:* 300 g l^{-1} ZnSO$_4$ · 7 H$_2$O + 30 g l^{-1} Al$_2$(SO$_4$)$_3$ · 18 H$_2$O + 15 g l^{-1} NaCl + 30 g l^{-1} H$_3$BO$_3$ + 3.0 g l^{-1} dextrin + 1.4 ml l^{-1} furfural) were very close to reflection characteristics of the zinc coating obtained from solution containing the dextrin/salicyl aldehyde mixture. The surface of this zinc coating is relativelly smooth (Fig. 31), covered by zinc crystals, the top planes of which are flat and mutually parallel (Fig. 32). These zinc crystals were smooth on the atomic scale. The estimated mean size of these atomically flat zinc crystals was approximately (30 x 40) nm^2 according to the ACVF function and approximately 31.35 nm by the PSD function.[13]

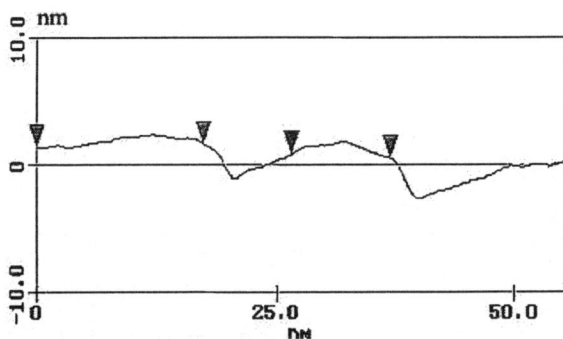

Figure 30.The line section analysis from the portion of the STM surface shown in Fig. 21. (Reprinted from Ref.[19] with permission from Union of Engineers and Technicians for Protecting of Materials of Serbia.)

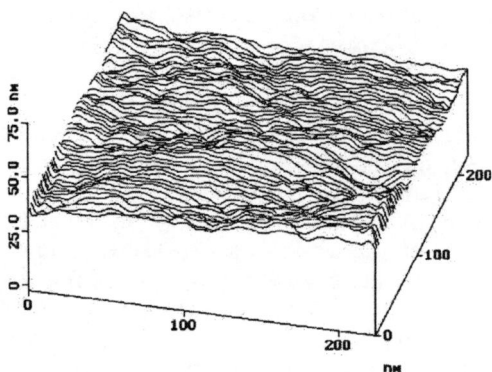

Figure 31. STM image of zinc coating obtained from solution containing the dextrin/furfural mixture (*solution Zn II*). Scan size: (300 x 300) nm. (Reprinted from Ref.[13] with permission from Elsevier.)

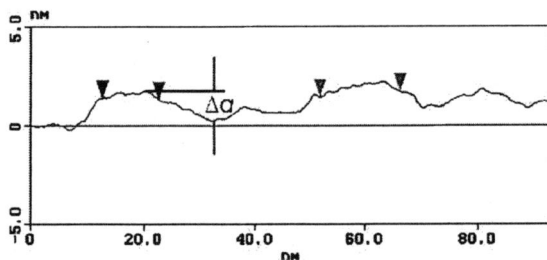

Figure 32. The line section analysis from the portion of the STM surfaces shown in Fig. 31. The vertical distance: $\Delta a = 1.379$ nm. (Reprinted from Ref.[13] with permission from Elsevier.)

(iv) *Nickel Coatings*

(a) *Reflection analysis*

The ideal reflectance of nickel[10] and the degrees of total reflection as a function of wavelength of visible light for a nickel coating electrodeposited in the presence of the basic brightening addition agent and the nickel coating electrodeposited in the presence of both the basic and the top brightening addition agents[20] are shown in Fig. 33a. The degrees of mirror and diffuse reflections of the same nickel coatings as a function of wavelength are shown in Fig. 33b.

It can be seen from Fig. 33b that the addition of the top brightening addition agent to a nickel plating bath containing the basic brightening additive led to an increase of the degree of mirror reflection and a decrease of the degree of diffuse reflection. The degree of mirror reflection of the nickel coating obtained in the coating obtained in the presence of the basic brightening addition agent only. Also, it can be observed from Fig. 33 that the degree of mirror reflection of the nickel coating obtained in the presence of the basic and top-brightening-addition agents approaches very nearly the ideal reflectance of nickel. Although the degree of mirror reflection of the nickel coating, obtained in the presence of the basic and top brightening addition agents, was only 3 - 6 % greater and the degree of diffuse reflection was only 4 % smaller than the same degrees of the nickel coating obtained in the presence of the basic brightening addition agent alone, a visual difference between these nickel coatings was apparent.

(b) *Structural analysis*

The addition of the top brightening addition agent to the plating bath containing only the basic brightening addition agent led to a decrease of the roughness of the nickel coating, as illustrated by Fig. 34.

a)

b)

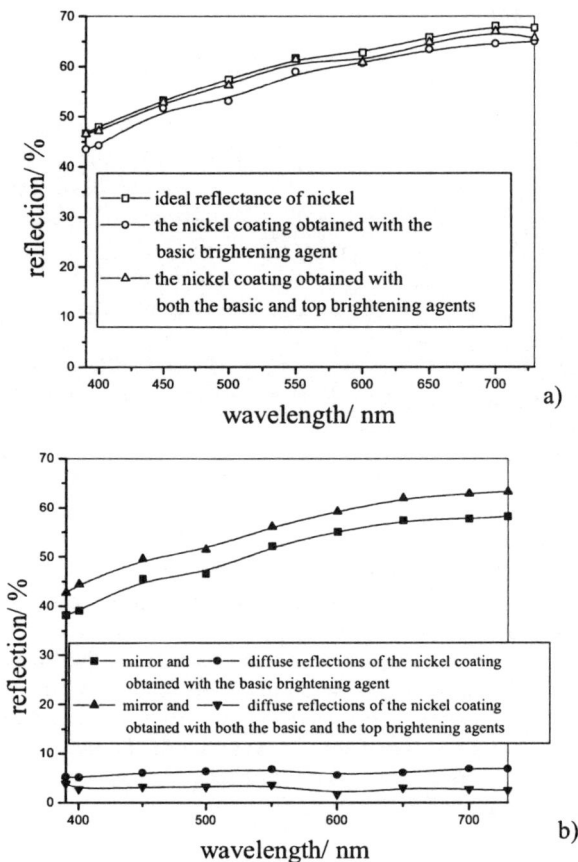

Figure 33. The dependence of the degrees of reflection on the wavelength of visible light: a) the ideal reflectance of nickel (□) and the total reflections of the nickel coating electrodeposited in the presence of the basic brightening addition agent (O) and the nickel coating electrodeposited in the presence of both the basic and the top brightening addition agents (Δ); b) mirror (■) and diffuse (•) reflections of the nickel coating electrodeposited in the presence of the basic brightening addition agent alone; mirror (▲) and diffuse (▼) reflections of the nickel coating electrodeposited in the presence of both the basic and the top brightening addition agents.[20] (Reprinted from Ref.[20] with permission from the Serbian Chemical Society.)

Relatively flat parts of the surface can be observed both for the nickel coating obtained in the presence of the basic brightening addition agent only and the nickel coating obtained in the presence of both the basic and top brightening addition agents (Fig. 35). From Fig.

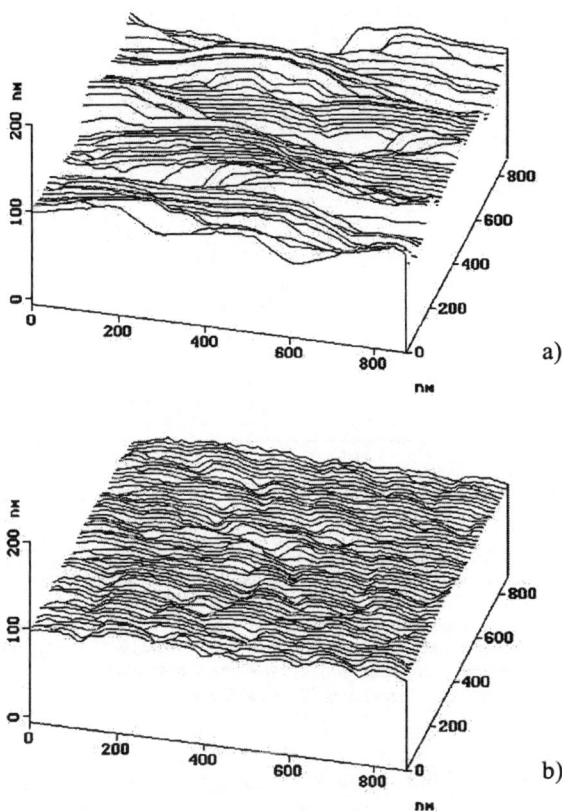

Figure 34. 3D STM images of the nickel coatings electrodeposited with: a) basic brightening addition agent, b) both basic and top brightening addition agents. Scan size: (880 x 880) nm. (Reprinted from Ref.[20] with permission from the Serbian Chemical Society.)

a)

b)

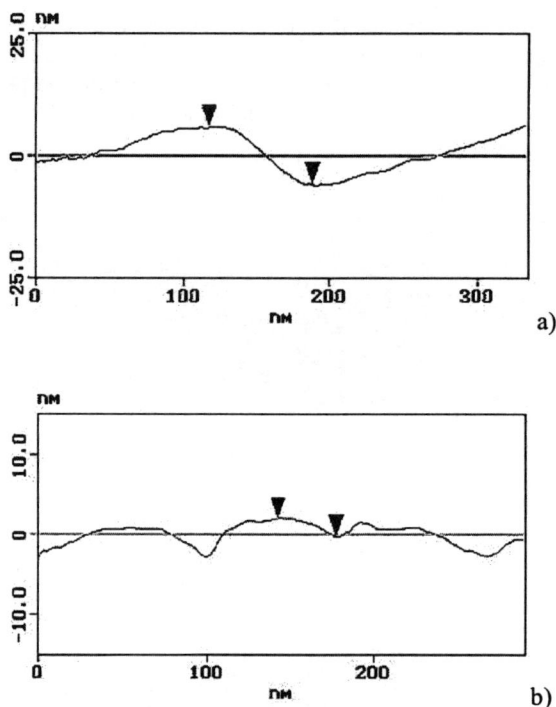

Figure 35. The line sections analysis from the portion of the STM surfaces shown in Fig. 34: a) the nickel coating obtained in the presence of the basic brightening addition agent, b) the nickel coating obtained in the presence of both the basic and the top brightening addition agents. The distances between the markers represent: a) 11.650 nm; b) 2.335 nm. (Reprinted from ref. 20 with permission from the Serbian Chemical Society.)

35b it can be seen that the nickel coating obtained in the presence of both the basic and top brightening addition agents consisted of flat and mutually parallel parts of the surface. It was shown by the STM software that the distances between adjacent flat parts of the nickel coating obtained in the presence of the basic brightening addition agent only are several times greater than the same distances of the nickel coating obtained in the presence of both the basic and the top brightening addition agents. The distance between adjacent flat parts of the nickel coating obtained in the presence of both the basic and the top brightening addition agents is several atomic diameters of nickel.[11] The same distance for the nickel coating obtained in the presence of the basic brightening addition agent only is approximately 40 atomic diameters of nickel.

The line section analysis of the flat part of the nickel coating obtained in the presence of both the basic and the top brightening addition agents is shown in Fig. 36. The roughness of this flat part of the surface is very small and less than the atomic diameter of nickel. For this reason, it can be said that the flat parts of the surface are smooth on the atomic level.

(v) Discussion of presented results and the model

On the basis of reflection and structural analyses of the silver mirror and other different metal surfaces, it can be concluded that the light from flat parts of surface is mirror reflected light. The diffuse reflection arises from the parts of the surface between the flat parts of surface. Hence, these parts of a surface scatter light.

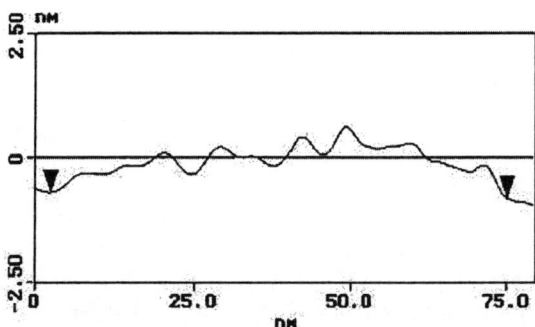

Figure 36. The line section analysis of the flat part of the nickel coating obtained in the presence of both the basic and the top brightening addition agents. The roughness of the observed surface was 0.220 nm. (Reprinted from Ref.[20] with permission from the Serbian Chemical Society.)

Reflection analysis of a silver mirror surface taken as a reference standard showed that the reflection of light from this surface is mostly mirror reflection and that the degrees of mirror reflection are very close to the ideal reflectance of silver. The structural characteristics of this surface, which provide a high degree of mirror reflection, are flat and mutually parallel parts which are smooth on the atomic level, with adjacent flat parts being separated by several atomic diameters of silver.

Hence, the conditions which must be fulfilled in order for metal surfaces to be mirror bright are: (i) flat parts of the surface which are smooth on the atomic level and (ii) distances between adjacent flat parts are comparable with the distances between the adjacent flat parts of a silver mirror.

Similarity of structural characteristics were shown by the the copper surface polished both mechanically and electrochemically, the copper coatings obtained from *solutions Cu I* and *Cu II*, the zinc coatings obtained from *solutions Zn I* and *Zn II* and the nickel coating obtained in the presence of both the basic and the top brightening addition agents (see 2.2.1–2.2.4). For that reason, these metal surfaces can be classified as mirror bright. On the other hand, the copper coating polished mechanically only and the nickel coating obtained in the presence of the basic brightening addition agent only, can be considered as semi-bright metal surfaces.

The variety of results presented can be illustrated by a simple model which treats brightness from the point of view of geometrical optics only.[12] The proposed model will be valid if the following assumptions are fulfilled:

- Metal surfaces are divided into equal elementary parts for which the surface area is n;
- the flat parts are smooth;
- the light falls onto the surface at a determined angle, and is reflected from the surface only in the direction making an angle with the normal equal to the angle of incidendce;
- *the upper flat parts of a surface area* ($k_u n$ parts) reflect light completely; and
- the parts of *the lower flat parts of a surface area* ($k_d n$ parts) do not reflect light completely because they are screened by the height of *the lateral parts of a surface area* ($k_l n$ parts). The screened zones are made by both incident and reflected lights.

In total, the brightness of a surface is determined by the ratio of flat parts to the surface between adjacent flat parts. Increasing the distance between adjacent flat parts leads to a decreasing brightness of the surface.[12]

The parts of the surface which are not able to reflect light, e_i, depend on the angle of incidence angle and the height of the lateral part of the surface. The effect of these factors is illustrated in Figs. 37a, 37b and 37c show that there is an additivity of the parts of the surface which are not

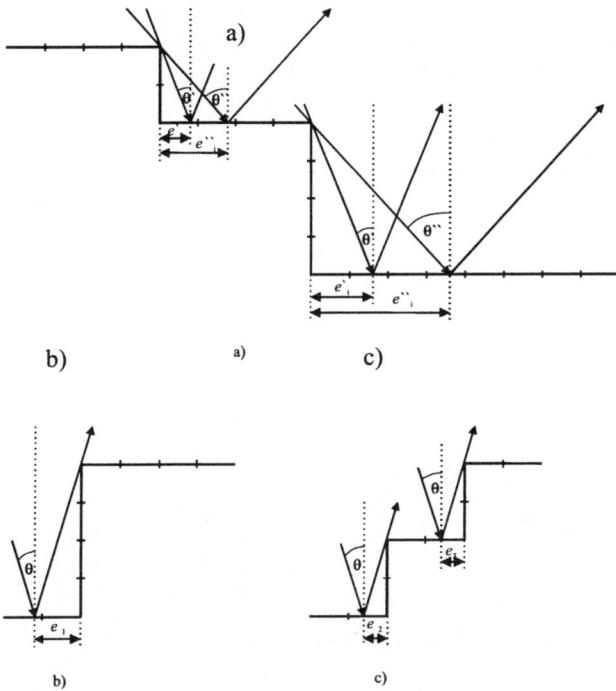

Figure 37. Schematic representation of line section showing the effects of different factors on the reflection of light from a surface: a) the angle of incidence θ and the height of a lateral part of a surface, b) and c) additivity of the parts of a surface which do not have the ability to reflect light. (Reprinted from Ref.[12] with permission from Elsevier.)

able to reflect light, *i.e.,* the screened part of the surface shown in Fig. 37b (denoted as e_1) is equal to the sum of screened parts of the surface shown in Fig. 37c (denoted as e_2 and e_3). Hence,

$$e = \sum_1^\infty e_i \tag{1}$$

where $i = 0, 1, 2, 3.....$

Then, mirror bright surfaces can be simulated by Fig. 38a, semibright surfaces by Fig. 38b and a surface which would show a small degree of mirror reflection (*i.e.,* a metal coating obtained from a solution in the absence of brightening addition agents) by Fig. 38c (see Figs. 8a, 11, 19, 21, 27, 30, 32 and 35).

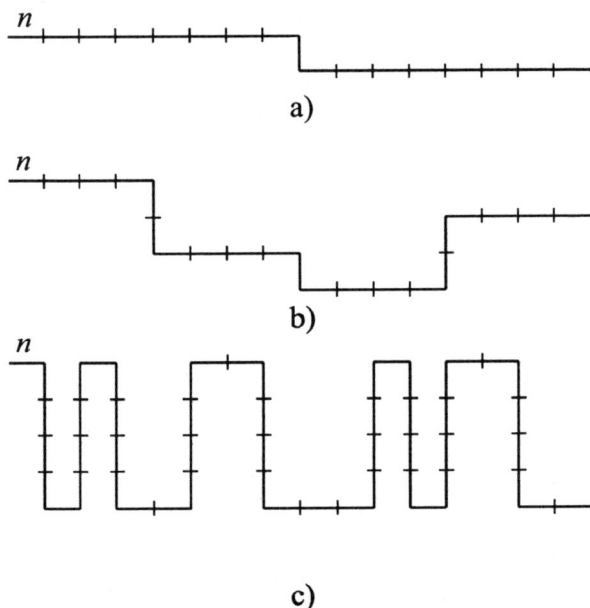

a)

b)

c)

Figure 38. Schematic presentation of the different metal surfaces: a) mirror bright metal surfaces, b) semi-bright metal surfaces, c) a surface which would show a small degree of mirror reflection. (Reprinted from Ref.[12] with permission from Elsevier.)

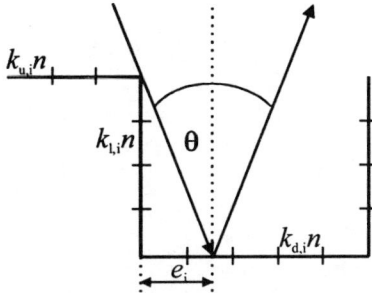

Figure 39. Line-section simulation of one surface of unit a metal surface.

These metal surfaces can be represented by one equivalent surface which consists of row elmentary surface units. One such surface unit is represented in Fig. 39.

It can be shown by simple mathematics that the part of the surface which is screened, e_i, and so cannot reflect light is given, for an angle of incidence, θ, by Eq. (2),

$$e_i = k_{l,i} \cdot n \cdot \text{tg}\theta \qquad (2)$$

where $i = 0, 1, 2, 3 \ldots\ldots$

Then, having in mind fact that the screened zones are made by both incident and reflected lights, the brightness of a surface, defined as the ratio of the geometrical minus the screened surface and the geometrical surface, can be given by Eq. (3):

$$\text{brightness} = \frac{\displaystyle\sum_1^\infty k_{u,i} \cdot n + \left(\sum_1^\infty k_{d,i} \cdot n - 2\sum_1^\infty e_i \right)}{\displaystyle\sum_1^\infty k_{u,i} \cdot n + \sum_1^\infty k_{d,i} \cdot n} \qquad (3)$$

$$= \frac{k_u \cdot n + \left(k_d \cdot n - 2 \cdot e\right)}{k_u \cdot n + k_d \cdot n} = \frac{k_u \cdot n + \left(k_d \cdot n - 2 \cdot \sum_1^\infty k_{l,i} \cdot n \cdot tg\theta\right)}{k_u \cdot n + k_d \cdot n}$$

$$= \frac{k_u + \left(k_d - 2k_l tg\theta\right)}{k_u + k_d} = 1 - 2\frac{k_l}{k_u + k_d}tg\theta$$

where

$$k_u = \sum_1^\infty k_{u,i}, \; k_d = \sum_1^\infty k_{d,i}, \; k_l = \sum_1^\infty k_{l,i} \tag{4}$$

Hence, according to Eq. (3), the brightness of a surface is a function of the incident angle θ, and the ratio between the flat parts, (k_u + k_d) and the lateral parts, k_l. Increasing the lateral parts of a surface, k_l leads to a decrease in the brightness of the surface. Therefore, when $k_l \to 0$, then the brightness of the surface approaches that of a mirror.

The boundary conditions of Eq. (3) depend on the expression in the bracket in the following way:

$$k_{d,i} - 2k_{l,i}tg\theta = \begin{cases} k_{d,i} - 2k_{l,i}tg\theta & tg\theta < \dfrac{k_{d,i}}{2k_{l,i}} \tag{4a} \\[3em] 0 & tg\theta \geq \dfrac{k_{d,i}}{2k_{l,i}} \tag{4b} \end{cases}$$

Hence, *the lower flat parts of a surface area* will reflect light only if the condition (4a) is fulfilled, *i.e.*, if tangent of an incident angle of light is smaller than the ratio ($k_{d,i}/2 \, k_{l,i}$). On the other hand, for larger incident angles or an incident angle equal to this ratio, *the lower flat parts of a surface area* will not be able to reflect light (the condition given by Eq. 4b).

Then, the consequences will be:

$$\theta = 0 \qquad\qquad \text{brightness} = 1 \qquad\qquad (4.c)$$

and

$$\text{tg}\,\theta \geq \frac{k_{d,i}}{2k_{l,i}} \qquad\qquad \text{brightness} = \frac{k_u}{k_u + k_d} \qquad\qquad (4d)$$

This means that the brightness of a surface decreases from 1 to some value given by Eq. (4d) with increase of the angle of incident light from 0 to some critical value.

It is necessary to note that an increase of the lateral parts of a surface, k_l is accompanied by an increase of the "degree of development of the surface." The later quantity, S', defined as the ratio of the real (actual) to geometrical surface, is given by Eq. (5):

$$S\,` = \frac{k_u \cdot n + k_d \cdot n + 2k_l \cdot n}{k_u \cdot n + k_d \cdot n} = 1 + 2\frac{k_l}{k_u + k_d} \qquad\qquad (5)$$

Then, according to Eqs. (3) and (5), the brightness of a surface can be defined by Eq. (5),

$$\text{brightness} = 1 - S\,\text{tg}\,\theta \qquad\qquad (6a)$$

or, in %, by Eq. (6a)

$$\text{brightness (in \%)} = 100 - S\,(\text{in \%}) \cdot \text{tg}\,\theta \qquad\qquad (6b)$$

where $S = (S\,' - 1)$ – a quantity experimentally measurably by STM data processing as *the surface area diff.*

The degrees of development of a surface, determined by the STM analysis as *the surface area diff.*, for the silver mirror surface, the copper surface polished mechanically only and the mirror-bright copper coating obtained from *solution Cu II* of 25 μm thickness, calculated from an area (880 x 880) nm^2, are given in Table 2.[12]

The results presented and the model can be verified in the following way. According to this model, the brightness of a surface depends on the angle of incidence and the ratio between the flat, $(k_u + k_d)$ and the lateral parts, k_l.

Table 2
The Values *The Surface Area Diff* (*S*, %) for Copper Surfaces from an (880 x 880) nm² area. (Reprinted from Ref.[12] with permission from Elsevier.)

Surface area diff.	Silver mirror surface	Copper surface polished mechanically	The mirror bright copper coating obtained from solution Cu II of 25 μm thickness
S, %	3.2 ± 0.2	94.0 ± 3.0	4.1 ± 0.3

The effect of the incident angle of light is illustrated by the results of the measurements of the mirror-reflected light from copper surfaces which were illuminated at 5° and 30° with respect to the normal to the surface. It should be noted that these results cannot be compared directly with the ones from Figs. 1, 6, 9, 15, 16, 24 and 33. Table 3 gives the values of the amount of mirror reflected light, in %, with respect to a mirror as the reference standard, from the copper coating obtained from *solution Cu II* of 25 μm thickness and the copper surface polished only mechanically. A value of 100 % means that the reflection of light approaches that from a mirror. Smaller values of the amount of mirror reflected light correspond to a decrease of the reflection of light with respect to that from a mirror.

Table 3
The Amount of Mirror-Reflected Light and Brightness of Surface Determined According to Eq. (6b). (Reprinted from Ref.[12] with permission from Elsevier.)

Surface	The amount of mirror reflected light, in %, for incident angles		Brightness calculated according to Eq. (6b), in %, for incident angles	
	5°	30°	5°	30°
Silver mirror	100	100	99.72	98.15
Coating obtained from *solution Cu II* of 25 μm thickness	100	91.2	99.64	97.63
Surface polished mechanically	88.9	51.5	91.78	45.73

It can be seen from Table 3 that increasing the angle of incidence leads to a decrease of the degree of specular reflection of a surface. The reflection of light from the copper coating obtained from *solution Cu II* of 25 μm thickness approaches that from a mirror. On the other hand, the reflection of light from the copper surface polished only mechanically is smaller than that from a mirror and the difference in the light reflection increases strongly with increasing incident angle.

Brightness of copper and silver mirror surfaces, calculated according to Eq. (6b), using values of degrees of development of surfaces determined by STM data processing (Table 2) and the degree of development of the silver mirror surface, for the same incident angles are given in Table 3. Table 3 shows that there is agreement within the limits of the experimental error, between experimentally obtained values of mirror reflected light and those of brightness obtained according to the proposed model (Eq. 6b).

The effect of the ratio between the flat and lateral parts can be verified easily by comparative analysis of reflection and structural characteristics of metal surfaces that were considered in Sections II.1 and II.2. Increasing the ratio of flat to lateral parts of surface areas from values corresponding to mat surfaces to those of metal surfaces denoted as mirror bright lead to increasing degrees of mirror reflection, *i.e.*, the reflection of light approaches that of a mirror. This obviously cannot be estimated from measurements of reflection at low light incident angle, because as $\theta \to 0$ the role of the lateral parts of surfaces becomes negligible. Hence, in order to define a metal surface as mirror bright by measurements of reflection, it is necessary to perform the determination of reflection at $\theta = 0$ in order to determine the quality of the flat parts of surfaces and at least at one relatively large angle (as, for example, 30°) in order to determine the role of the lateral parts in light reflection. Hence, if reflection at both incident angles (0 and, for example, 30°) approaches that of a mirror at the same incident angles, the surface can be denoted as mirror bright. Hence, these results represent a good semi-quantitative affirmation of the proposed mathematical model.

Also, the brightness of a metal surface depends on the mean size of atomically flat parts. The difference in the maximum degrees of mirror reflection between the copper coatings (above 85 %) and the zinc coatings (below 85 %) can be ascribed to the different mean size of atomically flat parts of the copper and zinc coatings (see Figs. 22 and 29). The smaller mean size of atomically flat parts of a surface, the greater is the ratio of screened parts which do not have the ability to reflect light. In both cases, the distances between adjacent flat parts are approximately same, and comparable with the same distances between adjacent flat parts of a silver mirror surface.[15]

Finally, the different preferred orientations of copper coatings, as well as the copper surface polished both mechanically and electrochemically, which are characterized as mirror bright metal surfaces, clearly indicate that mirror brightness of a surface is not associated with a preferred orientation. Also, the preferred orientations of the copper coatings did not depend on type of electrodes used.[21]

The properties which determine whether the metal surface is mirror bright are precisely determined by STM investigations. It is also shown that mirror bright metal surfaces can be obtained only by electrochemical polishing or electrochemical deposition in the presence of brightening addition agents. Hence, the next step in investigations of bright metal surfaces should be the determination of the mechanisms by which they are formed during corresponding electrochemical processes. For example, the mechanism of formation of bright copper surfaces can probably be easily done by correlating the results of Nichols and coworkers[5] on the structure of the copper surfaces obtained in the presence of additives, with their reflection characteristics.

ACKNOWLEDGMNETS

This work was supported by the Ministry of Sciences, Technologies and Development of the Republic of Serbia under the research projects "Electrodeposition of Metal Powders at a Constant and at a Periodically Changing Rate" 1806/2002) and "Surface Science and Thin Films" (2018/2002).

REFERENCES

[1] J.O`M. Bockris, G. A. Razumney, "Fundamental Aspects of Electrocrystallization", Plenum Press, New York, 1967.
[2] R. Weil, R. A.Paquin, *J.Electrochem.Soc.*, **107** (1960) 87.
[3] J.K.Dennis, T.E.Such, Nickel and Chromium plating, Woodhead Publ. Ltd, Cambridge England, 1993.
[4] R. Weil, H. C. Cook, *J.Electrochem.Soc.*, **109** (1962) 295.

[5] R. J. Nichols, C. E. Bach, H. Meyer, *Ber. Bunsenges. Phys. Chem.*, **97** (1993) 1012.

[6] F. Czerwinski, K. Kondo, J. A. Szpunar, *J. Electrochem. Soc.*, **144** (1997) 481.

[7] N.D.Nikolić, Z.Rakočević, K.I.Popov, *J. Serb. Chem. Soc.*, **66** (2001) 723.

[8] N.D.Nikolić, Z.Rakočević, K.I.Popov, *Materials Protection*, **42** (2001) 25 (in Serbian).

[9] N.D.Nikolić, E.R.Stojilković, K.I.Popov, M.G.Pavlović, *J. Serb. Chem. Soc.*, **63** (1998) 877.

[10] A.D.Rakić, A.B.Djuričić, J.M.Elazar, M.L.Majewski, *Applied Optics*, **37**(1998)5271.

[11] D. Grdenić, Molekule i kristali, Školska knjiga, Zagreb, 1987 (in Croatian).

[12] N. D. Nikolić, Z. Rakočević, K. I. Popov, *J.Electroanal.Chem.* **514** (2001) 56.

[13] N.D. Nikolić, G. Novaković, Z. Rakočević, D. R. Đurović, K. I. Popov, *Surface and Coating Technology*, **161** (2002)188.

[14] K. I. Popov, M. G. Pavlović, Z. Rakočević, and D. Škorić, *J. Serb. Chem. Soc.*, **60** (1995) 873.

[15] N. D. Nikolić, Z. Rakočević, K. I. Popov, *J. Solid State Electrochemistry*, submitted for publication.

[16] N. D.Nikolić, G. Novaković, Z. Rakočević, K. I. Popov, *Materials Protection*, **44** (2003)17 (in Serbian).

[17] N. D. Nikolić, B. Starčević, Z. Rakočević, K. I. Popov, unpublished data.

[18] N. D. Nikolić, K. I. Popov, Z. Rakočević, D. R. Đurović, M. G. Pavlović, and M. Stojanović, *J. Serb. Chem. Soc.* **65** (2000) 819.

[19] N. D. Nikolić, K. I. Popov, Z. Rakočević, M. G. Pavlović, M. Stojanović, *Materials Protection*, **41** (2000) 27 (in Serbian).

[20] N. D. Nikolić, Z. Rakočević, D. R. Đurović, K. I. Popov, *J. Serb. Chem. Soc.*, **67** (2002) 437.

[21] N. D. Nikolić, E. R. Stojilković, D. R. Đurović, M.G. Pavlović, V. R. Knežević, *Materials Science Forum*, **352** (2000) 73.

6

Electroplating of Metal Matrix Composites by Codeposition of Suspended Particles

Arjan Hovestad and Leonard J. J. Janssen

Faculty of Chemical Engineering, Eindhoven University of Technology, Eindhoven, The
Netherlands

I. INTRODUCTION

It is well known that insoluble substances present in an electroplating bath
codeposit with the metal and become incorporated in the deposits.[1] In
conventional electroplating processes various bath purification methods,
like continuous filtering and anode bagging, are employed to avoid
incorporation of insoluble anode debris (oxide, alloying elements),
airborne dust and solid drag-out from pre-treatment baths. These
incorporated substances generally have a strong adverse effect on the
deposit properties. The advantages of second phase material incorporated
in metallic deposits were realized only later.

In electrochemical composite plating inert particles are deliberately
added to the plating bath to obtain metal matrix composite coatings.
Figure 1 shows an example of metal matrix composite coating of
electroless nickel-phosphorous in which silicon carbide particles are
incorporated. The particle materials used should be inert to the bath in the

Modern Aspects of Electrochemistry, Number 38, edited by B. E. Conway *et al.* Kluwer
Academic/Plenum Publishers, New York, 2005.

sense that they do not dissolve into the bath. Different types of particles with a variety of properties, for example pure metals, ceramics and polymers, can be used. Combined with the variety of metals, which can be electrodeposited, electrochemical composite deposition enables the production of a wide range of composite materials. Compared to the plain metal coatings the composite coatings have improved physical and (electro)chemical properties.

Electrochemical composite plating combines the advantages of metal electroplating with those of composite materials. It requires only minor adjustments of proven and economical viable electroplating technology. Ideally, addition of particles to a standard electroplating bath suffices. Particularly in the field of composite coatings it compares favorably to other methods to produce composite materials with a metallic matrix. The most widely used methods are powder metallurgy, metal spraying and internal oxidation or nitridation.[2-4] Disadvantages of these techniques are that they need to be performed at high temperatures and often necessitate complicated and expensive equipment, which are not viable on industrial scale for numerous applications.

In recent years several excellent review papers on electrochemical composite plating have been published by Celis et al.[5-8] The uses and mechanism of the composite plating process are the main topics discussed

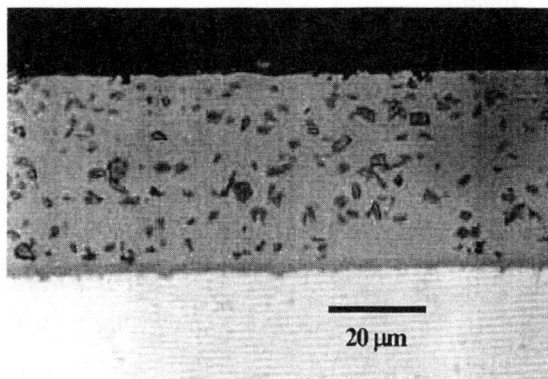

20 μm

Figure 1. Cross-section of an electroless nickel-phosphorous coating on aluminium 6063-T6 with incorporated silicon carbide particles.

in these papers. Hovestad and Janssen[9] published a review paper in 1995, which also includes the experimental facts reported in literature. The present chapter is an updated and extended revision of this earlier paper.

II. PROPERTIES AND APPLICATIONS

The first application of electrochemically deposited composites dates back to the beginning of this century. Sand particles held by a nickel matrix were utilized as anti-slip coatings on ship stairs. In 1928 Fink and Prince[10] investigated the possibility of using electrochemical composite deposition to produce self-lubricating copper-graphite coatings for use in car engines. Apart from these early attempts until about forty years ago little research was done in the field of electrochemical composite plating. In the early sixties the interest in the technique grew and new applications of electrodeposited composites were found. Particularly, the use of Ni-SiC and Ni-PTFE coatings in the automotive industry has accelerated the research in the last 15 to 20 years. The increasing demand for new materials having precisely defined properties offers a promising perspective for further applications of electrodeposited composites.

Applications of electrodeposited composites are generally determined by properties exhibited by the particles. The metal matrix merely serves as a dispersing medium for the particles. Frequently used particle materials like SiC, Al_2O_3 and diamond can be applied as a single coating on a metal substrate by vacuum deposition techniques, but it is difficult to deposit uniform and cohesive coatings and to coat complex shaped products. The difference in physical properties, like the coefficient of thermal expansion, between ceramics and metals often requires one or more intermediate layers are required to obtain a good adhesion to the metallic substrate. In a metal matrix composite the metal provides the adhesion to the substrate and keeps the particles together. The applications of electrochemically deposited composites can be divided into three main categories: Dispersion hardening, wear resistance and electrochemical activity.

1. Dispersion Hardening

Electrochemically deposited composites containing particles of refractory compounds, like oxides,[4,11-16] nitrides,[15,17,18] carbides[14,15,19-23] or borides[24] are dispersion hardened compared to the plain metal as evidenced by improved mechanical properties. The micro hardness, ultimate tensile strength and yield strength are considerable higher and the elongation percentage is reduced. For example, the micro hardness of a composite of Ni with 2.4 vol% Al_2O_3 particles is a factor of 1.5 higher than that of pure Ni.[19] Note that the hardness of the composites is still of the same order as that of the metal and that the particle materials are very much harder.[16] The incorporation of soft materials, like MoS_2, graphite[14] or PTFE[25] has the reverse effect on the mechanical properties, that is it reduces hardness and strength.

Dispersion hardening or strengthening of a material means an increased resistance to deformation. The movement of dislocations in the metal facilitates metal deformation. Incorporated particles block the dislocation movement and thus strengthen the metal.[4,11,12,21] Grain refinement of the metal due to the codeposition of particles has also been thought to contribute to the hardening effect, but this is not supported by experimental evidence. For several composites it was found[4,12,13,26] that the grain structure of the metal matrix was not altered by the codeposition of particles.

In most investigation[4,11,15-18,20,22,23] enhanced hardening was observed with an increase in the volume fraction incorporated particles,[4,11,20] a decrease in particle size[4,11] and a reduction in particle agglomeration.[4] A linear relationship between the composite hardness and the square root of the volume fraction incorporated particles has been reported.[4,22] The effect of composite deposition conditions, like particle bath concentration, current density or pH, on the dispersion strengthening can all be related to changes in the particle composite content.[4,15-18,22,23,25] Hence the effectiveness of the dislocation movement blockage increases with decreasing (effective) interparticle distance.[4,15] A high volume fraction of small particles finely dispersed through the metal matrix will therefore yield optimal hardening. It should be mentioned that there is no unlimited increase in dispersion strengthening with the volume fraction incorporated particles.[13,14] Brown and Gow[13] found a maximum ultimate tensile strength and hardness for Fe-Al_2O_3 composite at approximately 10 vol% Al_2O_3.

A heat treatment of as-plated composites is sometimes[5,19] considered necessary to achieve maximum hardening. Reported investigations[4,11,12,23] do not show a significant improvement in the mechanical properties of annealed composites, but composites do retain their strength up to higher temperatures. This makes the composites suitable for high-temperature applications. Nickel looses it mechanical strength when annealed above 700 K, whereas the strength of Ni-Al$_2$O$_3$ composites is lost above 1300 K.[12] Just as the strengthening, the strength retention of composites can be attributed to the blockage of dislocation movement by incorporated particles.[12] The loss of strength of a metal at high temperatures is due to the recrystallization of the metal, which is accompanied by the annihilation of dislocations. Verelst et al.[11] showed that grain refinement during recrystallization and some unknown process are also involved.

Dispersion hardened composites are mainly applied as free-standing structures, like bearings, die cavities or nozzles[21] and not as coatings. Electrodeposition is therefore not the most suitable preparation method for dispersion hardened composites. In specialized applications, like the electroforming of hollow balls with high thermostructural stability as described by Verelst et al.,[11] electrochemical composite plating is a useful alternative. The potential use of electroformed parts in microtechnology and the increasing availability of nano-sized particles might lead to a renewed interest into electrodeposited dispersion hardened composites in the future.

2. Wear Resistance

Particularly in the field of wear resistant coatings appear the advantages of electrochemical composite plating. The actually realized industrial applications of composite coatings are found in this field.[2,3,5,6] Incorporation of particles of either hard or low-friction materials strongly enhances the wear resistance of a metal coating. A coating of electrodeposited composite can considerably extend the lifetime of surfaces of tools, engine parts and machine parts that are in moving contact. Gages, dowel pins, saw blades, wood chisels and stamping mandrels coated with Ni-SiC last up to 10 times as long as tools made of hard chrome plated tool steel.[21] The effectiveness and lifetime of dentist drills is considerably enhanced by Ni-diamond coatings. Abrasion resistant Ni-SiC coatings are employed as cylinder lining or on piston rings in motor blocks of aluminum alloys[22,27] and coatings of Ni-PTFE are applied among others to reduce the wear of threads and low water adhesion coatings in condenser pipes.

(i) Abrasion Resistance

Coatings of composites containing particles of hard materials, like BN,[17] diamond,[20,28] WC,[21] SiC,[21,22,27,29] Al$_2$O$_3$,[21,26,30] and TiC[23] were shown to have much higher resistance to abrasive wear than plain metal coatings. Composites of Ni-3vol%TiC have a factor of 4 lower weight loss than Ni in a cyclic Taber wear test.[23] Similar to the dispersion hardened composites wear resistant coatings require a high concentration of small particles.[17,23,27,30] A high and uniform surface coverage of particles provides a large contact area and the smaller the particle size the more difficult they are removed during abrasion. However, investigations[30] on Cr- and CrNi- Al$_2$O$_3$ coatings indicate that a too high amount of incorporated particles should be avoided, because the composites become brittle. In general a particle volume fraction of 0.1 can be considered optimal. Composite coatings containing hard particles require careful and extensive grinding and polishing. Hard particles sticking out from the composite surface can dramatically increase the wear of opposite surfaces.

(ii) Lubrication

Another way of reducing the wear of metal surfaces in moving contact is through augmented lubrication. Nickel-SiC coatings have a factor 2 to 3 higher lubricated wear resistance than Ni, because the protruding SiC particles retain an oil film on the composite surface.[22] Otherwise, particles of low friction materials like PTFE,[19,25,31] graphite[29], BN or MoS$_2$ [29] included in a metal coating are used to reduce the friction between sliding metal surfaces.

Electrodeposited Cu-graphite coatings show a substantial lower coefficient of friction and rate of wear than Cu/Sn alloys or sintered bronze PTFE composites.[29] In frictional contact Ni-PTFE composites act as self-lubricating coatings, which slowly erode away thus releasing incorporated particles, which are smeared out over the surface.[25] In a similar manner a recent development[32] in electrochemical composite plating allows wet lubrication of metal surfaces. Microcapsules containing a liquid lubricant are incorporated in a metal. During use the microcapsules gradually wear away, thereby releasing the liquid lubricant. Since, electrochemical composite plating is practically the only method where liquids can be incorporated in metals, it opens up a new range of potential applications.[5]

3. Electrochemical Activity

A prerequisite for any application of a metal is a sufficient resistance to corrosive attack under operating conditions. Since a good resistance against corrosion does not naturally accompany favorable functional properties of a material, protective coatings are routinely applied. Certain composite coatings are suited for corrosion protective coatings of, among others, automotive body panels.[33] The improved corrosion resistance is achieved either directly by the composite itself or indirectly, that is as part of multiple-component coating system. In the first case corrosion resistance of the composite is due to the dispersed phase changing the electrochemical activity of the metal matrix. Otherwise composites containing (electro)chemically active particles could be used as catalytic electrodes in, for example, fuel cells.

(i) Corrosion Resistance

Composites of Ni and Al_2O_3,[33,34] Cr_2O_3,[15] SiC,[15] Si_3Ni_4,[15] TiO_2[33] and SiO_2[33] show decreased corrosion rates under certain conditions compared to Ni. The dissolution rate of Ni-6 vol% Si_3Ni_4 is a factor 3 lower than that of Ni in 100 mol m^{-3} H_2SO_4.[15] The electrochemical mechanism responsible for this reduced corrosion has not yet been elucidated. The effects of composite composition and the type of corrosive environment on the corrosion resistant of a composite are not straightforward. For some composites[33,35] the corrosion rate decreases with particle content, whereas for others[17,33] the opposite is observed. Ramesh Bapu et al.[36] reported that at pH 3 in a NaCl solution a Ni-V_2O_5 composites corrodes faster than Ni, but at pH 6.5 the corrosion rate is similar. The electrochemical activity of the particles, the presence of a metal/particle interface and changes in the metal matrix structure due to particle codeposition seem to be the main factors contributing to the corrosion resistance of composites.

Enhanced oxidation resistance was also found at elevated temperatures for Co-Cr_3C_2,[35] Ni-Cr_2O_3[15] and Ni-Si_3N_4[15] composites. In contrast Ni-SiC[15] and Ni-TiC[23] composites have a higher hot oxidation rate than nickel. During hot oxidation porous metal oxide scales are formed at the metal-air interface. At elevated temperature interdiffusion between the particles and the metal in composites affects the formation of these scales. The break down of TiC particles in Ni-TiC composites accelerates corrosion by favoring the formation of nickel oxide.[23] In

Co-Cr$_3$C$_2$[35] composites the carbide particles supply chromium to a mixed cobalt/chromium oxide scale, which improves corrosion protection. If these composites are heat-treated in the absence of oxygen, a homogeneous cobalt-chromium alloy is formed showing an even better corrosion resistance than conventional cobalt-chromium alloys.[35] Similarly, Stainless steel type anti-corrosion coatings are obtained by subjecting Ni-Cr[37] or Fe/Ni-Cr[38] composites to a homogenizing heat-treatment.

Takahashi et al.[39] used SiO$_2$ particles to obtain Zn/Cr-SiO$_2$ coatings exhibiting excellent corrosion resistance. The SiO$_2$ particles themselves hardly affect the corrosion rate, but they promote codeposition of chromium, which reduces the susceptibility to corrosion. Tomaszewski et al.[34] reported enhanced corrosion protection by nickel composites, with various submicrometer particles, covered by a thin chromium finish. The fine dispersion of particles in the nickel matrix makes the composite porous and this porosity is retained in the thin chromium layer. Numerous tiny chromium cathodes surrounded by numerous tiny Ni anodes are formed, which reduces the net corrosion current. Finally, a composite coating can be used to reduce corrosion by enhancing the adhesion of a protective lacquer to a metal substrate.[5] On the metal substrate the composite, a primer and the lacquer are successively applied. The metal matrix provides the adhesion to the substrate and holds the particles. Functional groups of the primer adhere to the lacquer and the particles.

(ii) Electrocatalysis

In chloralkali cells, electro-organic oxidation processes and batteries there is a need for electrodes with a high electroactive area. Relatively new is the use of electrochemical composite plating in the preparation of these electrocatalytically active electrodes.[40,41] Powder of the electroactive material is kept together by the metal matrix, which serves as the current collector. A Ni-LaNiO$_3$ composite[41] was found to be an effective and stable electrocatalyst for hydrogen evolution. Compared to sintered or electrodeposited Ni the polarization curve in alkaline solutions is shifted to lower potentials. If this due to an increase in surface area or an electrocatalytic effect of the LaNiO$_3$ is not clear. Keddam et al.[40] prepared Ni-Ni(OH)$_2$ composites to investigate the electrochemical activity of Ni(OH)$_2$ particles used in Ni-Cd battery. Since here the particles and not the composite are the object of study, it was verified that the activity of a

Ni-NiOH$_2$ composite electrode is fully determined by the NiOH$_2$ particles. These two examples show that composites can be suitable electrocatalytic electrodes. Though it is clear that more research has to be done to find out the full potential of electrochemical composite deposition in the preparation of electroactive electrodes.

III. PROCESS PARAMETERS

The amount of incorporated particles is the parameter characterizing a metal matrix composite. As discussed in the previous section it largely determines the composite properties. In order to obtain a composite exhibiting certain properties, the effect of process parameters on the particle composite content has therefore to be known. Apart from the practical significance knowledge of these effects is also a prerequisite for the understanding of the mechanism underlying particle codeposition.

Through the years it has been found that numerous process parameters directly or indirectly affect the particle composite content. These parameters can be divided into three main categories:

1) particle properties:
   - particle material
   - particle size
   - particle shape
2) bath composition:
   - constituents
   - pH
   - additives
   - aging
3) deposition variables:
   - particle bath concentration
   - current density
   - electrolyte agitation
   - temperature

A straightforward effect of a single parameter on the particle composite content can not always be given, because the influence of several parameters is interrelated. The fact that some parameters have been investigated extensively, whereas others were hardly examined even adds to this difficulty. If possible a general effect of a certain parameter will be given, else the various effects reported will be discussed.

The mechanism of particle incorporation is treated extensively in the next section, but a generalized mechanism is given here to better comprehend the effects of the process parameters. Particle incorporation in a metal matrix is a two step process, involving particle mass transfer from the bulk of the suspension to the electrode surface followed by a particle-electrode interaction leading to particle incorporation. It can easily be understood that electrolyte agitation, viscosity, particle bath concentration, particle density etc affect particle mass transfer. The particle-electrode interaction depends on the particle surface properties, which are determined by the particle type and bath composition, pH etc., and the metal surface composition, which depends on the electroplating process parameters, like pH, current density and bath constituents. The particle-electrode interaction is in competition with particle removal from the electrode surface by the suspension hydrodynamics.

1. Particle Properties

The particle properties are the least controllable process parameters. The choice of particle material is limited by the desired composite properties. The chosen particle material and (commercial) availability again restrict particle shape and size. Consequently the particle properties set the limits for the attainable particle composite contents.

(i) Particle Material

Comparison of different particle materials is very difficult, because generally particle shape and size will also vary, but it is clear that differences in particle density and surface composition will affect particle incorporation. Greco and Baldauf[4] noticed that three times as much TiO_2 is incorporated in a Ni matrix as Al_2O_3 under the same deposition conditions. Moreover, SiC[42] and Al_2O_3 [43,44] particles of different crystal structure yield composites with different particle content. The apparent impossibility of incorporating Al_2O_3 particles in Cr[45] and γ–Al_2O_3 in Cu[43] was overcome by changing the particles surface composition by means of dry grinding and calcining respectively.[46] Similarly, for Ni-SiC composites[42] it was shown that the surface composition of the particles that is the SiO/SiC ratio at the surface plays determines the particle codeposition.

Due to their hydrophobic character BN particles strongly aggregate in electrolyte solutions as is shown for an electroless Ni(P) bath in Figure 2.[107] Particle aggregation results in low particle incorporation in the Ni(P) coating. Particle codeposition is inhibited either by a reduction in particle bath content, because of particle flotation accompanied by strong foaming on top op the plating bath or dominant particle removal by hydrodynamics at the composite surface due the large size of the aggregates. Addition of a surfactant (see Section III.2.*iii*) facilitates suspension of BN particles and increases the composite particle content from around 1 to 15 vol%. However, Fig. 2 shows that also an oxide modification of the BN-particle surface can prevent particle aggregation and result in a Ni(P) composite

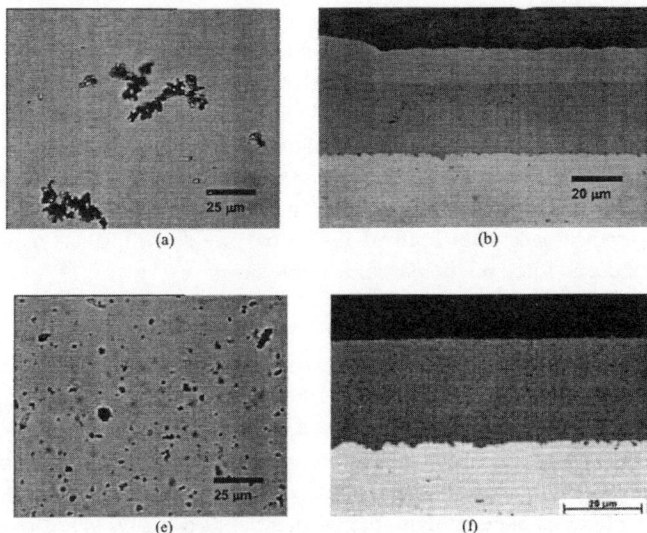

Figure 2. Optical microscope pictures of an electroless Ni(P) bath with suspended BN particles of 3 μm mean diameter (a,c) and a cross-section of the composite deposited from these baths (b,d);; untreated (a,b) and oxide treated (c,d) particels.[107]

with 15 – 20 vol% of finely distributed particles. The oxide modification changes the surface composition of the particle from mainly BN with about 3% B_2O_3, to a high oxide content.

The discussed examples clearly point out that the different interaction of for example particle surface oxides compared to particle surface nitrides and carbides with the electrolyte and the metal surface significantly affects particle incorporation. Note that, although it is often observed, it should not be concluded from the presented examples that oxide or oxide-covered particles will always codeposit more easily than non-oxides. Changing the surface of the BN particles to a silicon oxide does not prevent aggregation or enhance codeposition.[107]

Though it was stated that composite plating involves particles inert to the bath, particles do always interact with the electrolyte. Chemical and physical adsorption of electrolyte ions onto the particle occurs.[47] This adsorption and the initial particle surface composition determine the particle surface charge, which induces a double layer of electrolyte ions around the particle. In electrolytes double layers play a major role in the interactions between particles and between particles and the electrodes. According to the DLVO theory[48,49] surfaces in electrolytes interact through the competitive action of attractive and repulsive forces. Overlap of double layers results in the electro-osmotic force, which is repulsive for surfaces of like charge and attractive for surfaces of unlike charge. In an electroplating bath the applied electrical field will also exert an electrophoretic force on the particle double layer.

In view of this Tomaszewski et al.[50] proposed that particles with a positive charge codeposit more easily, because they are attracted to the negatively charged cathode. It was noticed that in a sodium sulfate electrolyte negatively charged SiO_2 particles move much more difficult to the cathode than the positively charged Al_2O_3 particles. Lee and Wan[51] quantified this by determining the ζ-potential, which is a measure for the double layer interactions. Corresponding to their respectively high and low particle composite content, the ζ-potential of α-Al_2O_3 is positive and that of γ-Al_2O_3 is negative in a dilute copper sulfate bath. In contrast measurements under practical conditions, that is in concentrated copper sulfate[52] and chromium/nickel chloride[53] baths, yield a negative ζ-potential for α-Al_2O_3 particles due to strong anion adsorption. Despite the negative charge α-Al_2O_3 codeposit readily from these baths.

As will also be shown later for charged surfactants adsorbed on codepositing particles, it can be concluded that the particle material influences the particle composite content through the particle surface composition, but that this is not determined by the particle charge. Other interaction forces, like the London-Van der Waals force or hydration force, dominate and are responsible for the effect of particle material on particle composite content as will be discussed further in the Section IV.

Apart from the surface composition the bulk properties of a particle material will affect composite deposition. Particle mass transfer and the particle-electrode interaction depend on the particle density, because of gravity acting on the particles. Since the particle density can not be varied without changing the particle material, experimental investigations on the effect of particle density have not been performed. However, it has been found that the orientation of the plated surface to the direction of gravity combined with the difference in particle and electrolyte density influences the composite composition. In practice it can be difficult to deposit composites of homogeneous composition on products where differently oriented surfaces have to be plated.

A horizontal cathode facing upward results in a high particle composite contents of particles denser than the electrolyte, because gravity causes the particles to settle on the cathode.[54] Using a mathematical model (Section IV.3.*iii*) Fransaer[55,56] calculated a maximum particle density of 1240 kg m^{-3} for a rotating disc electrode facing downward, because gravity imparts heavier particles to come in contact with the electrode. In reality denser particle can be codeposited on a RDE, because particle-particle collisions allow heavier particles to reach the electrode and become incorporated.

Tacken *et al.*[57] showed that magnetically charged Ni particles retain their magnetization, when they are suspended in a zinc deposition electrolyte. This remanent magnetization attracts the particles to the steel cathode, where up to three times as much magnetically charged as uncharged particles are incorporated. Figure 3 clearly shows that the magnitude of the remanent magnetization determines the amount of incorporated Ni particles. The electrical conductivity of particles can not be controlled, but conductive particles behave different in composite plating from isolating ones. Conductive particles tend to agglomerate on the composite surface during deposition.[58] Rough and porous deposits are obtained,[14,57] because as soon as the particles adsorb on the deposit surface metal deposition and hydrogen evolution occurs on them.

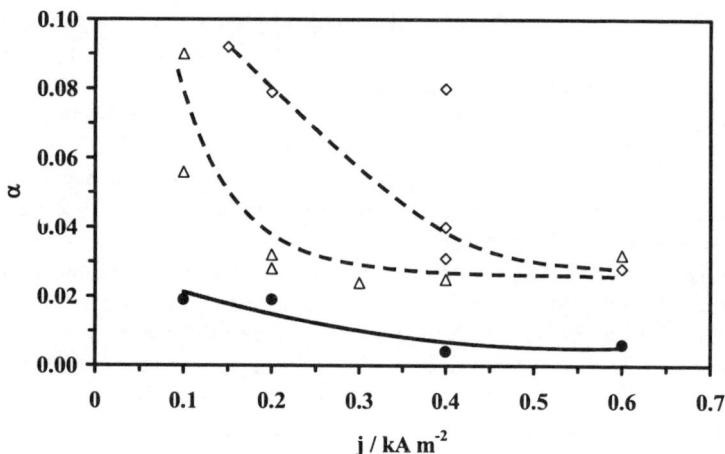

Figure 3. Mass fraction of Ni particles incorporated in Zn against the current density at $\phi_{Ni} = 0.0001$ and three remanent magnetizations of the particles: 0 (●); 3300 A m^{-1} (△) and 8600 A m^{-1} (◊).[57]

(ii) Particle Size

Regarding the effect of particle size on codeposition various results have been reported. For Ni-Al$_2$O$_3$,[45] Ni-SiC[42] and Ni-Cr[58] an increase in the particle composite content was reported, if the particle size was increased. It was shown[59] that the amount of P codeposited with Cu increases linearly with the median volume size of the particles. Yet negligible influence of particle size for Ni-Al$_2$O$_3$ was observed[11] and for Ag-Al$_2$O$_3$ [60] and Ni-V$_2$O$_5$ [36] a lower deposition ratio for larger particles was reported. The latter behavior agrees with the obvious thought that smaller particles are more easily incorporated.

The inconsistency in these various investigations is possibly due to the choice of the units in which the particle content is expressed. Just as ion concentrations are expressed in moles, that is the number of ions per volume, the parameter to be considered is the number of particles suspended and incorporated.[61] Volume or weight particle bath and composite content in fact yield an erroneous comparison of data obtained at different particle sizes. It was found[62] that the weight percentage of SiC particles incorporated in Ni increases, whereas the number of SiC particle incorporated decreases with increasing particle size. Even in this investigation the particle bath concentration is not expressed in numbers, however. It can be calculated from the presented data that if this done the difference between the SiC particle sizes considered is negligible. Further investigations, which take into account the particle number contents are required to unravel the exact effect of particle size on particle codeposition.

A negligible effect of the particle size would confirm the theoretical work of Bozzini et al.,[63] where it is predicted that instead of the absolute particle size the particle size distribution determines the particle composite content. This theoretical work is based on the discovery[53,63,64] of preferential codeposition due to particle parking problems. Parking problems arise, because particles can not deposit onto particles already present on the cathode surface. Depending on the particle surface coverage there is a limiting particle size above which newly arriving particles can not find a site to deposit. Consequently a fraction of the particles does not contribute to particle incorporation. For Cr-Ni-Al$_2$O$_3$[53] the finer 50 % of the particles in the bath accounts for at least 80 % of the particle composite content. In current practice the particle size distribution is generally neglected, but a careful choice of the particle size distribution could be an prerequisite for operating a successful composite plating process.

(iii) Particle Shape

To the knowledge of the authors only one investigation on the influence of the particle shape on particle codeposition has been reported in the open literature.[108] It was found that codeposition of Al$_2$O$_3$ particles in electroless Ni(P) increases in the order fiber, irregular and spherical shaped particles. Again it should be noted that these results were obtained for similar particle mass bath load and not particle number bath load. The fiber shaped particles are the largest in size and therefore had the lowest number content, which could also explain their lower codeposition. The particle shape influence was not further investigated, but considering the particle incorporation mechanism several possible effects of particle shape can be imagined.[65]

The particle shape determines the particles specific surface area and the observed order in increasing codeposition of the Al$_2$O$_3$ particle shapes corresponds to a decrease in the specific surface area.[108] The amount of electrolyte ions adsorbed on the particles and thereby the strength of the cathode-particle adsorption will differ for the different particle shapes. Referring to the detailed discussion of the mechanism in section IV this might be related to a higher ratio of the hydration force to particle adhesion force for a larger particle specific surface area. Although it has not been reported until now anisotropy of particles could give rise to a preferred orientation of incorporated particles. For example, needle-like particles are expected to adsorb perpendicular and not normal to the growing metal surface.

2. Bath Composition

Although less constricted than the particle properties the electrolyte composition is also largely determined by the desired composite. The bath constituents and pH can be varied only within certain limits to ensure a metal matrix of sufficient quality. Additives present an effective way of regulating the particle composite content , but can have adverse effects on the deposit quality. Consequently, also the electrolyte composition poses restrictions on the attainable particle composite contents

(i) Bath Constituents

The influence of the main bath constituents on the incorporation of particles is evident. For different types of baths different incorporation rates are reported, when the same kind of particles are used. For example, Al_2O_3 particles codeposit in Cu- and Ni-baths, but not in a Cr-bath[45] and Cu-Al_2O_3 deposits could be produced in a copper cyanide bath, but not in a copper sulfate bath.[45] Quantitative investigations[29,66] show that an increase in concentration of the main metal salt results in enhanced particle codeposition. Likewise, SiO_2 incorporation in Zn-Co, Zn-Ni or Zn-Fe alloy increases with increasing concentration of the Fe group metal salt.[39]

Since the particle surface composition is determined by the adsorption of electrolyte ions, changes in surface composition of the particles are expected to play a role in the effect of the bath constituents. Kariapper and Foster[67] found that the amount of metal ions adsorbed on a particle increases with increasing metal ion concentration in the electrolyte. For SiC particles this was again related to the ζ-potential of the particles, because it was found[68] that the ζ-potential increases with increasing electrolyte concentration. It should be mentioned that this was not a metal plating electrolyte. So a higher metal ion concentration is expected to increase the ζ-potential and consequently enhance the particle-cathode attraction and particle incorporation. As already discussed an electro-osmotic or electrophoretic attraction between the particles and the cathode, due to cation adsorption, is not essential for particle incorporation. In fact for α-Al_2O_3 in a copper sulfate bath it was found that about 7 times as much SO_4^{2-} is adsorbed on the particles as Cu^{2+}. It was concluded that indeed Cu^{2+} ions are adsorbed on the α-Al_2O_3 particles, but that their positive charge is more than compensated for by adsorbed SO_4^{2-} ions, resulting in a negative ζ-potential.

Figure 4 shows that in an electroless Ni(P) bath the $H_2PO_2^-$ bath concentration reduces on addition of SiC or Al_2O_3 particles. The Ni^{2+} bath concentration and pH, not shown in Fig. 4, in contrast is not affected by particle addition. Very likely a negative charge is inferred on the particles

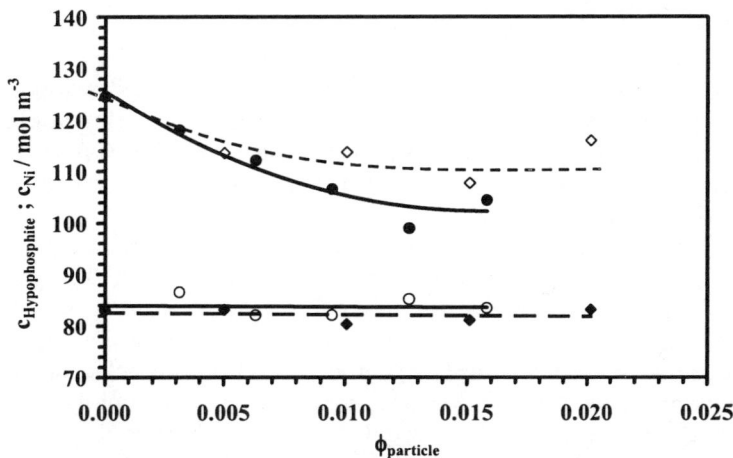

Figure 4. Nickel ($\circ$; $\blacklozenge$) and sodium hypophosphite ($\lozenge$; $\bullet$) concentration in an electroless nickel plating solution at pH 5 as a function of the volume fraction suspended SiC ($\circ$; $\bullet$) and Al_2O_3 ($\lozenge$; $\blacklozenge$) particles.[107]

by strong adsorption of $H_2PO_2^-$ anions instead of Ni^{2+} cations. The ζ-potential of the growing Ni(P) layer is unknown, but taking into account the fairly cathodic potential of -0.5 V vs. NHE measured for Ni(P) deposition it seems unlikely that the high particle incorporation, *i.e.*, $10 - 20$ vol%, even at low particle bath contents, is caused by electro-osmotic interactions.

An adverse effect of adsorption of $H_2PO_2^-$, which is the reducing agent in the bath, is a decrease in plating speed in the presence of suspended particles. Additionally, adsorption of stabilizers results in a reduced stability of electroless bath containing suspended particles. Typically, the life-time of a Ni(P) bath is reduced from $10 - 15$ metal turn-overs in a particle-free bath to around 5 in a particle-containing bath. Hence, through adsorption bath constituents do not only affect particle incorporation, but suspended particles also influence the metal deposition.

In metal plating baths often brighteners or wetting agents are present to improve the appearance of the deposit. Some of these additives act as surfactants, which can strongly affect particle incorporation as will be discussed in Section III.2.*iii*. Tomaszewski *et al.*[34] stated that brighteners can have several effects, but generally enhance particle codeposition. Greco and Baldauf[4] confirmed this and named the use of wetting agents as a tool to increase the particle composite content. On the other hand, a decrease in particle content on addition of wetting agents has also been observed.[44,45] It was suggested[45] that this is due to an increase in micro-throwing power, which leads to metal deposition behind the particles adsorbed at the cathode surface.

The metal surface properties also change with the bath constituents and thereby affect the particle-electrode interaction. Metal deposition constitutes a multi-step reaction mechanism that depends on the bath composition. In quite a number of reaction mechanism adsorbed intermediates, *e.g.* the presence of chromium and catalyst polyoxides on the metal surface during chromium plating, are involved. Not the metal surface, but the adsorbed intermediates will determine the particle-electrode interaction and might even compete for adsorption sites on the electrode surface with the particle. Although the reverse, *i.e.*, the change in metal deposition mechanism due to the presence of particles has been investigated (see Section 3.*ii*), no studies on the effect of the deposition mechanisms on particle codeposition have been reported.

(ii) pH

Investigations[11,12,16,17,34,36,44,45,47,50,53,62,69] concerning the effect of bath pH on the particle content of various composites give comparable results. Particle incorporation decreases sharply below pH 2 to 3 and is practically constant or decreases slightly above this value. An exception is the deposition of Ni-TiC composite,[23] where TiC codeposition reduces continuously with increasing pH. As evidenced by changes in bath pH on particle addition, adsorption of H^+ ions on the particles is considerable,. Together with the other electrolyte ions adsorbed H^+ ions, determine the particle surface composition and thus particle adsorption to the electrode. Besides the objections raised elsewhere, particle charge effects are certainly not involved. It was shown[68,70] that in the absence of metal ion adsorption the iso-electric point, where the ζ-potential equals zero, of SiC particles is reached at pH 2. Although the iso-electric point will be different in a plating bath due to the adsorption of other ions, the ζ-potential increases with decreasing pH, whereas particle incorporation decreases.

The amphoteric nature of oxide particles or oxide covered particles allows both H^+ adsorption and desorption. Due to the relatively high surface area of suspended particles H^+ adsorption or desorption can significantly alter the bath pH. Depending on the initial bath pH and the pretreatment of the particles, that is their initial surface composition, the bath pH either decreases or increases with time on addition of SiC[71] or Al_2O_3 [47,52] particles. The magnitude of the change in bath pH on particle addition decreases with increasing metal ion concentration due to the competitive adsorption of metal ions and H^+ ions.[47,71] Since adsorbed metal ions are by some (Section IV) considered essential for particle incorporation the decrease in particle composite content is attributed[47] to prevalence of H^+ adsorption at low bath pH. It has indeed been found[70] that metal ion adsorption increases with increasing bath pH.

Otherwise changes in metal deposition behavior with pH could be involved. Due to the competition between reduction of metal ions and hydrogen ions at the cathode the pH affects metal deposition. The current efficiency[70] of nickel deposition was seen to decrease markedly below pH 2 in the presence of SiC particles. Unfortunately, it was not determined if this effect is accompanied by a decrease in particle content below pH 2.

(iii) Additives

In order to enhance particle incorporation numerous additives for composite plating electrolytes were investigated. The addition of small amounts of monovalent cations, like Tl^+, Ce^+, Rb^+ and NH_4^+, or amines, like tetra-ethylene pentamine (TEPA), alanine and ethylenediamine tetra-acetic acid (EDTA) promotes particle codeposition. [44,50,51,67] The $BaSO_4$ particle content in a copper matrix increases from 0.5 to 4.5 wt% on addition of 25g l^{-1} EDTA.[50] Although the promoting effect is caused by the monovalent cations the anions of the added salts also play a role. For example, NH_4Cl yields a higher $BaSO_4$ composite content than NH_4F. An additional advantage of these additives is that they are not incorporated in the composite.[44,50,72] The addition of Tl^+, TEPA or EDTA raises the amount metal ions adsorbed on the particles,[67] but the additives themselves are not adsorbed on the particles.[44,72]

It is assumed that the these additives catalyze particle incorporation by enhancing metal ion adsorption. The relation between adsorbed metal ions and particle codeposition is still controversial (Section IV), so it can not be excluded that other processes play a role. The additives will also affect the metal deposition behavior through complexation of metal ions (EDTA, NH_4^+) or adsorption at the metal surface (Tl^+, amines). These processes have to be investigated to obtain a definite explanation for the promoting effect of these additives.

Another class of additives, which were found[10,30,31,42,65,68,73,74] to promote particle incorporation, are surface-active molecules or surfactants. Surfactants are usually employed to stabilize suspensions that are to prevent particle aggregation. In highly concentrated electrolytes the double layer of suspended particles is strongly compressed by the electrolyte ions. The repulsive electro-osmotic force between particles becomes negligible compared to the attractive London-van der Waals force and particles aggregate. Surfactants preferentially adsorb on particles and through mutual electro-osmotic or steric repulsion oppose particle aggregation. Despite the high electrolyte concentrations used in composite deposition baths particle aggregation was hardly found to be a problem. Only certain hydrophobic particles, like graphite,[10] polystyrene[54,75,76], BN^{107} and $PTFE^{31,73,74}$ require surfactant to prevent their agglomeration in a composite deposition bath.

For PTFE incorporation in Ni non-ionic fluorosurfactants were used to obtain a agglomerate-free suspension. Combined with a cationic fluorosurfactant they allow the deposition of composites containing up to

70 vol% PTFE particles.[31,73,74] Here, the use of surfactants is particular successful, because PTFE particles are very hydrophobic and the affinity of fluorocarbon surfactants to the fluoropolymer PTFE is large. However, Helle[31,73] showed that cationic fluorocarbon surfactant also produce a dramatic increase in particle composite content for SiC and diamond particles. It is assumed that the beneficial effect cationic surfactant is due to the positive charge they infer on the particles, which results in an electro-osmotic and electrophoretic attraction to the negatively charged cathode. It was already discussed though that the significance of these interactions in composite plating is doubtful. Hu et.al.[74] measured that the ζ-potential of PTFE becomes increasingly positive on addition of cationic fluorosurfactants, but contrary to their conclusions the obtained PTFE composite content does not vary correspondingly.

Other investigations also prove the assumption of an influence of surfactant charge wrong. The incorporation of polymeric microcapsules in Ni[32] is enhanced by addition of sodium dodecyl sulfate, which is an anionic surfactant. Figure 5(a) and 5(b) show that for polystyrene incorporation in zinc[54,75,76] in the presence of surfactant concentrations lower than 0.005 mol per kg of particles the volume fraction of incorporated polystyrene is equal to that without surfactant, namely 0.06 ± 0.02, independent of surfactant type. In the presence of the cationic surfactant cetylpyrridinium chloride a strong increase in polystyrene incorporation is found at higher concentrations.

At higher concentration of other cationic surfactants, cetylammonium chloride and bromide, the anionic surfactant sodium dodecylsulphate or the nonionic surfactant nonylphenol(ethoxylate)$_{28}$(propoxylate)$_{13}$ polystyrene incorporation is reduced and becomes practically zero. It is clear that there is no correlation between the change in polystyrene incorporation and the surfactant charge. Increased wetting of the particles due to adsorbed surfactants does also not contribute to promotion or inhibition of particle codeposition by the surfactants. Aggregation of polystyrene particles is prevented at high surfactant concentrations,[54,75,76] for all investigated surfactants but nonylphenol(ethoxylate)$_{28}$ (propoxylate)$_{13}$.

The surfactants only slightly affect polystyrene incorporation up to the surfactant concentration where surfactant adsorption on the particles is maximal,[54,76] that is 0.02 mol kg^{-1}. At higher surfactant concentrations the amount of free surfactant that is surfactants not adsorbed on the particles

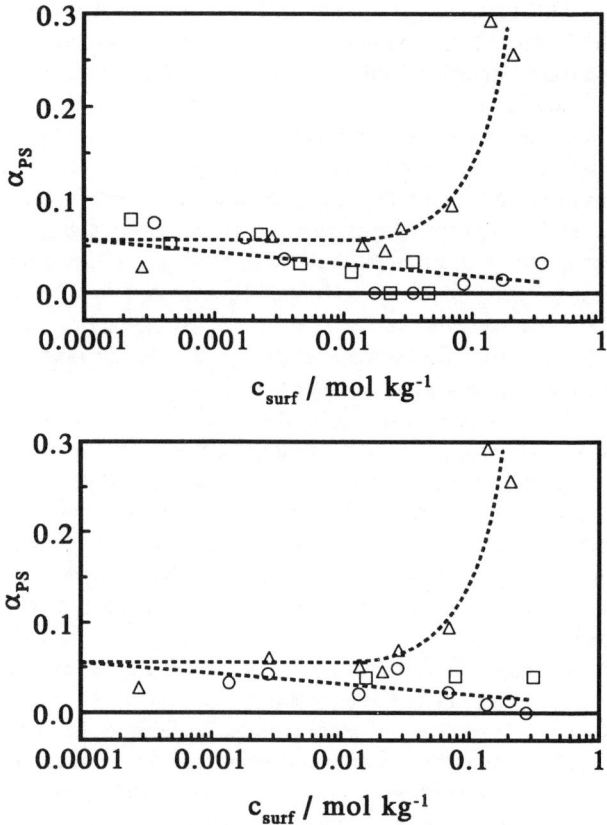

Figure 5. Volume fraction of polystyrene (PS) particles in zinc against the concentration of various surfactants at $j = 0.5$ kA m^{-2}, $\phi_{PS} = 0.02$;[54,76] (a): Cetylpyrridinium chloride ($\triangle$); Sodium dodecylsulphate ($\circ$) and Nonylphenol(ethoxylate)$_{28}$(propoxylate)$_{13}$ ($\square$); (b): Cetylpyrridinium chloride ($\triangle$); Cetyltrimethylammonium bromide ($\circ$) and Cetyltrimethylammonium chloride ($\square$).

becomes significant. It was found[54,76] that at these high concentrations the free surfactant alters the morphology of the deposited zinc matrix. Therefore it was proposed[54,76] that the effect of surfactants on the polystyrene incorporation in zinc is related to the changes in surface roughness of the deposit. In Section IV.3.*iv* this will be discussed in more detail. Further investigations are necessary to see if such correlations are also found for other composites.

A peculiar effect of composite electrodeposition in the presence of surfactant is the formation of a so-called "white layer". The exact conditions, which lead to white layer formation have not yet been established, but it does require hydrophobic particles and surfactants. The white layer is a layer of particles that remains on the surface of the composite after removal from the composite plating bath. The layer adheres to the composite surface, but is not strongly bonded and can be rinsed off. Figure 6 shows that BN particles from a white layer can become incorporated in a Cu layer deposited on top of a Ni(P)-BN composite from a particle-free bath. The distribution of the particles through the Cu matrix suggests that the white layer is not overgrown by the Cu layer. The Cu deposits below the white layer, which remains on the surface steadily releasing particles into the growing Cu layer. The white layer formed on Ni-PTFE composites has been used to create PTFE film on the composites by sintering the particles in the white layer particles in a post heat-treatment.[31,73]

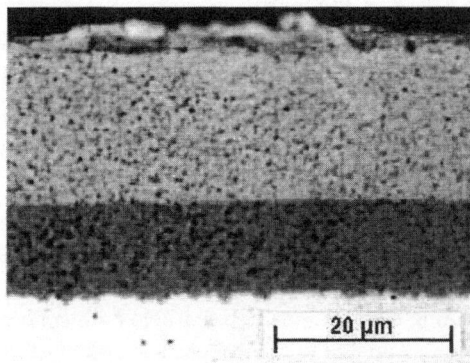

Figure 6. Cross-section of a copper layer deposited from a particle-free bath on an electroless nickel-phosphorous coating with incorporated BN particles.[107]

(*iv*) *Aging*

In practice aging of the composite deposition electrolyte will be a major factor in deciding on the feasibility of industrially producing a certain composite. Aging was found to influence only particular composite deposition baths. Narayan and Chattopadhyay[46] reported no aging effect up to 18 days for Cr-Al$_2$O$_3$, except under certain conditions, where the particle composite content increases until it reaches a limiting value after 10 days. For Cu-SiC composites[2,77,78] the SiC composite content fluctuates between 1 and 0.3 wt% during a period of 50 days. It was suggested that this is due to a time-dependent adsorption-desorption process of the electrolyte ions on the particles. However it was also noticed[78] that the changes in SiC composite content are accompanied by a change in surface morphology of the deposit.

The continuous decrease in Al$_2$O$_3$ incorporation in copper is also explained by a change in particle surface composition.[43] Chloride present as an impurity forms CuCl, which adsorbs on the particles and thereby inhibits the adsorption of copper ions on the particles. This is an effect characteristic for a copper sulfate bath and was not found in nickel or cobalt baths. A different type of aging was reported for the codeposition of aggregated polystyrene particles with zinc.[54,76] Polystyrene incorporation increased continuously in successive experiments, where the rotation speed of a cylinder electrode was randomly varied. Thixotropic viscous behavior of the aggregated suspension causes changes in aggregate size and suspension viscosity with rotation speed of the same time-scale as the experiments.

3. Deposition Variables

The deposition variables are the process parameters most suited to regulate the particle composite content within the limits set by the particle properties and plating bath composition. Particle bath concentration is the most obvious process variable to control particle codeposition. Within the limits set by the metal plating process and the practical feasibility also current density, bath agitation and temperature can be used to obtain a particular composite. Consequently the deposition process variables are the most extensively investigated parameters in composite plating. The models and mechanisms discussed in Section IV almost exclusively try to explain and model the relation between these process parameters and the particle codeposition rate.

(i) Particle Bath Concentration

Sautter[16] found that the volume percentage of Al_2O_3 particles in a Ni-matrix increases with particle bath concentration. In further investigations[4,9,11,23,26-30,33,42,44,46,52,53,55,56,58-60,62,63,66,69,72,75,76,79-82] this behavior was confirmed for a wide range of metal particle systems (Fig. 7). With a few exceptions the particle composite content increases with decreasing rate until a limiting value is reached at high particle bath concentration. As will be discussed in the next section this behavior points to particle adsorption at the electrode surface according to a Langmuir adsorption isotherm (Section IV).

For incorporation of polystyrene particles in copper[55,56] and zinc[54,75,76] a deviation from the Langmuir adsorption behavior at high particle bath concentration was observed (Fig. 7). The obtained particle composite content is higher than expected from the extrapolated curve at

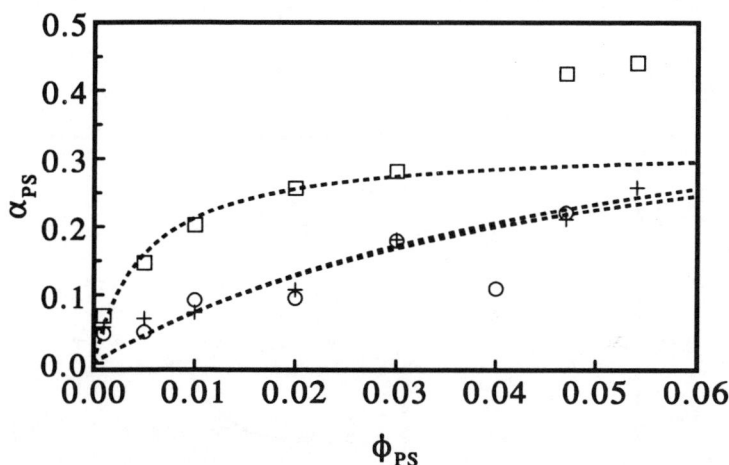

Figure 7. Volume fraction of incorporated polystyrene particles against the volume fraction of suspended particles at j = 0.5 kA m^{-2} and several concentrations of surfactant (cetylpyrridinium chloride): 0 (○); 0.02 mol kg^{-1} (+) and 0.2 mol kg^{-1} (□); Dotted lines are fits with a Langmuir isotherm.[54,76]

lower particle bath concentration. Hydrodynamic effects,[55,56] due to particle-particle interactions becoming significant at high particle bath concentrations, or additional particle adsorption[54,75,76] at sites, which were not accessible at low particle concentrations, were named as cause for the augmented particle composite content. Occasionally, a decrease in particle composite content is reported[17,23,36,69] at higher particle bath concentrations due to settling and agglomeration of the particles.

(ii) Current Density

Composites are deposited using both electroless[62-64] and electrolytic plating processes. In the latter case composite deposition occurs in the presence of an applied electrical field, which is characterized by the cathodic potential or current density. The current density is the most extensively investigated process parameter. Roughly two types of current density dependencies can be distinguished. The particle composite content against current density curve either decreases or increases continuously [4,11,16,24,27,30,42,59,66,67,79,82,83] or exhibits one or two peaks[14,17,22,29,33,41,42,44,46,54,67,69,72,75,76,80,84-87] (Figs. 4 and 8). It can not be

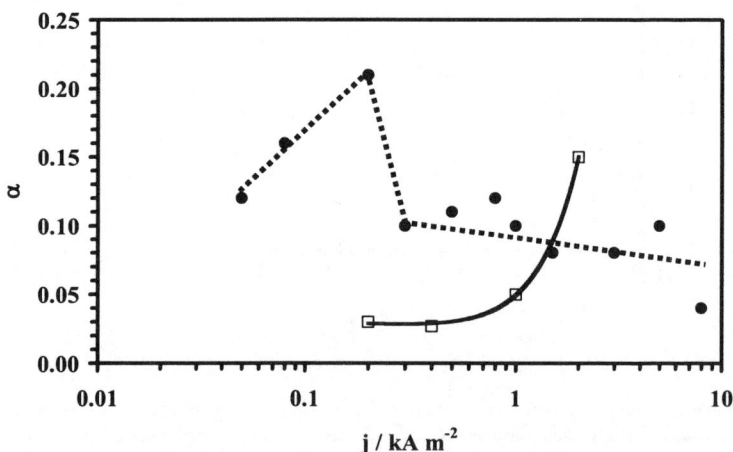

Figure 8. Current density variation of Si particle incorporation in Fe[54,83] at $\phi_{Si} = 0.086$ (□) and of polystyrene (PS) incorporation in Zn[54,75,76] at $\phi_{PS} = 0.02$ (•).

excluded that in the first cases peaks were not detected due to a too low accuracy of the data, a limited number of measurements or a too small current density range considered. The shape of the particle composite content against current density curves depends on other process parameters. The peak height and position change with agitation,[42,81] particle bath concentration,[14,46,69,80,81] and particle type.[42] For Ni-SiC composites[42] the curve of the SiC composite content against the current density shifts from a continuous decreasing curve to a curve with a maximum when the electrode rotation speed or particle size is decreased.

Recently, it was found[88] that the use of pulse-reverse plating can significantly increase the incorporation of nano-sized γ-Al_2O_3 particles in a copper matrix. Instead of depositing at a constant cathodic current density the current density is periodically set at an anodic value, resulting in composite dissolution. The highest particle composite content is obtained when the thickness of metal deposited in one cycle equals the particle diameter. Four times as much Al_2O_3 particles are incorporated as compared to a constant current density. So pulse-reverse plating could present a very effective way of enhancing the particle composite content. Though it has to be established if the method is applicable for other composites, particularly when larger particles are used. An explanation for the codeposition enhancement has not yet been found. Particle parking problems related to preferential codeposition might play a role here.

The nature of the current density dependence of particle codeposition is the most disputed aspect in the mechanism of composite plating (Section IV). In the simplest case the particle deposition rate is not affected by the current density, either because of particle mass transfer limitations or a current density independent particle-electrode interaction. Since the metal deposition rate increases with current density, this results in a continuously decreasing particle composite content. In other cases the particle-electrode interaction has to be current density dependent. An unambiguous explanation for this dependence has not yet been found, but it is apparent that the metal deposition behavior is involved.

The peaks in particle incorporation often[55,56,77,85,89] occur at the same current density as kinks in the polarization curve for metal deposition. For Au-Al_2O_3 composite deposition[77] the peaks and kinks also correlate with the preferred orientation of the Au crystallites. Similarly, for zinc-polystyrene composites[54,76] the peak in polystyrene codeposition corresponds to a change in morphology of the zinc deposit. Polarization curves were not recorded, but the peak appears around the current density, where Wiart et al.[90,91] found a 'S'-type kink in galvanostatic polarization curves for zinc deposition.

The presence of particles in the plating bath also changes the polarization behavior of the metal deposition. For silver deposition in the presence of Al_2O_3 particles[92] the polarization curve shifts to lower current densities at low overpotentials and to higher current densities at high overpotentials. Generally, for the same cathodic potential a higher current density, that is a depolarization, is found[44,58,77,83,85,89,93] in the presence of particles. In contrast for Ni-Al_2O_3 composite deposition a large polarization accompanied by a sharp decrease in the nickel deposition current efficiency was observed[89] at the current density, where a peak in Al_2O_3 incorporation occurs. It should be mentioned that contrary to other investigations the particles used were very fine, that is 0.01 μm. Such small particles could disturb the electrical double layer at cathode and the reduction reactions occurring there.[42] Moreover, the results could be obscured by the presence of surfactant in the plating bath, a fact which was not considered by the authors.

A depolarization up to 20 mV is shown in Fig. 9 for iron deposition in the presence of Si particle.[83] The depolarization is accompanied by an

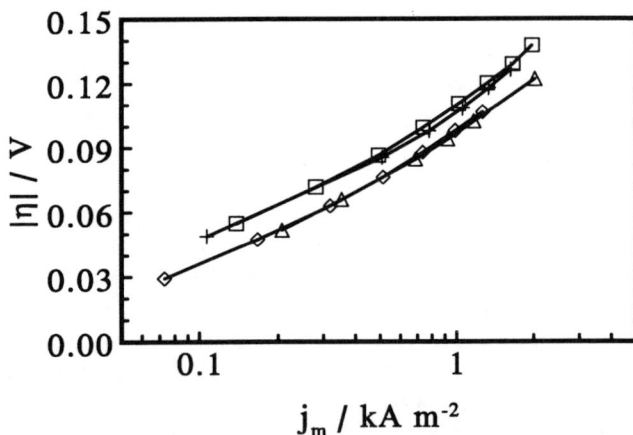

Figure 9. Modified Tafel plots for Fe deposition on a Pt-RDE from a 3 M FeCl$_2$ solution at 363 K containing various amounts of Si particles. ω = 100 s^{-1}. ϕ_{Si}: 0 (□), 0.004 (+), 0.086 (◊), 0.172 (Δ).[54,83]

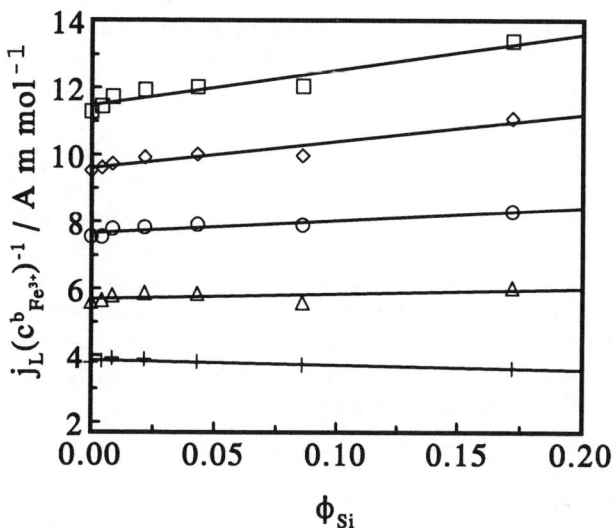

Figure 10. Concentration normalized limiting current density for Fe^{3+} reduction at a Pt-RDE in a 3 M $FeCl_2$ solution at 363 K as a function of ϕ_{Si} at different RDE rotation speeds, ω: 25 (+), 60 (Δ), 100 ($\circ$), 160 ($\diamond$) and 225 s^{-1} ($\square$).[54,83]

increase in the current efficiency for iron deposition from 90 to 95%. The magnitude of the depolarization changes with the Si particle bath concentration, but not with the applied current density. The slope of the Tafel curve in Fig. 9 is independent of the particle bath concentration, whereas the exchange current density increases. Since, in Fig. 8 the particle composite content is seen to increase strongly with current density, it is concluded[83] that suspended Si particles and not the actually codeposited particles enhance the rate of iron deposition. The mechanism behind this catalytic effect was not elucidated.

For Ag-Al_2O_3 deposition the polarization found at low overvoltages is attributed[92] to blockage of the cathode surface by the adsorbed particles and the depolarization observed at high overvoltages is explained[92] by mass transfer enhancement of the Ag^+ ions by the particles. Yet investigations on other composite systems do not support these explanations. A simple calculation shows that in Ni-Al_2O_3 deposition the observed polarization leads to unrealistically high particle cathode coverages close to 100%, when blockage is assumed. Mass transfer enhancement of ions by suspended particles is a well-known phenomenon,[94-97] but in metal plating baths, due to the high metal ion concentrations, mass transfer limitation of the metal ions becomes significant only at very high current densities. Figure 10[54,83] shows the variation in limiting current density for ferric ion reduction in a 3 M $FeCl_2$ iron plating bath as function of the amount of suspended Si particles. If the diffusion coefficient of a ferrous ion is taken equal to that of a ferric ion it can be calculated from Fig. 10 that the limiting current density for iron deposition is reached around 25 kA m^{-2}. A depolarization is already found at much lower concentrations (Fig. 9). Furthermore, Si particles only marginally enhance mass transfer of ferric ions. Only at high rotation rate of the RDE a slight increase in the limiting current density for Fe^{3+} reduction can be seen in Fig. 10.

From impedance measurement[70,89,93] in nickel composite baths it was concluded that SiC and Cr particles cause a depolarization by catalyzing the formation of adsorbed nickel intermediates. Under certain conditions[70,89] formation of adsorbed hydrogen intermediates is catalyzed even more, resulting in a decrease in current efficiency for nickel deposition. On the other hand the current efficiency for iron deposition increases with increasing Si particle bath concentration.[54,83] Figure 9 shows that the Tafel curve[54,83] for Fe deposition corrected for the current efficiency shifts to lower potentials in the presence of Si particles, but that the slope is not affected. Based on these facts it was suggested[54,83] that the Si particles increase the iron deposition rate, but do not interfere in the deposition mechanism. The nature of this catalytic effect is however not clear.

(iii) Electrolyte Agitation

The primary purpose of electrolyte agitation is to keep the particles suspended and prevent them from settling or floating. Agitation is achieved by stirring, air bubbling, recirculation of the electrolyte and, on labscale, by a rotating electrode. The rate of agitation affects particle codeposition in two opposite ways. Increased agitation results in a larger particle composite content,[4] because particle transfer from the bulk of the electrolyte to the cathode surface is augmented. Too much agitation[29] decreases the particle composite content, because the particles are ejected

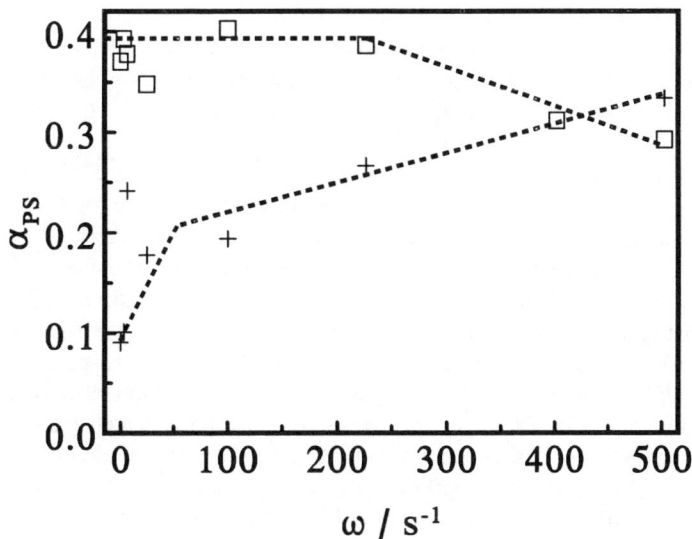

Figure 11. Volume fraction of polystyrene particles incorporated in Zn against the rotation speed of a cylinder electrode at $j = 0.5$ kA m^{-2}, $\phi = 0.054$ and 0.02 mol kg^{-1} cetylpyrridinium chloride.[54,76]

from the cathode surface before being incorporated. This is evidenced by the occurrence of a maximum in the particle composite against agitation rate curve.[51,60]

The balance between particle mass transfer and particle removal depends on the agitation type and the particle properties as is shown for Fe-Si and Zn-polystyrene composites in Table 1. Under the turbulent flow conditions of the stirrer and the RCE particle removal is stronger than under the laminar flow conditions of the RDE and consequently Si codeposition is lower. For the polystyrene particles the agitation type has less effect, because particle removal is low at the agitation rates considered and particle mass transfer is not rate determining. Figure 11 [54,75,76] shows that at low rotation speeds of the RCE the effect of electrode rotation speed on the polystyrene incorporation is negligible and at high rotation speeds polystyrene codeposition is reduced due to particle ejection. The Al$_2$O$_3$ content of a Cu matrix composite deposited on a rotating disc electrode (RDE) varies in a similar way, that is no effect at low rotation rates and a decrease in Al$_2$O$_3$ content at high rotation rates.[77,92] Here the two cases correspond to different flow regimes, respectively laminar and turbulent flow. In the transition flow between these regimes the particle content successively decreases and increase with increasing rotation speed. The formation and incorporation of particle agglomerates explain this.

As for an electrode in channel flow[98] and a RDE[42,51,55,56] the hydrodynamic conditions vary over the electrode surface, particles do not become homogeneously dispersed through the metal matrix. Different zones of hydrodynamic flow along a channel electrode[98] give rise to a distribution of particle contents over the cathode. On a RDE the centrifugal force, which is the removal force, exerted on a particle at the electrode surface increases radially. Accordingly at high rotation speeds and high particle bath concentration, where particle removal dominates, the SiC incorporation in Ni[42] decreases going from the center to the edges of the electrode. The mean size of incorporated particles also decreases radially, because larger particles are removed more easily. In contrast under mass transfer control that is at low rotation speeds and low particle bath concentration SiC incorporation is constant over the electrode, due to the uniform mass transfer to a RDE. It can be concluded that the electrolyte agitation is one of the main process parameters governing particle incorporation.

(iv) Temperature

Electrochemical composite deposition is usually performed at temperatures typical for the utilized metal plating bath. It has been shown that the bath temperature can have a pronounced effect on the particle composite content, but the effect varies considerably per composite. For

Table 1

Effect of type of agitation on Si incorporation in Fe ($\phi_{Si} = 0.05$) and polystyrene (PS) incorporation in Zn ($\phi_{PS} = 0.05$) at j = 500 A m^{-2} and $\omega = 100$ s^{-1} for the rotating electrodes.

Type of agitation	PS particles in Zn m%	Si particles in Fe m%
Magnetic stirrer	2.3 ± 0.5	0.05 ± 0.08
Rotating cylinder electrode	4.5 ± 0.5	0.10 ± 0.04
Rotating disc electrode	4.5 ± 0.5	2.8 ± 0.2

Ni-Al$_2$O$_3$ no effect of the temperature on the particle composite content was found.[11,16] On the other hand a continuos decrease of the particle content of copper matrix composite deposition[57] and a continuous increase of TiC incorporation in nickel[23] with increasing temperature have been reported. A maximum particle composite content at 50 °C was observed for Ni-PTFE,[69] Ni-V$_2$O$_5$ [36] and Ni-BN.[17] In a like manner a limiting particle incorporation in Cr above 50 °C preceded by either an increase[66] or a decrease[46] below this temperature was found.

This diversity in effects of bath temperature is a result of the various temperature dependent parameters affecting particle incorporation. The particle surface composition, electrolyte viscosity and density and metal deposition reaction all change with temperature. Ion adsorption on the particle and particularly lowered H$^+$ adsorption was named[46] as a cause of the reduction in Al$_2$O$_3$ incorporation in Cr with temperature. However, the effect of changes in ion adsorption are small compared to the changes in particle mass transfer due to lowered electrolyte density and viscosity[56] and changes in metal surface composition and morphology. The augmentation in TiC incorporation in Ni[23] with temperature is for example accompanied by the occurrence of dull and gray deposits.

IV. MECHANISMS AND MODELS

In the previous section it was shown that a generalized mechanism underlying particle incorporation in a metal matrix allows some insight into the effect of process parameters on the particle composite content. However, it is evident that a more elaborate mechanism is required to fully comprehend the processes involved. A detailed mechanism is also a prerequisite for the development of a mathematical model describing the particle codeposition behavior. Ideally, such a model should be able to predict the particle composite content from a given set of process parameters. This would facilitate screening composite types and optimization of process conditions for industrial applications.

Several attempts have been made to elucidate the mechanism and to develop models based hereupon. Although an increased insight into composite deposition has been obtained, there still remains some ambiguity. This is partly caused by the tedious and time-consuming work necessary to acquire a set of experimental data sufficient to validate a proposed model and the numerous interrelated process parameters involved. The evolution of the understanding of the mechanism through the years and the models proposed at certain stages are discussed in this section.

1. Early Mechanisms

Simultaneously with the growing interest in composite deposition the first explanations for particle codeposition with a metal were reported in the early sixties. Whithers[99] proposed that the particles having a positive surface charge are drawn to the cathode by electrophoresis. Williams and Martin[29] suggested that in addition the particles are transported to the cathode by bath agitation and are mechanically entrapped by the growing metal layer. Based on mechanical entrapment only Saifullin and Khalilova[100] presented the first model to calculate the amount of incorporated particles. The idea of mechanical entrapment was rejected by Brandes and Goldthorpe,[45] because it signifies that particle incorporation is independent of electrolyte composition and particle properties, in contrast to observations. They suggested the existence of an attractive force, for example an electro-osmotic one, holding the particles at the cathode surface long enough to be incorporated by the growing metal layer. Correspondingly Bazzard and Boden[58] proposed that particles collide with the cathode surface due to the bath agitation and should stay at the cathode surface a certain time to become incorporated. A simple equation to calculate the particle composite content was developed, but it was rightly stated that it lacks any physical significance. These first attempts can be summarized to the generalized two step mechanism involving particle mass transfer to the cathode followed by a particle cathode interaction.

2. Empirical Models

After the first initiatives, more extensive mechanisms and consequently more realistic models were developed. The break-through came with the model put forward by Guglielmi[82] in 1972. It presented the basis for various models, which have in common that they are highly empirical. A mechanism is deduced from experimental data and mathematical equations describing these data are developed. Like this relatively simple models containing several fit parameters of sometimes limited physical significance were obtained.

(i) Guglielmi[82]

From experimental data Guglielmi[82] inferred two fundamental phenomena comprising the particle electrode interaction. The resemblance between the particle composite content versus particle bath concentration curve and a Langmuir adsorption isotherm (Section III.3.i) implies adsorption of particles on the cathode. Additionally, to account for the observed current density dependence of the particle composite content the electrical field at the cathode has to play a role. An electrophoretic interaction is rejected, because the high electrolyte concentration in metal plating baths completely shields the particle charge. Therefore, a field-assisted adsorption mechanism consisting of two steps is proposed. In the first step, which is of a physical nature, particles approaching the cathode become loosely adsorbed on the cathode surface. The loosely adsorbed particles are still surrounded by a cloud of adsorbed ions. In the second step the particles loose the ionic cloud and become strongly adsorbed on the cathode. This step is thought to be of an electrochemical character, *i.e.,* it depends on the electrical field at the cathode. Finally, the strongly absorbed particle is incorporated in the growing metal layer.

The loose adsorption step is described by a Langmuir adsorption isotherm, taking into account the cathode area available for this loose adsorption:

$$\sigma = \frac{k\phi}{1+k\phi}(1-\theta) \tag{1}$$

where ϕ is the particle bath volume fraction, σ is the loose adsorption surface coverage, θ the strong adsorption surface coverage and k is a measure for the intensity of the particle cathode interaction. Obviously the second step depends on σ and Guglielmi considers this dependence linear. Together with a factor to describe its postulated dependence on the electrical field at the cathode, represented by the cathode overpotential (η), the following equation for the strong adsorption rate is obtained:

$$V_p = \sigma v_0 e^{B\eta} \tag{2}$$

where v_0 and B are constants.

The deposition rate of the metal is found using Faraday's law:

$$V_M = \frac{M_M j}{nF\rho_M} \tag{3}$$

Taking into account the area of the cathode available for metal deposition, that is $(1-\theta)$ for non-conducting particles the current density is related to the overpotential by the Tafel equation:

$$j = (1 - \theta) j_0 e^{A\eta} \tag{4}$$

where j_0 is the exchange current density and A is the Tafel slope. Assuming that the volume fraction of embedded particles $\alpha \approx \theta << 1$ these expressions give:

$$\frac{\phi}{\alpha} = \frac{M_M j_0^{B/A}}{nF\rho_M v_0} j^{(1-B/A)} \left(\frac{1}{k} + \phi \right) \tag{5}$$

The constants k, v_0 and B depend on the type of composite considered and have to be determined from experimental data using Eq. (5). A plot of ϕ/α against ϕ at constant j gives a straight line, from whose intercept at $\phi/\alpha = 0$ $1/k$ can be calculated. Plotting the logarithm of the slope of this line at different j against j will also yield a straight line, whose slope gives the ratio B/A and from whose intercept v_0 can be calculated. Other parameters can either be measured, that is A and j_0 , or are known constants, like ρ_M and M_M.

Guglielmi validated his model for Ni-TiO$_2$ and Ni-SiC composite deposition. From the obtained values for k it was found using Eq. (1) that $\theta << \sigma$ and it was concluded that the strong adsorption step is rate determining. The model was also successful in describing the variation of the particle composite content with the particle bath concentration and current density of several other composites.[57,59,60,66,79,84] For example,

Table 2

Values of k and σ obtained using Guglielmi's model for the codeposition of Ni particles with Zn at different remanent magnetizations of the Ni partilces.[57]

M_r / A m^{-1}	k	σ at ϕ_{Ni} = 0.0002
0	0.13	0.23
3300	0.24	0.34
8600-34000	0.90	0.64

Tacken et al.[57] adapted the model for conducting particles by leaving out the term $(1-\theta)$ in Eq. (4). From the modified Eq. (5) it was found that the increased incorporation of magnetically charged Ni particles in Zn is reflected in an increased k and σ (Table 2). Ramasubramanian et al.[101] used Guglielmi's mechanism for particle deposition to model the deposition of Fe-Ni-SiO$_2$ composites. Equation (2) was introduced in a material balance for the electrochemical reaction kinetics in order to describe the competition for adsorption sites on the cathode surface between SiO$_2$ particles and intermediates in the alloy deposition reactions. Like this the reduction in SiO$_2$ incorporation with increasing deposition potential and the decrease in Ni and Fe partial current densities with increasing SiO$_2$ bath concentration were successfully modeled.

Despite these successes, important process parameters, like bath agitation, bath constituents and particle type are disregarded. The constants k, v_0 and B inherently account for these constants, but they have to be determined separately for every set of process parameters. Moreover, the postulated current density dependence of the particle deposition rate, that is Eq. (2), is not correct. A peak in the current density against the particle composite content curve, as often observed (Section III.3.ii), can not be described. The fact that the peak is often accompanied by a kink in the polarization curve indicates that also the metal deposition behavior can not be accounted for by the Tafel equation (Eq. 4). Likewise, the $(1 - \theta)$ term in this equation signifies a polarization of the metal deposition reaction, whereas frequently the opposite is observed (Section III.3,ii). It can be concluded that Guglielmi's mechanism presented an important step in the understanding of composite deposition, but that his model has only limited validity.

(ii) Kariapper and Foster[67]

Regarding the role of the ionic cloud surrounding the particles some obscurity can be noted in Guglielmi's model.[82] It is stated that loosely adsorbed particles are surrounded by an ionic cloud indicating that the particles are adsorbed on the cathode through their ionic cloud. This is contradictory to the definition of k as the intensity of the particle cathode interaction. Kariapper and Foster[67] noted the importance of adsorption of ions on the particles and concluded that adsorbed metal ions play a twofold role. Firstly, the positively charged metal ions cause an electrostatic attraction of the particles to the negatively charged cathode. Secondly, the adsorbed metal ions are reduced at the cathode and create a physical bond between the particle and the cathode. It is striking that Guglielmi[82] inherently took into account this second phenomenon by using a Tafel-type of equation to describe the particle deposition rate (Eq. 2).

Kariapper and Foster derived a simple model considering the effect of several process parameters. The particle deposition rate is again defined as a Langmuir adsorption isotherm, where the measure of the particle cathode interaction k depends on:

- The electrostatic interaction, which is determined by the charge q adsorbed on the particles and the potential field at the cathode E.
- The physical bond, which depends on the rate at which metal is deposited, that is the current density j. When L is the physical bond strength per unit area, the physical bond is a function of Lj^2.
- Mechanical factors, like the particle properties a and the agitation rate b.

Taking $N*$ as the number of particle collisions with the cathode suitable for particle incorporation, which is affected by the agitation rate, the particle deposition rate is given by:

$$V_p = \frac{N*k*(qE + Lj^2 - ab)\phi}{1 + k*(qE + Lj^2 - ab)\phi} \tag{6}$$

where k^* is a constant. The resulting equation was not verified with experimental data, but it was shown that theoretically it is able to predict a peak in the particle composite content versus current density curve. It is evident that the factors in this equation can hardly be measured or evaluated, and has to be fitted with experimental data. Due to the large number of fit parameters an extensive set of experimental data is necessary to obtain a reliable fit. Despite the academic value of the equation Kariapper and Foster introduced important mechanistic concepts.

(iii) Buelens and Celis et al.[72,77]

Celis et al.[44,72,77,84] also noticed the inability of Guglielmi's model to describe the variation in particle composite content with current density. Two current density ranges have to be distinguished to explain the peak in Al_2O_3 incorporation in copper, when using this model. In the low current density range the particle deposition rate increases faster with increasing overvoltage than the metal deposition rate, that is $B > A$ in Eq. (5), and the particle composite content, α, increases with j. In the high current density range on the other hand $B < A$ and α decreases with j.

At the transition between the two current density ranges, the polarization curve for Cu deposition starts diverging from the calculated Tafel curve. This divergence was attributed to the transition from charge transfer to concentration overvoltage control of the copper reduction. It was concluded from these results that the reduction at the cathode surface of metal ions adsorbed on the particles plays a fundamental role in the codeposition mechanism.

Based on this postulate and the pronounced effect of agitation on particle incorporation Buelens et al.[72,77] proposed a five-step mechanism for composite deposition. In the first step particles in the bulk of the electrolyte obtain an ionic cloud by adsorbing ions from the electrolyte. In the second and third step the particles are transported by bath agitation to the hydrodynamic boundary layer and by diffusion through the diffusion layer to the cathode surface. Finally, the particles adsorb on the cathode surface still surrounded by their ionic cloud and are incorporated by the reduction of some of the adsorbed ions. A model for the calculation of the weight percent of incorporated particles was developed consistent with this mechanism. The basic hypothesis of the model is that a certain amount, x, out of X ions adsorbed on a particle must be reduced at the cathode for the particle to become incorporated. Hence not all particles present at the cathode surface are incorporated, but a minimum residence time for a particle adsorbed on the cathode surface is assumed.

The mass fraction of embedded particles is defined as follows:

$$\beta = \frac{W_p N_p P}{\dfrac{M_M j}{nF} + W_p N_p P} \tag{7}$$

where W_p is the particle mass, N_p is the number of particles crossing the diffusion layer at the cathode per unit of time and surface area and P is the chance of a particle to become incorporated. Faraday's law gives the weight of deposited metal. From the basic hypothesis it follows that P depends on the probability $P_{(x/X,j)}$ that at least x out of X adsorbed ions are reduced. Hence, if p_j is the chance that one ion is reduced at current density j:

$$P_{(x/X,j)} = \sum_{z=x}^{X} C_z^X (1 - p_j)^{X-z} p_j^z \tag{8}$$

To calculate p_j a new assumption is made, that is no distinction is made between the adsorbed ions and free ions and thus:

$$p_j = \frac{\dfrac{j}{nF}}{\dfrac{(c_M^b + c_M^s)\delta}{2} + \dfrac{j}{nF}} \tag{9}$$

Where δ is the diffusion layer thickness and c_M^b and c_M^s are respectively the bulk concentration and the concentration at the cathode surface of the metal ions. It is not clear why in Eq. (9) a time factor, which would make it dimensionless, is neglected. Buelens et al.[72,77] just state that a negligible error is created. A factor H is introduced to take into account bath agitation:

$$P = H P_{(x/X,j)} \tag{10}$$

From experiments with varying bath agitation it was obtained that $H = 1$ under laminar flow conditions, $0 < H < 1$ under transition flow conditions and $H = 0$ under turbulent flow conditions.

Finally, N_p is related to the number of ions crossing the diffusion layer per unit time and surface area N_M and to the type of overvoltage control:

$$N_p = N_M \frac{C_p^*}{C_M^*} \left(\frac{j_{tr}}{j} \right)^\lambda \qquad (11)$$

where j_{tr} is the transition current density from charge transfer to concentration overvoltage control and C_p^* and C_M^* are respectively the number of particles and the number of ions in the bulk. Under charge transfer overvoltage control $\lambda = 0$, because the ion-reduction is rate determining, while the diffusion rate is high enough for both particles and free ions. However, $\lambda \neq 0$ under concentration overvoltage control, because diffusion of ions is rate determining, which is obviously much slower for adsorbed ions than for the free ions.

The authors obtain a good agreement between the model and experiments for Cu-Al$_2$O$_3$ and Au-Al$_2$O$_3$ composite deposition, but some assumptions in the model can be questioned. The cathodic overvoltage at j_{tr} corresponds to the value where Degrez and Winand[102] observe a change in the reduction mechanism of Cu^{2+}. It was found that the cathodic charge transfer coefficient is 0.5 in the low overpotential region and 0.1 in the high overpotential region. This indicates that the change in codeposition behavior at j_{tr} is associated with a change in the metal deposition behavior and not with a transition from charge transfer to mass transfer overvoltage control of metal deposition. In general the peak in particle incorporation appears at current densities smaller than about 500 Am^{-2}, which in the concentrated plating electrolytes is much smaller than the limiting current density for metal ion reduction. For example, the peak in polystyrene incorporation in zinc[54,75,76] is found at 0.2 kA m^{-2}, whereas composites can be deposited at least up to 8 kA m^{-2}. The reduction of x out of X adsorbed ions is also difficult to imagine[55,56] considering that the particles are a few orders larger in size than ions. Only very few adsorbed ions will be close enough to the electrode surface, *i.e.*, in the inner Helmholtz plain, to be reduced. Hence the ratio x/X is either very small or the metal has to grow around the particle to reach a larger ratio. In the latter case it seems improbable though that a partly incorporated particle will leave the electrode surface again if the required amount of reduced ions is not reached.

(iii) Hwang and Hwang[86]

Hwang and Hwang[86] proposed an improvement on Guglielmi's model by adapting it to the mechanism put forward by Buelens et al.[72,77] For three current density ranges the particle deposition rate is determined by the electrode reactions for ions adsorbed on the particles, whose rates are determined by kinetic and/or diffusion parameters. A diffusion layer and concentration profile equivalent to that at an electrode are thought to develop at the particle surface. Since, Co-SiC composite deposition was experimentally investigated, the starting point is the reduction of H^+ and Co^{2+} adsorbed on the particles. Three different current density ranges for the reduction of these ions are distinguished:

- Low current density were only H^+ ions are reduced
- Intermediate current density, where the H^+ reduction rate has reached its limiting value and also Co^{2+} is reduced.
- High current were for both ions the reduction rate is at its limiting value.

Similar to Guglielmi's model the metal deposition rate V_M is defined as:

$$V_M = \frac{M_M}{\rho_M nF} j\Gamma_M(1-\theta)$$ (12)

where Γ_M is the current efficiency. The factor v_0 in Eq. (2) is similar to the exchange current density j_0 in electrochemical reactions. Since, j_0 depends on the concentration of reacting species, v_0 depends on the concentration of reacting species in particle deposition that are the adsorbed ions. In the low current density range the particle deposition rate V_p is determined by the reduction of the adsorbed H^+ ions:

$$V_p = k_1 c_{H^+}^s \sigma e^{B_1 \eta}$$ (13)

where $c^s_{H^+}$ is the concentration of H^+ ions adsorbed on the particle surface, which decreases with increasing H^+ reduction:

$$c^s_{H^+} = \left(1 - \frac{V_p}{V_{p,H^+}}\right) c^b_{H^+} \qquad (14)$$

where V_{p,H^+} is the maximum particle deposition rate due to H^+ reduction and $c^b_{H^+}$ is the H^+ ion concentration in the bulk solution.

In the intermediate current density range the particle deposition rate due to H^+ reduction is at its limiting value V_{p,H^+}, whereas the contribution of the metal reduction is similar to that of H^+ in the low current density range. Consequently, the equation for V_p in this range is given by:

$$V_p = V_{p,H^+} + k_2\left(1 - \frac{V_p}{V_{p,M}}\right) c^b_M \sigma e^{B_2\eta} \qquad (15)$$

where $V_{p,M}$ is the limiting particle deposition rate due to the metal reduction and c^b_M is the metal ion concentration in the bulk. Finally, in the high current density range the particle deposition rate is solely determined by diffusion and is independent of the current density and the adsorbed ions concentration. Equation (13) is simplified to:

$$V_p = k_3\sigma \qquad (16)$$

The volume fraction of embedded particles can now be calculated using Eq. (12) and, depending on the current density range, Eq. (13), (15) or (16). The deposition of Co-SiC composites, including the peak in SiC incorporation, can indeed be described by the model.

The model presents an improvement of Guglielmi's model, but it also suffers from the same limitations. Process parameters, like bath agitation and particle properties, are not taken into account and even more fit parameters have been introduced. The reduction of adsorbed ions again leads to some debatable assumptions. Inherently the reduction of adsorbed ions is supposed to differ completely from that of free ions. The efficiency of the metal deposition, that is the competition between the reduction of free H^+ and Co^{2+}, is considered to be independent of current density, whereas for adsorbed ions different regimes are distinguished. The authors do not discuss the validity of this assumption. Besides it is difficult to imagine how reduction of adsorbed H^+ ions can create a bond between the metal matrix and a particle.

3. Advanced Models

Considering the mechanisms treated so far in view of the generalized two step mechanism it is noticed that the nature of the particle-electrode interaction is based on disputable hypotheses, particularly the necessary reduction of adsorbed ions for which only indirect evidence exists. The particle mass transfer step has just been globally treated. This is related to the empirical character of these mechanisms and models. From other fields of research detailed descriptions of mass transfer of solid particles and particle-surface interactions are known.[61,103] Recently, researchers' tried[55,56,79,104] to develop a model for composite deposition using such descriptions. In comparison to the earlier models these models are much more elaborated. They are building up of various often interrelated equations containing numerous parameters, which necessitate the use of extensive computer calculations. Although this renders it difficult to get an easy insight into the effect of a particular process parameter, the number of questionable assumptions and fit parameters without physical meaning is greatly reduced.

(i) Valdes[104]

In 1987 Valdes[104] developed a model for composite deposition at a RDE taking into account the various ways in which a particle is transported to the cathode surface. As starting point an equation of continuity for the particle number concentration, C^*_p, based on a differential mass balance was chosen, that is:

$$\frac{\partial C^*_p}{\partial t} + \frac{\partial}{\partial r} N_p = 0 \tag{17}$$

where the particle flux, N_p, is composed of expressions for the different mass transfer processes, which is Brownian diffusion and convection. The convection term takes into account all the forces and torques acting on a particle due to hydrodynamic migration, electromigration and diffusiomigration. Together with expressions for the local electrical field and the local electrolyte concentration for a binary electrolyte a highly coupled set of transport equations is obtained.

Next the difficulties in obtaining a good description of the particle electrode interaction are noticed. For non-electrochemical systems several particle surface interaction models exist of which the 'perfect sink', that is all particles arriving within a critical distance of the electrode are captured, is the simplest one. However, the 'perfect sink' condition can not be used, because it predicts a continuous increase in particle codeposition with increasing current density, which contradicts experimental observations. Therefore, an interaction model based on the assumption that the reduction of adsorbed ions is the determining factor for particle deposition is proposed. This electrode-ion-particle electron transfer (EIPET) model leads to a Butler-Volmer like expression for the particle deposition rate:

$$V_p = k_0 c_{ion}^s \left[\exp\left(\frac{-\alpha_T nF}{RT} \eta \right) - \exp\left(\frac{(1-\alpha_T)nF}{RT} \eta \right) \right] \quad (18)$$

where k_0 is an electrochemical rate constant, c_{ion}^s, is the concentration of ions adsorbed on the particles, η is the overpotential and α_T is the cathodic transfer coefficient. Using this model a peak in the particle inclusion versus current density curve is predicted, but it is found close to the limiting current density instead of at low current densities. It can be concluded that Valdes model uses a far better description of particle mass transfer than previous models, but the particle-electrode interaction again relies on the controversial necessity of the reduction of adsorbed ions

(ii) Guo et al.[79]

Guo et al.[79] proposed a model based on a description of the mass transport by so-called similitude numbers. Similitude numbers are dimensionless numbers determined by factors influencing mass transfer. A standard description of the Sherwood number for mass transfer of solid particles in a dilute suspension to a fixed plate was modified for composite deposition. If certain parameters, like temperature and bath constituents are considered to be constant, the particle deposition rate can be calculated from the similitude number Sh':

$$Sh' = CoRe^c Dm^d Sx^e Gq^f \tag{19}$$

where Re is the Reynolds number describing mass transfer and Co, c, d, e and f are constants, which have to be determined by fitting the model with experimental data. Co contains among others the Van der Waals attraction that is the physical adsorption of particles on the cathode. Electro-osmotic interactions between particles and the cathode are accounted for by the electrical double layer number Dm. The factor Sx is introduced for the effect of the particle bath concentration and comprises a Langmuir adsorption isotherm. The factor Gq describes the incorporation process of the particles in the metal matrix. It is determined by the ratio of the particle diameter and the thickness of metal deposited during the residence time of a particle at the cathode surface. This residence time is obtained from the electrolyte flow velocity at the center of a particle on the electrode surface. Hence, Gq accounts for the removal of particles from the electrode surface by electrolyte agitation (Section III.3.*iii*).

Satisfactory agreement with experimental data was obtained for Cu-SiC composite deposition in a channel flow. Because of the limited range of experimental data it is not clear if the model is also able to describe important features, like the peak in the particle composite content versus current density curve. In comparison to Valdes model, the particle mass transfer is poorly taken into account by using the Reynolds number. The particle-electrode interaction on the other hand is treated much more adequately by the balance between particle adsorption (Co, Sx and Dm) and particle ejection due to hydrodynamics (Gq). For example, a small value for d is obtained, indicating that, in accordance with experimental data (Section III), electro-osmotic interactions between particles and the cathode (Dm) are negligible.

(iii) Fransaer et al.[55,56]

Fransaer *et al.*[55,56] adapted both existing particle mass transfer and particle-electrode interaction descriptions to composite deposition at a RDE. A trajectory description for a particle was developed based on all the forces and torques acting on it. This comprises the forces due to fluid convection and particle motion and the forces acting directly on the particle. Expressions for all these forces were developed and lead to a set of equations describing the particle trajectory. The particle volume flux to the cathode is determined by calculating the limiting particle trajectory that is the particle trajectory separating the trajectories of particles reaching the electrode from those passing by.

Close to the electrode surface the trajectory description fails, because it leads to the 'perfect sink' condition, which was seen to be wrong. A reaction term characterizing the particle electrode interaction is introduced. A force balance on the particle gives an equation for the probability that a particle at the electrode surface is incorporated (Fig. 12). A

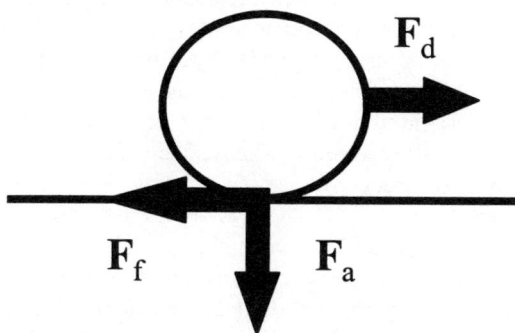

Figure 12. Forcticle adsorbed on the electrode; F_f is the friction force, F_a is the adhesion force and F_r is the removal force.[55,56]

friction force, F_f which results from forces responsible for particle adhesion, F_a, counteracts forces tending to remove particles from the electrode, F_r. The adhesion forces consists of particle-electrode interaction forces, like the London-van der Waals force, the electro-osmotic force and the electrophoretic force. Depending on the electrode geometry, forces acting only on the particles, like gravity, buoyancy and hydrodynamic forces, contribute to F_a. The removal forces, F_r, are of hydrodynamical origin, for example due to electrode rotation. Using the adhesion force determined from Cu-polystyrene codeposition data, the trajectory model gives a good description of the variation of the polystyrene composite content with the polystyrene bath concentration.

The trajectory model does not predict the maximum in the particle inclusion versus current density curve. Experiments indicate that this maximum is related to the potential of zero charge of the copper electrode, which suggests that the electro-osmotic force is responsible for the maximum. Calculations of the adhesion force dependence on the current density from experimental data do not support this last suggestion. It is concluded that the electro-osmotic and electrophoretic force do not affect particle incorporation in agreement with several experimental data discussed in Section III. The electrical double layers of the particles and the electrode are strongly compressed due to the high electrolyte concentration. Another strong repulsive force between the particle and the electrode has to be present at short distances. Therefore, the structural or hydration force is introduced, which is a short range repulsive force arising from the work required to remove the ordered hydration layers at the solid/liquid interfaces of solids coming into close contact in concentrated electrolytes.[105,106] This hydration force will be minimal if the electric field at the electrode is minimal, that is at the potential of zero charge (p.z.c.). The occurrence of a maximum in the particle inclusion versus current density curve is attributed to changes in the ordering of the water dipoles due to changes in electrode charge. Plausible explanations for the effect of particle type, monovalent cations and surfactants on the codeposition rate can be given using the hydration force.

In contrast to earlier models the trajectory model is based on widely accepted descriptions of particle mass transfer and particle-substrate interactions and does not heavily rely on assumptions, like the reduction of adsorbed ions, for which only indirect evidence exists. The extensive mathematical calculations and the complexity of an adaptation to other electrode geometries than a RDE have prevented further quantitative investigations. In spite of that from a mechanistic point of view the trajectory model offers a very good description for gaining a better insight into composite deposition. Particularly, the force balance depicted in Fig. 12 is a very powerful, but relatively simple, tool for the understanding of particle codeposition phenomena. It was, for example, successful in qualitatively explaining experimental data for Ni-SiC[42] and Zn-polystyrene[54,76] composite depositions. Nevertheless the particle-electrode interaction forces and their relative importance remain a point of discussion.

(iv) Hovestad et al.[54,76]

A description of the effect of surfactants on zinc polystyrene composite deposition[54,76] using the hydration force is not entirely satisfactory. Fransaer proposed that the changes in hydrophobicity of the particles, and thus the hydration force, due to surfactant adsorption determine the effect of surfactants on particle codeposition. As discussed in Section III.2.iii up to the concentration of maximum surfactant adsorption (0.02 mol kg^{-1})[54,76] surfactants hardly affect polystyrene codeposition, although the particles become increasingly hydrophilic. Figure 13 indicates that the variation in polystyrene incorporation in zinc due to the surfactant addition correlates with changes in the zinc appearance and surface morphology.[54,76]

Figure 14 shows a generalized picture of the main types of growth morphologies which can be encountered for zinc deposition from acid solutions.[109,110] Zinc deposits as hexagonal platelets, which are oriented at different angels to the substrate depending on the plating conditions. At intermediate current densities the platelets are stacked at random angles to the substrate as observed for deposition from a particle-free electrolyte. At high current densities a vertical type of deposit is formed where the platelets make an angle of 90° with the electrode surface. A basal type of deposit characterized by the platelets lying parallel to the electrode surface is formed at low current densities.

A basal type of deposit is also obtained in the presence of impurities like Co, Ni and Sb, which deposit in between the stacks of platelets resulting a nodular structure.[109,110] Such morphology is found in the presence of high concentrations cetylpyrridinium chloride. Cetylpyrridinium chloride produces basal type deposits (Fig. 13), because

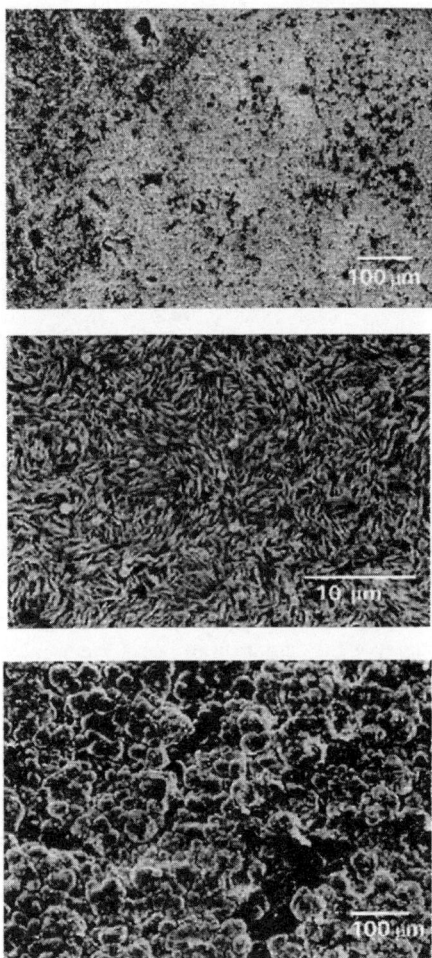

Figure 13. Electron microscope picture of the surface morphology of zinc polystyrene composites deposited from , from top to bottom, surfactant-free, $\phi_{PS} = 0.05$; 0.3 mol kg^{-1} CTAC, $\phi_{PS} = 0.02$ and 0.2 mol kg^{-1} CPC.

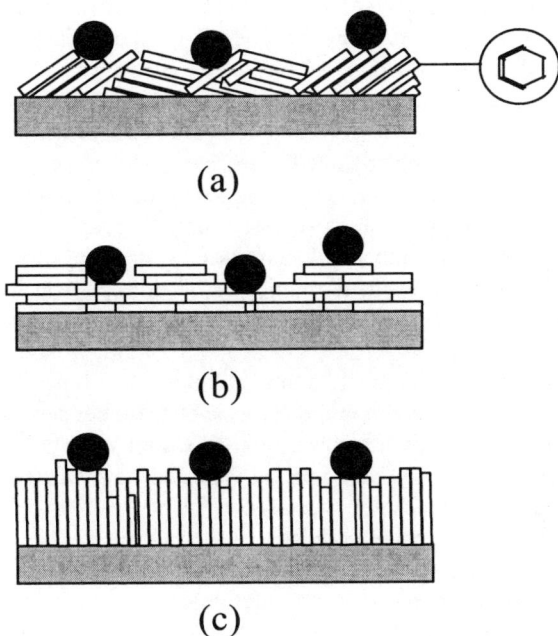

(a)

(b)

(c)

Figure 14. Growth morphologies of zinc deposits; (a) intermediate type, (b) basal type, (c) vertical type; Black spots represent particles.

it can be reduced at the cathode to a dimer, which deposits in between the nodules. Addition of organics, like glue, changes the morphology to the vertical type. Cetylammonium chloride has a similar effect on the composite deposits (Fig. 13), although the platelets are changed into needle-like crystallites. In the presence of polystyrene particles and low surfactant concentrations the intermediate type of growth is observed (Fig. 13).

It is obvious that there is a difference in surface roughness between the different morphology types. Therefore, it was proposed[54,76] that a change in surface roughness is responsible for the differences in polystyrene codeposition. For example glue is used, because the vertical type morphology yields smooth deposits. The friction force, which prevents a particle from being removed from the surface before being incorporated depends on the local surface roughness around the particle.

Particles are depicted in Fig. 14 to show how they can be adsorbed onto the deposit. A particle adsorbed in a recessed area has a much larger probability of becoming included than one adsorbed on a flat surface. Particularly if particles are able to move along the surface due to shearing forces or Brownian rotation to recessed areas, a rough surface leads to higher amounts of embedded particles. Comparing the roughness associated with the growth morphologies of zinc, particle incorporation will decrease in the order: basal, intermediate and vertical type. Correspondingly polystyrene codeposition is the largest at high concentrations of cetylpyrridinium chloride, the lowest with high concentrations of cetylammonium chloride or the other surfactants and in between for suspensions containing up to 0.02 mol kg-1 surfactant.

A similar reasoning could explain the peaks at low current density in the polystyrene codeposition versus current density curves (Fig. 8). A deposit consisting of mossy nodules was found at the peak in the presence of 0.02 mol kg^{-1} cetylpyrridinium chloride. Similar to composites prepared at high cetylpyrridinium chloride concentrations this is a basal type of morphology, where particles deposit in between nodules. Consequently increased codeposition compared to the intermediate type of morphology obtained at higher current densities is expected. For Au-Al$_2$O$_3$[77] codeposition an equivalent correlation between the orientation of Au crystallites and peaks in codeposition with current density were reported. Also the maximum in the particle incorporation versus current density curve for Cu-matrix composites is accompanied by a morphological change of the Cu deposit.[102] As discussed in Section III.3.*ii* these peaks occur at the same current density as kinks in the polarization curves. Similarly, Wiart *et. al.*[90,91] found a kink in polarization curves at low current density for Zn deposition from acidic ZnSO4 electrolyte, where the morphology changes from the mossy basal type to the compact intermediate type.

LIST OF SYMBOLS

a	particle properties parameter (-)
A	constant in Tafel equation for metal deposition (V^{-1})

b	agitation rate parameters (-)
B, B_1, B_2	constant in Tafel equation for particle deposition (V^{-1})
c_i	concentration of species i (mol m^{-3})
c_{surf}	amount of surfactant per unit weight of particles (mol kg^{-1})
C^*	ion or particle number concentration (m^{-3})
Co	dimensionless constant (-)
C_z^X	binomial constant (-)
Dm	double layer dimensionless number
E	electrode potential (V)
F	Faraday constant (C mol^{-1})
$\mathbf{F}$	force (N)
Gq	particle incorporation dimensionless number (-)
H	hydrodynamic coefficient (-)
j	current density (A m^{-2})
J_L	limiting current density (A m^{-2})
j_0	exchange current density (A m^{-2})
j_{tr}	transition current density (A m^{-2})
k	Langmuir adsorption constant (-)
k_0	constant (-)
k^*	electrochemical rate constant (m^4 mol^{-1} s^{-1})
k_1, k_2, k_3	rate constants for particle deposition (m^4 mol^{-1} s^{-1})
L	physical bond strength (m^{-2})
M	molecular weight (kg mol^{-1})
n	number of electrons transferred for the oxidation or reduction of an ion (-)
N	flux of ions or particles (m^{-2} s^{-1})
N^*	number of particle collisions (m s^{-1})
P	particle incorporation probability (-)
p_j	probability for an ion to be reduced at current density j (-)
q	charge (C)
Re	Reynolds number (-)
R	gas constant (J K^{-1} mol^{-1})
r	radial distance (m)
Sh'	modified Sherwood number (-)
Sx	dimensionless number for particle bath concentration (-)
t	time (s)

V	deposition rate (m s^{-1})
v_0	constant for particle deposition (m s^{-1})
W	weight (kg)
X	amount of ions adsorbed on a particle (-)
x	amount of ions adsorbed on a particle that need to be reduced (-)

Greek letters

α	volume fraction of incorporated material (-)
α_T	charge transfer coefficient (-)
β	mass fraction of incorporated material (-)
δ	diffusion layer thickness (m)
ϕ	volume fraction of suspended particles (-)
Γ	current efficiency (-)
η	overpotential (V)
λ	measure of interaction between free and adsorbed ions due to current density (-)
θ	strong adsorption coverage (-)
ρ	density (kg m^{-3})
σ	loose adsorption coverage (-)
ω	angular velocity of rotating electrode (s^{-1})

Subscripts

a	adsorbed
f	friction
M	metal
p	particle
r	removal

Superscripts

b	bulk
s	surface

REFERENCES

1 E. H. Lyons Jr., in *Modern Electroplating*, 3rd ed., Ed. by F.A. Lowenheim, J. Wiley & Sons, New York, 1974, pp. 38-40.

2 J. R. Roos and J. P. Celis, "Is The Electrolytic Codeposition of Solid Particles A Reliable Coating Technology ?," in *Proceedings of the 71st Annual Technical Conference of the American Electroplaters Society*, paper 0-1, New York, 1984, p. 1.

3 J. R. Roos, "A New Generation of Electrolytic and Electroless Composite Coatings," in: *Proceedings INCEF'86*, Bangalore, 1988, p. 382.

4 V. P. Greco and W. Baldauf, *Plating* 55 (1968) 250.

5 J. R. Roos, J. P. Celis, J. Fransaer and C. Buelens, *J. Metals* 42 (1990) 60.

6 J. P. Celis, J. R. Roos, C. Buelens and J. Fransaer, *Trans. Inst. Met. Finish.* 69 (1991) 133.

7 C. Buelens, J. Fransaer, J. P. Celis and J. R. Roos, *Bull. Electrochem.* 8 (1992) 371.

8 J. Fransaer, J. P. Celis and J. R. Roos, *Met. Finish.* 91 (1993) 97.

9 A. Hovestad and L. J. J. Janssen, *J. Appl. Electrochem.* 25 (1995) 519.

10 C. G. Fink and J. D. Prince, *Trans. Am. Electrochem. Soc.* 54 (1928) 315.

11 M. Verelst, J. P. Bonino and A. Rousset, *Mat. Sci. Eng.* A135 (1991) 51.

12 V. O. Nwoko and L. L. Shreir, *J. Appl. Electrochem.* 3 (1973) 137.

13 D. S. R. Brown and K. V. Gow, *Plating* 59 (1972) 437.

14 V. D. Stankovic and M. Gojo, *Surf. and Coat. Technol.* 81 (1996) 225.

15 N. Periene, A. Cesuniene and L. Taicas, *Plat. Surf. Finish.* 80 (1993) 73.

16 F. K. Sautter, *J. Electrochem. Soc.* 110 (1963) 557.

17 G. K. N. Ramesh Bapu, *Plat. Surf. Finish.* 82 (1995) 70.

18 N. S. Ageenko and V. P. Gavrilko, *J. Appl. Chem. USSR* 57 (1984) 2091.

19 J. K. Dennis and T. E. Such, *Nickel and Chromium plating*, 2nd ed., Butterworth & Co, Cambridge, 1986, pp.281-283.

20 E. A. Lukashev, *Russ. J. Electrochem.* 30 (1994) 87.

21 A. E. Grazen, *Iron Age* 183 (1959) 94.

22 S. H. Yeh and C. C. Wan, *Mat. Sci. Technol.* 11 (1995) 589.

23 G. K. N. Ramesh Bapu, *Surf. Coat. Technol.* 67 (1994) 105.

24 P. Fellner and P. K. Cong, *Surf. Coat. Technol.* 82 (1996) 317.

25 M. Pushpavanam, N. Arivalagan, N. Srinivasan, P. Santhakumar and S. Suresh, *Plat. Surf. Finish.* 83 (1996) 72.

26 O. Berkh, S. Eskin and J. Zahavi, *Plat. Surf. Finish.* 82 (1995) 72.

27 F. Mathis, B. Pierragi, B. Lavelle and B. Criqui, "Deposition processes and characterisation of Ni-SiC composite coating," in *Proceedings 24th ISATA International symposium on automotive technology and automation*, Florence, 1991, p. 171.

28 E. A. Lukashev, *Russ. J. Electrochem.* 30 (1994) 83.

29 R. V. Williams and P. W. Martin, *Trans. Inst. Met. Finish.* 42 (1964) 182.

30 Y. S. Chang and J. Y. Lee, *Mater. Chem. Phys.* 20 (1988) 309.

31 K. Helle, "Electroplating with inclusions," in *Proceedings of the 4th International Conference in Organic Coatings Science and Technology*, Vol. 2, Athens, 1979, p. 264.

32 S. Alexandridou, C. Kiparissides, J. Fransaer and J. P. Celis, *Surf. Coat. Technol.* 71 (1995) 267.

33 M. Kimoto, A. Yakawa, T. Tsuda and R. Kammel, *Metall* 44(12) (1990) 1148.

530   A. Hovestad and L.J. J. Janssen

34 T. W. Tomaszewski, R. J. Clauss and H. Brown, *Proc. Am. Electroplaters Soc.* **50** (1963) 169.
35 B.P. Cameron, J. Foster and J.A. Carew, *Trans. Inst. Met. Finish.* **57** (1979) 113.
36 G.N.K. Ramesh Bapu and M. Mohammed Yusuf, *Mat. Chem. Phys.* **36** (1993) 134.
37 R. Bazard and P. J. Boden, *Trans. Inst. Met. Finish.* **50** (1972) 207.
38 G. R. Smith, J. E. Allison and W. J. Kolodrubetz, *Electrochem. Soc. Ext. Abstr.* **85-2** (1985) 326.
39 A. Takahashi, Y. Miyoshi and T. Hada, *J. Electrochem. Soc.* **141** (1994) 954.
40 M. Keddam, S. Senyarich, H. Takenouti and P. Bernard, *J. Appl. Electrochem.* **24** (1994) 1037.
41 A. Anani, Z. Mao, S. Srinivasan and A. J. Appleby, *J. Appl. Electrochem.* **21** (1991) 683.
42 G. Maurin and A. Lavanant, *J. Appl. Electrochem.* **25** (1995) 1113.
43 G. R. Lakshminarayanan, E. S. Chen and F. K. Sautter, *Plat. Surf. Finish.* **63** (1976) 38.
44 J. P. Celis, *Elektrolytische depositie van koper-aluminiumoxyde deklagen*, Phd-thesis, Catholic University of Leuven (1976).
45 E. A. Brandes and D. Goldthorpe, *Metallurgia* **76** (1967) 195.
46 R. Narayan and S. Chattopadhay, *Surf. Technol.* **16** (1982) 227.
47 M. J. Bhagwat, J. P. Celis and J. R. Roos, *Trans. Inst. Metal Finish.* **61** (1983) 72.
48 K. Strenge and K. Mühle, in *Coagulation and Flocculation*, Ed. by B. Dobiáš, Surfactant Science Series 47, Marcel Dekker Inc., New York, 1993, pp. 265-320, 355-390.
49 R. F. Probstein, *Physicochemical Hydrodynamics*, 2nd ed., J. Wiley & Sons, New York, 1994, pp.237-256, 277-302.
50 T. W. Tomaszewski, L. C. Tomaszewski and H. Brown, *Plating* **56** (1969) 1234.
51 C. C Lee and C. C. Wan, *J. Electrochem. Soc.* **135** (1988) 1930.
52 H. Hayashi, S. Izumi and I. Tari, *J. Electrochem. Soc.* **140** (1993) 362.
53 O. Berkh, S. Eskin, S. Berner and A. Zahavi, *Plat. Surf. Finish.* **82** (1995).
54 A. Hovestad, *Electrochemical Deposition of Metal Matrix Composites*, PhD thesis, Eindhoven University of Technology (1997).
55 J. Fransaer, J. P. Celis and J. R. Roos, *J. Electrochem. Soc.* **139** (1992) 413.
56 J. Fransaer, *Study of the behaviour of particles in the vicinity of electrodes*, PhD thesis, Catholic University of Leuven (1994).
57 R. A. Tacken, P. Jiskoot and L. J. J. Janssen, *J. Appl. Electrochem.* **26** (1996) 129.
58 R. Bazard and P. J. Boden, *Trans. Inst. Met. Finish.* **50** (1972) 63.
59 J. W. Graydon and D. W. Kirk, *J. Electrochem. Soc.* **137** (1990) 2061.
60 Y. Suzuki and O. Asai, *J. Electrochem. Soc.* **134** (1987) 1905.
61 Z. Adamczyk, *Colloids and Surfaces* **35** (1989) 283.
62 O. Berkh, A. Bodnevas and J. Zahavi, *Plat. Surf. Finish.* **82** (1995) 62.
63 B. Bozzini, G. Giovanelli and P. L. Cavallotti, *J. Microscopy* (1996) .
64 B. Bozzini, G. Giovannelli, L. Nobili and P.L. Cavallotti, *AIFM Galvanotecnica e Nuove Finiture* **5** (1995) 92.
65 P. K. N. Bartlett, *Industrial training report AKZO*, Arnhem, (1980) pp. 10-39.
66 R. Narayana and B. H. Narayana, *J. Electrochem. Soc.* **128** (1981) 1704.
67 A. M. J. Kariapper and J. Foster, *Trans. Inst. Met. Finish.* **52** (1974) 87.
68 K. Meguno, T. Ushida, T. Hiraoka and K. Esumi, *Bull. Chem. Soc. Jpn.* **60** (1987) 89.
69 G. N. K. Ramesh Bapu and S. Mohan, *Plat. Surf. Finish.* **82** (1995) 86.
70 S. H. Yeh and C. C. Wan, *J. Appl. Electrochem.* **24** (1994) 993.

71 B. Szczygieł, *Plat. Surf. Finish.* **84** (1997) 62.
72 J. P. Celis, J. R. Roos and C. Buelens, *J. Electrochem. Soc.* **134** (1987) 1402.
73 K. Helle, *Codeposition of particle with a metallic matrix*, Report AKZO Research, Arnhem (1993).
74 X. Hu, C. Dai, J. Li and D. Wang, *Plat. Surf. Finish.* **84** (1997) 51.
75 A. Hovestad, R. J. C. H. L. Heesen and L. J. J. Janssen, *Trans. Met. Finish. Ass. India* **6** (1997) 93.
76 A. Hovestad, R. J. C. H. L. Heesen and L. J. J. Janssen, *J. Appl. Electrochem.* submitted for publication (1997).
77 C. Buelens, *A Model for the Codeposition of Inert Particles with a Metal*, Phd-thesis, Catholic University of Leuven (1984).
78 H. Kelchtermans, J. P. Celis and J. R. Roos, *Oberfläche-Surface* **23** (1982) 10.
79 H-T. Guo, Q-X. Qin and A-M. Wang, *Proc. Electrochem. Soc.* **88-18** (1988) 46.
80 J. Foster, B. P. Cameron and J. A. Carew, *Trans. Inst. Met. Finish.* **63** (1985) 115.
81 S. H. Yeh and C. C. Wan, *Plat. Surf. Finish.* **84** (1997) 54.
82 N. Guglielmi, *J. Electrochem. Soc.* **119** (1972) 1009.
83 A. Hovestad, R. Ansink and L. J. J. Janssen, *J. Appl. Electrochem.* **27** (1997) 756.
84 J. P. Celis and J. R. Roos, *J. Electrochem. Soc.* **124** (1977) 1508.
85 C. Buelens, J. P. Celis and J. R. Roos, *J. Appl. Electrochem.* **13** (1983) 541.
86 B. J. Hwang and C. S. Hwang, *J. Electrochem. Soc.* **140** (1993) 979.
87 P. R. Webb and N. L. Robertson, *J. Electrochem. Soc.* **141** (1994) 669.
88 E. J. Podlaha and D. Landolt, *J.Electrochem. Soc.* **144** (1997) L200.
89 S. W. Watson, *J. Electrochem. Soc.* **140** (1993) 2235.
90 I. Epelboin, M. Ksouri and R. Wiart, *J. Electrochem. Soc.* **122** (1975) 1206.
91 J. Bressan and R. Wiart, *J. Appl. Electrochem.* **9** (1979) 43.
92 Y. Suzuki, M. Wajima and O. Asai, *J. Electrochem Soc.* **133** (1986) 259.
93 S. W. Watson and R. P. Walters, *J. Electrochem. Soc.* **138** (1991) 3633.
94 M. Marie de Ficquelmont-Loizos, L. Tamisier and A. Caprani, *J. Electrochem. Soc.* **135** (1988) 626.
95 A. Caprani, M. Marie de Ficquelmont-Loizos, L. Tamisier and P. Peronneau, *J. Electrochem. Soc.* **135** (1988) 635.
96 P. J. Sonneveld, W. Visscher, E. Barendrecht, *J. Appl. Electrochem.* **20** (1990) 563.
97 D. W.Gibbons, R. H. Muller and C. W. Tobias, *J. Electrochem. Soc.* **138** (1991) 3255.
98 J. W. Graydon and D. W. Kirk, *Can. J. Chem. Eng.* **69** (1991) 564.
99 J. C. Withers, *Prod. Fin.* **26** (1962) 62.
100 R. S. Saifullin and R. G. Khalilova, *J. Appl. Chem. USSR.* **43** (1970) 1274.
101 M. Ramasubramanian, S. N. Popova, B. N. Popov and R. E. White, *J. Electrochem. Soc.* **143** (1996) 2164.
102 M. Degrez and R. Winand, *Electrochim.Acta* **29** (1984) 365.
103 Z. Adamczyk and J. Petlicki, *J. Colloid. Interface Sci.* **118** (1987) 20.
104 J. L. Valdes, *J. Electrochem. Soc.* **134** (1987) 223C.
105 R. M. Pashley and J. N. Israelachvili, *J. Colloid Interface Sci.* **101** (1984) 511.
106 R. M. Pashley and J. N. Israelachvili, *Colloids and Surfaces* **2** (1981) 169.
107 A. Hovestad, M. I. van der Meulen, Unpublished results, *Innovation Oriented Research Program on Surface Technology IOT97002*, TNO Institute of Industrial Technology (2001).
108 I. Apachitei, Synthesis and Characterisation of Autocatalytic Nickel Composite Coatings on Aluminium, PhD-thesis, Delft University of Technology (2001).

109 R.C. Kerby, in *Application of polarization measurements in the control of metal deposition*, Ed. by Proces Metallurgy 3, I.H. Waren, Elsevier Science Publishers B.V., Amsterdam, 1984, 111.

110 D.J. MacKinnon, J.M. Brannen and P.L. Fenn, *J. Appl. Electrochem.* **17** (1987) 1129.

Index

Abrasion resistance, 480
Absolute potential differences, 358
Acid / base chemistry in oxides, 20
Activation, in electron transfer, 185
Activity coefficients, for solid-state system, 42
Adsorbed charges, screening of, 333
Adsorption reaction, with partial charge transfer, 305
Adsorption, from electrolyte solution, 348
Adsorption, from gas phase, 347
Alloy electrocatalysts, for H_2 oxidation, 414
Alloy electrocatalysts, ternary, 396
American ordered alloy-catalyst patents, tabulated, 401
Anderson and Newns model, for partial charge transfer, 343
Anion adsorption, and density of states, 343

Area loss at electrocatalysts, 395, 396, 406, 407
Arrhenius plots, for C corrosion, 413

Booth theory, of dielectric saturation, 211
Born–Oppenheimer approximation, 193
Boundaries, defect chemistry at, 48
Boundaries, in solids, 4
Bright metal surfaces, electroplating of, 425
Bright metal surfaces, nanostructure of, 425
Brightness, and mirror reflectivity, 470
Brightness, definitions of, 425
Brightness, equation for, 467
Brightness, treatment by geometrical optics, 464
Bromide adsorption, 325–330
Brønsted slopes, 283
Buelens and Celis et al., for composite plating, 513
Bulk processes, in solids, 95

Capacitance, space-charge, 48
Carbon specific areas, heat
    effects on, 409–411
Carbon support materials, for
    electrocatalysts, 373
Carbon support, Pt intercrystal
    distance on, 382
Carbon, electrocatalyst
    supports, 404
Carbonaceous fuel molecules,
    reforming of, 415–
    418
Catalysation of C supports,
    patents on, 389
Catalyst activity and
    interatomic distances
    in Pt alloys, diagram,
    391
Catalyst particles, ultra-fine,
    375
Catalytic activity, and Pt
    crystallite separation,
    383, 384, 388
Cathode catalyst patents,
    tabulated, 394
Charge carrier concentrations,
    10
Charge carrier equilibria, 10
Charge carriers, as point
    defects, 5
Charge transfer in solids,
    kinetics of, 84
Charge transfer, partial, 304,
    305
Charge-transfer, and solvation,
    262
Charge-transfer, non-Franck–
    Condon conditions,
    269

Chemisorption, with charge
    transfer, 304
Chloride adsorption, 325–330
Codeposition of suspended
    particles, 475
Composite plating, additive
    effects in, 494
Composite plating, advanced
    models, by named
    workers, 518
Composite plating, bath
    compositions for, 490
Composite plating, current-
    density factor in, 500
Composite plating, deposition
    variables for, 498
Composite plating, early
    mechanisms, 508
Composite plating, electroless,
    491–492, 500
Composite plating, electrolyte
    agitation in, 504
Composite plating, empirical
    methods, 508
Composite plating,
    mechanisms and
    models, 507
Composite plating, pH effects
    in, 493
Composite plating,
    temperature factor,
    506
Composites, electroplated, 526
Conductance effects, with
    space-charge, 54
Conductance, due to interfaces
    in solids, 65
Conductivity, static and
    dynamic effects in
    solids, 114

Continuum energy, 246
Continuum media, 196, 198
Continuum, ion interaction with, 219
Copper coatings, reflectivity of, 439
Copper mirror, STM image of, 431
Copper mirror, STM line analysis, 432
Copper surface, polished, XRD pattern, 438
Copper surfaces, electrodeposited from various solutions, 449
Copper, mechanically and electrochemically polished, 438
Copper, mechanically polished, STM image, 435
Corrosion of C, Tafel slopes for, 410-412
Corrosion rates at C, Arrhenius plots for, 413
Corrosion rates at C, water vapor pressure effect on, 414
Corrosion rates, in hot phosphoric acid, 404, 409
Corrosion resistance, of electroplated composites, 481
Corrosion, in hot phosphoric acid, 404
Crystallite separation, and Pt catalyst activity, 382
Crystallite separation, Bregoli work, 382

Crystallite separation, Stonehart theory of, 382
Crystallites, hemispherical diffusion to, 385

Defect chemistry, at boundaries, 48
Defect chemistry, freezing-in of, 32
Defect chemistry, tuning of, 29
Defect concentrations, in pure solids, 13
Defect crystals, Gibbs energies for, 13
Defect reactions, in solids, 10
Defect, defect interactions, 36
Density of states, and anion adsorption, 343
Dielectric constant and displacement, 212
Dielectric constant, at high fields, 207, 211
Dielectric continuum, models, 180, 196
Dielectric saturation, in polar media, 202, 225
Diffusion, at crystallites, calculations on, 385
Diffusion, chemical, 106, 117
Diffusion, hemispherical, 385
Diffusion, tracer studies of, 103
Diffusion-layer, at separated crystallites, 387
Dipole moment, and electrosorption valency, 347
Dipole moments, in hard-sphere models, 351

Dipole oscillations, in electron transfer, 186
Dipole terms, and electrosorption valency, 341, 347
Dipole/dipole neighbors, in solvation, 235
Dispersion hardening, in electroplated composites, 477
DLVO theory, 486
Doping effects, in solids, 22
Double-layer, model, 306

Electrical conduction, in bulk solids, 95
Electrocatalysis rates, reason for low values, 376
Electrocatalyst design, for fuel cells, 373
Electrocatalyst supports, carbon, 404
Electrocatalyst, in phosphoric acid fuel-cells, 374
Electrochemical activity, of electroplated composites, 481
Electrochemical potentials, in solids, 16
Electrochemistry, solid state, 1
Electrode kinetics, in solids, 145
Electrode potential, affecting Pt solubility, 381
Electron excitation, in solids, 9
Electron transfer reactions, 175
Electron transfer, and Franck–Condon principle, 175
Electron transfer, continuum model, 180

Electron transfer, historical, 176
Electron transfer, inner and outer sphere processes, 179
Electron transfer, reorganizational energy in, 183, 184
Electron transfer, Tafel slopes for, 194
Electronegativities, and electrosorption valency, 327
Electrons and holes, in solids, 9
Electroplated composites, electrochemical activity of, 481
Electroplating, of bright metal surfaces, 425
Electroplating, of metal matrix composites, 475
Electrosorption valency, 303
Electrosorption valency, and dipole moment, 347
Electrosorption valency, and dipoles, 334, 341
Electrosorption valency, and electronegativities, 327
Electrosorption valency, and partial charge transfer, 333
Electrosorption valency, definition, 305
Electrosorption valency, determination, 324
Electrosorption valency, experimental plots of, 339, 340

Electrosorption valency,
    extrathermodynamic
contributions, 314
Electrosorption valency,
    themodynamics of,
    308
Electrosorption, Gibbs
    equation for, 309
Electrostatic polarization in a
    continuum, 198
Equations, for brightness of
    metals, 467–469
Equilibria of charge carriers,
    10
Equivalent circuits for solid-
    state processes, 147
Esin and Markov coefficient,
    321

Finishing, of metal surfaces,
    432
Force constants, and inner
    sphere rearrangement,
    191
Franck–Condon
    approximation, 193
Franck–Condon principle, 175
Franck–Condon proton
    transfer, 265
Franck–Condon redox
    processes, 262,
Fransaer et al., for composite
    plating, 521
Freezing-in, of defects, 32
Frenkel disorder, 10
Frenkel reaction, 5
Fuel-cell catalyst patents,
    tabulated, 389
Fuel-cell lifetimes, 374

Fuel-cells, electrocatalyst for,
    373
Fuel-cells, 373

Geometrical optics, of
    reflectivity, 464
Gibbs equation, for
    electrosorption, 309
Grain boundaries, 4
Growth morphologies, in zinc
    composite deposition,
    523, 526
Guglielmi method, for
    composite plating,
    509
Guo et al., for composite
    plating, 519

H_3O^+ ion, solvation of, 253,
    259
Hardening effects, in
    electroplating of
    composites, 478
Heat treatment effect on
    carbon areas, 392,
    409–411
Hemispherical diffusion to a
    crystallite, 385
High fields, and dielectric
    constants, 207
Historical development, of
    electron-transfer
    treatments, 176
Holes and electrons, in solids,
    9
Hovestad et al., for composite
    plating, 523
Hwang and Hwang, for
    composite plating,
    516

Hydrogen molecule oxidation, alloy catalysts for, 414

Induction effects, in polarization, 221
Inner and Outer sphere concepts, 179
Inner and Second-sphere energies, 230
Inner sphere rearrangement, 186
Inner-sphere solvation, at multivalent cations, 228
Integral electrosorption valency, 355
Interactions, between defects, 36
Interfacial defects, thermodynamics of, 70
Interphase, 309
Iodide adsorption, 325, 327–330
Ion solvation and dielectric saturation, 196
Ion solvation, in polar media, 196
Ion transfer, and electron transfer, 175
Ion, solvation energies, 248
Ionic conductivity, in solids, 5
Ionic states, in silver chloride, 5

Japanese alloy-catalyst performances, tabulated, 399

Kariapper and Foster, for composite plating, 512
Kinetics of charge transfer, in solids, experimental, 84
Kinetics, of surface processes in solids, 113

Lifetimes, of fuel-cell electrodes, 374
Lorenz and Salié, partial charge transfer coefficient, 316
Lubrication, 480

Mass activity of catalysts, and specific surface areas, 377, 378
Mass fraction, of composite components, 488
Mat, metal surfaces, 430
Materials strengthening, in codeposition of composites, 478
Matrices, for phosphoric acid fuel-cells, 400
Mechanisms, of composite plating, 507
Mercury / water, absolute potential difference at, 358
Mercury, thiol layers at, 352
Metal composites, electroplating of, 475
Metal finishing, systems, 432
Metal surface, of silver, 426
Metal surfaces, mat, 430
Metal surfaces, nanostructure of, 425

Metal surfaces, reflectivity of, 425

Mirror reflectivity, and brightness, 470

Mobility, of states in solids, 96

Multivalent cations, inner-sphere interactions at, 228

Nanodimensional Pt catalysts, 420

Nanoionics, 75

Nanostructure of bright metal surfaces, 425

Nickel coatings, 459

Non-Franck–Cordon redox electron transfer, 270

Non-linear processes, in solids, 152

Optics, geometrical, of reflectivity, 464

Orientation of adsorbates, and electrosorption valency, 339

Oxides, redox chemistry in, 20

Oxygen catalyst activities, temperature effects on, 402

Oxygen reduction at Pt, in phosphoric acid, 375

Oxygen reduction catalyst patents, tabulated, 394

Oxygen reduction Pt, crystallite size effects, 379

Oxygen-reduction catalysts, Pt-alloy, 390

Partial charge transfer coefficient, definition, 303, 316

Partial charge transfer, 304

Partial charge transfer, basis of, 333

Partial charge transfer, relation to electrosorption valency, 333

Particle bath cocentration, 499

Particle properties, in composite plating, 483–484

Particle shape, in composite plating, 489

Particle size factor, in composite plating, 480, 488

Patents, for catalysation of C by Pt, 389

Patents, for cathode catalysts, 394

Patents, Japanese, for phosphoric acid fuel cells, 397

PEM, fuel-cell, membrane type, 373

Phosphoric acid fuel cells, Japanese patents for, 397

Phosphoric acid fuel -cells, matrices for, 400

Phosphoric acid fuel cells, operating temperatures of, 419

Phosphoric acid, fuel-cell type, 373

Phosphoric acid, hot, corrosion in, 404

Phosphoric acid, Pt solubility
    in hot, 380
Platinum alloy catalysts, for
    oxygen reduction, 390
Platinum alloy electrocatalysts,
    various works on, 392
Platinum catalyst, in
    phosphoric acid, 375
Platinum loading, for fuel-cell
    catalysts, 389
Platinum loading, optimization
    of, 389
Platinum particles,
    intercrystallite
    separation on C, 382
Platinum particles, on C
    support, 375
Platinum solubility,
    temperature effect on,
    381
Platinum, complexation by
    ligands, 380
Platinum, solubility in hot
    phosphoric acid, 380
Platinum, solubility,
    dependence on
    potential, 381
Point defects, as charge
    carriers, 5
Polar solvents, ion solvation
    in, 196
Polarization, "ut" type, 221
Polarization, electrostatic, 202
Polystyrene, as a plated
    composite, 494–499,
    501, 505, 506, 515,
    522-6
Potential distribution, in
    double-layer, 306

Process parameters, in
    composite plating,
    483
Proton transfer and solvation,
    274
Proton transfer, Franck–
    Condon condition,
    265
Proton transport, in solids, 99

Quadrupole interactions, 224

Rate constants, for transfers
    processes in solids, 89
Reactions, in solid phases, 149
Redox chemistry, 20
Redox processes, Franck–
    Condon, 262
Redox processes, Tafel plots
    for, 280
Reflectivity factors,
    diagrammatic, 465,
    466
Reflectivity of metals, wave-
    length effect, 440
Reflectivity, components of, at
    silver, 428
Reflectivity, diffuse, at metals,
    426
Reflectivity, of metal, 425
Reflectivity, of nickel coatings,
    460
Reflectivity, specular, at
    metals, 426
Reforming of carbonaceous
    fuels,415–418
Reforming, autothermal
    processes, 417

Refractory compounds, composite plating of, 477
Reorganizational energy, 183

Screening, of adsorbed charges, 333
Self-assembled thiol monolayers, 352
Silicon carbide, electroplated composite, 476
Silver chloride, ionic states in, 5
Silver mirror surface, 426
Silver mirror, STM image of, 428, 430
Silver mirror, STM line analysis, 429
Sintering effect, with small Pt particles, 382
Size-effects, in nanoionics, 75
SO, solid-oxide fuel cell type, 373
Solid state electrochemistry, 1
Solid vs liquid state, comparisons, 2
Solids, bulk processes in, 95
Solids, charge-transfer kinetics in, 84
Solids, conductance at interfaces of, 65
Solids, defect concentrations in, 10, 13
Solids, doping effects in, 22
Solids, electrical conduction in bulk, 95
Solids, electrochemical potentials for, 16
Solids, electrode kinetics in, 145

Solids, equivalent circuits for processes in, 147
Solids, non-linear processes in, 152
Solids, proton transport in, 99
Solids, rate constants for transfer processes, 89
Solids, reactions in or between, 149
Solids, surface kinetics at, 113
Solid-state reactions, 149
Solvation shells, outer, 241
Solvation energies, of ions, 248
Solvation of H_3O^+ ion, 253, 259
Solvation of ions, and charge transfer, 262
Solvation shell assembly, 269
Solvent, screening effect of, 339
Space-charge capacitance, 48
Space-charge conductance, 54
Space-charge profiles, 48
Specific adsorption, model for, 322
Specular reflexion, 426
STM image, of copper electrodeposited from various solutions, 449
STM image, of thick, electrodeposited film, 442
STM images, of electrodeposited nickel, 461
STM images, of electrodeposited zinc, 453, 456, 458

STM imaging of copper, mechanically and electrochemically polished,
STM imaging, line analysis in, 429, 432, 436
STM imaging, of metal imaging, 428, 430, 431, 435
Stonehart theory, of crystallite separation, 382
Structural analysis, of surfaces, 427, 430, 433, 439, 452, 459
Structure, of water, 251
Sulfide adsorption, 327–330
Surface diffusion on Pt electrocatalysts, 396
Surface excess, in adsorption, 309
Surface irregularities, and reflectivity, 426
Surface kinetics, in solid-state processes, 113
Surfactant effect, in composite plating, 522–526
Suspended particles, composite electroplating of, 475
Symbols, in solid state electrochemistry, 162

Tafel plots, for redox processes, 280
Tafel plots, linear, 283
Tafel plots, non-linearity of, 195
Tafel slopes, and electron transfer, 194

Temperature effects in composite plating, 506
Ternary alloy electrocatalysts, 396
Ternary electrocatalysts, lifetimes, 402, 420
Thermodynamics, of interfacial defects, 70
Thiol monolayers, electrosorption valency in, 352, 355, 361
Thiols at Hg drop, experimental, 363
Third shell, modeling, 244
Time-scales, in Franck– Condon principle, 267
Topography, digital imaging of, 426
Tracer diffusion methods, 103
Transport kinetics, in solids, 84
Transport processes in solids, formalism for, 87
Transport, thermodynamics of irreversible transfer processes, 95

Underpotential deposition, and electrosorption valency, 332

Valdes, for composite plating, 518
Vetter and Schultze, definition of electrosorption valency, 335

Water molecule orientation, 288
Water structure, modeling, 254
Water, dielectric constant of, 208
Water, structure of, 251
Wave-length effect, on reflectivity of metals, 440
Wear resistance, by composite electroplating, 479

XRD pattern, of polished copper surface, 438
XRD patterns, of thick copper deposits, 450

Zinc coatings, 450
Zinc coatings, wave-length dependence of reflectivity, 450

139.50
S.O.